METHODS IN MOLECULAR BIOLOGY™

Series Editor
John M. Walker
School of Life Sciences
University of Hertfordshire
Hatfield, Hertfordshire, AL10 9AB, UK

For further volumes:
http://www.springer.com/series/7651

Signal Transduction Protocols

Edited by

Louis M. Luttrell

Departments of Medicine and Biochemistry & Molecular Biology, Medical University of South Carolina, Charleston, SC, USA; Charleston VA Medical Center, Charleston, SC, USA

Stephen S.G. Ferguson

The J. Allyn Taylor Centre for Cell Biology, Robarts Research Institute, The University of Western Ontario, London, ON, Canada; Department of Physiology & Pharmacology, The University of Western Ontario, London, ON, Canada

Humana Press

Editors
Louis M. Luttrell MD, PhD
Departments of Medicine and
Biochemistry & Molecular Biology
Medical University of South Carolina
Charleston, SC, USA
Charleston VA Medical Center, Charleston
SC, USA
luttrell@musc.edu

Stephen S. G. Ferguson PhD
The J. Allyn Taylor Centre for Cell Biology
Robarts Research Institute
The University of Western Ontario
London, ON, Canada
Department of Physiology & Pharmacology
The University of Western Ontario
London, ON, Canada
ferguson@robarts.ca

ISSN 1064-3745 e-ISSN 1940-6029
ISBN 978-1-4939-6108-5 ISBN 978-1-61779-160-4 (eBook)
DOI 10.1007/978-1-61779-160-4
Springer New York Dordrecht Heidelberg London

Printed on acid-free paper

Humana Press is part of Springer Science+Business Media (www.springer.com)

Preface

Signal transduction is the process whereby a physical or chemical stimulus in the extracellular environment is detected by a receptor on the plasma membrane or in the cytosol or nucleus of a sensitive cell and translated into a chemical or electrochemical signal that produces a change in cellular metabolism. Rather than representing a series of simple linear cascades, it is increasingly clear that signal transduction is a highly organized and integrated process. Extensive crosstalk between signaling cascades, communicated directly through receptor oligomerization or indirectly through the activation of autocrine and paracrine feedback loops, enables one type of receptor to modulate activity in multiple intracellular pathways. Additional factors impose spatial or temporal constraints on signaling that influence the final cellular response by determining where within the cell, and for how long, the signal persists.

This volume focuses on experimental approaches to understand the complexity of signal transduction. Introductory chapters have been included to provide perspective on several of the challenges in signal transduction research and guidance on selecting the best approaches to various types of questions. The individual chapters provide detailed experimental protocols, beginning with the effects of ligand binding on receptor conformation and effector coupling, then moving inside the cell to capture the spatial and temporal characteristics of signaling events.

We would like to express our deepest appreciation to the coauthors of this publication. We hope that *Signal Transduction Protocols – Third Edition* will prove to be a valuable resource for future progress in the field of signal transduction research.

Charleston, SC — *Louis M. Luttrell*
London, ON — *Stephen S.G. Ferguson*

Contents

Contributors

LAMIA ACHOUR • *Institut Cochin, Université Paris Descartes, Paris, France*
BENJAMIN AGUILA • *Hormones and Cancer Research Unit, Department of Medicine, McGill University Health Center Research Institute, McGill University, Montréal, QC, Canada*
SYED M. AHMED • *Department of Pharmaceutical Sciences & Biochemistry, Leslie Dan Faculty of Pharmacy, University of Toronto, Toronto, ON, Canada*
CHRISTOPHE ALTIER • *Department of Physiology and Pharmacology, Hotchkiss Brain Institute, University of Calgary, Calgary, AB, Canada*
STÉPHANE ANGERS • *Department of Pharmaceutical Sciences, Leslie Dan Faculty of Pharmacy, University of Toronto, Toronto, ON, Canada; Department of Biochemistry, Faculty of Medicine, University of Toronto, Toronto, ON, Canada*
NICOLAS AUDET • *Department of Pharmacology, Faculty of Medicine, University of Montreal, Montreal, QC, Canada; Centre de Recherche du CHU Ste-Justine, Bureau, Montreal, QC, Canada*
LARRY S. BARAK • *Department of Cell Biology, Duke University, Durham, NC, USA*
CATHERINE H. BERLOT • *Weis Center for Research, Geisinger Clinic, Danville, PA, USA*
BROCK F. BINKOWSKI • *Promega Corporation, Madison, WI, USA*
WAYNE CHADWICK • *Receptor Pharmacology Unit, National Institute on Aging, National Institutes of Health, Baltimore, MD, USA*
RAELENE A. CHARBENEAU • *Department of Pharmacology, The University of Michigan Medical School, Ann Arbor, MI, USA*
SHENG CHEN • *Department of Neuroscience, Centre for Addiction and Mental Health, University of Toronto, Toronto, ON, Canada*
LAËTITIA COMPS-AGRAR • *Institut de Génomique Fonctionnelle, University of Montpellier 1 and 2, Montpellier, France*
LIANNE B. DALE • *The J. Allyn Taylor Centre for Cell Biology, Robarts Research Institute, The University of Western Ontario, London, ON, Canada; Department of Physiology & Pharmacology, The University of Western Ontario, London, ON, Canada*
AVAIS M. DAULAT • *Department of Pharmaceutical Sciences, Leslie Dan Faculty of Pharmacy, University of Toronto, Toronto, ON, Canada*
CHARLENE DEPRY • *Department of Pharmacology and Molecular Sciences, The Johns Hopkins University School of Medicine, Baltimore, MD, USA*
PO HIEN EAR • *Département de Biochimie, Université de Montréal, Montréal, QC, Canada*
ERIC TRINQUET • *Cisbio Bioassays, Bagnols-sur-Cèze Cedex, France*
FRANK FAN • *Promega Corporation, Madison, WI, USA*

Stephen S.G. Ferguson • *The J. Allyn Taylor Centre for Cell Biology, Robarts Research Institute, The University of Western Ontario, London, ON, Canada; Department of Physiology & Pharmacology, The University of Western Ontario, London, ON, Canada*
Lisa L. Gallegos • *Department of Pharmacology, University of California San Diego, La Jolla, CA, USA*
Randy A. Hall • *Department of Pharmacology, Emory University School of Medicine, Atlanta, GA, USA*
Terence E. Hébert • *Department of Pharmacology and Therapeutics, McGill University, Montréal, QC, Canada*
Thomas R. Hynes • *Weis Center for Research, Geisinger Clinic, Danville, PA, USA*
Ralf Jockers • *Institut Cochin, Université Paris Descartes, Paris, France*
Maud Kamal • *Institut Cochin, Université Paris Descartes, Paris, France*
Kuljeet Kaur • *Department of Pharmacology, The University of Michigan Medical School, Ann Arbor, MI, USA*
Jason M. Kehrl • *Department of Pharmacology, The University of Michigan Medical School, Ann Arbor, MI, USA*
Terry P. Kenakin • *Department of Pharmacology, University of North Carolina, School of Medicine, Chapel Hill, NC, USA*
Jane E. Lamerdin • *Odyssey Thera Incorporated, San Ramon, CA, USA*
Christian Landry • *Département de Biochimie, Université de Montréal, Montréal, QC, Canada*
Stéphane A. Laporte • *Hormones and Cancer Research Unit, Departments of Medicine and Pharmacology and Therapeutics, McGill University Health Center Research Institute, McGill University, Montréal, QC, Canada*
Shupeng Li • *Department of Neuroscience, Centre for Addiction and Mental Health, University of Toronto, Toronto, ON, Canada*
Fang Liu • *Departments of Neuroscience and Psychiatry, Centre for Addiction and Mental Health, University of Toronto, Toronto, ON, Canada*
Louis M. Luttrell • *Departments of Medicine and Biochemistry & Molecular Biology, Medical University of South Carolina, Charleston, SC, USA; Charleston VA Medical Center, Charleston, SC, USA*
Mohan K. Malleshaiah • *Département de Biochimie, Université de Montréal, Montréal, QC, Canada*
Bronwen Martin • *Metabolism Unit, National Institute on Aging, National Institutes of Health, Baltimore, MD, USA*
Stefano Marullo • *Institut Cochin, Université Paris Descartes, Paris, France*
Stuart Maudsley • *Receptor Pharmacology Unit, National Institute on Aging, National Institutes of Health, Baltimore, MD, USA*
Damien Maurel • *Institut de Génomique Fonctionnelle, University of Montpellier 1 and 2, Montpellier, France*
Vincent Messier • *Département de Biochimie, Université de Montréal, Montréal, QC, Canada*

STEPHEN W. MICHNICK • *Département de Biochimie, Université de Montréal, Montréal, QC, Canada*
RICHARD R. NEUBIG • *Departments of Pharmacology and Internal Medicine, The University of Michigan Medical School, Ann Arbor, MI, USA*
ALEXANDRA C. NEWTON • *Department of Pharmacology, University of California San Diego, La Jolla, CA, USA*
SUNG-SOO PARK • *Receptor Pharmacology Unit, National Institute on Aging, National Institutes of Health, Baltimore, MD, USA*
DARLAINE PÉTRIN • *Department of Pharmacology and Therapeutics, McGill University, Montréal, QC, Canada*
JEAN-PHILIPPE PIN • *Institut de Génomique Fonctionnelle, University of Montpellier 1 and 2, Montpellier, France*
GRACIELA PIÑEYRO • *Department of Psychiatry, Faculty of Medicine, University of Montreal, Montreal, QC, Canada; Centre de Recherche du CHU Ste-Justine, Bureau, Montreal, QC, Canada*
LAURENT PRÉZEAU • *Institut de Génomique Fonctionnelle, University of Montpellier 1 and 2, Montpellier, France*
STEFANIE L. RITTER • *Department of Pharmacology, Emory University School of Medicine, Atlanta, GA, USA*
MÉLANIE ROBITAILLE • *Leslie Dan Faculty of Pharmacy, University of Toronto, Toronto, ON, Canada*
PHILIPPE RONDARD • *Institut de Génomique Fonctionnelle, University of Montpellier 1 and 2, Montpellier, France*
ALI SALAHPOUR • *Department of Pharmacology and Toxicology, University of Toronto, Toronto, ON, Canada*
SUDHA K. SHENOY • *Departments of Medicine and Cell Biology, Duke University Medical Center, Durham, NC, USA*
MAY SIMAAN • *Hormones and Cancer Research Unit, Department of Medicine, McGill University Health Center Research Institute, McGill University, Montréal, QC, Canada*
ROGER K. SUNAHARA • *Department of Pharmacology, University of Michigan Medical School, Ann Arbor, MI, USA*
GISSELLE A. VÉLEZ-RUIZ • *Department of Pharmacology, University of Michigan Medical School, Ann Arbor, MI, USA*
JEAN-PIERRE VILARDAGA • *Laboratory for GPCR Biology, Department of Pharmacology and Chemical Biology, University of Pittsburgh, School of Medicine, Pittsburgh, PA, USA*
MARK VON ZASTROW • *Departments of Psychiatry and Cellular & Molecular Pharmacology, University of California San Francisco, San Francisco, CA, USA*
LIYUN WANG • *Receptor Pharmacology Unit, National Institute on Aging, National Institutes of Health, Baltimore, MD, USA*
YU TIAN WANG • *Brain Research Center, University of British Columbia, Vancouver, BC, Canada*
JOHN K. WESTWICK • *Odyssey Thera Incorporated, San Ramon, CA, USA*
KEITH V. WOOD • *Promega Corporation, Madison, WI, USA*
EVAN A. YOST • *Weis Center for Research, Geisinger Clinic, Danville, PA, USA*

STACY M. YOST • *Weis Center for Research, Geisinger Clinic, Danville, PA, USA*
GUILLERMO A. YUDOWSKI • *Departments of Psychiatry and Cellular & Molecular Pharmacology, University of California San Francisco, San Francisco, CA, USA*
GERALD W. ZAMPONI • *Department of Physiology and Pharmacology, Hotchkiss Brain Institute, University of Calgary, Calgary, AB, Canada*
JIN ZHANG • *Department of Pharmacology and Molecular Sciences, The Johns Hopkins University School of Medicine, Baltimore, MD, USA; Solomon H. Snyder Department of Neuroscience, The Johns Hopkins University School of Medicine, Baltimore, MD, USA; Department of Oncology, The Johns Hopkins University School of Medicine, Baltimore, MD, USA*
YU ZHOU • *Receptor Pharmacology Unit, National Institute on Aging, National Institutes of Health, Baltimore, MD, USA*

Part I

Overviews

Chapter 1

Refining Efficacy: Allosterism and Bias in G Protein-Coupled Receptor Signaling

Louis M. Luttrell and Terry P. Kenakin

Abstract

Receptors on the surface of cells function as conduits for information flowing between the external environment and the cell interior. Since signal transduction is based on the physical interaction of receptors with both extracellular ligands and intracellular effectors, ligand binding must produce conformational changes in the receptor that can be transmitted to the intracellular domains accessible to G proteins and other effectors. Classical models of G protein-coupled receptor (GPCR) signaling envision receptor conformations as highly constrained, wherein receptors exist in equilibrium between single "off" and "on" states distinguished by their ability to activate effectors, and ligands act by perturbing this equilibrium. In such models, ligands can be classified based upon two simple parameters; affinity and efficacy, and ligand activity is independent of the assay used to detect the response. However, it is clear that GPCRs assume multiple conformations, any number of which may be capable of interacting with a discrete subset of possible effectors. Both orthosteric ligands, molecules that occupy the natural ligand-binding pocket, and allosteric modulators, small molecules or proteins that contact receptors distant from the site of ligand binding, have the ability to alter the conformational equilibrium of a receptor in ways that affect its signaling output both qualitatively and quantitatively. In this context, efficacy becomes pluridimensional and ligand classification becomes assay dependent. A more complete description of ligand–receptor interaction requires the use of multiplexed assays of receptor activation and screening assays may need to be tailored to detect specific efficacy profiles.

Key words: Agonist, G protein-coupled receptor, Heterotrimeric guanine nucleotide-binding protein, Pharmaceutical chemistry, Pharmacodynamics, Signal transduction

1. Introduction

Most of the basic tenets of receptor pharmacology predate our understanding of the molecular structure of receptors themselves. When Stephenson defined efficacy in 1956, he was studying the acetylcholine-like effects of a series of alkyl-trimethyl

Louis M. Luttrell and Stephen S.G. Ferguson (eds.), *Signal Transduction Protocols*, Methods in Molecular Biology, vol. 756,
DOI 10.1007/978-1-61779-160-4_1,

ammonium salts on the contraction of guinea pig ileum (1). In this work, the readout of receptor activation was a relatively simple bioassay. Although the intervening 50 years have seen an explosion in our knowledge of receptor structure and mechanisms of intracellular signaling, even today most drug discovery efforts rely on using a single readout, often in a highly artificial system engineered for high throughput automated screening, as the basis for classifying the effect of ligand binding on receptor activity. Within such systems, where receptor density is constant and activity is measured either as an integrated whole cell or tissue response, e.g., muscle contraction, or a single molecular event, e.g., influx of cytosolic calcium, the relationship between the ligand concentration and receptor activation can be adequately described by two terms; affinity, the equilibrium dissociation constant of the ligand–receptor complex; and the maximal response that can be observed (2, 3), which is a function of efficacy. In this paradigm affinity and efficacy are largely independent functions, i.e., a ligand may have high affinity but low efficacy or vice versa, and ligands are classified as full agonists if they can elicit a maximal response from the system, partial agonists if they can only generate a submaximal response, and antagonists if they lack intrinsic efficacy but interfere with the ability of agonists to evoke a response.

Although these principles provide the framework that has guided signal transduction and drug discovery research for decades, advances in our understanding of the complexity of signal transduction networks and the evolution of technology to measure receptor activation in many dimensions have unambiguously demonstrated that the nature of efficacy is far more complex than originally envisioned, and a more general model is needed to explain the action of ligands on receptors (4). Rather than functioning like simple switches that transition between tightly constrained "off" and "on" states, receptors are highly dynamic proteins capable of adopting a large number of conformational states, some subset of which is capable of coupling to variable sets of downstream effectors. Viewed in this way, it is evident that any ligand, small molecule, or other protein that contacts the receptor in a manner that alters its conformational equilibrium may initiate, attenuate, or even qualitatively change signaling. Orthosteric ligands, allosteric modulators, even other proteins contacting the receptor in the lipid bilayer or on its cytosolic face, all work in essentially the same way. In the sections that follow, we will review the changing concepts of efficacy, their implications for drug development, and the challenges arising from the need to incorporate a more complete characterization of ligand action into experimental and industrial research.

2. Two-State Models of Receptor Activation

When only a single readout of receptor activation is considered, receptors can be described as existing in either an empty "off" state that is silent in the assay or an agonist-bound "on" state that elicits a measurable response. The early model of "induced fit" advanced by Koshland in 1958 to describe enzymatic catalysis, proposed that the interaction between a substrate and amino acid residues within the active site of an enzyme changes the structure of the enzyme so as to bring the catalytic groups into proper alignment (5). In other words, for an enzyme (or receptor) that exists in a preferred low energy "inactive" state and must transition to a higher energy "active" state to function, substrate (or ligand) binding facilitates the transition by contributing energy that makes the "active state" become the new preferred low energy state. The alternative concept of "conformational selection" arises from the Monod–Wyann–Changeux model of allostery, which proposes that proteins exist in spontaneous equilibrium between different conformations and that a molecule that binds to a specific conformation will stabilize it, shifting the conformational population toward that favored state (6). The use of such allosteric models to describe membrane receptor function began in the late 1960s (7, 8). The assumption is that the probability that an unbound receptor will exist in the active state is very low, but that stabilization of this state upon ligand binding drives the equilibrium toward the "on" state by interfering with the transition back to the "off" state.

While molecular simulations favor conformational selection models for the binding of small molecules to proteins (9), selection of a relatively rare pre-existing conformation would thermodynamically resemble conformational induction (10), leaving little need to choose between them in modeling two-state receptor behavior. Structural and biophysical data demonstrate that GPCRs vary widely in their degree of conformational flexibility. One extreme is the visual photoreceptor, rhodopsin, which for many years was the only GPCR for which high-resolution X-ray crystallographic structure was available (11, 12). Given its function, it is not surprising that rhodopsin is completely inactive toward transducin in the dark adapted state, i.e., it has evolved to function as an "on–off" switch. To achieve this, it is tightly constrained in the "off" position by intramolecular interactions between the transmembrane helices, notably an "ionic lock" linking the highly conserved E/DRY sequence found at the cytoplasmic end of TM3 in 70% of class A GPCRs, to the NPxxY motif located in TM6. More recent structures of light-activated rhodopsin and of opsin, the ligand-free form of rhodopsin, bound to a C-terminal fragment of transducin, demonstrate that the upon activation the ionic lock is released, allowing a outward turn of TM6 that exposes the transducin-binding site (13, 14).

Studies of the β_2 adrenergic receptor, which unlike rhodopsin catalyzes a low level of G protein activation even in the absence of agonist, are perhaps more representative of a "typical" GPCR. Fluorescence lifetime spectroscopy of fluorescently labeled β_2 receptors demonstrates that the receptor spontaneously oscillates around a single preferred conformation. Such oscillation admits the possibility of spontaneous, but rare, adoption of an active conformation. Antagonist binding does not change the preferred conformation but does reduce the extent of oscillation, while agonist binding results in the appearance of a distinct conformational population that presumably reflects stabilization of the otherwise rare active state (15, 16). The crystallographic structure of the receptor provides a physical basis for this enhanced flexibility (17–19). In the β_2 receptor TM3 and TM6 are farther apart than in rhodopsin and the salt bridge that comprises the ionic lock is "broken," permitting greater conformational freedom (20).

The far end of the conformational flexibility spectrum is illustrated by constitutively active receptors; engineered or naturally occurring GPCRs that exhibit a high degree of spontaneous G protein activation (21–24). The finding that substitution of Ala293 located near the IC3-TM6 interface in the α_{1B} adrenergic receptor with any of the 19 other possible amino acids results in some degree of constitutive activity (25), suggests the existence of "hot spots" where any change that disrupts the normal helical packing can destabilize conformational constraints and confer constitutive activity. Indeed, biochemical and spectroscopic analysis of purified constitutively active β_2 adrenergic receptors reveals greater structural instability and an exaggerated conformational response to drug binding (26).

The accidental discovery of constitutively activating GPCR mutations led to the finding that some ligands that appear as antagonists in the setting of low basal receptor activity actually possess the ability to suppress constitutive activity, while others do not (22, 27). The behavior of such "inverse agonists" prompted a revision of the classic allosteric model of GPCR activation. The "extended ternary complex" model envisions the receptor existing in spontaneous equilibrium between two states (active: R*; inactive: R) that differ in their ability to activate G proteins (22). In the model, the intrinsic efficacy of a ligand is a reflection of its ability to alter the equilibrium between R and R*. *Full agonists* stabilize the R* conformation, pulling the equilibrium toward the active state to generate a maximal response; *Partial agonists* have lower intrinsic efficacy than full agonists, thus producing a submaximal system response and potential attenuation of full agonist activation; *Antagonists* bind indiscriminately to both R and R*, producing no physiological response but blocking the response to agonists; *Inverse agonists* act as antagonists in non-constitutively-active systems, but have the added property of reducing receptor-mediated constitutive activity by binding preferentially to R and pulling the equilibrium toward the inactive state (Fig. 1a).

Thus, a more precise formulation of efficacy must account for factors that affect receptor conformation other than ligand binding, such as intrinsic activity of the unliganded receptor. The cubic ternary complex model, for example, allows that receptors exist in a native conformational ensemble, within which only certain conformations, e.g., R*–G and H–R*–G, are "active," meaning that they produce a measurable response (28). A ligand is efficacious only the extent that it changes production of the active species relative to what is observed in the native state, i.e., efficacy must be defined in terms of net stimulus. In this case, even the direction of efficacy can be system dependent, and one can accommodate the behavior of "protean agonists," ligands that appear as partial agonists in systems with low basal activity and inverse agonists in systems with high basal activity, if one assumes that the ligand has intrinsic efficacy that is greater than the basal state of the low activity system but less than that of the constitutively active system (29).

Despite their utility in describing positive and negative efficacy, two-state models have limitations. With only two possible states, the receptor alone is the determinant of information flow across the membrane. Ligand binding may alter the fraction of receptors in the "on" state, but cannot qualitatively change the nature of that state. To hold true, the classification of a ligand as an agonist, antagonist, or inverse agonist must be independent of the assay used to detect receptor activation, and the relative order of potency for a series of ligands cannot vary when two or more assays are employed (Fig. 1b). Deviations from these principles can only be explained using strength-of-signal arguments, which posit that receptors coupling to different downstream effectors may do so with different efficiencies, such that the most efficiently coupled response will be activated first, followed by less efficiently activated processes. Indeed, new signaling responses commonly emerge as the level of receptor expression increases (30, 31). Similar phenomena arise from changes in the expression levels of the participating G proteins (32). In experimental systems, an agonist activating a GPCR that stimulates multiple G proteins frequently elicits signals downstream of each G protein with differing efficacy and/or potency (33). In this case, variation in receptor density can create the illusion of unique functional states. For example, the muscarinic receptor agonist oxotremorine is twofold more potent than carbachol in promoting contraction of guinea pig ileum. When receptor density is lowered through alkylation with phenoxybenzamine, the response to oxotremorine disappears, but the response to carbachol, while reduced, is still present. The reason for this apparent reversal of potency is that oxotremorine is a high affinity but low efficacy

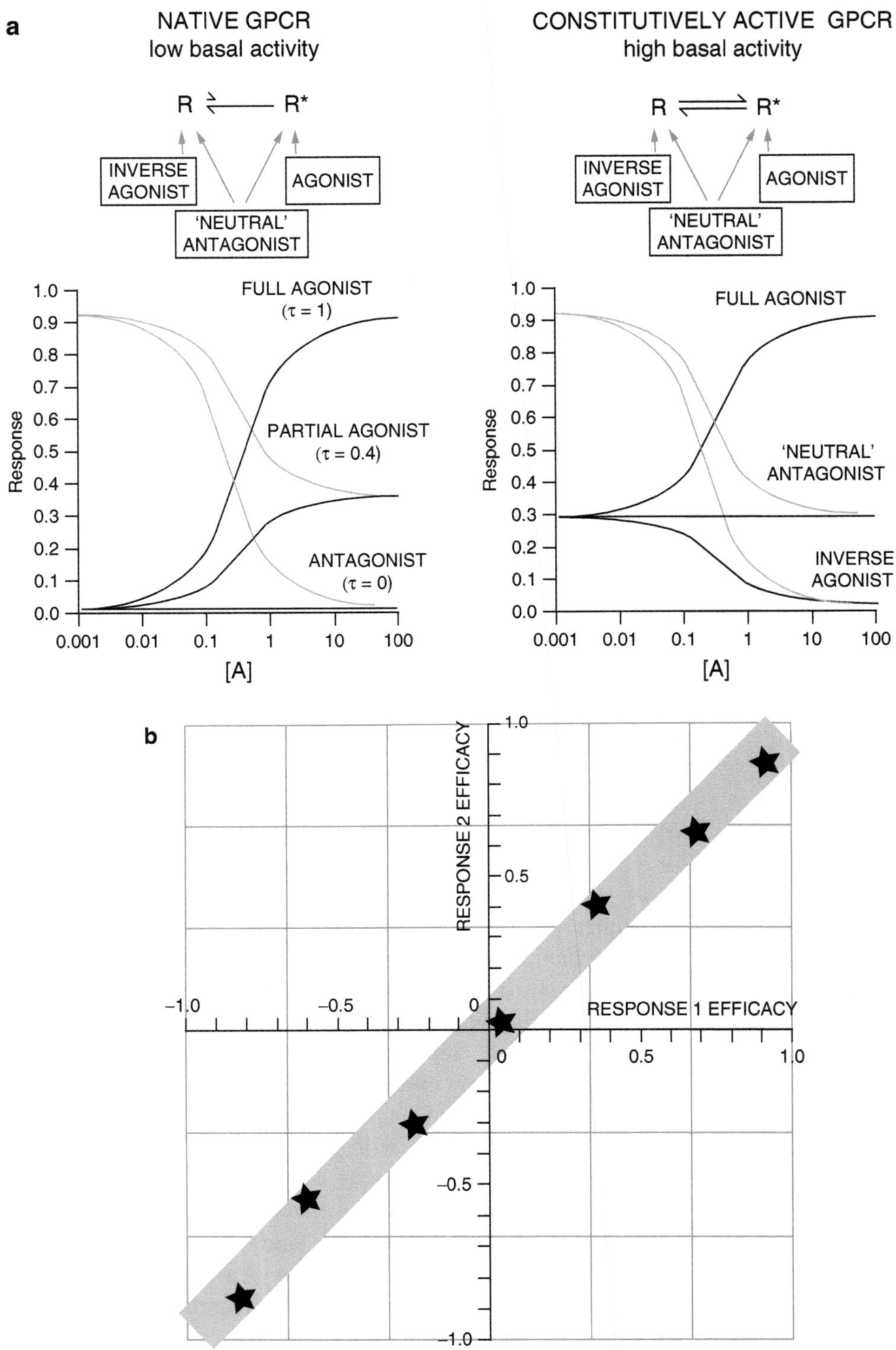
a
NATIVE GPCR
low basal activity
R
R*
INVERSE AGONIST
AGONIST
'NEUTRAL' ANTAGONIST
FULL AGONIST (τ = 1)
PARTIAL AGONIST (τ = 0.4)
ANTAGONIST (τ = 0)
Response
[A]
CONSTITUTIVELY ACTIVE GPCR
high basal activity
R
R*
INVERSE AGONIST
AGONIST
'NEUTRAL' ANTAGONIST
FULL AGONIST
'NEUTRAL' ANTAGONIST
INVERSE AGONIST
Response
[A]
b
RESPONSE 2 EFFICACY
RESPONSE 1 EFFICACY

agonist, hence more sensitive to decreased receptor number, while carbachol is a low affinity but high efficacy agonist, which is less affected by the loss of tissue sensitivity. Although oxotremorine and carbachol clearly produce opposite effects in high and low receptor density systems, these findings do not require the postulate of separate agonist-induced receptor active states (34).

3. Multistate Models and Functional Selectivity

While there is nothing inherent in two-state models of GPCR activation that precludes the possibility of multiple active states, they are limited to describing the conformational equilibrium of unliganded receptors, and their characterization of efficacy is based on the assumption that ligand binding affects only the proportion of receptors in the "active" state. But if receptors are conformationally flexible there is no a priori reason to assume that the active conformation stabilized by a ligand will be identical either to the spontaneously formed active state or that produced by a structurally distinct ligand. As techniques were developed to measure efficacy in different ways it became apparent that the relative activity of agonists did not always adhere to the predictions of simple receptor theory. Reversal of agonist potency, which cannot occur in a two-state model, has been described for several GPCRs that activate more than one G protein species, including the serotonin $5HT_{2c}$, pituitary adenylate cyclase-activating polypeptide (PACAP), dopamine D_2, neurokinin NK_1, CB_1 cannabinoid, and type 1 parathyroid hormone (PTH_1) receptors (35–40). An early and striking example was found upon comparison of the ability of two PACAP analogues, $PACAP_{1-27}$ and $PACAP_{1-38}$, to

Fig. 1. Efficacy in a two-state system. (**a**) Most native GPCRs exhibit low basal activity, i.e., the equilibrium between the "off" state of the receptor (R) and the "on" state (R*) heavily favors R. Ligands with agonist activity preferentially stabilize R* pulling the equilibrium toward the "on" state. The intrinsic efficacy of an agonist (τ) is a reflection of its ability to stabilize R*, hence "full" agonists are highly selective for R* while partial agonists exhibit less selectivity. Antagonists lack intrinsic efficacy, but both antagonists and partial agonists will competitively reduce receptor activity measured in the presence of an agonist. In systems with low basal activity, ligands that preferentially bind R cannot be distinguished from ligands that bind equivalently to both states. However, in systems with high basal activity, e.g., constitutively active GPCRs, a detectable quantity of R* exists in the absence of agonist. In this setting it is possible to demonstrate that some ligands, termed "inverse agonists," are selective for R, enabling then to lower the basal activity of the system. A true "neutral antagonist" would bind equivalently to R and R*, hence would have no effect on basal activity, but would reduce activity measured in the presence of an agonist ligand with intrinsic efficacy greater than the basal activity of the system. (**b**) Since in a two-state system, efficacy reflects only the ability to influence the R-R* equilibrium, ligand classification should be independent of the assay used to detect the response. Hence, a plot of intrinsic efficacy measured for any two responses to a series of ligands (*stars*) in a single system should approximate the line of unity from full agonist activity (efficacy 1:1), through neutral antagonism (efficacy 0:0), to full inverse agonism (efficacy −1:−1). Significant deviations can result only from differences in signal strength, i.e., the intrinsic efficiency of coupling between the receptor and effectors 1 and 2 in the system.

stimulate cAMP and phosphatidylinositol production in LLC-PK1 cells transfected with the PACAP receptor (36). Whereas the relative potency of the two ligands in the cAMP assay was $PACAP_{1-27} > PACAP_{1-38}$, the order for inositol phosphate production was reversed. These data definitively demonstrated that the two agonists were not activating the receptor in the same way. Similar, but even more dramatic examples of ligand-dependent "bias" have been shown for the PTH_1 receptor. Whereas PTH_{1-34} activates both the protein kinase (PK)A and PKC pathways, PTH_{1-31} only stimulates cAMP production, while the N-terminally truncated analogue PTH_{3-38} activates PKC, but not PKA (39, 40). Other examples include the findings that certain antagonists of the $5HT_{2A}$, AT_{1A} angiotensin, and PTH_1 receptors can produce active receptor internalization in the absence of G protein activation (41–45). Conversely, G protein agonists can be "nondesensitizing," i.e., activate G protein signaling without producing receptor desensitization or internalization (44–46).

Biochemical and biophysical evidence further supports the hypothesis that ligands can stabilize distinct receptor conformations. Indeed, multiple G protein-coupled states of the β_2 adrenergic receptor have been distinguished using guanine nucleotide analogues (47). Similarly, some receptor mutations produce constitutive activity that is restricted to a single signaling pathway among those ordinarily activated by the receptor (48, 49), presumably by restricting receptor isomerization to a subset of conformations that promote selective G protein coupling. Fluorescence lifetime spectroscopy of β_2 adrenergic receptors fluorescently labeled at Cys265 shows that agonists select discrete arrays of receptor conformation, consistent with the induction of ligand-selective active states (16, 50). Other approaches, including plasmon-waveguide resonance spectroscopy, fluorescence resonance energy transfer (FRET), bioluminescence resonance energy transfer (BRET), circular dichroism, antibody binding, site-directed mutagenesis, and kinetic studies, have similarly yielded evidence of multiple receptor conformations (51–58).

If different ligands can produce different active receptor conformations, then the receptor alone cannot be the minimal recognition unit for the cytosolic elements involved in signaling. The first formal model to account for these digressions postulated that it is the ligand–receptor complex, not the receptor alone, that specifies the active state (34). In this case, the formation of agonist-selective active states can "bias" the coupling of the receptor to different signaling pathways. Many terms have been coined to describe this phenomenon, including "stimulus-trafficking," "functional dissociation," "biased agonism," "biased inhibition," "differential engagement," "discrete activation of transduction," and "functional selectivity" (34, 59–64).

Whatever term is applied, the implications for signal transduction are dramatic. Functional selectivity can range from relatively modest deviations from the predicted line of unity depicted in Fig. 1b (65), to frank reversal of efficacy, wherein the characterization of a ligand as agonist, antagonist or inverse agonist becomes assay dependent (66, 67).

Among the more dramatic examples of functional selectivity is the phenomenon of G protein-independent signaling arising from GPCR "coupling" to non-G protein effectors like arrestins (68, 69). The arrestins are a family of four GPCR-binding proteins involved in homologous desensitization and receptor endocytosis (70). Arrestins 1 and 4 are confined to visual sensory tissue, whereas arrestins 2 and 3 (β-arrestin1 and β-arrestin2) are ubiquitously expressed. Upon agonist binding, GPCRs are phosphorylated by G protein-coupled receptor kinases (GRKs), creating high-affinity binding sites for arrestins. Unlike the catalytic GPCR–G protein interaction, arrestins form stable bimolecular complexes with receptors, in which state they are sterically uncoupled from G protein activation. In addition, arrestins 2 and 3 function as adapters, physically linking the receptor to the endocytic machinery. It was the discovery that arrestins serve as adapters not only for GPCR sequestration, but also for linking GPCRs to other enzymatic effectors (71), that changed our view of GPCR signal transduction. A host of catalytically active proteins have been reported to bind arrestins and undergo recruitment to agonist-occupied GPCRs; among them Src family tyrosine kinases; components of the ERK1/2 and JNK3 mitogen-activated protein kinase cascades; the E3 ubiquitin ligase, Mdm2; the cAMP phosphodiesterases, PDE4D3/5; diacylglycerol kinase; the inhibitor of NF-κB, IκBα; the Ral-GDP dissociation stimulator, Ral-GDS; and the Ser/Thr protein phosphatase, PP2A. It is now generally accepted that that ligand binding elicits two mutually exclusive GPCR signaling modes; a transient G protein-coupled state that dominates early signaling, and an arrestin-coupled state in which signals originate from multi-protein receptor–arrestin "signalsomes" that continue to signal as the receptor internalizes (68, 69, 72).

Once it became clear that arrestins act as alternative GPCR signal transducers, it was logical to test whether ligands that promote arrestin-dependent GPCR internalization without G protein activation (41–45) might exhibit arrestin pathway-selective efficacy in signaling. Indeed, using small-interfering RNA to silence arrestin expression, is possible to show that one such ligand, the peptide AT_{1A} receptor antagonist, Sar^{1}-Ile^{4}-Ile^{8}, produces arrestin-dependent ERK1/2 activation under conditions where G protein activation in the system is undetectable (73). Even more dramatic complete reversal of efficacy is observed with (D-Trp^{12}, Tyr^{34}) PTH_{7-34}, an inverse agonist for PTH_1 receptor–Gs coupling, that acts as an arrestin-dependent agonist for ERK1/2 activation (45). Similarly, ICI118551, propranolol and carvedilol, β_2-adrenergic receptor ligands that act as inverse agonists with respect to Gs activation, function as partial agonists for the ERK1/2 pathway by engaging arrestins (74, 75). The existence of such "perfect bias," wherein a ligand is silent in one assay yet active in another, or exhibits diametrically opposing efficacy in different assays, suggests that most, if not all, GPCRs can assume multiple active states, and that GPCR ligands can be identified that select from a menu of receptor conformations to induce only a subset of the full response profile (Fig. 2).

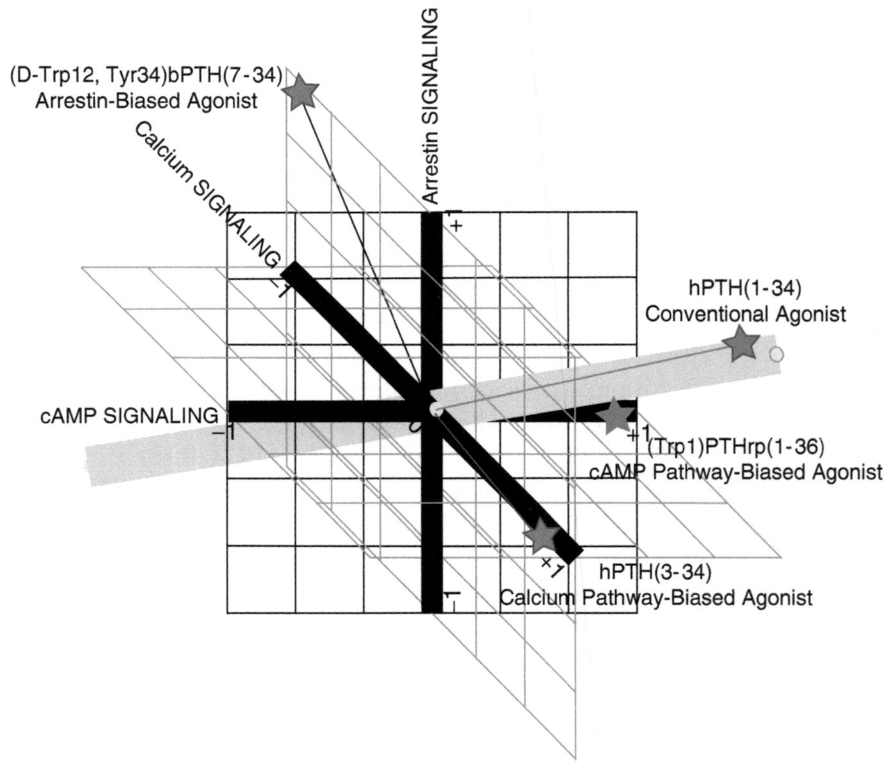

Fig. 2. Functional selectivity in a multistate system. If GPCRs adopt multiple "active" receptor conformations, each capable of coupling the receptor to a subset of possible effectors, then ligands may exert functional selectivity by stabilizing different conformational populations. In this case, ligands can exhibit significant deviations from the two-state "line of unity" and even demonstrate "perfect bias," i.e., positive efficacy in one assay with no efficacy, or reversal of efficacy, in another. In this case ligand classification becomes assay dependent. Shown is a conceptual plot of PTH_1 receptor agonism determined using three different signaling readouts of PTH1 receptor activity based on published data (39, 44); cAMP production, calcium signaling, and receptor sequestration/arrestin signaling. Whereas the conventional agonist PTH_{1-34} exhibits positive efficacy in all three assays, the cAMP-selective agonist (Trp1)$PTHrP_{1-36}$, and the calcium-selective agonist PTH_{3-34}, exhibit functional selectivity for Gs and Gq/11 coupling, respectively. The arrestin pathway-selective ligand, (D-Trp12, Tyr34)PTH_{7-34} exhibits true reversal of efficacy, activating arrestin pathways while functioning as an inverse agonist for PTH1 receptor–Gs coupling and a neutral antagonist of G*q/11* signaling.

4. GPCRs as Allosteric Proteins

Thus far we have limited the discussion to orthosteric ligands, molecules that modulate receptor behavior by interacting with the native ligand-binding pocket. But it has long been clear that other molecular interactions affect GPCR conformation and function. From the earliest in vitro reconstitution of agonist-regulated activation of G proteins (76), it was known that in the absence of guanine nucleotide, the receptor and heterotrimeric G protein form a stable complex that displays increased affinity for agonist binding. The "ternary complex" model developed to describe this behavior proposed the existence of two GPCR states; a high agonist affinity state representing the ternary complex between agonist (H), receptor (R), and heterotrimeric G protein (G); and, a low affinity (H–R) state observed in the presence of GTP, which allows receptor-catalyzed G protein activation and dissociation of the H–R–G complex (77). Although it considers only two conformations, the model captures the key point that GPCR conformation is influenced not just by ligands, but by other proteins in their environment. However, G proteins are not alone in exerting allosteric effects on receptors that affect ligand binding. Arrestin-bound receptors also demonstrate high agonist affinity, prompting the receptor–arrestin complex to be described as an "alternative ternary complex" that can be modeled similarly (78, 79).

It is perhaps more accurate to envision GPCRs as collections, or ensembles, of tertiary conformations (4). Receptors "sample" these different conformations according to changes in the thermal energy of the system, taking conformational excursions away from some canonical native structure. The probability that a given receptor will exist in a particular conformation, hence the fraction of the receptor population in that conformation at any instant, depends on the energy required to attain it. For any set of conditions there exists some number of nearly isoenergetic conformers associated with energy "wells" in the landscape that are frequented more often than random chance in the normal course of conformational sampling (80). If one of these conformations leads to a measurable outcome, i.e., biological response, it can be operationally defined as an "active" state, and the biological activity of the receptor under those conditions will reflect the energy-weighted contributions of the component microstates of the conformational ensemble (81). The more flexible the receptor, i.e., the more readily it can adopt new conformations, the more susceptible its biological activity is to allosteric modulation. Receptors must maintain a balance between thermodynamic stability to support specificity, and flexibility to undergo conformational change and catalyze biochemical reactions (82). Molecular dynamic analyses have shown that signaling proteins in general have an unusual amount of intrinsic disorder, making them ideal candidates for allosteric modulation (83). The power of allosterism emanates from the ability of the receptor to sense from sites other than the active site, or the site being modulated. Therefore, the active site is free to function until changes in the environment lead to change the energy landscape. Any molecular interaction that imparts energy, whether it involves ligand binding, interaction with other membrane or cytosolic proteins, or binding of a small molecule somewhere outside the ligand-binding pocket, can affect the conformational ensemble in a manner that may affect signaling (84). As allosteric proteins, GPCRs are thus susceptible to numerous inputs that modify their signaling properties. Apart from orthosteric ligand effects, two pharmacologically important factors are lateral allostery arising from protein–protein interactions within the plasma membrane or cytosol, and allosteric modulation arising from the interaction of small molecules with sites on the receptor outside the ligand-binding pocket (Fig. 3).

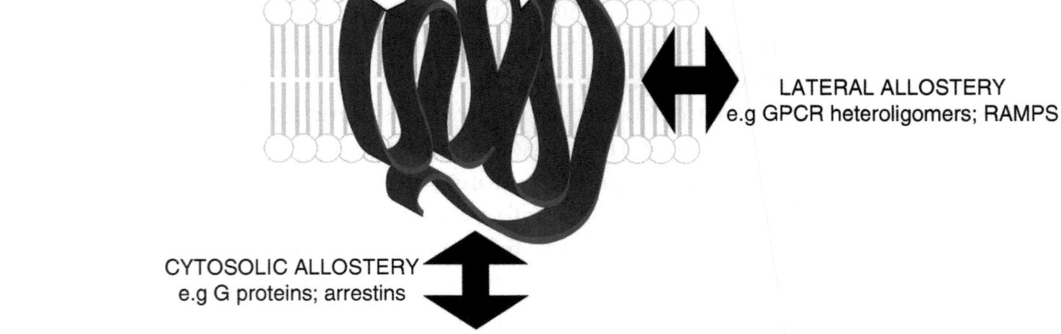

Fig. 3. GPCRs as allosteric proteins. Intermolecular interactions between GPCRs and other proteins or small molecules in their environment can alter the conformational equilibrium of the receptor in ways that change its reactivity toward guest probes, e.g., ligands or cytosolic effectors. In addition to orthosteric allostery exerted through the native ligand-binding pocket, protein–protein interactions within the plane of the plasma membrane (lateral allostery) or at the cytosolic interface (cytosolic allostery) can change receptor properties. Likewise, small molecule allosteric modulators can exert effects by binding to recognition sites outside of the orthosteric ligand site. The results can be changes in orthosteric ligand-binding affinity or selectivity, or altered coupling to cytosolic effectors, e.g., the imposition of functional selectivity.

5. Allosteric Modulation by Protein–Protein Interaction

The interaction between GPCRs and numerous other proteins modifies the specificity, selectivity, and time course of signaling by the minimal H–R–G module (67). These protein–protein interactions include the formation of GPCR dimers (85–87), the interaction of GPCRs with nonreceptor transmembrane proteins (88, 89), and the binding of PDZ domain-containing and non-PDZ domain scaffold proteins to the intracellular loops and C-termini of receptors (90–92).

Coprecipitation approaches, complementation studies using mutated or chimeric receptors, and FRET/BRET measurements all support the conclusion that many, if not most, GPCRs can exist as homodimers, heterodimers or higher order multimers. Indeed, FRET/BRET data suggest that many homodimeric or heterodimeric GPCR combinations are allowed (85–87). The clearest examples of dimerization involve Class C GPCRs (93), where dimer formation is required to assemble a functional receptor. The γ-amino butyric acid $(GABA)_B$ receptor is such an obligatory dimer (94, 95). The $GABA_BR1$, which contains the structural determinants necessary for ligand binding but not for G protein coupling, fails to traffic to the plasma membrane unless it is coexpressed with a second $GABA_B$ receptor transcript, the $GABA_BR2$. The $GABA_BR2$ alone can reach the cell surface and is capable of G protein coupling, but cannot bind ligand. Dimerization of the two receptors, mediated via their C-terminal tails, masks an endoplasmic reticulum retention sequence located in the tail of the $GABA_BR1$, permitting the $GABA_BR2$ to chaperone for $GABA_BR1$ to the plasma membrane (96). Perhaps importantly, GPCR dimerization enables receptor partners to exert lateral allosteric effects within the plane of the plasma membrane through contact between their transmembrane domains. In some cases, dimer formation has been shown to modulate ligand binding or to enable an orthosteric ligand of one receptor to modify the signaling of the other. For example, positive cooperativity has been reported for ligand binding to δ and κ opioid receptors when coexpressed (97). Conversely, negative cooperativity in dopamine D_2 receptor agonist binding in the presence of an adenosine A_2 receptor agonist has been observed (98). In the context of $\mu - \delta$ opioid receptor dimers, antagonist occupancy of δ receptors enhances μ opioid receptor agonist binding and signaling in vitro, and δ opioid antagonists enhance morphine-induced analgesia in vivo (99). Similarly, dimerization between angiotensin AT_{1A} and bradykinin B_2 receptors increases the potency and efficiency of angiotensin II, a vasopressor, while decreasing that of bradykinin, a vasodilator (100). Allosteric antagonism within GPCR heterodimers is also possible. In murine cardiomyocytes, antagonism of β-adrenergic receptors inhibits angiotensin AT1a receptor-mediated contractility and vice versa (101). This phenomenon arises within β_2-adrenergic-AT_{1A} receptor heterodimers, wherein each receptor is uncoupled from its cognate G proteins when its partner is bound to an orthosteric antagonist. Similar allosteric antagonism occurs within μ opioid-CB1 cannabinoid receptor heterodimers (102). To the extent that GPCR heterodimers comprise pharmacologically unique entities with tissue- or disease-specific expression patterns, lateral allostery opens the potential for trans-facilitation or inhibition of signaling or even the development of dimer-specific agonist or antagonist ligands.

Other examples of lateral allostery include GPCR complexes with nonreceptor proteins that modify ligand binding, signaling or trafficking, even to the extent of creating altogether "new" receptors. Receptor activity modifying proteins (RAMPs) are a family of three single membrane-spanning glycoproteins with large extracellular domains and short cytoplasmic domains (88). RAMPs form complexes with the calcitonin receptor-like receptor (CRLR) and calcitonin receptor, and it is the RAMP–CRLR complex, not the receptor per se, that determines ligand specificity. The CRLR–RAMP1 complex functions as the receptor for calcitonin gene-related peptides, a pleiotropic family of neuropeptides with homology to calcitonin, amylin, and adrenomedullin. When CRLR is complexed with RAMP2 or RAMP3 it serves as an adrenomedullin receptor. Similarly, complexes between a naturally occurring splice variant of the calcitonin receptor and RAMP1 or RAMP3 yields a functional amylin receptor. RAMP expression changes under various forms of physiologic stress and in response to glucocorticoids, suggesting that cellular responsiveness to certain hormones may be regulated through control of accessory protein expression. Melanocortin 2 (MC_2) receptor accessory protein (MRAP) is another example (89). MRAP binding to the MC_2, or adrenocorticotrophic hormone (ACTH), receptor facilitates nascent MC_2 receptor trafficking to the plasma membrane and is required for ACTH binding and activation of adenylyl cyclase. Humans lacking MRAP are ACTH-resistant and deficient in glucocorticoid production.

Interactions with cytosolic proteins similarly modify GPCR signaling (89–92). A good example is the Na^+/H^+ exchanger regulatory factors (NHERF) 1 and 2, PDZ domain scaffolding proteins with restricted tissue distribution. NHERF1/2 bind to a consensus PDZ binding motif at the C-terminus of the PTH_1 receptor, linking the receptor to specific effectors like phospholipase Cβ1 (103). Whereas uncomplexed PTH_1 receptors robustly stimulate adenylyl cyclase by activating Gs, the PTH_1 receptor–NHERF2 complex exhibits enhanced phospholipase C activation and inhibition of adenylyl cyclase arising from coupling to Gi/o proteins. Thus, PTH_1 receptor signaling in cells that express NHERF, like renal tubular epithelium, is qualitatively changed.

6. Refining Efficacy Through Allosteric Modulation

The current GPCR pharmacopeia consists almost entirely of drugs that target the orthosteric ligand-binding pocket. Nonetheless, it is unsurprising that GPCRs can be affected by small molecules that bind outside of the ligand-binding site. Such allosteric modulators (AMs) are ligands that bind receptor domains that are topographically distinct from the orthosteric site, leading to an increase or decrease in the ability of the orthosteric ligand to interact with the receptor and/or modulate its ability to stabilize the active conformation of the receptor. Additionally, AMs may engender collateral efficacy by biasing the stimulus, thus leading to signaling-pathway-selective allosteric modulation (104, 105). The broad range of effects that can be achieved through allosteric modulation offers significant promise for the development of new classes of GPCR-targeted drugs.

AMs have the ability to change orthosteric ligand affinity, efficacy, or both. The effect of an AM on orthosteric ligand affinity is commonly described in terms of a cooperativity factor (α), which specifies the strength and direction of the change in affinity for one site when the other is occupied (2, 106, 107). AMs can be broadly grouped as either positive AMs ($\alpha > 1$) or as negative AMs ($\alpha < 1$). For example, binding of the orthosteric antagonist N-methylscopolamine to the M_2 muscarinic acetylcholine receptor (mAChR) is allosterically enhanced by alcuronium but is allosterically inhibited by gallamine, even though both AMs bind to a common site on the receptor (108, 109). Cinacalcet, a positive allosteric modulator of the calcium-sensing receptor, increases its affinity for calcium and enhances calcium-induced inhibition of PTH secretion by the parathyroid glands (110). This property led its approval by the Food and Drug Administration for the management of secondary hyperparathyroidism in chronic renal failure and parathyroid carcinoma. AMs may also change the intrinsic efficacy of the receptor–orthosteric ligand complex by governing the transition of the receptor between its resting and activated states independent of effects on orthosteric ligand binding. Such effects are specified by an efficacy cooperativity factor (β), wherein $\beta > 1$ confers increased efficacy, and $\beta < 1$ reduced efficacy, in the presence of the AM. For example, the allosteric modulator Org27569 enhances binding of the orthosteric agonist CP55940 to cannabinoid CB_1 receptors ($\alpha > 1$), while simultaneously reducing its efficacy ($\beta < 1$) (111).

Although most AMs are pharmacologically silent in the absence of an orthosteric ligand, some, termed "ago-allosteric" modulators, possess intrinsic agonist-like activity. Such allosteric agonists further expand the repertoire of possible effects, because they have the potential to initiate signaling in their own right in addition to modulating orthosteric ligand pharmacology. A relative efficacy factor (ϕ), representing the fraction of the maximal efficacy of the reference orthosteric agonist (τ) produced by the allosteric ligand, can be applied to quantify these effects. For example, a series of substituted quinoxalines act both as allosteric activators ($\phi > 0$) of the human glucagon-like peptide-1 (GLP-1) receptor and as allosteric modulators of GLP-1 affinity (112). Similarly, McN-A-343 and AC-42 are allosteric partial agonists of mAChRs. In addition to their partial agonist effects, they inhibit the binding N-methylscopolamine to rat M_2 (McNA-A-343) and human M_1 (AC-42) mAChRs, while retarding NMS dissociation (113, 114).

The situation is further complicated by the fact that AMs may affect receptor conformation so as to favor certain active states or change the interaction of the receptor with proteins, introducing bias into the signal output generated by orthosteric ligands. In cortical astrocytes, 5, N-{4-chloro-2-((1,3-dioxo-1,3-dihydro-2H-isoindol-2-yl) methyl) phenyl}-2-hydroxybenzamide (CPPHA), an AM of the type 5 metabotropic glutamate receptor (mGluR), potentiates calcium mobilization by the orthosteric agonist 3, 30-difluorobenzaldazine but decreases the maximal ERK activation stimulated by the same agonist (115). Similarly, binding of the allosteric agonist peptide ASLW to the CXCR4 chemokine receptor induces a stronger chemotactic response than the orthosteric ligand, CXCL12, but does not promote receptor internalization like CXCL12 (116). Thus, it is clear that AMs possess the same capacity to engender functional selectivity in GPCR signaling as orthosteric ligands, offering the potential for "selecting" desired pharmacological effects and excluding nondesired effects.

Among the more dramatic examples of allosteric effects on GPCR conformation are small molecule AMs of the type 5 chemokine receptor (CCR_5) (117, 118). CCR_5 acts as the cell surface co-receptor for the HIV-1 viral coat protein gp120, and binding is essential for viral entry and replication. CCR_5 and gp120 make contact at numerous points, and mutational studies have shown that the regions of the receptor involved in binding the endogenous ligands, chemokine ligand $(CCL)_3$ and CCL_5, differ from those that bind gp120. As a result, small molecules targeting the orthosteric-binding site do not inhibit HIV-1 entry. Nonetheless, structurally diverse AMs of CCR_5 (aplaviroc, maraviroc, vicriviroc, TAK-779 and TAK-220) that bind to a common allosteric site are able to produce global changes in CCR5 conformation that interfere with the interaction between CCR_5 and gp120 (119–121). The strategy is sufficiently effective at reducing HIV infectivity and reducing systemic viral load that maraviroc has been approved by the Food and Drug Administration as salvage therapy in advanced HIV disease.

The ability to modulate GPCR signaling via allosteric effects exerted by small molecules binding outside the orthosteric site offers potential advantages in pharmaceutical design. One is enhanced subtype selectivity. Most GPCRs cluster into families of closely related receptors that share a common endogenous ligand, e.g., M_1–M_5 mAChR and mGluR1–mGluR8. While selectivity between families that bind structurally distinct ligands is usually achievable, it is often difficult to obtain subtype selectivity between members of an individual family by targeting the orthosteric site. In contrast, AMs can exhibit exquisite selectivity between closely related receptors (122, 123). One reason may be that allosteric sites are under less evolutionary pressure with respect to conservation of function and thus display wider protein sequence divergence across receptor subtypes relative to orthosteric sites (124). AMs with little inherent intrinsic activity that act by enhancing or attenuating the response elicited by the endogenous ligand offer several potential advantages over conventional agonists and antagonists. First, AM effects are saturable and therefore less likely to elicit adverse effects from overdose. Second, their effects are exerted primarily in the presence of the orthosteric ligand. Thus, AM activity is tied to the temporal pattern of endogenous ligand release, such that they only amplify or reduce the receptor signal when the hormone or neurotransmitter is released. Third, the lack of chronic receptor activation may limit tachyphylaxis, overcoming the problem of diminishing therapeutic efficacy seen with many chronically administered orthosteric agonists. Fourth, AMs can bias signal output in favor of only part of the receptor response profile by imposing conformational constraints that limit the receptor's ability to engage effector/accessory proteins. In such cases, functionally biased AMs may be useful in restoring signal balance in systems where disease has altered downstream signaling, or even establish "new" functional receptor systems with unique signaling capability.

7. Quantifying Efficacy and Bias in an Allosteric World

In allosteric systems, it is useful to consider the receptor as a *conduit*, through which the energy imparted by the binding of a *modulator* leads to changes in the behavior of the *guest*, a second molecule interacting with the receptor at a different site. Although modulators are usually orthosteric or allosteric ligands, it is important to recognize that other proteins, e.g., RAMPs, can also act as modulators. Similarly, guests may be ligands, signaling proteins, e.g., G proteins, or other receptors. It is also important to recognize that these allosteric effects are reciprocal, in that the guest imparts the same energy through the conduit back to the modulator. From the thermodynamic perspective, the modulator and guest are interchangeable, in that the effect of each on the other is identical (125).

The simplest model to quantify functional allosteric effects is derived from the Ehlert allosteric model (106) and the Black/Leff operational model (3), describing the response to an agonist (A) in the presence of an allosteric modulator (B). The capacity of B to affect the response to A is reflected in the affinity and efficacy cooperativity factors, α and β, respectively. The equation below is an elaboration of the model that further incorporates the relative efficacy (ϕ) factor for B ligands that possess intrinsic efficacy ($\phi = \tau_B/\tau_A$), K_A and K_B are the equilibrium dissociation constants for A and B, respectively, and E_M is the maximum response capability of the system (126, 127):

$$\text{Response} = \frac{\left(\frac{[A]}{K_A}\left(1+\frac{\alpha\beta[B]}{K_B}\right)+\frac{\phi[B]}{K_B}\right)\tau A\, E_M}{\frac{[A]}{K_A}\left(1+\tau A+\frac{\alpha[B]}{K_B}(1+\beta\tau A)\right)+\frac{[B]}{K_B}(1+\phi\tau A)+1}.$$

As shown in Fig. 4, changes in the relative values of α, β, and ϕ can produce marked changes in dose–response curves generated using ligand A. Experimentally, values for α, β, and ϕ, along with the equilibrium dissociation constant for ligand B (K_B) can be derived by fitting dose response data to this equation, although the number of parameters demands the largest possible dataset to avoid ambiguity. Alternatively, complete allosteric datasets can be analyzed using the method of Ehlert (125). This approach is valid for dose–response data described by curves with Hill coefficients of unity and requires the generation of multiple dose–response curves. The technique uses the "relative activity" of ligands; a ratio of the products of the maxima of the allosterically modulated dose–response curve (MAX_{AB}) multiplied by the EC_{50} of the control curve (EC_{50A}), divided by the maxima of the control curve (MAX_A) curve multiplied by the EC_{50} of the modulated curve ($EC50_{AB}$) [$RA = (MAX_{AB} \times EC_{50A})/(MAX_A \times EC_{50AB})$], to estimate allosteric parameters.

Three important behaviors of AMs that emerge from these models are as follows: (1) their effects are saturable; (2) they can

Fig. 4. Allosteric modulation of GPCR function. In theory, AMs have the potential to independently change orthosteric ligand affinity or efficacy, and may themselves possess intrinsic efficacy. Shown are conceptual plots illustrating the range of possible AM effects on a reference agonist dose–response curve (*gray* line in each panel) based on the allosteric model incorporating direct allosteric agonism shown in the text. The cooperativity factors for agonist affinity (α) and efficacy (β) are assumed to vary independently. The intrinsic efficacy of the allosteric agonist is represented by the relative efficacy factor (ϕ). **(a-c)** Effect of varying the efficacy factor β of an AM with an affinity factor $\alpha<1$. The result is reduced agonist potency with either enhanced or diminished agonist efficacy. (d-f) Effect of varying the efficacy factor β of an AM with an affinity factor $\alpha>1$. The result is enhanced agonist potency with either enhanced or diminished agonist efficacy. (g-i) Introduction of allosteric agonism ($\phi>0$) to an AM with an affinity factor $\alpha<1$ and variable efficacy factor β. (j-l) Allosteric agonism ($\phi>0$) superimposed onto an AM with an affinity factor $\alpha>1$ and variable efficacy factor β.

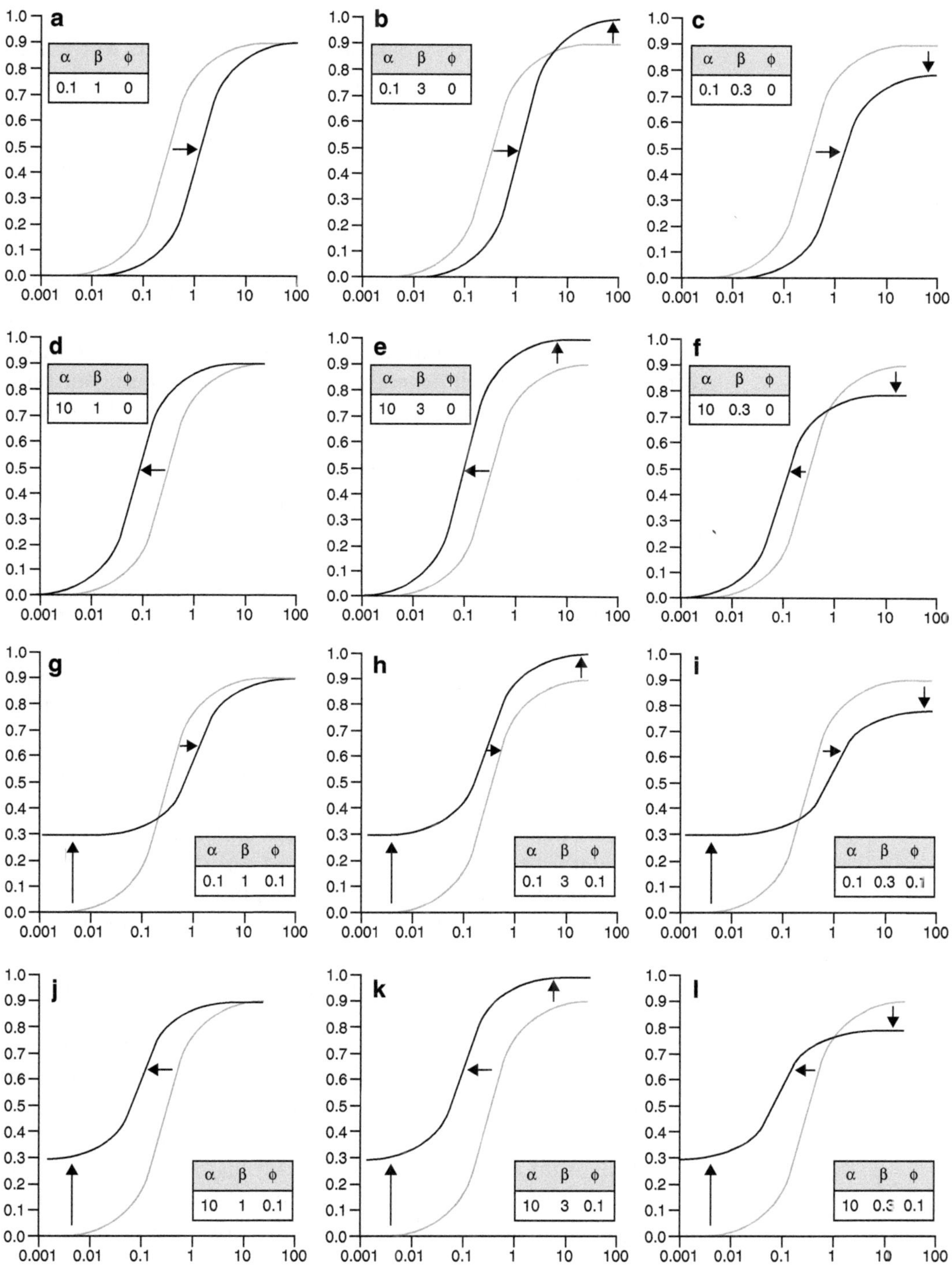
a
α β ϕ
0.1 1 0
b
α β ϕ
0.1 3 0
c
α β ϕ
0.1 0.3 0
d
α β ϕ
10 1 0
e
α β ϕ
10 3 0
f
α β ϕ
10 0.3 0
g
α β ϕ
0.1 1 0.1
h
α β ϕ
0.1 3 0.1
i
α β ϕ
0.1 0.3 0.1
j
α β ϕ
10 1 0.1
k
α β ϕ
10 3 0.1
l
α β ϕ
10 0.3 0.1
0.0 0.1 0.2 0.3 0.4 0.5 0.6 0.7 0.8 0.9 1.0
0.001 0.01 0.1 1 10 100

be probe-dependent; and, (3) they can have differential effects on affinity and efficacy (104). Saturability of effect refers to the fact that no further modulation can be achieved once the allosteric binding site on the receptor is fully occupied. Whereas orthosteric antagonists can eventually surmount the effect of a fixed concentration of agonist, allosteric antagonists have a maximum effect beyond which further increases in concentration have no effect on the agonist response. This is because the allosterically modulated receptor is not necessarily inactive; rather it is conformationally constrained such that it exhibits altered reactivity toward guest probes. AMs with moderate values of α ($0.1 < \alpha < 1$) may produce a right shift in the dose–response curve, while those with β ($0.1 < \beta < 1$) will decrease the maximal response. But only at values of α or β approaching 0 will agonist responsiveness be abolished.

Probe dependence results from the fact that AMs alter the tertiary conformations available to receptor and there is no a priori reason that this "new" receptor will behave like the unmodulated receptor. A probe can be any molecule interacting with the receptor whose behavior can be measured, e.g., orthosteric ligand binding, G protein activation, etc., and it may be expected that the modulated receptor will exhibit different reactivity toward different probes. For example, the CCR_5 AM aplaviroc inhibits the actions of both CCL_3 and CCL_5 and blocks the binding of CCL_3, yet has minimal effect on the binding of CCL_5 (119). With respect to cytosolic probes, the neurokinin NK_2 receptor agonist neurokinin A normally activates Gs and Gq, but in the presence of the AM LP1805 it produces enhanced Gs activation ($\beta_{Gs} > 1$) without Gq activation ($\beta_{Gq} << 1$) (128). Similarly, Nα-tosyltryptophan, an AM of the corticotrophin-releasing hormone CRH_2 receptor, causes the natural agonist prostaglandin D2 to alter its activation profile from coupling to both Gi and arrestin to sole activation of Gi (129). Thus probe dependence opens the possibility of modulating receptor conformation in a manner that biases the cytosolic response to receptor activation by endogenous agonists, i.e., to impose functional selectivity on orthosteric ligands that do not normally possess it.

Antagonism of orthosteric agonists in the presence of an AM can result from a reduction in either agonist affinity or efficacy (α or $\beta < 1$), while increases in either affinity or efficacy can potentiate the response (α or $\beta > 1$). The fact that the cooperativity factors α and β can vary independently means that AMs may modulate affinity and efficacy separately, leading to a wide range of possible effects. AMs that decrease agonist affinity but do not change efficacy ($\alpha < 1$; $\beta = 1$) can produce surmountable antagonism that will resemble the effect of a competitive antagonist save for the saturability of its effect. Conversely, AMs that change efficacy ($\beta <$ or > 1) can either increase or decrease the maximal response independent of effects on ligand affinity, creating the potential for mixed effects. An illustrative example would be an AM that blocks agonist efficacy ($\beta = 0$), but increases agonist affinity ($\alpha > 1$). Because the transfer of energy between the allosteric- and orthosteric-binding sites is reciprocal, such a ligand would exhibit the interesting property of increasing its affinity for the receptor as agonist concentration rises, i.e., it would "sense" the degree to which the system is being driven by the endogenous agonist and adjust its potency accordingly (104). Ifenprodil, an allosteric antagonist of the N-methyl D-aspartate receptor, is such a "smart antagonist," demonstrating increased potency in the presence of increasing concentrations of the agonist, *N*-methyl D-aspartate (130). When one adds the further factor of allosteric agonism ($\phi > 0$), it becomes clear that allosteric modulation of GPCR signaling can produce a wide range of cellular effects (Fig. 4).

Functional selectivity adds yet another dimension to the quantitative description of ligand effects. Because of probe dependence, the factors K_A and τ that describe agonism in the Black/Leff operational model (3) cannot be assumed to be the same for each signaling pathway downstream of the receptor. Indeed, if efficacy is truly "pluridimensional" (131), it must be defined in terms of the specific signaling pathway being used to probe the activation state of the receptor. Still, if one views the functionally selective ligand as "modulator" and the signaling protein as "guest," there is no need for a new model. All that is required is a means to quantitatively monitor different signaling pathways activated by the receptor.

A practical approach to quantify ligand bias using the operational model is to determine ratios of Log (τ/K_A) (4, 104) for each agonist and each pathway. This can be done by fitting the dose–response curve for each agonist to the operational model in the form below, where τ_A is the efficacy of the agonist for the pathway, K_A is the equilibrium dissociation constant for the agonist (A), and E_M is the maximum response capability of the system:

$$\text{Response} = \frac{[\text{A}]^n \tau^n E_\text{M}}{[\text{A}]^n \tau^n + ([\text{A}] + K_\text{A})^n}.$$

The term τ encompasses both the intrinsic efficacy of the agonist and system-dependent factors such as receptor density and coupling efficiency. Since the latter factors are constant for any dose–response curve determined in the same cell for any given signaling pathway, the ratio of τ values for any two agonists in the same system will yield a ratio of intrinsic efficacy for activation of the pathway that is independent of receptor number or coupling efficiency. Because allosteric systems can impose effects on both affinity and efficacy, it cannot be assumed that K_A values for a given agonist will be constant for all signaling pathways, hence functional selectivity must be quantified with τ/K_A ratios for each pathway. Once determined for each agonist/pathway of interest, τ/K_A ratios can be used to describe bias relative to a reference agonist, e.g., the endogenous ligand (104).

An advantage of the operational model is its capability to quantify the full range of agonism from submaximal effects to effects in very sensitive systems with receptor reserve (2), meaning that separate scales, such as relative potency for full agonists and relative efficacy for partial agonists need not be used. Further, τ/K_A ratios determined in one system should be applicable to all systems without the need to quantify functional selectivity in all systems. Variation in receptor density and coupling efficiency between systems might change the ability of all agonists targeting a given receptor to activate a particular pathway, but it will not change the pathway selective "bias" of different ligands relative on one another.

8. The Pitfalls and Perils of Assay Design

The evolution of our concept of GPCRs from simple molecular switches to allosteric proteins whose function is modified through contact with orthosteric and allosteric ligands, as well as other proteins, has immediate implications for the design of assays to characterize GPCR signaling in the research or drug discovery settings (132–134). In particular, the phenomena of probe dependence and pluridimensional efficacy present significant challenges, since the use of unidimensional assays to characterize ligand effects may miss critical properties. For example, screens based on cAMP production would characterize the PTH analog (D-Trp^{12}, Tyr^{34})PTH(7–34) as a PTH_1 receptor antagonist (135). If a constitutively activated PTH_1R were used to elevate basal cAMP levels in the assay, it would appear as an inverse agonist for PTH_1R-Gs coupling (44). If assayed based on arrestin recruitment, its agonist efficacy for arrestin binding and arrestin-dependent PTH_1R sequestration would emerge (43, 136). Whereas the agonist activity of the conventional PTH_1R agonist PTH_{1-34} would emerge from any of these screens, only when activity is compared across cAMP, arrestin recruitment, and ERK activation assays would (D-Trp^{12}, Tyr^{34})PTH(7–34) be identified as an arrestin pathway-selective biased agonist for the PTH_1R (45). Even this would miss the important finding that in vivo, (D-Trp^{12}, Tyr^{34})PTH(7–34) promotes osteoblastic bone formation without stimulating Gs-dependent bone resorption or hypercalcuria like the conventional agonist (137). The emergence of such potentially beneficial therapeutic effects, in this case the ability of the biased ligand to dissociate the desired property of increased bone formation from deleterious effects on bone resorption and calcium excretion, could not have been predicted a priori from the in vitro efficacy profile.

How then, should we proceed when studying ligand effects on signal transduction or establishing a drug discovery platform? For the present, the answer depends on the objective. Discovering AMs presents particular problems. For one, since many allosteric sites lie outside of the physiological ligand-binding domain, there is far less structural conservation than for orthosteric sites that evolved to confer fidelity in receptor activation by endogenous ligands. Although the recent elucidation of high-resolution GPCR structures (20) may eventually reveal commonalities among AM binding pockets and enable virtual screening, at present the discovery of AMs is primarily through the use of functional assays. But because of their structural diversity, AMs have much greater potential to produce off-target effects. For example, the clinically useful diuretic amiloride, which produces its therapeutic effects via blockade of renal tubular epithelial sodium channels, also exerts an allosteric effect at $\alpha_{2A/23}$ adrenergic receptors (138). The greater potential for cross-reactivity inherent in small molecules needs to be considered both when screening AMs for activity and when attempting to ascribe physiologic effects on complex systems to a specific receptor target, and well-defined experimental criteria must be employed in ascribing insurmountable drug effects to allosteric modulation (139). Probe dependence also comes into play, since allosteric effects on orthosteric ligand affinity may differ when different orthosteric probes are used, making it desirable, albeit not always practical, to use the native ligand as the probe in such screens (4).

If the objective is an unbiased screen for whether compounds exert any form of activity against a receptor, then technology designed to detect integrated whole cell pharmacologic responses may provide the first step. Resonant waveguide grating technology has led to the development of optical biosensors that can measure dynamic mass redistribution (DMR) signals in living cells (140, 141). This technology can detect interactions of GPCRs with cytosolic signaling molecules at a depth of 150–200 nM and also detect receptor internalization. The resulting DMR signal is a noninvasive cell-based technology that can measure virtually any receptor activation in any cell type in real time. The approach has been applied to the detection and quantification of functional selectivity in intact cells (142). Another technology that can be used for the same purpose involves measuring changes in the electrical impedance of cell monolayers caused by receptor-mediated changes in cell mass redistribution (143, 144). The principal limitation of these "label-free" systems is that since whole cell responses represent the integration of numerous pathways, it may be difficult or impossible to deconvolute the kinetic pattern of the response in a manner that allows identification of the specific signaling pathway(s) being affected. As a result, functionally selective ligands can exhibit highly variable agonist potency ratios in different cell types due to changes in receptor coupling to downstream pathways caused by variation in the abundance of different effectors and regulatory proteins, e.g., G protein and arrestin isoforms. On the other hand, label-free systems can be used with primary cells, circumventing the potential for signaling artifacts arising from promiscuous coupling by overexpressed receptors in engineered systems, and allowing ligands to be screened in the most therapeutically relevant cell type.

If the goal is to define efficacy in a limited number of known dimensions, e.g., G protein and arrestin coupling, then multiplexing assays using different probes can provide a more complete efficacy profile and permit identification/quantification of functional selectivity (145). Beyond traditional assays based on ligand binding, G protein activation, or second messenger generation, a number of recent technological advances have expanded the tools available to measure different aspects of receptor activation. High content assays based on imaging techniques that utilize fluorescent signals can provide both temporal and spatial information about signaling events (146, 147). Considering one such event, the interaction of GPCRs with arrestins, responses can be monitored through direct visualization of GPCR/arrestin–green fluorescent protein complexes (148, 149), with bioluminescence resonance energy transfer (150), with enzyme fragment complementation (151), or with protease-activated transcriptional reporter genes (152). Greater efficiency can be obtained by multiplexing green fluorescent protein- and immunofluorescence-based assays within the same cell to provide simultaneous readouts of multiple signaling pathways within the same cell (153). The principal challenge is less in designing assays to capture the pluridimensionality of signaling, than in determining which aspects of signaling are relevant to the problem at hand; in other words which assays to multiplex. Our current understanding of GPCR signaling, particularly the physiological relevance of recently discovered G protein-independent signals (69), is often inadequate to allow us to predict the ligand efficacy profile most likely to elicit a desired physiological effect.

Yet another consideration is the impact of cell type on assay results. Because GPCRs often display promiscuous coupling at the high levels of receptor expression commonly employed in cell-based screening assays (30–32), use of highly engineered cells to characterize ligand efficacy could lead to the detection of effects that have no relevance in the physiologic setting. In addition, some potentially desirable signaling events are cell background specific, since they are dependent upon not only receptor density but also the relative abundance of effector proteins and the influence of lateral allosteric effects. For example, PTH_1 receptor coupling to Gi is detectable only in renal tubular epithelium, since it is dependent on expression of NHERF1/2 (103). If the goal is to find compounds that produce a specific efficacy profile in a specific tissue, then there are clear advantages to screening in native cell systems. Advances in "label-free" assay systems make such approaches more feasible. Still, primary cells are difficult to produce in quantity needed for screening efforts and often exhibit dramatic batch-to-batch variability, making them far from ideal for industrial purposes. Engineered systems, on the other hand, offer many advantages, despite their propensity to reveal non-physiologic responses. One significant advantage is the ability to "bias" the system to increase sensitivity to detect the signal of interest. For example, varying the level of G protein α subunit expression in α_{2B} adrenergic receptor-expressing Sf9 cells allows for the detection of differences between noradrenaline and the synthetic agonist UK14304 in their ability to couple the receptor to Gi and Gs (32). Similarly, overexpressing GRK2 and arrestin3 in HEK293 cells increases sensitivity for detecting ligand-related differences in the ability of morphine and herkinorin to induce μ opioid receptor internalization (154). At the end of the day it is clear that assays can detect only what they are designed to detect, so judicious choice of both assay and cell background is necessary if the results of screening are to be relevant.

9. Conclusions

The pluridimensionality GPCR signaling, as illustrated by the phenomena of guest allostery and functional selectivity, mandates a change from the traditional approach to signal transduction research and pharmaceutical development. Classical models of affinity and efficacy often fail to adequately describe the behavior of receptors, and the classification of ligand activity must be viewed and system and assay dependent. As our knowledge of GPCR signaling expands along with the assays to detect these events, we are able to generate a more detailed picture of the effect of ligands on receptors and even to tailor ligands to elicit specific signal profiles. At the same time it is apparent that the traditional nomenclature for classifying drugs as agonists, partial agonists, antagonists, etc., based on their activity in cellular systems is inadequate to capture the complexity arising from ligand–receptor interaction. It is increasingly necessary to define ligand effects using multiple readouts of receptor activation and to describe ligand bias in quantitative terms. Rather than a hindrance to therapeutic design, this additional complexity offers the prospect of new generations of GPCR-targeted therapies that exploit functional selectivity and allosteric modulation to achieve more specific therapeutic effects or diminish toxicity. The greatest challenge at present is not in designing assays to detect these phenomena, but in determining what efficacy profile is needed to produce the optimal treatment response for any given disease.

Acknowledgments

The Luttrell laboratory is supported by National Institutes of Health Grant DK55524 and the Research Service of the Charleston SC Veterans Affairs Medical Center.

References

1. Stephenson, R. P. (1956) A modification of receptor theory. *Br J Pharmacol* **11**, 379–93.
2. Black, J. W., and Leff, P. (1983) Operational models of pharmacological agonist. *Proc R Soc Lond [Biol]* **220**, 141–62.
3. Black, J. W., Leff, P., and Shankley, N.P. (1985) An operational model of pharmacological agonism: The effect of E/[A] curve shape on agonist dissociation constant estimation. *Br J Pharmacol* **84**, 561–71.
4. Kenakin, T., and Miller, L. E. (2010) Seven transmembrane receptors as shapeshifting proteins: the impact of allosteric modulation and functional selectivity on new drug discovery. *Pharmacol Rev.* **62**, 265–304.
5. Koshland, D. E., Jr. (1998) Conformational changes: How small is big enough? *Nature Med* **4**, 1112–4.
6. Monod, J., Wyann, J., and Changeux, J-P. (1965) On the nature of allosteric transitions: a plausible model. *J Mol Biol* **12**, 88–118.
7. Karlin, A. (1967). On the application of "a plausible model" of allosteric proteins to the receptor for acetylcholine. *J Theor Biol* **16**, 306–320.
8. Thron, C.D. (1973) On the analysis of pharmacological experiments in terms of an allosteric receptor model. *Mol Pharmacol* **9**, 1–9.
9. Okazaki, K-I., and Takada, S. (2008) Dynamic energy landscape view of coupled binding and protein conformational change: Induced-fit versus population shift models. *Proc Natl Acad Sci USA* **105**, 11182–7.
10. Kenakin, T. P. (1996) Receptor conformational induction versus selection: All part of the same energy landscape. *Trends Pharmacol Sci.* **17**, 190–1.
11. Palczewski, K., Kumasaka, T., Hori, T., Behnke, C. A., Motoshima, H., Fox, B. A., Trong, I. Le., Teller, D. C., Okada, T., Stenkamp, R. E., Yamamoto, M., and Miyano, M. (2000) Crystal structure of rhodopsin: a G protein-coupled receptor. *Science* **289**:739–45.
12. Okada, T., Sugihara, M., Bondar, A. N., Elstner, M., Entel, P., and Buss, V. (2004) The retinal conformation and its environment in rhodopsin in light of a new 2.2 Å crystal structure, *J Mol Biol* **342**, 571–83.
13. Salom, D., Lodowski, D. T., Stenkamp, R. E., Trong, I. Le, Golczak, M., Jastrzebska, B., Harris, T., Ballesteros, J. A., and Palczewski, K. (2006) Crystal structure of a photoactivated deprotonated intermediate of rhodopsin. *Proc Natl Acad Sci U S A* **103**, 16123–8.
14. Scheerer, P., Park, J. H., Hildebrand, P. W., Kim, Y. J., Krauss, N., Choe, H. W., Hofmann, K. P., and Ernst, O. P. (2008) Crystal structure of opsin in its G-protein-interacting conformation. *Nature* **455**, 497–502.
15. Gether, U., Lin, S., and Kobilka, B.K. (1995) Fluorescent labeling of purified β2 adrenergic receptor. Evidence for ligand-specific conformational changes. *J Biol Chem* **270**, 28268–75.
16. Ghanouni, P., Gryczynski, Z., Steenhuis, J.J., Lee, T.W., Farrens, D.L., Lakowicz, J.R., and Kobilka, B.K. (2001) Functionally different agonists induce distinct conformations in the G protein coupling domain of the β2 adrenergic receptor. *J Biol Chem* **276**, 24433–6.
17. Cherezov, V., Rosenbaum, D. M., Hanson, M. A., Rasmussen, S. G. F., Thian, F. S., Kobilka, T. S., Choi, H-J., Kuhn, P., Weis, W. I., Kobilka, B. K., and Stevens, R. C. (2007)

High-resolution crystal structure of an engineered human β_2-adrenergic G protein-coupled receptor. *Science.* **318**, 1258–65.

18. Rosenbaum, D. M., Cherezov, V., Hanson, M. A., Rasmussen, S. G. F., Thian, F. S., Kobilka, T. S., Choi, H-J., Yao, X-J., Weis, W. I., Stevens, R. C., and Kobilka, B. K. (2007) GPCR engineering yields high-resolution structural insights into β_2-adrenergic receptor function. *Science.* **318**, 1266–73.
19. Rasmussen, S. G. F., Choi, H-J., Rosenbaum, D. M., Kobilka, T. S., Thian, F. S., Edwards, P. C., Burghammer, M., Ratnala, V. R. P., Sanishvili, R., Fischetti, R. F., Schertler, G. F. X., Weis, W. I., and Kobilka, B. K. (2007) Crystal structure of the human β_2 adrenergic G-protein-coupled receptor. *Nature* **450**, 383–7.
20. Weis, W. I., and Kobilka, B. K. (2008) Structural insights into G-protein-coupled receptor activation. *Curr Opin Struct Biol* **18**, 734–40.
21. Allen, L. F., Lefkowitz, R. J., Caron, M. G., and Cotecchia, S. (1991) G-protein-coupled receptor genes as protooncogenes: constitutively activating mutation of the alpha 1B-adrenergic receptor enhances mitogenesis and tumorigenicity. *Proc Natl Acad Sci U S A* **88**, 11354–8.
22. Samama, P., Cotecchia, S., Costa, T., and Lefkowitz, R. J. (1993) A mutation-induced activated state of the beta 2-adrenergic receptor. Extending the ternary complex model. *J Biol Chem* **268**, 4625–36.
23. Kosugi, S., Shenker, A., and Mori, T. (1994) Constitutive activation of cyclic AMP but not phosphatidylinositol signaling caused by four mutations in the 6th transmembrane helix of the human thyrotropin receptor. *FEBS Lett* **356**, 291–4.
24. Kosugi, S., Van Dop, C., Geffner, M. E., Rabl, W., Carel, J. C., Chaussain, J.L., Mori, T., Merendino, J. J. Jr., and Shenker, A. (1995) Characterization of heterogeneous mutations causing constitutive activation of the luteinizing hormone receptor in familial male precocious puberty. *Hum Mol Genet* **4**, 183–8.
25. Kjelsberg, M. A., Cotecchia, S., Ostrowski, J., Caron, M. G., and Lefkowitz, R. J. (1992) Constitutive activation of the alpha 1B-adrenergic receptor by all amino acid substitutions at a single site. Evidence for a region which constrains receptor activation. *J Biol Chem* **267**, 1430–3.
26. Gether, U., Ballesteros, J. A., Seifert, R., Sanders-Bush, E., Weinstein, H., and Kobilka, B. K. (1997) Structural instability of a constitutively active G protein-coupled receptor. Agonist-independent activation due to conformational flexibility. *J Biol Chem* **272**, 2587–90.
27. Costa. T., and Herz, A. (1989) Antagonists with negative intrinsic activity at delta opioid receptors coupled to GTP-binding proteins. Proc Natl Acad Sci U S A **86**, 7321–5.
28. Weiss, J. M., Morgan, P. H., Lutz, M. W., and Kenakin, T. P. (1996) The cubic ternary complex receptor-occupancy model. III. resurrecting efficacy. *J Theor Biol* **181**, 381–97.
29. Kenakin T. (1995) Pharmacological proteus? *Trends Pharmacol Sci* **16**, 256–8.
30. Cordeaux, Y., Briddon, S. J., Megson, A. E., McDonnell, J., Dickenson, J. M., and Hill, S. J, (2000) Influence of receptor number on functional responses elicited by agonists acting at the human adenosine A(1) receptor: evidence for signaling pathway-dependent changes in agonist potency and relative intrinsic activity. *Mol Pharmacol* **58**, 1075–84.
31. Zhu, X., Gilbert, S., Birnbaumer, M., and Birnbaumer, L. (1994) Dual signaling potential is common among Gs-coupled receptors and dependent on receptor density. *Mol Pharmacol* **46**, 460–9.
32. Nasman, J., Kukkonen, J. P., Ammoun, S., and Akerman, K. E. (2001) Role of G-protein availability in differential signaling by α2-adrenoceptors. *Biochem Pharmacol* **62**, 913–22.
33. Offermanns, S., Wieland, T., Homann, D., Sandmann, J., Bombien, E., Spicher, K., Schultz, G., and Jakobs, K. H. (1994) Transfected muscarinic acetylcholine receptors selectively couple to Gi-type G proteins and Gq/11. *Mol Pharmacol* **45**, 890–898.
34. Kenakin, T. (1995) Agonist-receptor efficacy. II. Agonist-trafficking of receptor signals. *Trends Pharmacol Sci* **16**, 232–38.
35. Meller, E., Puza, T., Diamond, J., Lieu, H. D., and Bohmaker, K. (1992) Comparative effects of receptor inactivation, 17 β-estradiol and pertussis toxin on dopaminergic inhibition of prolactin secretion in vitro. *J Pharmacol Exp Ther* **263**, 462–9.
36. Spengler, D., Waeber, C., Pantaloni, C., Holsboer, F., Bockaert, J., Seeburg, P. H., and Journot, L. (1993) Differential signal transduction by five splice variants of the PACAP receptor. *Nature* **365**, 170–5.
37. Berg, K. A., Maayani, S., Goldfarb, J., Scaramellini, C., Leff, P., and Clarke, W. P. (1998) Effector pathway-dependent relative efficacy at serotonin type 2A and 2 C receptors: evidence for agonist-directed trafficking of receptor stimulus. *Mol Pharmacol* **54**, 94–104.

38. Sagan, S., Chassaing, G., Pradier, L., and Lavielle, S. (1996) Tachykinin peptides affect differently the second messenger pathways after binding to CHO-expressed human NK-1 receptors. *J Pharmacol Exp Ther* **276**, 1039–48.

39. Takasu, H., Gardella, T. J., Luck, M. D., Potts, J. T., and Bringhurst, F. R. (1999) Amino-terminal modifications of human parathyroid hormone (PTH) selectively alter phospholipase C signaling via the type 1 PTH receptor: Implications for design of signal-specific PTH ligands *Biochemistry* **38**, 13453–60.

40. Mohan, S., Kutilek, S., Zhang, C., Shen, H. G., Kodama, Y., Srivastava, A. K., Wergedal, J. E., Beamer, W. G., and Baylink, D. J. (2000) Comparison of bone formation responses to parathyroid hormone(1–34), (1–31), and (2–34) in mice. *Bone* **27**, 471–8.

41. Gray, J. A., and Roth, B. L. (2001) Paradoxical trafficking and regulation of 5-HT(2A) receptors by agonists and antagonists. *Brain Res Bull* **56**, 441–51.

42. Holloway, A. C., Qian, H., Pipolo, L., Ziogas, J., Miura, S., Karnik, S., Southwell, B. R., Lew, M. J., and Thomas, W. G. (2002) Side-chain substitutions within angiotensin II reveal different requirements for signaling, internalization and phosphorylation of type 1A angiotensin receptors. *Mol Pharmacol* **61** 768–77.

43. Sneddon, W. B., Magyar, C. E., Willick, G. E., Syme, C. A., Galbiati, F., Bisello, A., and Friedman, P. A. (2004) Ligand-selective dissociation of activation and internalization of the parathyroid hormone (PTH) receptor: conditional efficacy of PTH peptide fragments. *Endocrinology* **145**, 2815–23.

44. Bisello, A., Chorev, M., Rosenblatt, M., Monticelli, L., Mierke, D. F., and Ferrari, S. L. (2002) Selective ligand-induced stabilization of active and desensitized parathyroid hormone type 1 receptor conformations. *J Biol Chem* **277**, 38524–30.

45. Gesty-Palmer, D., Chen, M., Reiter, E., Ahn, S., Nelson, C. D., Wang, S., Eckhardt, A. E., Cowan, C. L., Spurney, R. F., Luttrell, L. M., and Lefkowitz, R. J. (2006) Distinct conformations of the parathyroid hormone receptor mediate G protein and beta-arrestin dependent activation of ERK1/2. *J Biol Chem* **281**, 10856–64.

46. Walters, R. W., Shukla, A. K., Kovacs, J. J., Violin, J. D., DeWire, S. M., Lam, C. M., Chen, J. R., Muehlbauer, M. J., Whalen, E. J., and Lefkowitz, R. J. (2009) beta-Arrestin1 mediates nicotinic acid-induced flushing, but not its antilipolytic effect, in mice. *J Clin Invest* **119**, 1312–21.

47. Seifert, R., Gether, U., Wenzel-Seifert, K., and Kobilka, B. K. (1999) Effects of guanine, inosine, and xanthine nucleotides on β(2)-adrenergic receptor/G(s) interactions: evidence for multiple receptor conformations. *Mol Pharmacol* **56**, 348–58.

48. Perez, D. M., Hwa, J., Gaivin, R., Mathur, M., Brown, F. and Graham, R. M. (1996) Constitutive activation of a single effector pathway: evidence for multiple activation states of a G protein-coupled receptor. *Mol Pharmacol* **49**, 112–22.

49. Barroso, S., Richard, F., Nicolas-Etheve, D., Kitabgi, P., and Labbe-Jullie, C. (2002) Constitutive activation of the neurotensin receptor 1 by mutation of Phe(358) in helix seven. *Br J Pharmacol* **135**, 997–1002.

50. Swaminath, G., Xiang, Y., Lee, T. W., Steenhuis, J., Parnot, C., and Kobilka, B. K. (2004) Sequential binding of agonists to the beta2 adrenoceptor. Kinetic evidence for intermediate conformational states. *J Biol Chem* **279**, 686–91.

51. Hruby, V. J., and Tollin, G. (2007) Plasmon-waveguide resonance (PWR) spectroscopy for directly viewing rates of GPCR/G-protein interactions and quantifying affinities. *Curr Opin Pharmacol* **7**, 507–14.

52. Vilardaga, J. P., Bunemann, M., Krasel, C., Castro, M., and Lohse, M. J. (2003) Measurement of the millisecond activation switch of G-protein-coupled receptors in living cells. *Nat Biotechnol* **21**, 807–12.

53. Swaminath, G., Deupi, X., Lee, T. W., Zhu, W., Thian, F. S., Kobilka, T. S., and Kobilka, B. (2005) Probing the β2-adrenoceptor binding site with catechol reveals differences in binding and activation by agonists and partial agonists. *J Biol Chem* **280**, 22165–71.

54. Zurn, A., Zabel, U., Vilardaga, J-P., Schindlein, H., Lohse, M. J., and Hoffmann, C. (2009) Fluorescence resonance energy transfer analysis of α2a-adrenergic receptoractivation reveals distinct agonist-specific conformational changes. *Mol Pharmacol* **75**, 534–41.

55. Galandrin, S., Oligny-Longpre, G., Bonin, H., Ogawa, K., Gales, C., and Bouvier, M. (2008) Conformational rearrangements and signaling cascades involved in ligand-based mitogen-activated protein kinase signaling through the β1-adrenergic recepor. *Mol Pharmacol* **74**, 162–72.

56. Baneres, J-L., Mesnier, D., Martin, A., Joubert, L., Dumuis, A., and Bockaret, J. (2005) Molecular characterization of a purified 5-HT4 receptor. *J Biol Chem* **280**, 20253–60.

57. Tutor, A., Penela, P., and Mayor, F. (2007) Anti-β1-adrenergic receptor autoantibodies are potent stimulators of the ERK1/2 pathway in cardiac cells. *Cardiovasc Res* **76**, 51–60.

58. Pellissier, L. P., Sallander, J., Campillo, M., Gaven, F., Queffeulou, E., Pillot, M., Dumuis, A., Claeysen, S., Bockaert, J., and Pardo, L. (2009) Conformational toggle switches implicated in basal constitutive and agonist-induced activated states of 5-hydroxytryptamine-4 receptors. *Mol Pharmacol* **75**, 982–90.

59. Whistler, J. L., Chuang, H. H., Chu, P., Jan, L.Y., and von Zastrow, M. (1999) Functional dissocation of mu opioid receptor signaling and endocytosis: implications for the biology of opiate tolerance and addiction. *Neuron* **23**, 737–46.

60. Jarpe, M. B., Knall, C., Mitchell, F. M., Buhl, A. M., Duzic, E., and Johnson, G. L. (1998) [D-Arg1,D-Phe5,D-Trp7,9,Leu11]Substance P acts as a biased agonist toward neuropeptide and chemokine receptors. *J Biol Chem* **273**, 3097–104.

61. Kudlacek, O., Waldoer, M., Kassack, M. U., Nickel, P., Salmi, J. I., Freissmuth, M., and Nanoff, C. (2002) Biased inhibition by a suramin analogue of A1-adenosine receptor/G protein coupling in fused receptor/G protein tandems: the A1 adenosine receptor is predominantly coupled to Goalpha in human brain. *Naunyn Schmiedeberg's Arch Pharmacol* **365**, 8–16.

62. Manning, D. R. (2002) Measures of efficacy using G proteins as endpoints: differential engagement of G proteins through single receptors. *Mol Pharmacol* **62**, 451–52.

63. Gurwitz, D., Haring, R., Heldman, E., Fraser, C. M., Manor, D., and Fisher, A. (1994) Discrete activation of transduction pathways associated with acetylcholine M1 receptor by several muscarinic ligands. *Eur J Pharmacol* **267**, 21–31.

64. Lawler, C. P., Prioleau, C., Lewis, M. M., Mak, C., Jiang, D., Schetz, J. A., Gonzalez, A. M., Sibley, D. R., and Mailman, R. B. (1999) Interactions of the novel antipsychotic aripiprazole (OPC-14597) with dopamine and serotonin receptor subtypes. *Neuropsychopharmacology* **20**, 612–27.

65. Drake, M. T., Violin, J. D., Whalen, E. J., Wisler, J. W., Shenoy, S. K., and Lefkowitz, R. J. (2008) beta-Arrestin-biased agonism at the beta2-adrenergic receptor. *J Biol Chem* **283**, 5669–76.

66. Kenakin, T. P. (2007) Functional selectivity through protean and biased agonism: who steers the ship? *Mol Pharmacol* **72**, 1393–401.

67. Maudsley, S., Martin, B., and Luttrell, L. M. (2005) The origins of diversity and specificity in G protein-coupled receptor signaling. *J Pharmacol Exp Ther* **314**, 485–94.

68. Gesty-Palmer, D., and Luttrell, L. M. (2008) Heptahelical terpsichory. Who calls the tune? *J Recept Signal Transduct Res* 28, 39–58.

69. Luttrell, L. M., and Gesty-Palmer, D. (2010) Beyond desensitization: physiological relevance of arrestin-dependent signaling. *Pharmacol Rev* **62**, 305–30.

70. Ferguson, S. S. (2001) Evolving concepts in G protein-coupled receptor endocytosis: the role in receptor desensitization and signaling. *Pharm Rev* **53**, 1–24.

71. Luttrell, L. M., Ferguson, S. S. G., Daaka, Y., Miller, W. E., Maudsley, S., Della Rocca, G. J., Lin, F-T., Kawakatsu, H., Owada, K., Luttrell, D. K., Caron, M. G., and Lefkowitz, R. J. (1999) β-Arrestin-dependent formation of β2 adrenergic receptor/Src protein kinase complexes. *Science* **283**, 655–61.

72. Lefkowitz, R. J., and Shenoy, S. K. (2005) Transduction of receptor signals by beta-arrestins. *Science.* **308**, 512–7.

73. Wei, H., Ahn, S., Shenoy, S. K., Karnik, S., Hunyady, L., Luttrell, L. M., and Lefkowitz, R. J. (2003) Independent G protein and beta-arrestin2 mediated activation of ERK by angiotensin. *Proc Natl Acad Sci USA* **100**, 10782–7.

74. Azzi, M., Charest, P. G., Angers, S., Rousseau, G., Kohout, T., Bouvier, M., and Pineyro, G. (2003) Beta-arrestin-mediated activation of MAPK by inverse agonists reveals distinct active conformations for G protein-coupled receptors. *Proc Natl Acad Sci USA* **100**, 11406–11.

75. Wisler, J. W., DeWire, S. M., Whalen, E. J., Violin, J. D., Drake, M. T., Ahn, S., Shenoy, S. K., and Lefkowitz, R. J. (2007) A unique mechanism of beta-blocker action: carvedilol stimulates beta-arrestin signaling. *Proc Natl Acad Sci USA* **104**, 16657–62.

76. Ross, E. M., and Gilman, A. G. (1980) Biochemical properties of hormone-sensitive adenylate cyclase. *Annu Rev Biochem.* **49**, 533–64.

77. DeLean, A., Stadel, J.M., and Lefkowitz, R.J. (1980) A ternary complex model explains the agonist specific binding properties of the adenylate cyclase-coupled beta-adrenergic receptor. *J Biol Chem* 255, 7108–17.

78. Gurevich, V. V., Pals-Rylaarsdam, R., Benovic, J. L., Hosey, M. M., and Onorato, J. J. (1997) Agonist-receptor-arrestin, an alternative ternary complex with high agonist affinity. *J Biol Chem* **272**, 28849–52.

79. Key, T. A., Bennett, T. A., Foutz, T. D., Gurevich, V. V., Sklar, L. A., and Prossnitz, E. R. (2001) Regulation of formyl peptide receptor agonist affinity by reconstitution with arrestins and heterotrimeric G proteins. *J Biol Chem* **276**, 49204–12.
80. Kumar, S., Ma, B., Tsai, C-J., Sinha, N., and Nussinov, R. (2000) Folding and binding cascades: Dynamic landscapes and population shifts. *Prot Sci* **9**, 10–9.
81. Hilser, V. J., Garcia-Moreno, E. B., Oas, T. G., Kapp, G., and Whitten, S. T. (2006) A statistical thermodynamic model of protein ensembles. *Chem Rev* **106**, 1545–58.
82. Livesay, D. R., Dallakyan, S., Wood, G. G., and Jacobs, D. J. (2004) A flexible approach for understanding protein stability. *FEBS Lett.* **576**, 468–76.
83. Hilser, V. J., and Thompson, E. B. (2007) Intrinsic disorder as a mechanism to optimize allosteric coupling in proteins. *Proc Natl Acad Sci USA* **104**, 8311–5.
84. Kobilka, B. K., and Deupi, X. (2007) Conformational complexity of G-protein-coupled receptors. *Trends Pharmacol Sci* **28**, 397–406.
85. Devi, L. (2001) Heterodimerization of G-protein-coupled receptors: Pharmacology, signaling and trafficking. *Trends Pharmacol Sci* **22**, 532–37.
86. Milligan, G. (2001) Oligomerisation of G-protein-coupled receptors. *J Cell Sci* **114**, 1265–71.
87. Angers, S., Salahpour, A., and Bouvier, M. (2002) Dimerization: An emerging concept for G protein-coupled receptor ontogeny and function. *Ann Rev Pharmacol Toxicol* **42**, 409–35.
88. Foord S. M., and Marshall, F. H. (1999) RAMPs: accessory proteins for seven transmembrane domain receptors. *Trends Pharmacol Sci* **20**, 184–7.
89. Hinkle, P. M., and Sebag, J. A. (2009) Structure and function of the melanocortin2 receptor accessory protein (MRAP). *Mol Cell Endocrinol* **300**, 25–31.
90. Brady, A. E., and Limbird, L. E. (2002) G protein-coupled receptor interacting proteins: emerging roles in localization and signal transduction. *Cell Signal* **14**, 297–309.
91. Bockaert, J, Marin, P, Dumuis, A and Fagni, L (2003) The "magic tail" of G protein-coupled receptors: An anchorage for functional protein networks. *FEBS Lett* **546**, 65–72.
92. Weinman, E. J., Hall, R. A., Friedman, P. A., Liu-Chen, L. Y., and Shenolikar, S. (2006) The association of NHERF adaptor proteins with G-protein-coupled receptors and receptor tyrosine kinases. Annu Rev Physiol 68, 491–505.
93. Fredriksson, R., Lagerstrom, M. C., Lundin, L. G., and Schioth, H. B. (2003) The G-protein-coupled receptors in the human genome form five main families. Phylogenetic analysis, paralogon groups, and fingerprints. *Mol. Pharmacol* **63**, 1256–72.
94. Jones, K. A., Borowsky, B., Tamm, J. A., Craig, D. A., Durkin, M. M., Dai, M., Yao, W. J., Johnson, M., Gunwaldsen, C., Huang, L. Y., Tang, C., Shen, Q., Salon, J. A., Morse, K., Laz, T., Smith, K. E., Nagarathnam, D., Noble, S. A., Branchek, T. A., and Gerald, C. (1998) GABA(B) receptors function as a heteromeric assembly of the subunits GABA(B) Rl and GABA(B)R2. *Nature* **396**, 674–9.
95. Kaupmann, K., Malitschek, B., Schuler, V., Heid, J., Froestl, W., Beck, P., Mosbacher, J., Bischoff, S., Kulik, A., Shigemoto, R., Karschin, A., and Bettler, B. (1998) GABA(B)-receptor subtypes assemble into functional heteromeric complexes. *Nature.* **396**, 683–7.
96. Margeta-Mitrovic, M., Jan, Y. N., and Jan, L. Y. (2000) A trafficking checkpoint controls GABA(B) receptor heterodimerization. *Neuron* **27**, 97–106.
97. Jordan, B. A., and Devi, L. A. (1999) G-protein-coupled receptor heterodimerization modulates receptor function. *Nature* **399**, 697–700.
98. Franco, R., Ferre, S., Agnati, L., Torvinen, M., Gines, S., Hillion, J., Casado, V., Lledo, P., Zoli, M., Lluis, C., and Fuxe, K. (2000) Evidence for adenosine/dopamine receptor interactions: indications for heteromerization. *Neuropsychopharmacol* **23**, S50–9.
99. Gomes, I., Jordan, B. A., Gupta, A., Trapaidze, N., Nagy, V., and Devi, L. A. (2000) Heterodimerization of μ and δ opioid receptors: A role in opiate synergy. *J Neurosci* **20**, RC110.
100. AbdAlla, S., Lother, H., and Quitterer, U. (2000) AT1-receptor heterodimers show enhanced G-protein activation and altered receptor sequestration. *Nature* **407**, 94–8.
101. Barki-Harrington, L., Luttrell, L. M., and Rockman, H. A. (2003) Dual inhibition of beta-adrenergic and angiotensin II receptors by a single antagonist: A functional role for receptor-receptor interaction in vivo. *Circulation.* **108** 1611–8.
102. Rios, C., Gomes, I., and Devi, L. A. (2006) mu Opioid and CB1 cannabinoid receptor interactions: reciprocal inhibition of receptor signaling and neuritogenesis. *Br J Pharmacol* **148**, 387–95.
103. Mahon, M. J., Donowitz, M., Yun, C. C., and Segre, G. V. (2002) Na^+/H^+ exchanger regulatory factor 2 directs parathyroid hormone 1 receptor signalling. *Nature* **417**, 858–61.

104. Kenakin, T. P. (2009) 7TM receptor allostery: putting numbers to shapeshifting proteins. *Trends Pharmacol Sci* **30**, 460–9.

105. Wang, L., Martin, B., Brenneman, R., Luttrell, L. M., and Maudsley, S. (2009) Allosteric modulators of G protein-coupled receptors: future therapeutics for complex physiological disorders. *J Pharmacol Exp Ther* **331**, 340–8.

106. Ehlert F. J. (1988) Estimation of the affinities of allosteric ligands using radioligand binding and pharmacological null methods. *Mol Pharmacol* **33**, 187–94.

107. Christopoulos, A., and Kenakin, T. (2002) G protein-coupled receptor allosterism and complexing. *Pharmacol Rev* **54**, 323–74.

108. Avlani, V., May, L. T., Sexton, P. M., and Christopoulos, A. (2004) Application of a kinetic model to the apparently complex behavior of negative and positive allosteric modulators of muscarinic acetylcholine receptors. *J Pharmacol Exp Ther* **308**, 1062–72.

109. Lanzafame, A., Christopoulos, A., and Mitchelson, F. (1997) Three allosteric modulators act at a common site, distinct from that of competitive antagonists, at muscarinic acetylcholine M2 receptors. *J Pharmacol Exp Ther* **282**, 278–285.

110. Nemeth, E. F., Heaton, W. H., Miller, M., Fox, J., Balandrin, M. F., Van Wagenen, B. C., Colloton, M., Karbon, W., Scherrer, J., Shatzen, E., Rishton, G., Scully, S., Qi, M., Harris, R., Lacey, D., and Martin, D. (2004) Pharmacodynamics of the type II calcimimetic compound cinacalcet HCl. *J Pharmacol Exp Ther* **308** 627–35.

111. Price, M. R., Baillie, G. L., Thomas, A., Stevenson, L. A., Easson, M., Goodwin, R., McLean, A., McIntosh, L., Goodwin, G., Walker, G., Westwood, P., Marrs, J., Thomson, F., Cowley, P., Christopoulos, A., Pertwee, R. G., and Ross, R. A. (2005) Allosteric Modulation of the Cannabinoid CB1 Receptor. *Mol Pharmacol* **68**, 1484–95.

112. Knudsen, L. B., Kiel, D., Teng, M., Behrens, C., Bhumralkar, D., Kodra, J. T., Holst, J. J., Jeppesen, C. B., Johnson, M. D., de Jong, J. C., Jorgensen, A. S., Kercher, T., Kostrowicki, J., Madsen, P., Olesen, P. H., Petersen, J. S., Poulsen, F., Sidelmann, U. G., Sturis, J., Truesdale, L., May, J., and Lau, J. (2006) Small molecule agonists for the glucagon-like peptide 1 receptor. *Proc Natl Acad Sci USA* **104**, 937–42.

113. Langmead, C. J., Fry, V. A., Forbes, I. T., Branch, C. L., Christopoulos, A., Wood, M. D., and Herdon, H. J. (2006) Probing the molecular mechanism of interaction between 4-n-butyl-1-[4-(2-methylphenyl)-4-oxo-1-butyl]-piperidine (AC-42) and the muscarinic M1 receptor: direct pharmacological evidence that AC-42 is an allosteric agonist. *Mol Pharmacol* **69**, 236–46.

114. Valant, C., Gregory, K. J., Hall, N. E., Scammells, P. J., Lew, M. J., Sexton, P. M., and Christopoulos, A. (2008) A novel mechanism of G protein-coupled receptor functional selectivity. Muscarinic partial agonist McN-A-343 as a bitopic orthosteric/allosteric ligand. *J Biol Chem* **283**, 29312–21.

115. Zhang, Y., Rodriguez, A., and Conn, P. J. (2005). Allosteric potentiators of metabotropic glutamate receptor subtype 5 have differential effects on different signaling pathways in cortical astrocytes. *J Pharmacol Exp Ther* **315**, 1212–9.

116. Sachpatzidis, A., Benton, B. K., Manfredi, J. P., Wang, H., Hamilton, A., Dohlman, H. G., and Lolis, E. (2003) Identification of allosteric peptide agonists of CXCR4. *J Biol Chem* **278**, 896–907.

117. Kondru, R., Zhang, J., Jim, C., Mirzadegan, T., Rotstein, D., Sankurati, S., and Dioszegi, M. (2008) Molecular interactions of CCR5 with major classes of small molecule anti-HIV CCR5 antagonists. *Mol Pharmacol* **73**, 789–800.

118. Maeda, K., Nakata, H., Koh, Y., Miyakawa, T., Ogata, H., Takaoka, Y., Sbibayama, S., Sagawa, K., Fukushima, D., Moravek, J., Koyanagi, Y., and Mitsuya, H. (2004) Spirodiketopiperazine based CCR5 inhibitor which preserves CC-chemokine/CCR5 interactions and exerts potent activity against R5 human immunodeficiency virus type 1 in vitro. *J Virol* **78**, 8654–62.

119. Watson, C., Jenkinson, S., Kazmierski, W., and Kenakin, T. P. (2005) The CCR5 receptor-based mechanism of action of 873140, a potent allosteric noncompetitive HIV entry-inhibitor. *Mol Pharmacol* **67**, 1268–82.

120. Muniz-Medina, V. M., Jones, S., Maglich, J. M., Galardi, C., Hollingsworth, R. E., Kazmierski, W. M., Ferris, R. G., Edelstein, M. P., Chiswell, K. E., and Kenakin, T. P. (2009) The relative activity of "function sparing" HIV-1 entry inhibitors on viral entry and CCR5 internalization: Is allosteric functional selectivity a valuable therapeutic property? *Mol Pharmacol* **75**, 490–501.

121. Wood A and Armour D. (2005). The discovery of the CCR5 receptor antagonist, UK-427,857, a new agent for the treatment of HIV infection and AIDS. *Prog Med Chem.* **43**: 239–71.

122. Lazareno, S., Popham, A., and Birdsall, N. J. M. (1998) Muscarinic interactions of bisindolylmaleimide analogues. *Eur J Pharmacol* **360**, 281–4.

123. Ellis, J., and Seidenberg, M. (2000) Interactions of alcuronium, TMB-8 and other

allosteric ligands with muscarinic acetylcholine receptors: Studies with chimeric receptors. *Mol Pharmacol* **58**, 1451–60.

124. Bridges, T. M., and Lindsley, C. W. (2008) G-protein-coupled receptors: from classical modes of modulation to allosteric mechanisms. *ACS Chem Biol* **3**, 530–41.

125. Ehlert, F. J. (2005) Analysis of Allosterism in Functional Assays. *J Pharmacol Exp Ther* **315**, 740–54.

126. Kenakin, T. (2007) Allosteric agonist modulators. *J Recept Signal Transduct Res* **27**, 247–59.

127. Leach, K., Sexton, P. M., and Christopoulos, A. (2007) Allosteric GPCR modulators: taking advantage of permissive receptor pharmacology. *Trends Pharmacol Sci* **28**, 382–9.

128. Maillet, E. L., Pellegrini, N., Valant, C., Bucher, B., Hibert, M., Bourguignon, J. J., and Galzi, J. L. (2007) A novel, conformation-specific allosteric inhibitor of the tachykinin NK2 receptor (NK2R) with functionally selective properties. *FASEB J* **21**, 2124–34.

129. Mathiesen, J. M., Ulven, T., Martini, L., Gerlach, L. O., Heinemann, A., and Kostenis, E. (2005) Identification of indole derivatives exclusively interfering with a G protein-independent signaling pathway of the prostaglandin D2 receptor CRTH2. *Mol Pharmacol* **68**, 393–402.

130. Kew, J. N., Trube, G., and Kemp, J. A. (1996) A novel mechanism of activity-dependent NMDA receptor antagonism describes the effect of ifenprodil in rat cultured cortical neurones. *J Physiol* **497**, 761–72.

131. Galandrin, S., and Bouvier, M. (2006) Distinct signaling profiles of beta1 and beta2 adrenergic receptor ligands toward adenylyl cyclase and mitogen-activated protein kinase reveals the pluridimensionality of efficacy. *Mol Pharmacol* **70**, 1575–84.

132. Watson, C., Chen, G., Irving, P. E., Way, J., Chen, W-J., and Kenakin, T. P. (2000) The use of stimulus-biased assay systems to detect agonist-specific receptor active states: Implications for the trafficking of receptor stimulus by agonists. *Mol Pharmacol* **58**, 1230–8.

133. Woolf, P. J., Kenakin, T. P., and Linderman, J. J. (2001) Uncovering biases in high throughput screens of G-protein coupled receptors. *J Theor Biol* **208**, 403–18.

134. Chen, G., Way, J., Armour, S., Watson, C., Queen, K., Jayawickreme, C. K., Chen, W. J., and Kenakin, T. (2000) Use of constitutive G protein-coupled receptor activity for drug discovery. *Mol Pharmacol* **57**, 125–34.

135. Gardella, T. J., Luck, M. D., Jensen, G. S., Schipani, E., Potts, J. T. Jr., Juppner, H. (1996) Inverse agonism of amino-terminally truncated parathyroid hormone (PTH) and PTH-related peptide (PTHrP) analogs revealed with constitutively active mutant PTH/PTHrP receptor. *Endocrinology* **137**: 3936–3941.

136. Shukla, A. K., Violin, J. D., Whalen, E. J., Gesty-Palmer, D., Shenoy, S. K., and Lefkowitz, R. J. (2008) Distinct conformational changes in beta-arrestin report biased agonism at seven-transmembrane receptors. *Proc Natl Acad Sci USA* **105**, 9988–93.

137. Gesty-Palmer, D., Flannery, P., Yuan, L., Spurney, R., Lefkowitz, R. J., and Luttrell, L. M. (2009) A beta-arrestin biased agonist of the parathyroid hormone receptor (PTH1R) promotes bone formation independent of G protein activation. *Science – Transl Med* **1**:1ra1.

138. Leppik, R. A., and Birdsall, N. J. (2000) Agonist binding and function at the human α_{2A}-adrenoceptor: allosteric modulation by amilorides. *Mol Pharmacol* **58**, 1091–9.

139. Kenakin, T., Jenkinson, S., and Watson, C. (2006) Determining the potency and molecular mechanism of action of insurmountable antagonists. *J Pharmacol Exp Ther* **319**, 710–23.

140. Fang, Y., Ferrie, A. M., Fontaine, N. H., and Yuen, P. K. (2005) Characteristics of dynamic mass redistribution of EGF receptor signaling in living cells measured with label free optical biosensors. *Anal Chem* 77, 5720–5.

141. Cunningham, B. T., Li, P., Schultz, S., Lin, B., Baird, C., Gerstenmaier, J., Genick, C., Wang, F., Fine, E., and Laing, L. (2004) Label free assays on the BIND system. *J Biomolec Screen* **9**, 481–90.

142. Fang, Y., and Ferrie, A. M. (2008) Label-free optical biosensor for ligand-directed functional selectivity acting on β2-adrenoceptor in living cells. *FEBS Lett* **582**, 558–64.

143. McGuinness, R. (2007) Impedance-based cellular assay technologies: Recent advances. *Curr Opin Pharmacol* 7, 535–40.

144. Peters, M. F., and Scott, C. W. (2009) Evaluating cellular impedance assays for the detection of GPCR pleiotropic signaling and functional selectivity. *J Biomol Screen* **14**, 246–55.

145. Kenakin, T. (2005) New concepts in drug discovery: collateral efficacy and permissive antagonism. *Nat Rev Drug Discov* **4**, 919–27.

146. Milligan, G. (2003) High-content assays for ligand regulation of G-protein coupled receptors. *Drug Disc Today* **8**, 579–85.

147. Ross, D. A., Lee, S., Reiser, V., Xue, J., Alves, K., Vaidya, S., Kreamer, A., Mull, R., Hudak, E., Hare, T., Detmers, P. A., Lingham, R., Ferrer, M., Strulovici, B., and Santini, F.

(2008) Mulitplexed assays by high-content imaging for assessment of GPCR activity. *J Biomol Screen* **13**, 449–455.

148. Ghosh, R. N., DeBasio, R., Hudson, C. C., Ramer, E. R., Cowan, C. L., and Oakley, R. H. (2005) Quantitative cell-based high content screening for vasopressin receptor agonists using Transfluor® technology. *J Biomol Screen* **10**, 476–84.
149. Hanson, B. J., Wetter, J., Bercherer, M. R., Kopp, L., Fuerstenau-Sharp, M., Vedvik, K. L., Zielinski, T., Doucette, C., Whitey, P. J., and Revankar, C. (2009) A homogeneous fluorescence live-cell assay for measuring 7-transmembrane receptor activity and agonist functional selectivity through beta-arrestin recruitment. *J Biomol Screen* **14**, 798–810.
150. Hamdan, F. F., Audet, M., Garneau, P., Pelletier, J., and Bouvier, M. (2005) High throughput screening of G-protein-coupled receptor antagonists using bioluminescence resonance energy transfer 1-based β-arrestin2 recruitment assay. *J Biomol Screen* **10**, 463–75.
151. Zhao, X., Jones, A., Olson, K. R., Peng, K., Wehrman, T., Park, A., Mallari, R., Nebalasca, D., Young, S. W., and Xiao, S-H. (2008) A homogeneous enzyme fragment complementation-based β-arrestin translocation assay for high throughput screening of G-protein-coupled receptors. *J Biomol Screen* **13**, 737–47.
152. Barnea, G., Strapps, W., Herrada, G., Berman, Y., Ong, J., Kloss, B., Axel, R., and Lee, K. J. (2008) The genetic design of signaling cascades to record receptor activation. *Proc Natl Acad Sci USA* **105**, 64–9.
153. Henriksen, U., Fog, J., Loechel, F., and Praestegaard, M. (2008) Profiling of multiple signal pathway activities by multiplexing antibody and GFP-based translocation assays. *Comb Chem High Throughput Screen* **11**, 537–44.
154. Groer, C. E., Tidgewell, K., Moyer, R. A., Harding, W. W., Rothman, R. B., Prisinzano, T. E., and Bohn, L. M. (2007) An opioid agonist that does not induce mu opioid receptor-arrestin interactions or receptor internalization. *Mol Pharmacol* **71**, 549–557.

Chapter 2

Imaging-Based Approaches to Understanding G Protein-Coupled Receptor Signalling Complexes

Darlaine Pétrin and Terence E. Hébert

Abstract

In the last 10 years, imaging assays based on resonance energy transfer (RET) and protein fragment complementation have made it possible to study interactions between components of G protein-coupled receptor (GPCR) signalling complexes in living cells under physiological conditions. Here, we consider the history of such approaches, the current tools available and how they have changed our understanding of GPCR signalling. We also discuss some theoretical and methodological issues important when combining the different types of assay.

Key words: G protein-coupled receptor, Bioluminescence resonance energy transfer, G protein, Protein–protein interaction assays, Protein fragment complementation assays

1. Introduction

GPCR signalling complexes comprise a diverse set of stable, metastable, and transient protein–protein interactions that occur in distinct tissues, cells, and subcellular compartments. Further, the transience or stability of these interactions may in fact be modulated by the localization of the relevant partners or particular cellular or subcellular conditions. That is to say, the organization of these signalling complexes presents a spectrum from a series of transient interactions aimed at generating and amplifying cellular signals, such as in the mammalian visual system, to highly organized and stable complexes which may be assembled during biosynthesis, trafficked to the cell surface and remain together during a defined series of signalling events.

During the last 10 years or so, the field has undergone a renaissance most notably in our ability to interrogate the organization

Louis M. Luttrell and Stephen S.G. Ferguson (eds.), *Signal Transduction Protocols*, Methods in Molecular Biology, vol. 756, DOI 10.1007/978-1-61779-160-4_2, © Springer Science+Business Media, LLC 2011

and function of these signalling systems. We have gone from a basic functional understanding of distinct, individual GPCR signalling events measured using biochemical assays to a rich appreciation of the diversity and organization of signalling at a systems level. To a large extent, this new understanding of GPCR signalling is due to our ability to probe a large number of the relevant protein–protein interactions that occur during signalling events in living cells and in real-time. Here, we will examine the development of these approaches and discuss some theoretical and methodological considerations for their usage.

2. Historical Aspects of the Development of Imaging-Based Assays of GPCR Signalling

G protein signalling in the mammalian visual system, involving the retinal-bound rhodopsin, the heterotrimeric G protein, transducin, and cGMP phosphodiesterase, is based on the need for significant signal amplification (1). This necessitates an organization where one activated rhodopsin molecule must interact with and activate several transducin equivalents, that is the interactions must be transient. Of course, it was also believed that the receptor itself was monomeric. This model, developed for the visual system, was thought to be generally applicable to GPCR signalling systems in other cells and tissues as well. However, evidence in the literature, based on radiation inactivation experiments and thermodynamic considerations of ligand binding indicated that these complexes may actually have been somewhat larger than predicted by the data in the mammalian visual system (reviewed in (2)). The first direct demonstration of higher order structures for GPCRs was in a study in which I participated as a post-doctoral fellow in the laboratory of Michel Bouvier. Here, we showed, using what was novel at the time, i.e. differential epitope tagging and co-immunoprecipitation, that the β_2-adrenergic receptor (β_2AR) was in fact dimeric (3). There was understandable scepticism regarding these findings. Although the concept that GPCRs were dimeric (or even oligomeric) is accepted now (see (4–6) for review), several key findings were required to ultimately convince investigators of the validity of the initial observations. First, the discovery that the GABA-B receptor was composed of two distinct subunits in the context of a receptor heterodimer definitively proved that GPCR heterodimers (and GPCR dimers in general) existed (7–10). However, the existence and role of homodimeric receptors remained and to some extent remains an open question. In part, this was due to the initial reliance on co-immunoprecipitation as the principle technique demonstrating interactions between monomer equivalents. There was additional functional information supporting this notion ((11), see (2) for review) but the possibility that interactions detected were simply artifacts of membrane solubilization with detergent could not be avoided. Thus, another approach was needed that could be performed in living cells. Imaging techniques using antibodies or GFP fusion proteins have long been used to monitor the trafficking of GPCRs (reviewed in (12)). However, with the development of interaction assays based on fluorescence or bioluminescence such as fluorescence or bioluminescence resonance energy transfer (FRET and BRET, respectively), the stage was set for their general use to monitor protein–protein interactions as well. Thus, the first use of resonance energy transfer approaches revolutionized the study of GPCR signalling. In the year 2000, four groundbreaking studies appeared using either FRET and BRET. Two of these studies used photobleaching FRET with labelled ligands to demonstrate dimerization between somatostatin receptors (13) or heterodimerization between dopamine and somatostatin receptors (14), one used a classic FRET approach with CFP- and YFP-tagged yeast α-factor receptors (15) and one used BRET to demonstrate homodimerization of the β_2AR (16). Since then, as we will see below, these approaches have been used to probe many aspects of GPCR signalling, assembly and trafficking. One of life's little ironies is that in the 9 years or so that my lab has been using BRET, most journal reviewers now also ask for co-immunoprecipitation or other classical protein co-purification approaches, mainly in the untransfected native context. The take home message would be that both types of approaches have their value and are complementary.

3. Resonance Energy Transfer Approaches to GPCR Signalling

Since the publication of those four initial reports, a large number of studies have appeared confirming and extending the notion that GPCRs form homo- and heterodimers (see (4–6) for review). Further, a number of biosensors for various GPCR signalling pathways have also been developed (reviewed in (17–22)). Often, these assays are used simultaneously, to develop a broad picture of the signalling kinetics from receptor activation down to production of second messengers in multiple subcellular compartments (23, 24). The use of these techniques has also spread to other signalling receptor families and ion channels. For example, recent BRET and FRET studies have focused on the IL-5 cytokine receptor (25), receptor tyrosine kinases (26–28), KCNQ voltage-gated potassium channels (29), and cyclic nucleotide-gated HCN channels (30). The reviews I have cited are detailed summaries of these observations which I will not focus on here. Rather, I will discuss how variants in the generations of RET approaches and new assay formats have been developed to probe different GPCR signalling events and interactions. Many excellent reviews on the technical aspects of using these approaches have also appeared in the last few years (see for example (31–38)).

3.1. Basic Considerations Regarding FRET and BRET

Conceptual aspects regarding the use and interpretation of resonance energy transfer experiments have also been reviewed thoroughly (31–40). We will discuss them here briefly and refer the reader to the aforementioned reviews for more detail.

RET depends upon the acceptor and donor tags being in close proximity (<100 Å, Fig. 1a). Although the possibility of this occurring inadvertently is relatively low, it is a potential problem and these experiments cannot be correctly interpreted without the proper negative controls (see Box 1 and the references therein). If expression levels of the RET pair are high enough, crowding of acceptor and donor molecules can produce "bystander" RET even though the tagged proteins have no true affinity for each other (41). A simple approach to determining if bystander RET occurs is to assay for resonance energy transfer between the proteins being investigated and a tagged protein which does not normally interact with either of the RET pair, that is localized to the same subcellular compartment. When expressed at the same or higher levels, no RET should occur between the proteins that do not interact. Since bystander RET is most likely to occur when expression levels are high, it can be minimized by keeping expression levels of donor and acceptor as low as possible given the constraints of instrument sensitivity.

Box 1
Controls for RET Experiments

1. RET partners must be verified for localization and function.
2. Known interactors which localize to the same compartment as the proteins of interest must be used as positive controls.
3. Negative controls (i.e. RET pairs that do not interact with proteins of interest) also must be localized to same compartment as protein of interest.
4. Both positive and negative control constructs must be expressed at similar levels.
5. Specificity of interactions must be confirmed using RET saturation experiments (donor saturation) as well as competition with "cold" versions of each partner.
6. Where possible donor and acceptor moieties on each partner should be switched.

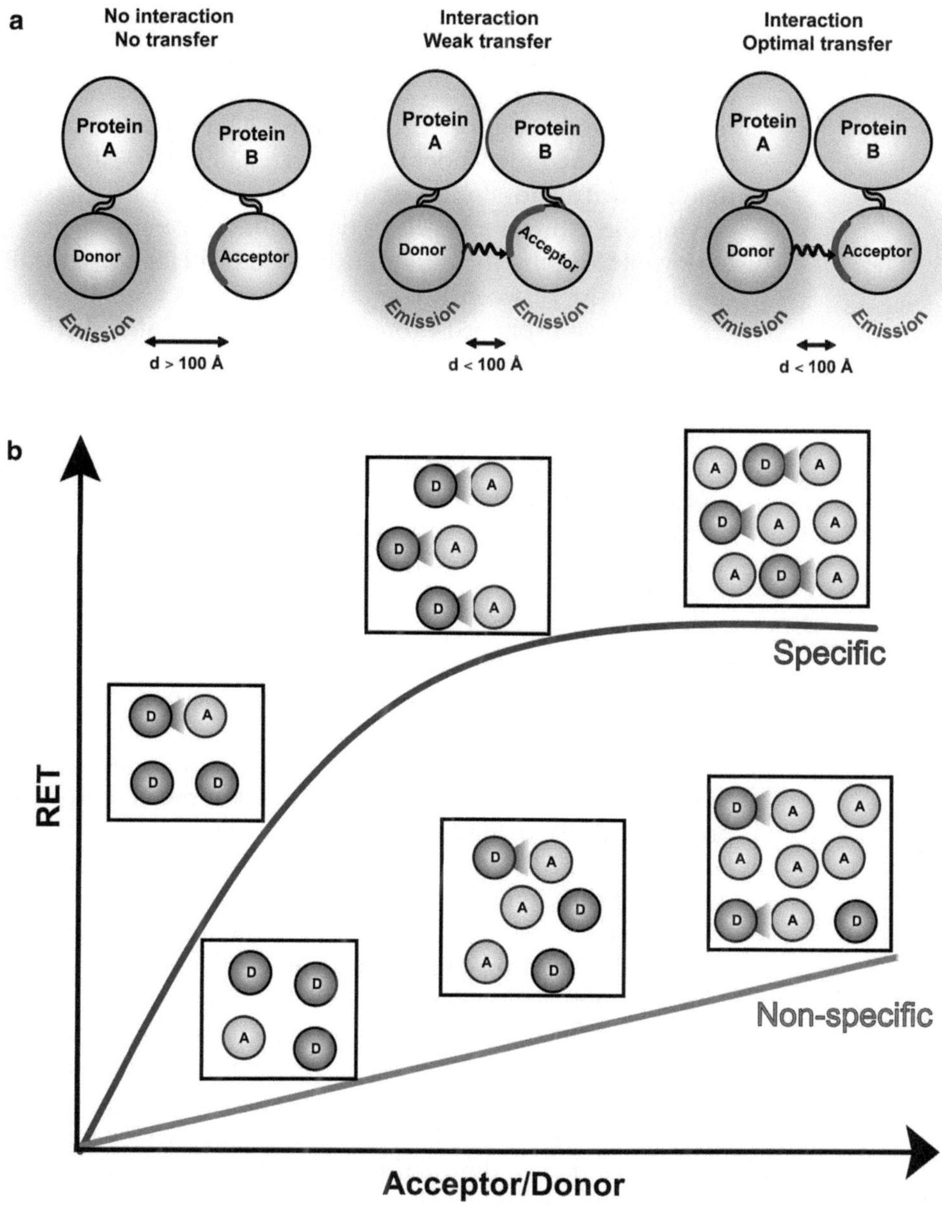

Fig. 1. Considerations for correct performance and interpretation of RET experiments. (**a**) RET depends on the distance between donor and acceptor molecules and is independent of the magnitude of the measured signal. The efficiency of energy transfer depends on each individual interaction and the relative orientation and distance of donor and acceptor. (**b**) Interactions that are specific, as measured by RET, result in a "saturable" RET signal in the sense that increasing the amount of acceptor against a constant background of donor molecules will lead to a plateau which can be fitted as a binding assay would. Non-specific interactions, based on random collisions of donor and acceptor generally result in smaller and non-saturating RET signals.

Published RET studies have sometimes been performed without carefully considering the ratio of expressed donor- to acceptor-tagged proteins or without other controls discussed below. Control over the donor/acceptor ratio is important for confirming the specificity and affinity of any given interaction. If the interaction is specific, then RET will be sensitive to the donor/acceptor ratio. On the other hand, if the RET is a consequence of non-specific interactions caused by random collisions between the donor and acceptor (i.e. bystander RET), resonance energy transfer will be independent of this ratio (39, 40). Alternatively, the specificity and affinity of an interaction can be assessed by varying the expression levels of donor- and acceptor-tagged proteins while keeping the ratio of expressed proteins constant. RET should persist between two specific interactors even at low concentrations of expressed proteins, but will be lost if the interaction between donor- and acceptor-tagged proteins is non-specific (39, 40). In effect, this approach is another way of determining if bystander RET occurs.

One advantage of using RET to study protein–protein interactions is that changes in the efficiency of energy transfer can reveal information regarding changes in the interaction between the proteins in question. For G protein-mediated signalling pathways, these changes often occur in response to ligand binding to a receptor. However, a change in RET may indicate one or a combination of two different things: (1) changes in the affinity of proteins for each other; or (2) changes in intermolecular distance or orientation (i.e. a conformational change) within a protein complex (Fig. 1a). One approach to distinguishing between these possibilities is to express varying amounts of the acceptor-tagged protein with fixed amounts of the donor-tagged protein. As discussed above, RET signals will be sensitive to the ratio of tagged proteins if they interact specifically with each other. By fixing the expression of the donor-tagged protein and increasing the expression of the acceptor-tagged protein, RET between them will increase, approaching a maximum value asymptotically, thus producing, in effect, a saturation curve (Fig. 1b). Receptor ligands, for example, that cause either an increase or a decrease in the amount of acceptor-tagged protein needed to attain half-maximal RET (the $BRET_{50}$ or $FRET_{50}$), reflect decreases or increases in the affinity of tagged proteins for each other, respectively (41).

An alternative method for distinguishing between a change in affinity, and a change in conformation, is to introduce the tags at different positions in donor or acceptor proteins that do not compromise biological function. If a receptor ligand, or expression of modulating proteins causes opposing changes in RET efficiency depending upon the location of the tag, this suggests that these changes represent different conformational states within a protein complex rather than changes in the affinity of the proteins for one another per se.

3.2. Measuring RET in Real-Time

As instrumentation for measuring RET has become faster and more sensitive, it has become possible to conduct RET experiments in real-time, that is, to tease out kinetic data regarding the interactions being measured in response to different stimuli and also to assess the stability of given interactions within the context of physiologically relevant signalling events. Real-time BRET or FRET has been used to detect changes in the conformation of receptor/G protein complexes (42), receptor/β-arrestin complexes (43), or many of these events simultaneously (23). Some specific considerations for these types of experiments are provided in one of the chapters in this volume (44).

3.3. Inter- and Intra-receptor Interactions Measured Using BRET and FRET

Dimerization has been demonstrated to be required for efficient surface localization of a number of GPCRs including the β_2AR (45, 46) and the α_{1B}AR (47), and this has been reviewed recently (48). Different ligands can also "induce" distinct conformations in receptor dimers. For example, it has been demonstrated that the direction of the change in BRET depends on the ligand used to modulate CXCR4 chemokine receptors (49). Similarly, FRET has been used to detect ligand-specific conformations in the 5-HT$_{2A}$ receptor (50). RET has been used to map out receptor/receptor interfaces delineated by site-directed mutagenesis (45, 51). BRET has also been used to measure dimerization of folding mutations of β_1AR which are trapped in the ER but which can be rescued through the use of pharmacological chaperones (52). RET approaches have also been used to detect signalling events unique to receptor heterodimers. For example, it has been demonstrated that by altering the position of tags used to measure BRET, heterodimeric GPCRs composed of MT$_1$ and MT$_2$ melatonin receptors undergo conformational changes in response to agonist without any change in affinity of the monomeric receptors for each other (53). Intermolecular FRET has also been used to examine conformational changes within receptor monomers in response to agonist stimulation. Here, both the donor and acceptor molecules are inserted into the primary structure of the receptor, usually in the conformationally flexible third intracellular loops and the C-tail (reviewed in (18)). In some cases, such as the α_2AR, the donor and acceptor have been CFP and YFP, respectively (54) and in others the FRET acceptor was based on a much smaller FlAsH reagents and tetracysteine motifs (55). In a recent study, the α_2AR was tagged with the tetracysteine motif at different positions in the third intracellular loop, and different full and partial agonists varied in their ability to modulate FRET depending on the position of the insertion (56). Consistent with data from the recent GPCR crystal structures (reviewed in (57–59)), full agonists caused changes in FRET independent of their position, while partial agonists could only cause changes in a subset of these reporters (55, 56). These approaches are likely to be even more useful when combined with approaches to simultaneously interrogate multiple interactions, as discussed below.

3.4. Interactions Between Receptors, G Proteins, and Effector Molecules

As discussed above, one of the advantages of the BRET and FRET approaches is that the tags can be engineered into different sites in the proteins of interest. Assuming that insertion at these distinct sites does not compromise the function of the protein, different sites of insertion offer different conformational vantage points from which to interrogate changes in the interaction between the two proteins. A number of studies have performed these types of experiments between GPCRs and heterotrimeric G proteins and shown that receptor/G protein complexes are preformed and undergo conformational rearrangement following agonist stimulation (42, 60, 61). The first two studies demonstrated the validity of tagging G proteins from different conformational vantage points and showed that agonists could either increase or decrease BRET depending on the orientation of the distinct donor/acceptor positions in the same molecules. The latter study showed that complexes containing δ-opioid receptors (DOR) and heterotrimeric G proteins are differentially sensitive to different DOR ligands, highlighting the utility of this system to study and understand efficacy. BRET was also used to demonstrate that the β_2AR forms a complex with heterotrimeric G proteins and effector molecules during biosynthesis which are subsequently trafficked as a complex to the cell surface (46, 62).

BRET and FRET have both been used to detect pre-assembled receptor/G protein complexes and to monitor changes in these interactions in response to ligand stimulation (42, 60). Although there does seem to be as solid case for stability of receptor/G protein interactions in the face of agonist activation, recent data suggest that a spectrum of relative stabilities of the G protein heterotrimer are possible depending on the Gα subunit of the heterotrimeric G protein in question. For example, as described below, it has recently been demonstrated that Go-containing heterotrimers show a markedly increased propensity to dissociate following agonist stimulation than Gs-containing heterotrimers ((63, 64); reviewed in (65)).

3.5. Newer Variants of RET and Alternative Strategies

Originally, BRET experiments targeting GPCRs used *Renilla* luciferase as the donor and yellow fluorescent protein (YFP) as the acceptor, though the enhanced variant of GFP (EGFP) and Venus can also be used (16). A second generation of BRET (BRET^2 as distinguished from the original BRET^1) was developed that uses a novel substrate for RLuc, a coelenterazine derivative called DeepBlueC, and a GFP variant, GFP^2, with greater spectral separation between excitation and emission wavelengths (reviewed in (31–38, 66)). In addition to the now standard BRET pairs described above, a number of new luminescent RET variants have been generated. These include, firefly luciferase and the GFP variant DsRED as donor and acceptor pair (67, 68). Also, a BRET pair made up of *Renilla* luciferase and *Renilla* GFP has been developed which shows a marked increase in the efficiency of energy transfer and has been validated for GPCR/β-arrestin interactions (69). In addition to BRET per se, there is also luminescence RET (LRET) which takes advantage of luminescent donors covalently attached to purified (70) and non-purified (71) proteins of interest. LRET has also been used to study the dynamics of voltage-gated ion channel opening in response to changes in membrane potential (72).

Techniques to perform FRET are evolving as well. For example, biarsenical (FlAsH) reagents binding to smaller, engineered tetracysteine motifs in target proteins can be used as FRET donors for acceptors covalently bound to engineered cysteine residues for example (55, 56, 73, 74). FRET between fluorescent dyes and engineered dihistidine motifs, so-called transition metal ion FRET, has also been used to monitor the conformational dynamics of HCN channels in response to membrane hyperpolarization (75). These small molecule-based approaches (reviewed in (76)) may ultimately allow the combination of imaging for observational purposes and release of caged compounds for tight control of where and when the imaged events occur (see (77) for review).

As an alternative to RET approaches, a number of investigators have begun using fluorescence recovery after photobleaching (FRAP) approaches to study inter-GPCR interactions and interactions with G proteins and effectors. Using techniques such as antibody cross-linking to immobilize a portion of the GPCRs expressed at the cell surface, this technique was used to demonstrate the differential stability of distinct G protein heterodimer combinations ((63, 64); reviewed in (65)), and to confirm tight co-localization of GPCR complexes that regulate Kir3 channels (78). FRAP experiments have also raised questions regarding the proportion of some GPCRs that actually exist as pre-coupled complexes with heterotrimeric G proteins (79) and the relative stability of GPCR dimers (80). FRAP can also be used to determine the relative stoichiometry of GPCR dimers and signalling complexes in addition to interrogating the stability of these complexes. For example, a recent study demonstrated that β_2AR were much more likely to form stable dimers and oligomers than the closely related β_1AR (81). FRAP studies also revealed changes in the GABA-B heterodimer in response to agonist binding. It was noted that few structural changes occurred within each monomer but rather that the distance between gb1A and gb2 subunits was altered (82). Other high resolution optical techniques such as near-field scanning optical microscopy (NSOM) have led to similar conclusions in intact cardiomyocytes regarding the oligomerization of β_2AR (83).

3.6. The Development of Protein Fragment Complementation Assays

Protein complementation assays (PCA) are based on the notion that an enzyme, fluorescent or luminescent protein can be reconstituted from fragments which can be fused to other proteins to detect bimolecular protein interactions. When these proteins are split in two, neither fragment is active, fluorescent or luminescent when exogenously expressed in cells, nor does simple co-expression result in the reconstitution of the active parent protein. If the complementary N- and C-terminal fragments are genetically fused to two proteins that associate to form a complex, the fragments are brought together and can fold to produce an active protein by complementation. The development of protein fragment complementation assays (PCA) has yielded a number of complementary tools that have been useful for studying GPCR signalling. We will not delve deeply into the history of these assays as they have been reviewed extensively (84–86), but will restrict our focus to their applications for GPCR signalling. The two most common types of complementation assay used in this regard have been those that are based on the reconstitution of GFP variants or luciferase enzymes. A number of GFP-based assays, interrogating a variety of biochemical and signalling pathways have simultaneously been used for high content screening (87–89). Interestingly, such large-scale analyses can identify both off target effects of drugs as well as new potential therapeutic targets (reviewed in (84)). One type of split GFP reconstitution, called bimolecular fluorescence complementation (BiFC; (90)) was used to study the assembly of Gβγ subunits (91). This technique was further refined using multicolour split-YFP and split-CFP constructs (92) to compare the abilities of different Gβ and Gγ to assemble *in cellulo* and to activate effectors (93, 94). The unique spectra produced when different N- and C-terminal CFP or YFP fragments assemble can be resolved allowing for a number of different Gβγ species to be quantitated simultaneously. BiFC has also been used to study interactions between GPCRs and β-arrestin (described in (95)). Interestingly, GFP reconstitution assays have even been performed where the two interacting partners bearing the split GFP constructs are expressed on the surface of pre- and postsynaptic neurons in *Caenorhabditis elegans* (96).

A number of considerations must be taken into account when using PCA to study protein–protein interactions. As in RET studies, correct localization of tagged proteins, and similar function must be confirmed for such tagged proteins. Ideally, a perfect PCA-based biosensor would be one that can be expressed and measured at levels significantly below endogenous levels. However, if interactions of proteins that also interact with endogenous proteins is being measured, the risk is that the assay proteins may get titrated out by interaction with the endogenous proteins. Some issues have been raised regarding PCAs with respect to the rates of false positives and false negatives. This is clearly important with GFP-based PCA, since the interactions are essentially irreversible after both halves of GFP in the two proteins of interest interact and GFP folds (reviewed in (84–86)). The relatively slow kinetics of chromophore maturation can also be an issue when trying to capture physiologically relevant protein–protein interactions. We will discuss the chromophore maturation issue later, but other strategies have been developed to overcome the issue with respect to irreversibility (i.e. using reversible split interactors such as luciferase).

In an analogous way to the split GFP-based assays, luciferases can also be split in two so that the individual fragments are no longer bioluminescent alone or when simply co-expressed. Again if the complementary N- and C-terminal fragments are fused to two proteins that associate *in cellulo*, a bioluminescent protein can be reconstituted. In particular, a number of PCA, based on different luciferases have been developed in the last few years. These include firefly (97–101), *Gaussia* (102, 103), and *Renilla* luciferases (104). These assays have already been used to study interactions between different GPCRs including chemokine receptors (105), D2 dopamine receptors (106), and the β_2AR (107) as well as between CXCR4 and β-arrestin (98, 99). Since these constructs allow reversible assembly and disassembly, split luciferase fusion proteins have also been used to study the dynamics of cellular signalling over multiple rounds of agonist stimulation, for example the activation of PKA (97, 104).

Deciding where to split these PCA constructs is critical for their optimization as reporters for protein–protein interaction (103, 108–111). The use of superfolder variants of GFP with improved maturation characteristics is also useful for in vivo interaction assays based on PCA, and although there are caveats, may also facilitate in vitro protein–protein interactions assays using purified proteins (108, 112, 113). Multiplexing different split GFP variants amenable to “mixing and matching” different colours and producing GFP variants with unique spectral properties when partner proteins interact increases the potential for resolving discrete interactions and understanding the specificity of interactions (92–94). The recent design of RFP variants such as mKate, which are also amenable to being split, will further increase the combinatorial possibilities for discriminating unique interactions (114). The design of GFP variants which change colour over time is also based on the nature and rate of chromophore maturation (115). These “fluorescent timer” variants can certainly be used to monitor time-dependent changes in protein localization or interaction status (as they are amenable to use in FRET studies). Similar opportunities will be available when photoswitchable GFP variants (i.e. whose emissions can be altered by exposure to light at particular wavelengths) are used in FRET applications (116, 117).

The use of GFP-based PCA, in particular, to interrogate GPCR signalling has been reviewed recently (95). Assays based on split GFP and split luciferase are simple, modular and amenable to scaling up for use in FACS analysis (118), screens with cDNA libraries for interacting proteins in mammalian cells (119), HTS screens (see (120) for review) or even for use in living animals (96, 99, 121, 122).

4. Combining Assay Formats to Study Multiple Protein–Protein Interactions

The development of fluorescent and luminescent PCA and other labelling strategies such as FlAsH (123), and SNAP- or CLIP-tagging, both based on O6-alkylguanine-DNA alkyltransferase (AGT; (124, 125)) has led to the attempts to combine these strategies with FRET and BRET in order to study multi-partner interactions (126, 127). These types of experiments have shed further light on the nature of GPCR signalling complexes and have also yielded subtle nuances in how data from these experiments can be interpreted. For example, by combining BRET with a split-GFP interaction pair as the reconstituted acceptor molecule, a number of groups demonstrated that simultaneous interactions occur between three partners in GPCR signalling systems. We first used split-GFP constructs to show three partner interactions between effector molecules such as adenylyl cyclase and Kir 3 inwardly rectifying potassium channels tagged with luciferase with Gβ and Gγ bearing half of YFP each (62). Reconstitution of three partners in PCA/BRET experiments has also been used to demonstrate complexes of G protein heterotrimers (46), dimeric calcitonin-like receptors and RAMPs (128), and a complex of adenylyl cyclase tagged with Rluc and split YFP reconstituted by the β_2AR and Gγ2 fusion proteins, showing that the entire basic GPCR signalling complex can remain intact during signalling (107).

A number of studies have suggested that GPCRs form higher-order complexes in addition to monomers or simple homo- or heterodimers (129, 130). FRET approaches have indicated similar higher-order structures for M2 muscarinic receptor and the β_2AR (131, 132). Protein complementation has now been used to confirm and extend our knowledge regarding dimerization and oligomerization of GPCRs. Not only has reconstitution of split luciferase (*Gaussia* or *Renilla*) and split GFP constructs shown that dimers of β_2AR (107) and D2 dopamine receptors (106) exist, complementing immunopurification and RET approaches, but that these approaches can be combined to detect and examine larger complexes. A number of investigators have used three partner PCA/RET to show that higher-order complexes of GPCRs such as the A_{2A}-adenosine receptor homo- and heteroligomers with CB1 cannabinoid/D2 dopamine receptors (133–136) and CXCR4 multimers (137) can be detected. Similar results have been obtained when combining BRET and FRET sequentially (so called SRET). Here, A_{2A}-adenosine receptor heteroligomers with CB1 cannabinoid/D2 dopamine receptors were detected by measuring Rluc/YFP BRET and the subsequent transfer of energy by FRET to CFP in the context of three tagged receptors (138). Four partner BRET/PCA interactions using split luciferase and GFP constructs has been used to directly detect

D2 dopamine receptor homo- and heterotetramers (106) and for β_2AR homotetramers (107). The former article is exemplary in its use of complementary techniques to demonstrate higher order structures for GPCRs- demonstrating by RET, PCA, PCA/RET, and standard biochemical techniques that these species exist. The use of these combined techniques has recently been reviewed (139). Combining PCA with FRET has also been demonstrated recently (140) and will likely add an interesting dimension to the study of GPCR signalling complexes. The more refined combination of RET approaches with PCA and other strategies will allow us to dissect the stoichiometry of homo- and hetero-oligomeric GPCRs as well as their associated signalling complexes.

The use of other tagging approaches has allowed use of other FRET imaging modalities that have provided information on the asymmetric organization of GPCR oligomers. Time-resolved FRET overcomes a significant limitation of standard FRET in that it avoids the overlap between the emissions from FRET donors and acceptors. Further, since the ligands can only interact with surface receptors, this approach avoids the confounding effects of internal pools of GPCRs. By engineering SNAP-tagged mGluR and GABA-B receptors, and labelling them with either europium cryptate as a donor or the fluorophore d2 as an acceptor (127), this group was able to demonstrate the specificity of heterodimer formation using a number of other GPCRs. More interestingly, these authors were able to demonstrate that mGluR1 receptors formed dimers rather than oligomers (summarized in Fig. 2a). Also, they showed that GABA-B receptors form dimers of dimers, i.e. that although both GABA-B1 and GABA-B2 receptors can form homodimers, the organization of the heterooligomer containing both subunits is asymmetric in that RET efficiency between the GABA-B1 equivalents is distinct from GABA-B2 equivalents in the heterooligomer. These asymmetries certainly play into the coupling of GABA-B receptors to G proteins, and likely will be important for other GPCRs as well (discussed in (141)). These findings have tremendous implications for future studies of receptor oligomerization and how GPCR signalling systems are organized in general. For instance, even using classic RET approaches, it should be possible to detect such asymmetric dimers (or asymmetric associations with signalling partners such as G proteins) using competition assays where one type of RET interaction serves to compete against a second. Receptors that share the same interface for dimer formation should compete with each other. If oligomeric receptors are arranged asymmetrically, then one could predict that two RET interactions could be detected simultaneously, i.e. the competition would be asymmetric as well (Fig. 2b). Here, RET and PCA/RET will be as useful as TR-FRET approaches since many of these interactions occur inside the cell as well as at the cell surface.

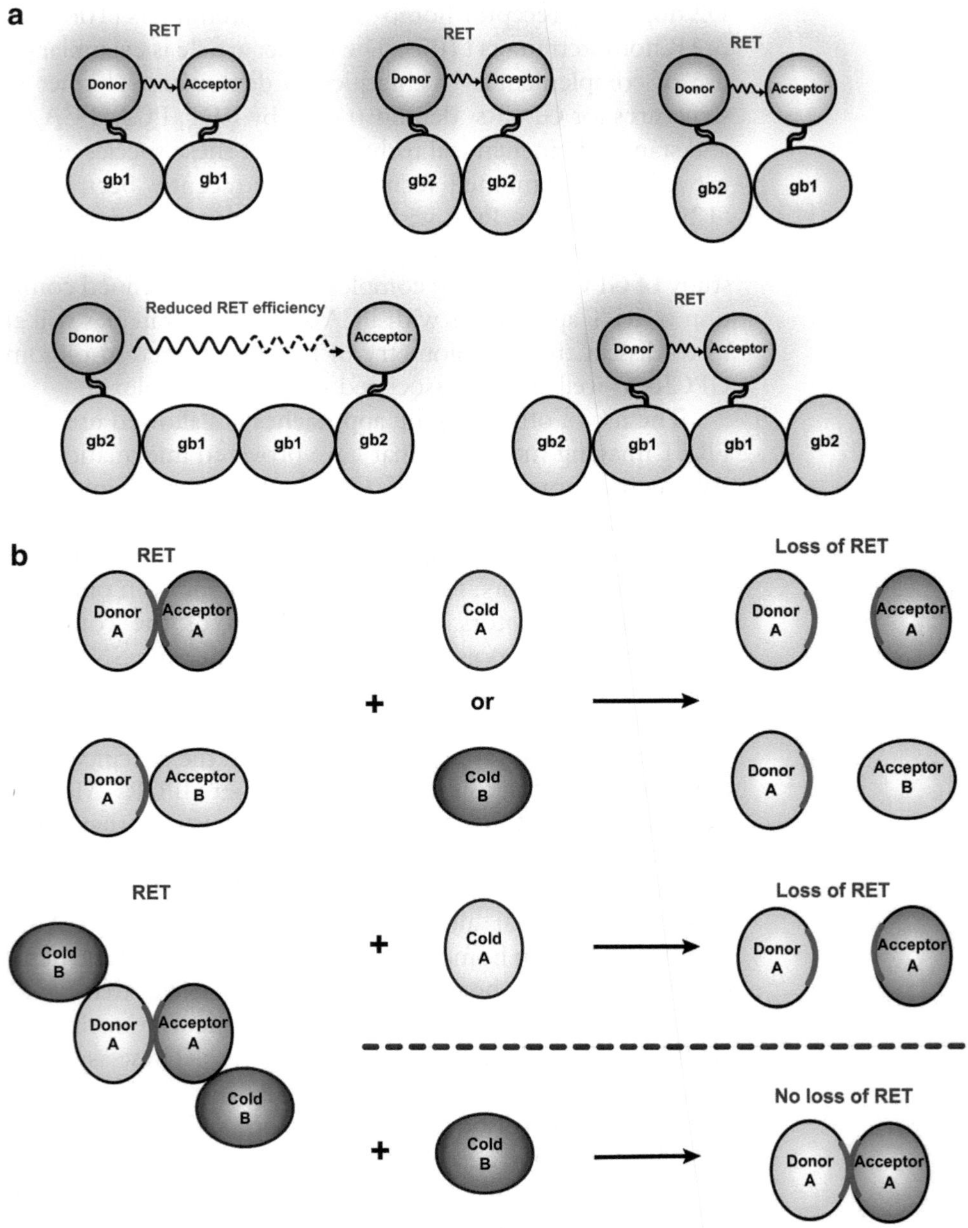

Fig. 2. Organizational complexities in GPCR oligomer organization revealed using imaging techniques. (**a**) Using a combination of SNAP- and epitope tagging and TR-FRET, it was shown that changes in FRET efficiency reflected the organization of a heteroligomeric GABA-B receptor with 2 gb1 and 2 gb2 subunits. In the top panel, it was demonstrated that both gb1 and gb2 subunits could form homodimers and heterodimers when each subunit is tagged with SNAP or a FLAG epitope and TR-FRET between either homo- or heterodimers was measured. In the lower panel, when both subunits are co-expressed, RET efficiency is much higher between gb1 equivalents than between gb2 equivalents in a heterotetramer, suggesting that the organization of the tetramer is asymmetric. Adapted from results in (127). (**b**) Organizational complexity in GPCR oligomers can be revealed by RET competition experiments. The "products" of competition reactions are only shown if they are tagged for RET in the figure. Untagged competitors could also be tracked by ELISA but are not shown here. If two proteins share an interface as either homo- or heterodimers, "cold," untagged versions of either will compete for a RET pair as in the top panel. However, in an asymmetric oligomer, as described in part A, cold competitors will have distinct effects on the RET pair (shown in pink and purple) depending on the different interfaces in the oligomer. One complication not considered here, is that A/A interface might also be affected by changes in the A/B interface by allosteric mechanisms.

RET, PCA, and TR-FRET approaches can also be used to distinguish cases where there are in fact interactions between GPCRs or when standard forms of molecular crosstalk are sufficient to explain how receptors and their signalling systems interact in a given cellular context ((142); reviewed in (143)).

5. The Current Tool Set

Although the discussion above has focused on the development of FRET, BRET, and PCA, imaging tools are continually being refined such that the these assays will become more robust, more efficient, and the fluorescent and luminescent tags more stable. This latter consideration is important for improving the kinetics of folding and chromophore maturation in PCA based on GFP reconstitution. Fluorescent protein technology is advancing steadily and has been thoroughly reviewed (144–147).

FRET has long been amenable to measurements in single cells (reviewed in (148, 149)). BRET would be useful in avoiding situations which require direct excitation by light such as when imaging animal tissues in situ. However, BRET has lagged in this regard although with the current common BRET vectors, some success has been achieved (53). The use of electron-multiplying CCD cameras which collect all available light (150–152) and the development of new *Renilla* luciferase variants such as Rluc8 and Rluc-M (153, 154) with improved quantum efficiency and stability will be of significant utility in this regard and have shown promise in both single cell (155) and whole animal BRET experiments (155, 156). Recent experiments with Rluc–YFP fusion proteins tagged with particular targeting sequences, may also lead to the development of BRET-based sensors for localizing structures and proteins in single cells and tissues (157). New BRET vectors will be particularly useful in performing BRET experiments in vivo. BRET using conventional vectors has recently been demonstrated in such an application in transgenic mice expressing β_2AR-Rluc and β-arrestin-GFP (158). These latter approaches will allow BRET to be measured in living animals using in vivo imaging systems such as the Caliper IVIS, capable of measuring fluorescence and luminescence (98, 99, 159). These FRET and BRET systems will be of significant utility in adapting current HTS screening assays for use in animals (120). Advances in microscopy will also benefit researchers wishing to adapt protein–protein interaction assays and/or signalling assays to either single cells, intact tissues or whole animals (for example (160, 161)).

Multifunctional tags are also being created that combine utility in imaging assays with use in protein purification protocols. This can be done by creating vectors encoding for sequential fluorescent and epitope tags (162) or by re-engineering fluorescent molecules. For example, a variant of GFP has been engineered that contains distinct molecular tags for purification inserted into one of the surface loops of the fluorescent protein (163). Mixing and matching such differential tagged constructs will allow combinatorial assortments for combining imaging and standard biochemical approaches.

6. Perspectives

This is an exciting time for research on GPCRs. The possibilities of combining imaging assays will lead to a much richer appreciation of the organization of GPCR signalling systems. We also have an unparalleled ability to interrogate the organization of signalling and the signalling events themselves, in homogenous formats and on similar time scales in living cells. Multiplexing the different types of assays will be possible, in high content, high throughput, and living animal contexts. It is also quite possible that the imaging techniques discussed here will also be applied to other receptor signalling systems as well as voltage- and ligand-gated ion channels.

Acknowledgments

This work was supported by grants from the Canadian Institutes of Health Research to T.E.H (MOP-36279) as well as the CIHR Team in GPCR Allosteric Regulation (CTiGAR). T.E.H. is a Chercheur National of the Fonds de la Recherche en Santé du Québec (FRSQ). We thank Vic Rebois (NIH) and the Hébert lab for helpful discussions.

References

1. Pugh, E. N., Jr., and Lamb, T. D. (1993) Amplification and kinetics of the activation steps in phototransduction. *Biochim Biophys Acta* **1141**, 111–49.
2. Hébert, T. E., and Bouvier, M. (1998) Structural and functional aspects of G protein-coupled receptor oligomerization. *Biochem Cell Biol* **76**, 1–11.
3. Hébert, T. E., Moffett, S., Morello, J. P., Loisel, T. P., Bichet, D. G., Barret, C., and Bouvier, M. (1996) A peptide derived from a β_2-adrenergic receptor transmembrane domain inhibits both receptor dimerization and activation *J Biol Chem* **271**, 16384–92.
4. Prinster, S. C., Hague, C., and Hall, R. A. (2005) Heterodimerization of G protein-coupled receptors: specificity and functional significance. *Pharmacol Rev* **57**, 289–98.
5. Bulenger, S., Marullo, S., and Bouvier, M. (2005) Emerging role of homo- and heterodimerization in G-protein-coupled receptor biosynthesis and maturation. *Trends Pharmacol Sci* **26**, 131–7.
6. Milligan, G. (2009) G protein-coupled receptor hetero-dimerization: contribution to pharmacology and function. *Br J Pharmacol* **158**, 5–14.
7. Ng, G. Y., Clark, J., Coulombe, N., Ethier, N., Hébert, T. E., Sullivan, R., Kargman, S., Chateauneuf, A., Tsukamoto, N., McDonald, T., Whiting, P., Mezey, E., Johnson, M. P., Liu, Q., Kolakowski, L. F., Jr., Evans, J. F.,

Bonner, T. I., and O'Neill, G. P. (1999) Identification of a GABAB receptor subunit, gb2, required for functional GABAB receptor activity. *J Biol Chem* **274**, 7607–10.
8. White, J. H., Wise, A., Main, M. J., Green, A., Fraser, N. J., Disney, G. H., Barnes, A. A., Emson, P., Foord, S. M., and Marshall, F. H. (1998) Heterodimerization is required for the formation of a functional GABA(B) receptor. *Nature* **396**, 679–82.
9. Jones, K. A., Borowsky, B., Tamm, J. A., Craig, D. A., Durkin, M. M., Dai, M., Yao, W. J., Johnson, M., Gunwaldsen, C., Huang, L. Y., Tang, C., Shen, Q., Salon, J. A., Morse, K., Laz, T., Smith, K. E., Nagarathnam, D., Noble, S. A., Branchek, T. A., and Gerald, C. (1998) GABA(B) receptors function as a heteromeric assembly of the subunits GABA(B) R1 and GABA(B)R2. *Nature* **396**, 674–9.
10. Kaupmann, K., Malitschek, B., Schuler, V., Heid, J., Froestl, W., Beck, P., Mosbacher, J., Bischoff, S., Kulik, A., Shigemoto, R., Karschin, A., and Bettler, B. (1998) GABA(B)-receptor subtypes assemble into functional heteromeric complexes. *Nature* **396**, 683–7.
11. Hébert, T. E., Loisel, T. P., Adam, L., Ethier, N., Onge, S. S., and Bouvier, M. (1998) Functional rescue of a constitutively desensitized β_2AR through receptor dimerization. *Biochem J* **330**, 287–93.
12. Bohme, I., and Beck-Sickinger, A. G. (2009) Illuminating the life of GPCRs. *Cell Commun Signal* 7, 16.
13. Rocheville, M., Lange, D. C., Kumar, U., Sasi, R., Patel, R. C., and Patel, Y. C. (2000) Subtypes of the somatostatin receptor assemble as functional homo- and heterodimers. *J Biol Chem* **275**, 7862–9.
14. Rocheville, M., Lange, D. C., Kumar, U., Patel, S. C., Patel, R. C., and Patel, Y. C. (2000) Receptors for dopamine and somatostatin: formation of hetero-oligomers with enhanced functional activity. *Science* **288**, 154–7.
15. Overton, M. C., and Blumer, K. J. (2000) G-protein-coupled receptors function as oligomers in vivo. *Curr Biol* **10**, 341–4.
16. Angers, S., Salahpour, A., Joly, E., Hilairet, S., Chelsky, D., Dennis, M., and Bouvier, M. (2000) Detection of β_2-adrenergic receptor dimerization in living cells using bioluminescence resonance energy transfer (BRET). *Proc Natl Acad Sci U S A* **97**, 3684–9.
17. Vilardaga, J. P., Bunemann, M., Feinstein, T. N., Lambert, N., Nikolaev, V. O., Engelhardt, S., Lohse, M. J., and Hoffmann, C. (2009) GPCR and G proteins: drug efficacy and activation in live cells. *Mol Endocrinol* **23**, 590–9.
18. Lohse, M. J., Hoffmann, C., Nikolaev, V. O., Vilardaga, J. P., and Bunemann, M. (2007) Kinetic analysis of G protein-coupled receptor signaling using fluorescence resonance energy transfer in living cells. *Adv Protein Chem* **74**, 167–88.
19. Lohse, M. J., Bunemann, M., Hoffmann, C., Vilardaga, J. P., and Nikolaev, V. O. (2007) Monitoring receptor signaling by intramolecular FRET. *Curr Opin Pharmacol* 7, 547–53.
20. Nikolaev, V. O., and Lohse, M. J. (2006) Monitoring of cAMP synthesis and degradation in living cells. *Physiology (Bethesda)* **21**, 86–92.
21. Elster, L., Elling, C., and Heding, A. (2007) Bioluminescence resonance energy transfer as a screening assay: Focus on partial and inverse agonism. *J Biomol Screen* **12**, 41–9.
22. Frommer, W. B., Davidson, M. W., and Campbell, R. E. (2009) Genetically encoded biosensors based on engineered fluorescent proteins. *Chem Soc Rev* **38**, 2833–41.
23. Ferrandon, S., Feinstein, T. N., Castro, M., Wang, B., Bouley, R., Potts, J. T., Gardella, T. J., and Vilardaga, J. P. (2009) Sustained cyclic AMP production by parathyroid hormone receptor endocytosis. *Nat Chem Biol* **5**, 734–42.
24. Lohse, M. J., Hein, P., Hoffmann, C., Nikolaev, V. O., Vilardaga, J. P., and Bunemann, M. (2008) Kinetics of G-protein-coupled receptor signals in intact cells. *Br J Pharmacol* **153 Suppl 1**, S125-32.
25. Zaks-Zilberman, M., Harrington, A. E., Ishino, T., and Chaiken, I. M. (2008) Interleukin-5 receptor subunit oligomerization and rearrangement revealed by fluorescence resonance energy transfer imaging. *J Biol Chem* **283**, 13398–406.
26. Tan, P. K., Wang, J., Littler, P. L., Wong, K. K., Sweetnam, T. A., Keefe, W., Nash, N. R., Reding, E. C., Piu, F., Brann, M. R., and Schiffer, H. H. (2007) Monitoring interactions between receptor tyrosine kinases and their downstream effector proteins in living cells using bioluminescence resonance energy transfer. *Mol Pharmacol* **72**, 1440–6.
27. Schiffer, H. H., Reding, E. C., Fuhs, S. R., Lu, Q., Piu, F., Wong, S., Littler, P. L., Weiner, D. M., Keefe, W., Tan, P. K., Nash, N. R., Knapp, A. E., Olsson, R., and Brann, M. R. (2007) Pharmacology and signaling properties of epidermal growth factor receptor isoforms studied by bioluminescence resonance energy transfer. *Mol Pharmacol* **71**, 508–18.
28. De Vries, L., Finana, F., Cachoux, F., Vacher, B., Sokoloff, P., and Cussac, D. (2010) Cellular BRET assay suggests a conformational rearrangement of preformed TrkB/Shc

complexes following BDNF-dependent activation. *Cell Signal* **22**, 158–65.

29. Bal, M., Zhang, J., Zaika, O., Hernandez, C. C., and Shapiro, M. S. (2008) Homomeric and heteromeric assembly of KCNQ (Kv7) K+channels assayed by total internal reflection fluorescence/fluorescence resonance energy transfer and patch clamp analysis. *J Biol Chem* **283**, 30668–76.
30. Whitaker, G. M., Angoli, D., Nazzari, H., Shigemoto, R., and Accili, E. A. (2007) HCN2 and HCN4 isoforms self-assemble and co-assemble with equal preference to form functional pacemaker channels. *J Biol Chem* **282**, 22900–9.
31. Pfleger, K. D. (2009) Analysis of protein–protein interactions using bioluminescence resonance energy transfer. *Methods Mol Biol* **574**, 173–83.
32. Ayoub, M. A., and Pfleger, K. D. (2010) Recent advances in bioluminescence resonance energy transfer technologies to study GPCR heteromerization. *Curr Opin Pharmacol* **10**, 44–52.
33. Bacart, J., Corbel, C., Jockers, R., Bach, S., and Couturier, C. (2008) The BRET technology and its application to screening assays. *Biotechnol J* **3**, 311–24.
34. Prinz, A., Reither, G., Diskar, M., and Schultz, C. (2008) Fluorescence and bioluminescence procedures for functional proteomics. *Proteomics* **8**, 1179–96.
35. Marullo, S., and Bouvier, M. (2007) Resonance energy transfer approaches in molecular pharmacology and beyond. *Trends Pharmacol Sci* **28**, 362–5.
36. Pfleger, K. D., and Eidne, K. A. (2006) Illuminating insights into protein–protein interactions using bioluminescence resonance energy transfer (BRET). *Nat Methods* **3**, 165–74.
37. Hébert, T. E., Galés, C., and Rebois, R. V. (2006) Detecting and imaging protein–protein interactions during G protein-mediated signal transduction in vivo and in situ by using fluorescence-based techniques. *Cell Biochem Biophys* **45**, 85–109.
38. Li, I. T., Pham, E., and Truong, K. (2006) Protein biosensors based on the principle of fluorescence resonance energy transfer for monitoring cellular dynamics. *Biotechnol Lett* **28**, 1971–82.
39. Kenworthy, A. K., and Edidin, M. (1999) Imaging fluorescence resonance energy transfer as probe of membrane organization and molecular associations of GPI-anchored proteins. *Methods Mol Biol* **116**, 37–49.
40. Kenworthy, A. K., and Edidin, M. (1998) Distribution of a glycosylphosphatidylinositol-anchored protein at the apical surface of MDCK cells examined at a resolution of <100 A using imaging fluorescence resonance energy transfer. *J Cell Biol* **142**, 69–84.
41. Mercier, J. F., Salahpour, A., Angers, S., Breit, A., and Bouvier, M. (2002) Quantitative assessment of β_1- and β_2-adrenergic receptor homo- and heterodimerization by bioluminescence resonance energy transfer. *J Biol Chem* **277**, 44925–31.
42. Galés, C., Van Durm, J. J., Schaak, S., Pontier, S., Percherancier, Y., Audet, M., Paris, H., and Bouvier, M. (2006) Probing the activation-promoted structural rearrangements in preassembled receptor-G protein complexes. *Nat Struct Mol Biol* **13**, 778–86.
43. Hamdan, F. F., Rochdi, M. D., Breton, B., Fessart, D., Michaud, D. E., Charest, P. G., Laporte, S. A., and Bouvier, M. (2007) Unraveling G protein-coupled receptor endocytosis pathways using real-time monitoring of agonist-promoted interaction between β-arrestins and AP-2. *J Biol Chem* **282**, 29089–100.
44. Pétrin, D., Robitaille, M., and Hébert, T. E. (2010) Real-time BRET assays to measure G protein/effector interactions. *Methods in Molecular Biology* see Chapter 2.
45. Salahpour, A., Angers, S., Mercier, J. F., Lagace, M., Marullo, S., and Bouvier, M. (2004) Homodimerization of the β_2-adrenergic receptor as a prerequisite for cell surface targeting. *J Biol Chem* **279**, 33390–7.
46. Dupré, D. J., Robitaille, M., Ethier, N., Villeneuve, L. R., Mamarbachi, A. M., and Hébert, T. E. (2006) Seven transmembrane receptor core signaling complexes are assembled prior to plasma membrane trafficking. *J Biol Chem* **281**, 34561–73.
47. Lopez-Gimenez, J. F., Canals, M., Pediani, J. D., and Milligan, G. (2007) The α_{1B}-adrenoceptor exists as a higher-order oligomer: effective oligomerization is required for receptor maturation, surface delivery, and function. *Mol Pharmacol* **71**, 1015–29.
48. Milligan, G. (2010) The role of dimerisation in the cellular trafficking of G-protein-coupled receptors. *Curr Opin Pharmacol* **10**, 23–9.
49. Berchiche, Y. A., Chow, K. Y., Lagane, B., Leduc, M., Percherancier, Y., Fujii, N., Tamamura, H., Bachelerie, F., and Heveker, N. (2007) Direct assessment of CXCR4 mutant conformations reveals complex link between receptor structure and G(α)(i) activation. *J Biol Chem* **282**, 5111–5.

50. Brea, J., Castro, M., Giraldo, J., Lopez-Gimenez, J. F., Padin, J. F., Quintian, F., Cadavid, M. I., Vilaro, M. T., Mengod, G., Berg, K. A., Clarke, W. P., Vilardaga, J. P., Milligan, G., and Loza, M. I. (2009) Evidence for distinct antagonist-revealed functional states of 5-hydroxytryptamine(2A) receptor homodimers. *Mol Pharmacol* **75**, 1380–91.
51. Harikumar, K. G., Pinon, D. I., and Miller, L. J. (2007) Transmembrane segment IV contributes a functionally important interface for oligomerization of the Class II G protein-coupled secretin receptor. *J Biol Chem* **282**, 30363–72.
52. Kobayashi, H., Ogawa, K., Yao, R., Lichtarge, O., and Bouvier, M. (2009) Functional rescue of β-adrenoceptor dimerization and trafficking by pharmacological chaperones. *Traffic* **10**, 1019–33.
53. Ayoub, M. A., Couturier, C., Lucas-Meunier, E., Angers, S., Fossier, P., Bouvier, M., and Jockers, R. (2002) Monitoring of ligand-independent dimerization and ligand-induced conformational changes of melatonin receptors in living cells by bioluminescence resonance energy transfer. *J Biol Chem* **277**, 21522–8.
54. Vilardaga, J. P., Steinmeyer, R., Harms, G. S., and Lohse, M. J. (2005) Molecular basis of inverse agonism in a G protein-coupled receptor. *Nat Chem Biol* **1**, 25–8.
55. Nikolaev, V. O., Hoffmann, C., Bunemann, M., Lohse, M. J., and Vilardaga, J. P. (2006) Molecular basis of partial agonism at the neurotransmitter α_{2A}-adrenergic receptor and Gi-protein heterotrimer. *J Biol Chem* **281**, 24506–11.
56. Zurn, A., Zabel, U., Vilardaga, J. P., Schindelin, H., Lohse, M. J., and Hoffmann, C. (2009) Fluorescence resonance energy transfer analysis of a α_{2A}-adrenergic receptor activation reveals distinct agonist-specific conformational changes. *Mol Pharmacol* **75**, 534–41.
57. Rosenbaum, D. M., Rasmussen, S. G., and Kobilka, B. K. (2009) The structure and function of G-protein-coupled receptors. *Nature* **459**, 356–63.
58. Mustafi, D., and Palczewski, K. (2009) Topology of class A G protein-coupled receptors: insights gained from crystal structures of rhodopsins, adrenergic and adenosine receptors. *Mol Pharmacol* **75**, 1–12.
59. Audet, M., and Bouvier, M. (2008) Insights into signaling from the β_2-adrenergic receptor structure. *Nat Chem Biol* **4**, 397–403.
60. Galés, C., Rebois, R. V., Hogue, M., Trieu, P., Breit, A., Hébert, T. E., and Bouvier, M. (2005) Real-time monitoring of receptor and G-protein interactions in living cells. *Nat Methods* **2**, 177–84.
61. Audet, N., Galés, C., Archer-Lahlou, E., Vallieres, M., Schiller, P. W., Bouvier, M., and Pineyro, G. (2008) Bioluminescence resonance energy transfer assays reveal ligand-specific conformational changes within preformed signaling complexes containing delta-opioid receptors and heterotrimeric G proteins. *J Biol Chem* **283**, 15078–88.
62. Rebois, R. V., Robitaille, M., Gales, C., Dupre, D. J., Baragli, A., Trieu, P., Ethier, N., Bouvier, M., and Hébert, T. E. (2006) Heterotrimeric G proteins form stable complexes with adenylyl cyclase and Kir3.1 channels in living cells. *J Cell Sci* **119**, 2807–18.
63. Digby, G. J., Lober, R. M., Sethi, P. R., and Lambert, N. A. (2006) Some G protein heterotrimers physically dissociate in living cells. *Proc Natl Acad Sci U S A* **103**, 17789–94.
64. Digby, G. J., Sethi, P. R., and Lambert, N. A. (2008) Differential dissociation of G protein heterotrimers. *J Physiol* **586**, 3325–35.
65. Lambert, N. A. (2008) Dissociation of heterotrimeric G proteins in cells. *Sci Signal* **1**, re5.
66. Milligan, G., and Bouvier, M. (2005) Methods to monitor the quaternary structure of G protein-coupled receptors. *FEBS J* **272**, 2914–25.
67. Arai, R., Nakagawa, H., Kitayama, A., Ueda, H., and Nagamune, T. (2002) Detection of protein–protein interaction by bioluminescence resonance energy transfer from firefly luciferase to red fluorescent protein. *J Biosci Bioeng* **94**, 362–4.
68. Yamakawa, Y., Ueda, H., Kitayama, A., and Nagamune, T. (2002) Rapid homogeneous immunoassay of peptides based on bioluminescence resonance energy transfer from firefly luciferase. *J Biosci Bioeng* **93**, 537–42.
69. Molinari, P., Casella, I., and Costa, T. (2008) Functional complementation of high-efficiency resonance energy transfer: a new tool for the study of protein binding interactions in living cells. *Biochem J* **409**, 251–61.
70. Du, M., Rambhadran, A., and Jayaraman, V. (2008) Luminescence resonance energy transfer investigation of conformational changes in the ligand binding domain of a kainate receptor. *J Biol Chem* **283**, 27074–8.
71. Gonzalez, J., Rambhadran, A., Du, M., and Jayaraman, V. (2008) LRET investigations of conformational changes in the ligand binding domain of a functional AMPA receptor. *Biochemistry* **47**, 10027–32.
72. Posson, D. J., and Selvin, P. R. (2008) Extent of voltage sensor movement during gating of *Shaker* K+ channels. *Neuron* **59**, 98–109.

73. Granier, S., Kim, S., Shafer, A. M., Ratnala, V. R., Fung, J. J., Zare, R. N., and Kobilka, B. (2007) Structure and conformational changes in the C-terminal domain of the β_2-adrenoceptor: insights from fluorescence resonance energy transfer studies. *J Biol Chem* **282**, 13895–905.
74. Glatz, R. V., Leifert, W. R., Cooper, T. H., Bailey, K., Barton, C. S., Martin, S., Aloia, A. L., Bucco, O., Waniganayake, L., Wei, G., Raguse, B., Wieczorek, L., and McMurchie, E. J. (2007) Molecular Engineering of G Protein-Coupled Receptors and G Proteins for Cell-Free Biosensing. *Aust. J. Chem.* **60**, 309–13
75. Taraska, J. W., Puljung, M. C., Olivier, N. B., Flynn, G. E., and Zagotta, W. N. (2009) Mapping the structure and conformational movements of proteins with transition metal ion FRET. *Nat Methods* **6**, 532–7.
76. Fernandez-Suarez, M., and Ting, A. Y. (2008) Fluorescent probes for super-resolution imaging in living cells. *Nat Rev Mol Cell Biol* **9**, 929–43.
77. Ellis-Davies, G. C. (2007) Caged compounds: photorelease technology for control of cellular chemistry and physiology. *Nat Methods* **4**, 619–28.
78. Lober, R. M., Pereira, M. A., and Lambert, N. A. (2006) Rapid activation of inwardly rectifying potassium channels by immobile G-protein-coupled receptors. *J Neurosci* **26**, 12602–8.
79. Qin, K., Sethi, P. R., and Lambert, N. A. (2008) Abundance and stability of complexes containing inactive G protein-coupled receptors and G proteins. *FASEB J* **22**, 2920–7.
80. Fonseca, J. M., and Lambert, N. A. (2009) Instability of a class a G protein-coupled receptor oligomer interface. *Mol Pharmacol* **75**, 1296–9.
81. Dorsch, S., Klotz, K. N., Engelhardt, S., Lohse, M. J., and Bunemann, M. (2009) Analysis of receptor oligomerization by FRAP microscopy. *Nat Methods* **6**, 225–30.
82. Matsushita, S., Nakata, H., Kubo, Y., and Tateyama, M. (2010) Ligand-induced rearrangements of the GABAb receptor revealed by fluorescence resonance energy transfer(FRET). *J Biol Chem* **285**, 10291–9.
83. Ianoul, A., Grant, D. D., Rouleau, Y., Bani-Yaghoub, M., Johnston, L. J., and Pezacki, J. P. (2005) Imaging nanometer domains of β-adrenergic receptor complexes on the surface of cardiac myocytes. *Nat Chem Biol* **1**, 196–202.
84. Michnick, S. W., Ear, P. H., Manderson, E. N., Remy, I., and Stefan, E. (2007) Universal strategies in research and drug discovery based on protein-fragment complementation assays. *Nat Rev Drug Discov* **6**, 569–82.
85. Kerppola, T. K. (2009) Visualization of molecular interactions using bimolecular fluorescence complementation analysis: characteristics of protein fragment complementation. *Chem Soc Rev* **38**, 2876–86.
86. Kerppola, T. K. (2008) Bimolecular fluorescence complementation (BiFC) analysis as a probe of protein interactions in living cells. *Annu Rev Biophys* **37**, 465–87.
87. MacDonald, M. L., Lamerdin, J., Owens, S., Keon, B. H., Bilter, G. K., Shang, Z., Huang, Z., Yu, H., Dias, J., Minami, T., Michnick, S. W., and Westwick, J. K. (2006) Identifying off-target effects and hidden phenotypes of drugs in human cells. *Nat Chem Biol* **2**, 329–37.
88. Michnick, S. W., MacDonald, M. L., and Westwick, J. K. (2006) Chemical genetic strategies to delineate MAP kinase signaling pathways using protein-fragment complementation assays (PCA). *Methods* **40**, 287–93.
89. Remy, I., and Michnick, S. W. (2007) Application of protein-fragment complementation assays in cell biology. *Biotechniques* **42**, 137, 139, 41 passim.
90. Hu, C. D., Chinenov, Y., and Kerppola, T. K. (2002) Visualization of interactions among bZIP and Rel family proteins in living cells using bimolecular fluorescence complementation. *Mol Cell* **9**, 789–98.
91. Hynes, T. R., Tang, L., Mervine, S. M., Sabo, J. L., Yost, E. A., Devreotes, P. N., and Berlot, C. H. (2004) Visualization of G protein βγ dimers using bimolecular fluorescence complementation demonstrates roles for both beta and gamma in subcellular targeting. *J Biol Chem* **279**, 30279–86.
92. Shyu, Y. J., Liu, H., Deng, X., and Hu, C. D. (2006) Identification of new fluorescent protein fragments for bimolecular fluorescence complementation analysis under physiological conditions. *Biotechniques* **40**, 61–6.
93. Hynes, T. R., Yost, E., Mervine, S., and Berlot, C. H. (2008) Multicolor BiFC analysis of competition among G protein β and γ subunit interactions. *Methods* **45**, 207–13.
94. Mervine, S. M., Yost, E. A., Sabo, J. L., Hynes, T. R., and Berlot, C. H. (2006) Analysis of G protein βγ dimer formation in live cells using multicolor bimolecular fluorescence complementation demonstrates preferences of β_1 for particular γ subunits. *Mol Pharmacol* **70**, 194–205.
95. Rose, R. H., Briddon, S. J., and Holliday, N. D. (2010) Bimolecular fluorescence complementation: lighting up seven transmembrane

domain receptor signalling networks. *Br J Pharmacol* **159**, 738–50.

96. Feinberg, E. H., Vanhoven, M. K., Bendesky, A., Wang, G., Fetter, R. D., Shen, K., and Bargmann, C. I. (2008) GFP Reconstitution Across Synaptic Partners (GRASP) defines cell contacts and synapses in living nervous systems. *Neuron* **57**, 353–63.
97. Fan, F., Binkowski, B. F., Butler, B. L., Stecha, P. F., Lewis, M. K., and Wood, K. V. (2008) Novel genetically encoded biosensors using firefly luciferase. *ACS Chem Biol* **3**, 346–51.
98. Luker, K. E., Gupta, M., and Luker, G. D. (2008) Imaging CXCR4 signaling with firefly luciferase complementation. *Anal Chem* **80**, 5565–73.
99. Luker, K. E., Gupta, M., Steele, J. M., Foerster, B. R., and Luker, G. D. (2009) Imaging ligand-dependent activation of CXCR7. *Neoplasia* **11**, 1022–35.
100. Villalobos, V., Naik, S., and Piwnica-Worms, D. (2008) Detection of protein–protein interactions in live cells and animals with split firefly luciferase protein fragment complementation. *Methods Mol Biol* **439**, 339–52.
101. Yang, K. S., Ilagan, M. X., Piwnica-Worms, D., and Pike, L. J. (2009) Luciferase fragment complementation imaging of conformational changes in the epidermal growth factor receptor. *J Biol Chem* **284**, 7474–82.
102. Remy, I., and Michnick, S. W. (2006) A highly sensitive protein–protein interaction assay based on Gaussia luciferase. *Nat Methods* **3**, 977–9.
103. Kim, S. B., Sato, M., and Tao, H. (2009) Split *Gaussia* luciferase-based bioluminescence template for tracing protein dynamics in living cells. *Anal Chem* **81**, 67–74.
104. Stefan, E., Aquin, S., Berger, N., Landry, C. R., Nyfeler, B., Bouvier, M., and Michnick, S. W. (2007) Quantification of dynamic protein complexes using Renilla luciferase fragment complementation applied to protein kinase A activities in vivo. *Proc Natl Acad Sci U S A* **104**, 16916–21.
105. Luker, K. E., Gupta, M., and Luker, G. D. (2009) Imaging chemokine receptor dimerization with firefly luciferase complementation. *FASEB J* **23**, 823–34.
106. Guo, W., Urizar, E., Kralikova, M., Mobarec, J. C., Shi, L., Filizola, M., and Javitch, J. A. (2008) Dopamine D2 receptors form higher order oligomers at physiological expression levels. *EMBO J* **27**, 2293–304.
107. Rebois, R. V., Robitaille, M., Petrin, D., Zylbergold, P., Trieu, P., and Hébert, T. E. (2008) Combining protein complementation assays with resonance energy transfer to detect multipartner protein complexes in living cells. *Methods* **45**, 214–8.
108. Cabantous, S., Terwilliger, T. C., and Waldo, G. S. (2005) Protein tagging and detection with engineered self-assembling fragments of green fluorescent protein. *Nat Biotechnol* **23**, 102–7.
109. Cabantous, S., Pedelacq, J. D., Mark, B. L., Naranjo, C., Terwilliger, T. C., and Waldo, G. S. (2005) Recent advances in GFP folding reporter and split-GFP solubility reporter technologies. Application to improving the folding and solubility of recalcitrant proteins from Mycobacterium tuberculosis. *J Struct Funct Genomics* **6**, 113–9.
110. Cabantous, S., and Waldo, G. S. (2006) In vivo and in vitro protein solubility assays using split GFP. *Nat Methods* **3**, 845–54.
111. Chen, Y., Li, S., Chen, T., Hua, H., and Lin, Z. (2009) Random dissection to select for protein split sites and its application in protein fragment complementation. *Protein Sci* **18**, 399–409.
112. Pedelacq, J. D., Cabantous, S., Tran, T., Terwilliger, T. C., and Waldo, G. S. (2006) Engineering and characterization of a superfolder green fluorescent protein. *Nat Biotechnol* **24**, 79–88.
113. Ottmann, C., Weyand, M., Wolf, A., and Kuhlmann, J. (2009) Applicability of superfolder YFP bimolecular fluorescence complementation in vitro. *Biol Chem* **390**, 81–90.
114. Chu, J., Zhang, Z., Zheng, Y., Yang, J., Qin, L., Lu, J., Huang, Z. L., Zeng, S., and Luo, Q. (2009) A novel far-red bimolecular fluorescence complementation system that allows for efficient visualization of protein interactions under physiological conditions. *Biosens Bioelectron* **25**, 234–9.
115. Subach, F. V., Patterson, G. H., Manley, S., Gillette, J. M., Lippincott-Schwartz, J., and Verkhusha, V. V. (2009) Photoactivatable mCherry for high-resolution two-color fluorescence microscopy. *Nat Methods* **6**, 153–9.
116. Andresen, M., Stiel, A. C., Folling. J., Wenzel, D., Schonle, A., Egner, A., Eggeling, C., Hell, S. W., and Jakobs, S. (2008) Photoswitchable fluorescent proteins enable monochromatic multilabel imaging and dual color fluorescence nanoscopy. *Nat Biotechnol* **26**, 1035–40.
117. Chudakov, D. M., Lukyanov, S., and Lukyanov, K. A. (2007) Tracking intracellular protein movements using photoswitchable fluorescent proteins PS-CFP2 and Dendra2. *Nat Protoc* **2**, 2024–32.

118. Morell, M., Espargaro, A., Aviles, F. X., and Ventura, S. (2008) Study and selection of in vivo protein interactions by coupling bimolecular fluorescence complementation and flow cytometry. *Nat Protoc* **3**, 22–33.
119. Ding, Z., Liang, J., Lu, Y., Yu, Q., Songyang, Z., Lin, S. Y., and Mills, G. B. (2006) A retrovirus-based protein complementation assay screen reveals functional AKT1-binding partners. *Proc Natl Acad Sci U S A* **103**, 15014–9.
120. Giacomotto, J., and Ségalat, L. (2010) High-throughput screening and small animal models, where are we? *Br J Pharmacol* **160**, 204–16
121. Luker, K., Gupta, M., and Luker, G. (2009) Bioluminescent CXCL12 fusion protein for cellular studies of CXCR4 and CXCR7. *Biotechniques* **47**, 625–32.
122. Hiatt, S. M., Shyu, Y. J., Duren, H. M., and Hu, C. D. (2008) Bimolecular fluorescence complementation (BiFC) analysis of protein interactions in Caenorhabditis elegans. *Methods* **45**, 185–91.
123. Adams, S. R., Campbell, R. E., Gross, L. A., Martin, B. R., Walkup, G. K., Yao, Y., Llopis, J., and Tsien, R. Y. (2002) New biarsenical ligands and tetracysteine motifs for protein labeling in vitro and in vivo: synthesis and biological applications. *J Am Chem Soc* **124**, 6063–76.
124. Keppler, A., Kindermann, M., Gendreizig, S., Pick, H., Vogel, H., and Johnsson, K. (2004) Labeling of fusion proteins of O6-alkylguanine-DNA alkyltransferase with small molecules in vivo and in vitro. *Methods* **32**, 437–44.
125. Gautier, A., Juillerat, A., Heinis, C., Correa, I. R., Jr., Kindermann, M., Beaufils, F., and Johnsson, K. (2008) An engineered protein tag for multiprotein labeling in living cells. *Chem Biol* **15**, 128–36.
126. Robitaille, M., Héroux, I., Baragli, A., and Hébert, T. E. (2009) Novel tools for use in bioluminescence resonance energy transfer (BRET) assays. *Methods Mol Biol* **574**, 215–34.
127. Maurel, D., Comps-Agrar, L., Brock, C., Rives, M. L., Bourrier, E., Ayoub, M. A., Bazin, H., Tinel, N., Durroux, T., Prezeau, L., Trinquet, E., and Pin, J. P. (2008) Cell-surface protein–protein interaction analysis with time-resolved FRET and snap-tag technologies: application to GPCR oligomerization. *Nat Methods* **5**, 561–7.
128. Héroux, M., Hogue, M., Lemieux, S., and Bouvier, M. (2007) Functional calcitonin gene-related peptide receptors are formed by the asymmetric assembly of a calcitonin receptor-like receptor homo-oligomer and a monomer of receptor activity-modifying protein-1. *J Biol Chem* **282**, 31610–20.
129. Ma, A. W., Pawagi, A. B., and Wells, J. W. (2008) Heterooligomers of the muscarinic receptor and G proteins purified from porcine atria. *Biochem Biophys Res Commun* **374**, 128–33.
130. Ma, A. W., Redka, D. S., Pisterzi, L. F., Angers, S., and Wells, J. W. (2007) Recovery of oligomers and cooperativity when monomers of the M2 muscarinic cholinergic receptor are reconstituted into phospholipid vesicles. *Biochemistry* **46**, 7907–27.
131. Pisterzi, L. F., Jansma, D. B., Georgiou, J., Woodside, M. J., Chou, J. T., Angers, S., Raicu, V., and Wells, J. W. (2010) Oligomeric size of the M2 muscarinic receptor in live cells as determined by quantitative fluorescence resonance energy transfer (FRET). *J Biol Chem.* **285**, 16723–38
132. Fung, J. J., Deupi, X., Pardo, L., Yao, X. J., Velez-Ruiz, G. A., Devree, B. T., Sunahara, R. K., and Kobilka, B. K. (2009) Ligand-regulated oligomerization of β_2-adrenoceptors in a model lipid bilayer. *EMBO J* **28**, 3315–28.
133. Vidi, P. A., Chemel, B. R., Hu, C. D., and Watts, V. J. (2008) Ligand-dependent oligomerization of dopamine D(2) and adenosine A(2A) receptors in living neuronal cells. *Mol Pharmacol* **74**, 544–51.
134. Vidi, P. A., Chen, J., Irudayaraj, J. M., and Watts, V. J. (2008) Adenosine A(2A) receptors assemble into higher-order oligomers at the plasma membrane. *FEBS Lett* **582**, 3985–90.
135. Gandia, J., Galino, J., Amaral, O. B., Soriano, A., Lluis, C., Franco, R., and Ciruela, F. (2008) Detection of higher-order G protein-coupled receptor oligomers by a combined BRET-BiFC technique. *FEBS Lett* **582**, 2979–84.
136. Navarro, G., Carriba, P., Gandia, J., Ciruela, F., Casado, V., Cortes, A., Mallol, J., Canela, E. I., Lluis, C., and Franco, R. (2008) Detection of heteromers formed by cannabinoid CB1, dopamine D2, and adenosine A2A G-protein-coupled receptors by combining bimolecular fluorescence complementation and bioluminescence energy transfer. *ScientificWorldJournal* **8**, 1088–97.
137. Hamatake, M., Aoki, T., Futahashi, Y., Urano, E., Yamamoto, N., and Komano, J. (2009) Ligand-independent higher-order multimerization of CXCR4, a G-protein-coupled

chemokine receptor involved in targeted metastasis. *Cancer Sci* **100**, 95–102.

138. Carriba, P., Navarro, G., Ciruela, F., Ferre, S., Casado, V., Agnati, L., Cortes, A., Mallol, J., Fuxe, K., Canela, E. I., Lluis, C., and Franco, R. (2008) Detection of heteromerization of more than two proteins by sequential BRET-FRET. *Nat Methods* **5**, 727–33.
139. Vidi, P. A., and Watts, V. J. (2009) Fluorescent and bioluminescent protein-fragment complementation assays in the study of G protein-coupled receptor oligomerization and signaling. *Mol Pharmacol* **75**, 733–9.
140. Shyu, Y. J., Suarez, C. D., and Hu, C. D. (2008) Visualization of ternary complexes in living cells by using a BiFC-based FRET assay. *Nat Protoc* **3**, 1693–702.
141. Rovira, X., Pin, J. P., and Giraldo, J. (2010) The asymmetric/symmetric activation of GPCR dimers as a possible mechanistic rationale for multiple signalling pathways. *Trends Pharmacol Sci* **31**, 15–21.
142. Rives, M. L., Vol, C., Fukazawa, Y., Tinel, N., Trinquet, E., Ayoub, M. A., Shigemoto, R., Pin, J. P., and Prezeau, L. (2009) Crosstalk between GABA(B) and mGlu1a receptors reveals new insight into GPCR signal integration. *EMBO J* **28**, 2195–208.
143. Pin, J. P., Comps-Agrar, L., Maurel, D., Monnier, C., Rives, M. L., Trinquet, E., Kniazeff, J., Rondard, P., and Prezeau, L. (2009) G-protein-coupled receptor oligomers: two or more for what? Lessons from mGlu and GABA(B) receptors. *J Physiol* **587**, 5337–44.
144. Tsien, R. Y. (2005) Building and breeding molecules to spy on cells and tumors. *FEBS Lett* **579**, 927–32.
145. Shaner, N. C., Steinbach, P. A., and Tsien, R. Y. (2005) A guide to choosing fluorescent proteins. *Nat Methods* **2**, 905–9.
146. Day, R. N., and Davidson, M. W. (2009) The fluorescent protein palette: tools for cellular imaging. *Chem Soc Rev* **38**, 2887–921.
147. Shaner, N. C., Patterson, G. H., and Davidson, M. W. (2007) Advances in fluorescent protein technology. *J Cell Sci* **120**, 4247–60.
148. Lymperopoulos, K., Kiel, A., Seefeld, A., Stohr, K., and Herten, D. P. (2010) Fluorescent probes and delivery methods for single-molecule experiments. *Chemphyschem* **11**, 43–53.
149. Joo, C., Balci, H., Ishitsuka, Y., Buranachai, C., and Ha, T. (2008) Advances in single-molecule fluorescence methods for molecular biology. *Annu Rev Biochem* **77**, 51–76.
150. Coulon, V., Audet, M., Homburger, V., Bockaert, J., Fagni, L., Bouvier, M., and Perroy, J. (2008) Subcellular imaging of dynamic protein interactions by bioluminescence resonance energy transfer. *Biophys J* **94**, 1001–9.
151. Perroy, J. (2010) Subcellular dynamic imaging of protein–protein interactions in live cells by bioluminescence resonance energy transfer. *Methods Mol Biol* **591**, 325–33.
152. Xu, X., Soutto, M., Xie, Q., Servick, S., Subramanian, C., von Arnim, A. G., and Johnson, C. H. (2007) Imaging protein interactions with bioluminescence resonance energy transfer (BRET) in plant and mammalian cells and tissues. *Proc Natl Acad Sci U S A* **104**, 10264–9.
153. Loening, A. M., Wu, A. M., and Gambhir, S. S. (2007) Red-shifted Renilla reniformis luciferase variants for imaging in living subjects. *Nat Methods* **4**, 641–3.
154. Loening, A. M., Fenn, T. D., Wu, A. M., and Gambhir, S. S. (2006) Consensus guided mutagenesis of Renilla luciferase yields enhanced stability and light output. *Protein Eng Des Sel* **19**, 391–400.
155. De, A., Loening, A. M., and Gambhir, S. S. (2007) An improved bioluminescence resonance energy transfer strategy for imaging intracellular events in single cells and living subjects. *Cancer Res* **67**, 7175–83.
156. De, A., Ray, P., Loening, A. M., and Gambhir, S. S. (2009) BRET3: a red-shifted bioluminescence resonance energy transfer (BRET)-based integrated platform for imaging protein–protein interactions from single live cells and living animals. *FASEB J* **23**, 2702–9.
157. Hoshino, H., Nakajima, Y., and Ohmiya, Y. (2007) Luciferase-YFP fusion tag with enhanced emission for single-cell luminescence imaging. *Nat Methods* **4**, 637–9.
158. Audet, M., Lagac, M., Silversides, D. W., and Bouvier, M. (2010) Protein–protein interactions monitored in cells from transgenic mice using bioluminescence resonance energy transfer. *FASEB J* **24**, 2829–38.
159. Mezzanotte, L., Fazzina, R., Michelini, E., Tonelli, R., Pession, A., Branchini, B., and Roda, A. (2010) In Vivo Bioluminescence Imaging of Murine Xenograft Cancer Models with a Red-shifted Thermostable Luciferase. *Mol Imaging Biol* **4**, 406–14.
160. Lippincott-Schwartz, J., and Patterson, G. H. (2009) Photoactivatable fluorescent proteins for diffraction-limited and super-resolution imaging. *Trends Cell Biol* **19**, 555–65.

161. Lippincott-Schwartz, J., and Manley, S. (2009) Putting super-resolution fluorescence microscopy to work. *Nat Methods* **6**, 21–3.
162. Deng, C., Xiong, X., and Krutchinsky, A. N. (2009) Unifying fluorescence microscopy and mass spectrometry for studying protein complexes in cells. *Mol Cell Proteomics* **8**, 1413–23.
163. Kobayashi, T., Morone, N., Kashiyama, T., Oyamada, H., Kurebayashi, N., and Murayama, T. (2008) Engineering a novel multifunctional green fluorescent protein tag for a wide variety of protein research. *PLoS One* **3**, e3822.

Chapter 3

Improving Drug Discovery with Contextual Assays and Cellular Systems Analysis

John K. Westwick and Jane E. Lamerdin

Abstract

Despite rapid growth in our knowledge of potential disease targets following completion of the first drafts of the human genome over 10 years ago, the success rate of new therapeutic discovery has been frustratingly low. In addition to the widely reported costs and single-digit success rate of the entire drug discovery and development process, it has recently been estimated that even the preliminary process of transitioning new targets to *preclinical* development succeeds in less than 3% of attempts [Vogel (ed.) *Drug Discovery and Evaluation: Pharmacological Assays.* 3rd ed. Springer, Berlin (2007)]. At these early stages of development, poor understanding of therapeutic mechanisms and lack of compound selectivity are often to blame for failed compounds. It is worth noting than the emerging class of nucleic acid-based therapeutics, including miRNA and RNAi, are likely to be even more prone to unexpected system-wide and off-target activities. For all therapeutic approaches, it is clear that discovery strategies permitting the assessment of drug targets in their native context are required. At the same time, these strategies need to retain the high throughput of current reductionist approaches to enable broad assessment of chemical space for small molecule and genetic therapeutics. We describe here an integrated system based on high-content cellular analysis combined with system-wide pathway interrogation. The platform can be applied to novel therapeutic target and drug candidate identification, and for providing detailed mechanistic and selectivity information at an early stage of development.

Key words: Signal transduction, Network biology, Chemical biology, Systems biology, Protein complex, High-content assay, Pathway analysis, Drug discovery, Drug profiling, G-protein-coupled receptor, Nuclear receptor, Proteasome, Protein-fragment complementation assay

1. Introduction

It is now widely appreciated that cellular signaling occurs via networks of interacting macromolecules, and disease states result from disturbances in the networks (1). The components of these networks – including potential therapeutic targets – do not physically exist as individual genes and proteins, but are comprised of large

Louis M. Luttrell and Stephen S.G. Ferguson (eds.), *Signal Transduction Protocols*, Methods in Molecular Biology, vol. 756, DOI 10.1007/978-1-61779-160-4_3, © Springer Science+Business Media, LLC 2011

and dynamic macromolecular complexes governed by diverse and dynamic interactions (2). These facts are at least partially responsible for the different levels of complexity seen in diverse organisms such as worms, humans, and flies that have similar numbers of genes (3). Therapeutic development should exploit these realities, but such an approach poses significant challenges. For example, the foregoing points would appear to indicate that definition of complex processes, such as a process of cellular differentiation or a specific disease state, would require monitoring the entire collection of proteins ultimately linked to that process, and perhaps the much larger number of their potential interactions. Given the total number of genes and the estimated number of protein interactions (650,000) (3), this would appear to be an intractable goal.

Fortunately, it is possible to categorize all cellular functions into as few as 16 defined signaling pathways (4), and functions related to complex disease states into as few as a dozen defined signaling pathways (5, 6). Cellular networks are not randomly organized, but instead exhibit scale free topology (7) and pathways within these networks contain essential relay points or "nodes" that can be used to further reduce complexity. These attributes can be exploited to simultaneously capture the inherent complexity and connectivity of signaling dynamics within a tractable technical framework. The approach requires, however, a strategy for subsequent de-convolution of the specific biochemical events responsible for observed higher-level changes. The follow-up strategy should also be agnostic as to the target class of the biochemical event in question, and should capture events throughout the cell (as opposed to, e.g. measurements that only identify changes in receptors or transcriptional read-outs).

We describe here an approach designed to satisfy these basic requirements. To enable high-throughput network analysis, the strategy is fundamentally pathway based. To enable dissection of the biochemical events in question, we employ automated microscopic analysis of diverse cellular and biochemical activities, including assessment of protein levels, posttranslational modifications, second messengers and other biochemical changes, transcriptional control, and protein complex dynamics by protein-fragment complementation assays (PCA). For the purposes of this overview, we will describe the application of the platform to analysis of protein complex dynamics.

2. Engineering a Platform for High-Throughput Protein Complex Analysis

To capture protein complex dynamics within cellular networks, and with a format amenable to high-throughput analysis of large compound or reagent panels, we have employed Protein-fragment Complementation Assay (PCA) technology. The details and

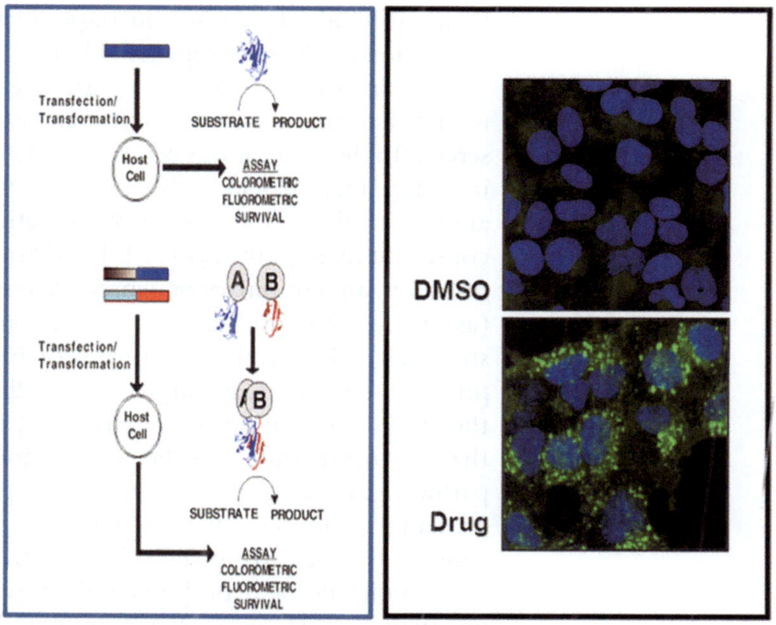

Fig. 1. PCA design. To engineer a PCA (*left panel*), a gene encoding a reporter protein is rationally dissected into two or more fragments. A test protein of interest (A) is fused in-frame to one of the reporter fragments, and the other test protein (B) is fused to the other reporter fragment. Assembly of the reporter protein from its fragments can only happen if the test proteins "A" and "B" interact. When the test proteins interact, the reporter fragments are brought into proximity, re-fold, and generate a detectable signal. An example of images from an automated fluorescence microscope is shown (*right panel*). The blue signal (Hoechst stain) corresponds to cell nuclei and the green is the PCA signal. In this example, the PCA signal is only evident after drug treatment.

examples of PCA have been previously described (8–10) and are illustrated in Fig. 1. In brief, PCA employs pairs of test proteins known or suspected to be involved in a protein complex. The cDNA for one of the test proteins is linked in-frame to a fragment of a reporter protein cDNA. The other test protein is linked in-frame to a sequence corresponding to the other portion of the same reporter protein. Reporter protein fragments are designed such that if the fusion proteins are co-expressed in a living cell, the fragments of the reporter protein can interact, re-fold, and generate a detectable signal only when the two test proteins are in close proximity (i.e., within a protein complex). The system therefore enables sensitive detection and measurement of specific protein complexes. An advantage of the assay technology is the inherent simplicity, which enables the use of the same assay format across target classes, regardless of the molecular or enzymatic mechanism of the target (8, 10).

One type of PCA utilizes synthesized fragments of inherently fluorescent proteins. Fluorescence PCA is particularly valuable because it can be combined with high-throughput automated fluorescence microscopy and image analysis. This strategy captures the dynamics of protein complexes – both their existence *and* their subcellular location – in response to cellular perturbation with drugs or genetic reagents (11, 12).

A common misconception regarding these assay technologies is that they amount to a mammalian version of a yeast 2-hybrid screen. In the applications described here, probe sets are expressed in living mammalian cells and exist within native pathways along with the endogenous proteins native to the complexes that constitute those pathways (12). In addition, the complexes being measured are not binary or expressed in unnatural compartments (as in yeast 2-hybrid analysis or previously described biosensor strategies). Numerous additional proteins, besides the two comprising the assay, create larger-order cellular complexes. Because the assays exist within native pathways, perturbation of a target that is "upstream" or otherwise connected to the measured pathway can be detected. If a drug or probe directly disrupts a protein interaction it will be detected, and the assays are ideal vehicles for identifying such disruptors. Indeed, the concept of small molecule or peptidomimetic targeting of protein–protein interaction interfaces is growing in popularity (13, 14). To date, however, the number of examples of direct interaction modulators is small, and the ability to capture pathway activities beyond those directly measured by a protein complex probe is essential.

To enable high-throughput analysis of these rich biological events, we engineered a platform consisting of automated cell and liquid handling, high-throughput confocal microscopy and cytometry. Data-intensive microscopy required development of a LIMS infrastructure consisting of on-the-fly automated image analysis, and database components capable of storing and analyzing large datasets. A schematic of this approach is shown in Fig. 2. The system is used to routinely capture of hundreds of thousands of images per day (each consisting of several hundred individual cells), and therefore enables screening of large chemical or target files.

3. Analysis of Diverse Targets

To capture the activity of cellular networks and to address target classes previously considered "un-drugable," a global pharmacology strategy should be agnostic as to protein pathway or target class. We therefore set out to test the breadth of target classes and cellular pathways that could be interrogated in high throughput with protein complex-based assays. The spectrum of target classes and cellular processes that we have successfully interrogated with PCA is shown in Fig. 3. Hundreds of unique PCAs were generated and tested, including examples of target classes that comprise the most common drug targets such as G-protein-coupled receptors and protein kinases (9–12). Even with high profile targets

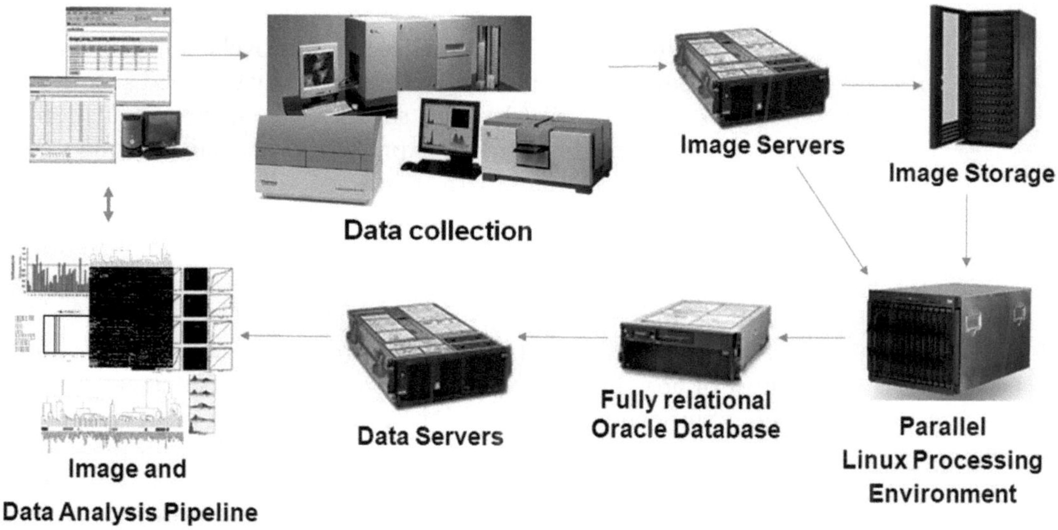

Fig. 2. Schematic of data information flow and customized IT infrastructure to support high-throughput, high-content protein complex analyses. Screening campaigns (small molecules, siRNA, etc.) performed on automated confocal microscopes and cytometry platforms (Evotec Opera; Perkin Elmer, Acumen eX3; TTP LabTech) generate ~130 GB of images per day, which are saved over a high speed fiber connection to a storage area network (SAN). On-the-fly image analysis is performed using highly parallelized custom algorithms on a suite of Linux blade centers to quantify changes observed in the relevant subcellular compartment(s) for each assay. Quantitative data (along with metadata from experimental templates) from each assay are loaded into a fully relational database (Oracle), enabling customized queries and integration with external statistical analysis and visualization tools.

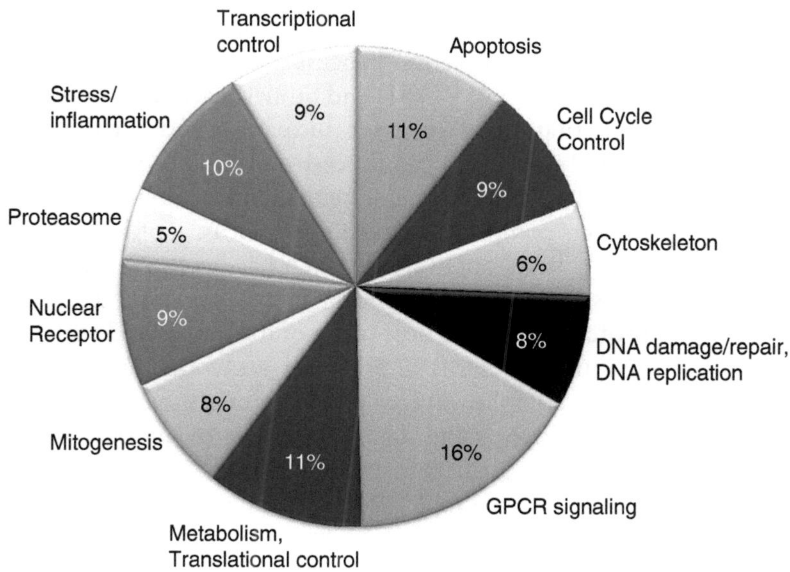

Fig. 3. Broad target class and pathway representation in engineered PCAs. 305 PCA assays targeting diverse signaling nodes or cellular processes were assigned to one of 11 categories (Apoptosis, Cell cycle control, Cytoskeleton, DNA damage & repair or DNA replication, GPCR signaling, Metabolism/Translational control, Mitogenesis, Nuclear receptor, Proteasome, Stress/inflammation, and Transcriptional control) based on the cellular properties of the protein complex as described in the literature. Assays involved in GPCR signaling are well-represented in the panel, with examples including GPCR:arrestin complexes as well as downstream effectors (e.g., Grk2 with effectors, etc). Kinase:effector pairs are found in many of the categories; e.g., the well-known Pdk1/Akt1 complex is found in the Apoptosis category, while Mnk1/eIF4E is classified under Metabolism/Translational control.

from these extensively studied classes, it is interesting to note that screens we have performed detected novel chemical hits and leads from chemical files that were previously screened against the same target in vitro (data not shown). The likely explanation is that a protein target, when expressed in its native context and existing in a complex with a host of additional proteins and other cellular structures, will be regulated by chemical matter that would be missed in screens utilizing purified components in a test tube. For this reason, we believe that the approach has significant promise as a drug re-indication strategy (11).

Beyond assays comprised of current drug targets, there are a host of proteins closely linked to various diseases that are considered "un-drugable," that is, the proteins do not possess inherent enzymatic activity or otherwise lack a facile strategy for measuring their activity. With the proliferation of genome-wide association studies, the number of disease-related yet un-drugable targets has become frustratingly large. Well-known examples include GTPases (such as Ras), various transcription factors, scaffolds, chaperones, and other cellular factors such as transport proteins and allosteric activators that rely on protein complex dynamics to effect signaling changes. Given the emerging importance of the control of protein levels and stability by the proteasome system, we provide examples of proteasome-related signaling events as an example of a target class which can be broadly and effectively analyzed by these methods (Fig. 4). For example, PCAs were generated that monitor different components of the ubiquitination and sumoylation processes, including interaction of the E3 ligases (Mdm2, PIAS1, and Smurf) with target proteins (Fig. 4a). Quantitation of complex formation and turn-over are demonstrated with ALLN-induced accumulation of the Mdm2/p53 complex and treatment with a novel compound that potently inhibits complex formation with an IC_{50} of 50 nM (Fig. 4b).

A useful application which has emerged from this expansion of assay space is multi-facet target analysis. Nearly every cellular protein contacts a host of other molecules and each of these interactions represents a potential assay that may be differentially regulated following cellular perturbation. It is intuitively clear that to understand a particular area of biology, it is necessary to assess that area from the standpoint of all components of that activity. An example of this strategy is shown in Fig. 5. In this example, the same cellular "target" (in this case the cell cycle regulator p27) is assayed in the context of five unique known co-regulators. We have used one of these assays measuring the direct ubiquitination of p27 to test the effect of the proteasome inhibitor ALLN. Inhibition of the proteasome would be expected to lead to accumulation of ubiquitinated substrates (such as p27), as we observed in Fig. 5b. We previously demonstrated a similar strategy for analyzing multiple aspects of the activity of the protein kinase Akt (9).

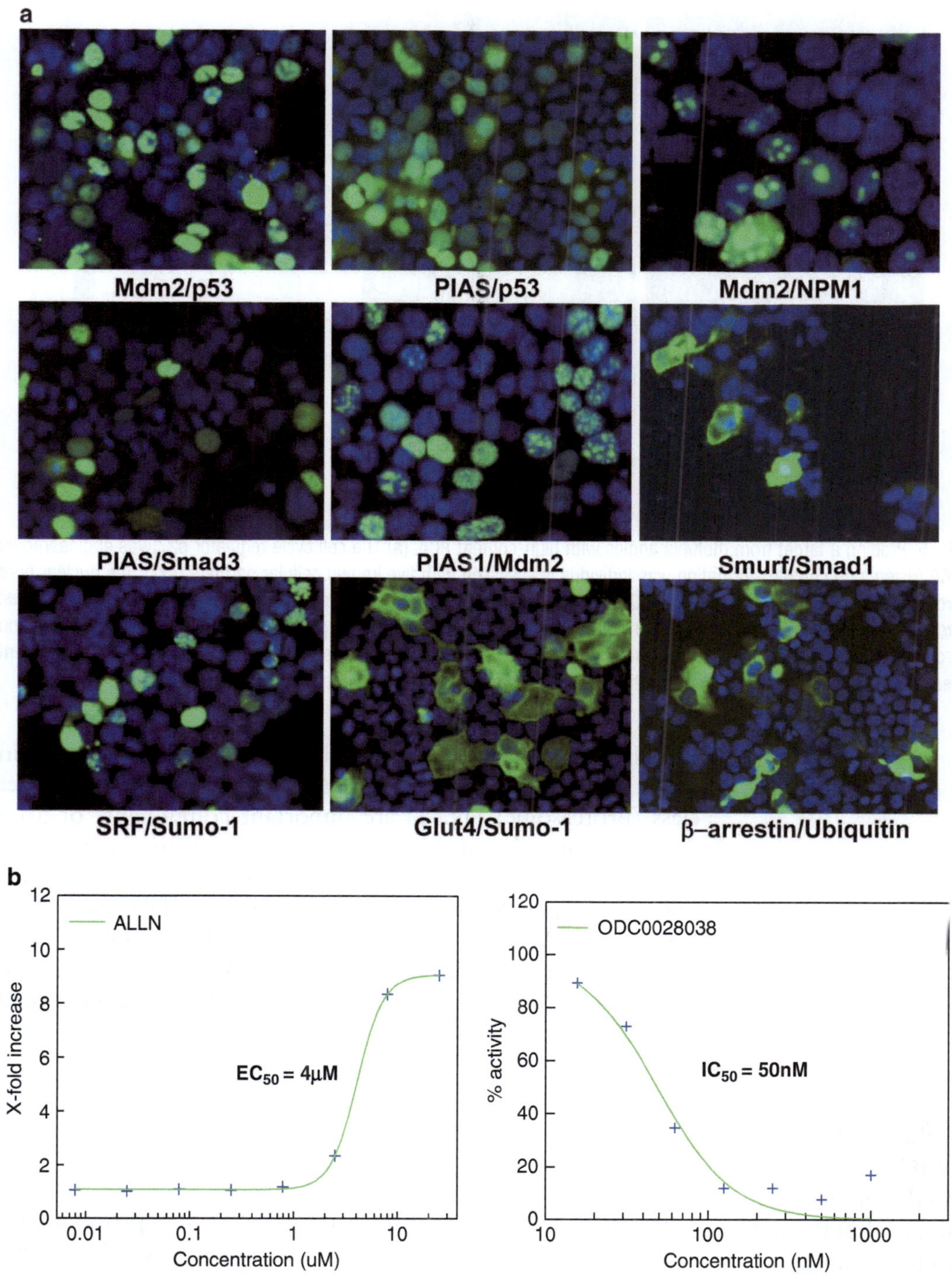

Fig. 4. Interrogating proteasome-regulated targets with PCA. (**a**) Examples of diverse proteasome-related activities as monitored by PCA. HEK cells were transiently transfected with DNA constructs encoding the indicated pairs of fusion proteins using FuGene. 48 h posttransfection, cells were fixed and stained with 33 μg/ml Hoechst 33342 in 2% formaldehyde for 10 min to identify nuclei, and images were captured on the Discovery 1 imaging system (Molecular Devices Corp) equipped with excitation and emission filters 470/35 and 535/60, respectively. In each panel, the nuclei are shown in *blue* (Hoechst) and the PCA signal is shown in *green* (YFP channel). (**b**) Proteasome-related PCAs were used to screen a collection of known proteasome inhibitors and novel compounds. As expected, incubation with ALLN for 24 h was found to increase levels of p53/Mdm2 complexes with an EC_{50} of 4 μM (*left panel*), while 8 h treatment with a novel compound led to decreased cellular complexes (*right panel*).

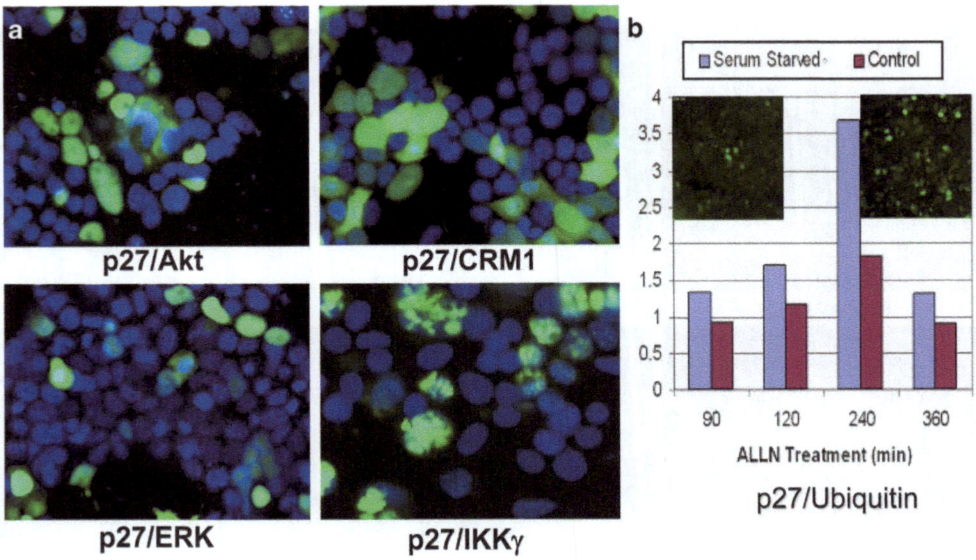

Fig. 5. Probing a target from multiple angles with high-content PCA. (**a**) The cell cycle regulator p27 was engineered as PCA fusion, and complex formation was individually assessed with five known cellular partners (Akt1, the nuclear pore protein CRM1, ERK, IKKγ, and Ubiquitin). Cells were transfected and imaged as described in the legend to Fig. 4. The *green signal* in each image corresponds to the PCA signal, representing protein complexes comprised of the indicated fusion proteins. (**b**) Images and time-course quantitation of the p27/Ubiquitin PCA response to ALLN (25 μM). The experiment demonstrates that treatment of cells with ALLN rapidly and transiently enhances levels of ubiquitinated p27.

The analysis of alternative signaling paths for a particular protein (p27; Fig. 5) and diverse proteins and mechanisms within a target class (proteasome; Fig. 4) are important components of drug discovery campaigns for identifying new therapeutic candidates, and for assessing the mechanism and specificity of drug leads. For example, currently marketed proteasome inhibitors such as Bortezomib (Velcade) are pan-inhibitors of the 26S proteasome. Despite their noted clinical success, it is likely that selective inhibitors targeting only a subset of proteasome client proteins would have significant clinical utility.

4. Putting It Together: Applications of a Global Cellular Pharmacology Platform

Systematic analysis of polypharmacology will improve understanding of drug activity, and may lead to improved therapeutic development and re-indication of existing therapeutics (15). This general concept may be particularly important for RNAi- and miRNA-based therapies, as these control mechanisms are likely to naturally impinge on multiple cellular targets. The diversity of assays described above suggests these strategies will be useful for discovery of multitargeted therapeutic agents. There is rapidly growing appreciation for polypharmacology – the concept that small molecule drugs can bind to more than one protein. We have previously reported on the identification of subsets of assays that can identify molecules with desired functional activity (11).

More broadly, the high-throughput capacity of the platform (11, 16), coupled with the knowledge that each high-content pathway-based assay in effect captures the activity of a much larger set of proteins connected by complex or pathway links, suggests that the approach could be a powerful strategy for profiling drug mechanisms, selectivity, and safety. To test this hypothesis, we screened a collection of over 9,000 chemical entities and genetic reagents against a panel of over a hundred live human cell high-content assays, each performed at multiple time points following probe addition. We found that every unique drug or probe elicits a unique "signature" of assay response across this panel. Even minor chemical modifications were found to generate changes in a signature, indicating that the approach could be used as an SAR-generating tool.

An example of drug signature creation is shown in Fig. 6a, with profiles for three known drugs. A signature is comprised of the quantitative data for a given compound – each red dot represents the fold change in a given assay for the test compound relative to vehicle controls. In this example, two of the known drugs, Astemizole and Terfenadine, display broad activity across the assay panel as evidenced by numerous assay hits visible in the profile plot, indicating widespread off-target activity. Notably, these drugs were withdrawn from clinical use due to cardiovascular toxicity. Fexofenadine, however, does not have this clinical liability and remains a widely used antihistamine. The profile plot for Fexofenadine (bottom panel; Fig. 6a) – even at a 10× higher concentration – shows no statistically significant off-target activity.

Two aspects of the approach have been found to be particularly useful for enhancing the pace and success rate of preclinical discovery. First, we and others have found that simply measuring compound selectivity across broad cellular networks can be a surprisingly useful early predictor of therapeutic candidate desirability (17). Following quantification of cellular images for a diverse panel of assays we calculate the number of statistically significant "hits" across our assay panel for each compound/dose combination. The resultant metric represents the cellular selectivity of that compound (an example of this calculation is displayed on the *Y*-axis of Fig. 6b). We have also found that it is useful to group the assays into target classes and pathways. In some cases, a compound may hit a relatively small percentage of total assays, but if these assay hits are broadly distributed across target classes and pathways that compound is likely to target a general cellular mechanism and is therefore unfavorable from a development standpoint.

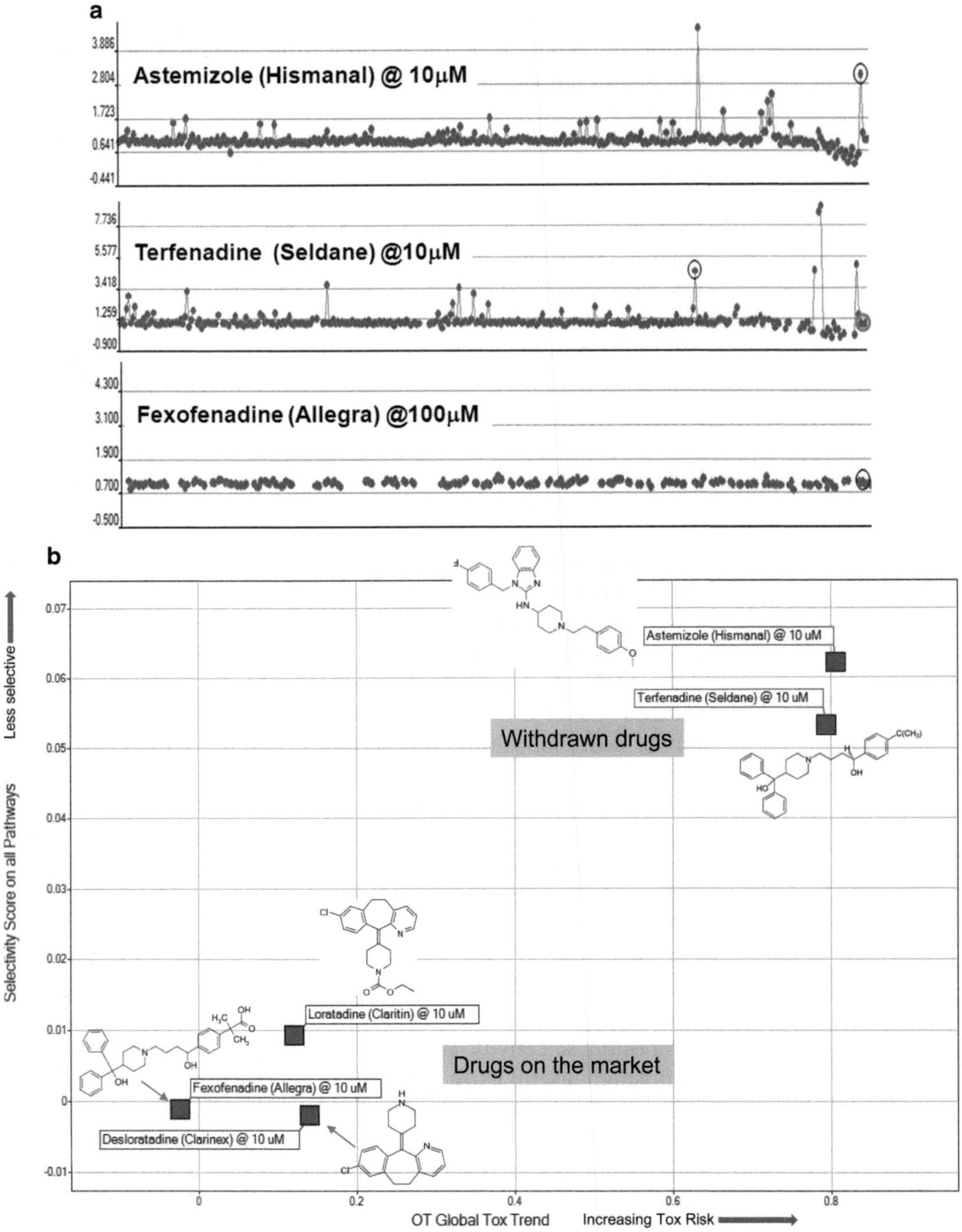

Fig. 6. Application of a high-content assay panel to compound profiling. (**a**) Every drug generates a unique "signature" across the assay panel. DMSO, Astemizole (10 μM), Terfenadine (10 μM), and Fexofenadine (100 μM), were used to treat cells engineered with a panel over 100 high-content assays. Following quantitative image analysis and data normalization, profile plots were created (Spotfire; Tibco). Each dot in a given profile represents the activity (fold change relative to DMSO control) for an individual assay/time point in the panel. Note difference in scale on *Y*-axis; even at 10× high concentration, Fexofenadine has much lower activity across the panel of assays, indicating higher cellular selectivity for this compound. (**b**) Selectivity and toxicity profiling effectively segregate successful and failed drugs. Selectivity scores were generated by adding statistically significant activities across the assay panel (*Y*-axis). Less selective compounds display a higher score on this metric. Test compounds were also scored for similarity to known toxicant signatures (*X*-axis); compounds with higher degrees of similarity to known toxicants have a higher score on this metric.

A second analytical strategy was engineered by analyzing a large panel of known toxicants across the same assay panel. Integrating the assay response patterns generated by specific classes of toxicants led to the development of "consensus signatures" of toxicant-stimulated assay responses. When new drugs or test agents are profiled across the assay panel, the signature they generate is automatically compared to toxicant signatures, and the degree of similarity plotted. An example plot is shown on the *X*-axis of Fig. 6b. The higher the similarity of a test agent signature to the toxicant signatures, the less desirable that agent is from a development perspective. Various subclassifications of cellular toxicity have been generated and employed, but for simplicity we display a combined metric here, which we term the Global Tox Trend (Fig. 6b).

As shown in Fig. 6b, by combining selectivity and known toxicant similarity analyses, a robust score is derived that can be used to triage large numbers of hits, leads and candidates at an early stage in the development process. The example shown here indicates that the combination of these two metrics effectively segregates marketed antihistamines from failed drugs in the same class (those which were pulled due to safety concerns). We have also found that these metrics can be effectively combined with various functional measurements in specialized cell types (e.g., stem cells, cardiomyocytes, hepatocytes) to further increase understanding of mechanisms and safety. Generally speaking, there is no simple numeric cut-off for test agent selectivity or toxicant similarity that defines the likelihood of toxicity, as many molecules are toxic by virtue of organ concentration or other issues. However, we have found that by scoring test agents with various combined metrics, the *relative* liabilities of particular chemical structures can easily be determined, enabling rank-ordering of candidate chemical structures and lead series at a very early stage in their development. The overall strategy has been validated to enhance understanding of compound mechanisms and selectivity throughout the process of preclinical development to IND (18).

5. Conclusions

Dwindling pipelines of novel medicines and the dismal success rates of both preclinical and clinical development programs indicate that new discovery and development strategies are needed (19). One solution is to expand the universe of potentially drug-like chemical matter, as dogma and rules and logistics have inadvertently limited the scope of chemistries represented in most compound libraries. Diversity-oriented synthetic approaches are likely to provide molecules required for novel targeting strategies (20).

Novel assay and screening technologies, both focused (e.g., specific protein–protein interface targeting (13), siRNA, and miRNA approaches) and more global (signature-based screening, pathway- and systems-based approaches) will certainly play a role. Pathway-based approaches in particular have gained momentum, with the realization that they can be exploited to simultaneously capture complexity and achieve high throughput (21). As described in this overview, the integration of diverse, contextual high-content assays into a pathway/systems-based analytical framework can provide deep target understanding in high-density formats suitable for interrogation of large chemical files. Although the approach described here involves an integrated platform of sophisticated instrumentation and significant investment in IT infrastructure, the basic strategy can be applied to focused analyses using instruments that are broadly available in academic core laboratories.

The strategy can also be used to address a key post-discovery challenge: the definition of drug candidate mechanisms, specificity, and safety. As described above, application of a broad high-content systems-based approach can be used to rank-order drug candidates "early and often" in the discovery and development process. The next frontier is the integration of the most relevant cell types with the analytical and systems strategies described above. The recent discovery successes achieved with simple functional analyses performed in cancer stem cells (22) and cells from defined genetic backgrounds (23), along with the development of robust gene transduction technologies (24), provides confidence that the combination of these strategies is within our grasp.

References

1. http://www.thehealthcareblog.com/the_health_care_blog/2009/08/rx-for-medical-research.html
2. Liddington, R. C. (2004) Structural basis of protein-protein interactions. *Methods Mol Biol* **261**, 3–14.
3. Stumpf, M. P., Thorne, T., de Silva, E., Stewart, R., An, H. J., Lappe, M., and Wiuf, C. (2008) Estimating the size of the human interactome. *Proc Natl Acad Sci USA* **105**, 6959–64.
4. Gerhart, J. and Kirschner, M.W. (1997) *Cells, Embryos, and Evolution*. Blackwell, Malden, MA.
5. Jones, S., Zhang, X., Parsons, D. W., Lin, J. C., Leary, R. J., Angenendt, P., Mankoo, P., Carter, H., Kamiyama, H., Jimeno, A., Hong, S. M., Fu, B., Lin, M. T., Calhoun, E. S., Kamiyama, M., Walter, K., Nikolskaya, T., Nikolsky, Y., Hartigan, J., Smith, D. R., Hidalgo, M., Leach, S. D., Klein, A. P., Jaffee, E. M., Goggins, M., Maitra, A., Iacobuzio-Donahue, C., Eshleman, J. R., Kern, S. E., Hruban, R. H., Karchin, R., Papadopoulos, N., Parmigiani, G., Vogelstein, B., Velculescu, V. E., and Kinzler, K. W. (2008) Core signaling pathways in human pancreatic cancers revealed by global genomic analyses. *Science* **321**, 1801–6
6. Jones, D. (2008) Pathways to cancer therapy. *Nat Rev Drug Discov* 7, 875–876.
7. Albert, R. J. (2005) Scale-free networks in cell biology. *J Cell Sci* **118**, 4947–57.
8. Michnick, S. W., Ear, P. H., Manderson, E. N., Remy, I., and Stefan, E. (2007) Universal strategies in research and drug discovery based on protein-fragment complementation assays. *Nat Rev Drug Discov* **6**, 569–82.
9. MacDonald, M. L., and Westwick, J. K. (2007) Exploiting Network Biology to Improve Drug

Discovery. In: *Methods in Molecular Biology, Vol 356: High Content Screening*. Lansing Taylor, Ed. Humana Press Inc., Totowa, NJ. 221–32;

10. Michnick, S. W., Macdonald, M. L., and Westwick, J. K. (2006) Chemical genetic strategies to delineate MAP kinase signaling pathways using protein-fragment complementation assays (PCA). *Methods* **40**, 287–93.
11. Macdonald, M. L., Lamerdin, J., Owens, S., Keon, B. H., Bilter, G. K., Shang, Z., Huang, Z., Yu, H., Dias, J., Minami, T., Michnick, S. W., and Westwick, J. K. (2006) Identifying Off-Target Effects and Hidden Phenotypes of Drugs in Human Cells. *Nat Chem Biol* **2**, 329–37.
12. Yu, H., West, M., Keon, B. H., Bilter, G. K., Owens, S., Lamerdin, J., and Westwick, J. K. (2003) Measuring drug action in the cellular context using protein-fragment complementation assays. *Assay Drug Dev Technol* **6**, 811–22.
13. Wells, J. A., and McClendon C. L. (2007) Reaching for high-hanging fruit in drug discovery at protein-protein interfaces. *Nature* **450**, 1001–9.
14. Westwick, J. K., and Michnick, S. W. (2005) Use of Protein-fragment Complementation Assays (PCA) in small GTPase research and drug discovery. *Methods Enzymol* **407**, 388–401.
15. Hopkins, A. L. (2008) Network pharmacology: the next paradigm in drug discovery. *Nat Chem Biol* **4,** 682–90.
16. Auld, D. S., Johnson, R. L., Zhang, Y. Q., Veith, H., Jadhav, A., Yasgar, A., Simeonov, A., Zheng, W., Martinez, E. D., Westwick, J. K., Austin, C. P., and Inglese, J. (2006) Fluorescent protein-based cellular assays analyzed by laser-scanning microplate cytometry in 1536-well plate format. *Methods Enzymol* **414**, 566–89.
17. Westhouse, R. A. (2010) Safety assessment considerations and strategies for targeted small molecule cancer therapeutics in drug discovery. *Toxicol Pathol* **38,** 165–8.
18. Murray, B. W., Guo, C., Piraino, J., Westwick, J. K., Lamerdin, J., Dagostino, E., Knighton, D., Zhang, C., Loi, C-M., Zager, M., Kraynov, E., Bouzida, D., Martinez, R., Karlicek, S., Bergqvist, S., Kephardt, S., Marakovits, J., Zhang, J., and Smeal, T. (2010) Small-molecule p21-activated kinase-4 inhibitor PF-3758309 is a potent inhibitor of oncogenic signaling and tumor growth. *Proc Natl Acad Sci U S A*. **107**, 9446–51.
19. Vogel, H. G., ed. (2007) *Drug Discovery and Evaluation: Pharmacological Assays*. 3rd Ed. Springer.
20. Shaw, J. T. (2009) Naturally diverse: highlights in versatile synthetic methods enabling target- and diversity-oriented synthesis. *Nat Prod Rep* **1**, 11–26.
21. Fishman, M. A., and Porter, J. A. (2005) A new grammar for drug discovery. *Nature* **437**, 491–493.
22. Raimondi, C., Cortesi, E., Gianni, W., and Gazzaniga P. (2010) Cancer Stem Cells and Epithelial-Mesenchymal Transition: Revisiting Minimal Residual Disease. *Curr Cancer Drug Targets*. **10**, 496–508.
23. Sur, S., Pagliarini, R., Bunz, F., Rago, C., Diaz, L. A. Jr., Kinzler, K. W., Vogelstein, B., and Papadopoulos, N. (2009) A panel of isogenic human cancer cells suggests a therapeutic approach for cancers with inactivated p53. *Proc Natl Acad Sci USA* **106**, 3964–9.
24. Maurisse, R., De Semir, D., Emamekhoo, H., Bedayat, B., Abdolmohammadi, A., Parsi, H., and Gruenert, D. C. (2010) Comparative transfection of DNA into primary and transformed mammalian cells from different lineages. *BMC Biotechnol* **10**, 1–9.

Chapter 4

RGS-Insensitive Gα Subunits: Probes of Gα Subtype-Selective Signaling and Physiological Functions of RGS Proteins

Kuljeet Kaur, Jason M. Kehrl, Raelene A. Charbeneau, and Richard R. Neubig

Abstract

The *R*egulator of *G* protein *S*ignaling (RGS) proteins were identified as a family in 1996 and humans have more than 30 such proteins. Their best known function is to suppress G Protein-Coupled Receptors (GPCR) signaling by increasing the rate of Gα turnoff through stimulation of GTPase activity (i.e., GTPase acceleration protein or GAP activity). The GAP activity of RGS proteins on the $G\alpha_i$ and $G\alpha_q$ family of G proteins can terminate signals initiated by both α and βγ subunits. RGS proteins also serve as scaffolds, assembling signal-regulating modules. Understanding the physiological roles of RGS proteins is of great importance, as GPCRs are major targets for drug development. The traditional method of using RGS knockout mice has provided some information about the role of RGS proteins but in many cases effects are modest, perhaps because of redundancy in RGS protein function. As an alternative approach, we have utilized a glycine-to-serine mutation in the switch 1 region of Gα subunits that prevents RGS binding. The mutation has no known effects on Gα binding to receptor, Gβγ, or effectors. Alterations in function resulting from the G>S mutation imply a role for both the specific mutated Gα subunit and its regulation by RGS protein activity. Mutant rodents expressing these G>S mutant Gα subunits have strong phenotypes and provide important information about specific physiological functions of $G\alpha_{i2}$ and $G\alpha_o$ and their control by RGS. The conceptual framework behind this approach and a summary of recent results is presented in this chapter.

Key words: G protein-coupled receptor, Heterotrimeric G protein, Regulator of G protein signaling protein, GTPase-activating protein, Signal transduction

1. RGS Proteins History and Importance

After the isolation of G proteins in 1981 (1), a clear understanding emerged about the role of GTP hydrolysis in the turnoff of GPCR signals (2). However, it became obvious by the late 1980s and

Louis M. Luttrell and Stephen S.G. Ferguson (eds.), *Signal Transduction Protocols*, Methods in Molecular Biology, vol. 756, DOI 10.1007/978-1-61779-160-4_4, © Springer Science+Business Media, LLC 2011

early 1990s that there was a discrepancy between the biochemical GTPase activity of Gα subunits and the turnoff rate of physiological signals. This was clearly documented in the visual system where the measured GTPase activity of transducin (Gαt) was about ten times slower than the turnoff of physiological responses to light (3). This discrepancy was attributed in part to effects of the phosphodiesterase (4) but other studies suggested the involvement of another protein (5). This was also about the time when GTPase accelerating proteins (GAPs) for the *ras* oncogene were discovered (6) leading to the suggestion that there might be similar GAPs for heterotrimeric G proteins. The solution to this conundrum ultimately came from studies in yeast and worms.

Pheromone signaling in the mating of yeast (*Saccharomyces cerevisiae*) utilizes GPCRs and its study has revealed a number of key insights into G protein signaling mechanisms including the role of Gβγ subunits as active signaling elements and the identification of RGS proteins (7). Haploid yeast secrete pheromones which act on GPCRs of yeast of the opposite mating type. This induces growth arrest and events that promote fusion of the two cells for mating. Chan and Otte (1982) discovered a mutant yeast strain (sst2) which was supersensitive to the α factor pheromone (8). The mutant strain also had a prolonged pheromone response; in continuous presence of α factor, the budding returned to baseline after about 4 h for the wild-type cells (WT) but in sst2 yeast the effect of α factor lasted more than 6 h after just 1 h of exposure. The effect of the Sst2 protein was through direct actions on the G protein α subunit (9, 10) and Sst2p is now known to represent the first member of the RGS family. At about the same time, studies in the model organism *Caenorhabditis elegans* identified the gene EGL-10 which suppresses serotonin signaling through the Gαo protein (11). This study and several others (12) defined a large mammalian RGS family that has sequence homology to the G protein regulators of the model organisms. Soon thereafter, the mechanism of action was shown to be GAP activity at Gi and Gq family Gα subunits (13). In the years that followed, there was a flurry of studies defining the biochemical properties, tissue-specific expression, protein complex formation, and regulation of RGS proteins. This work has been amply reviewed (14–18). One aspect of RGS action relates to the specificity of different RGS proteins for Gα subunits. Surprisingly, this specificity is relatively limited when examined in biochemical studies with purified RGS and Gα proteins. The Gi/o and Gq families are the primary targets. To date, there have not been convincing demonstrations of an RGS protein acting as a GAP on a Gs protein. Figure 1 summarizes the extensive literature on this point (14–19). Only a few RGS proteins show specificity for a Gα subunit at the biochemical level (e.g., RGS2, RGS6 & 7, and RGS20, see Fig. 1).

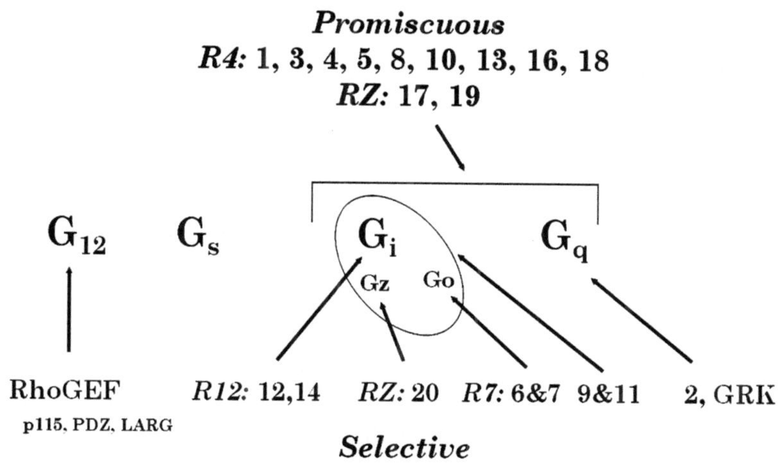

Fig. 1. Summary of the specificity of RGS proteins for Gα subunits in vitro. The four families of Gα subunits are illustrated as are all of the RGS proteins (designated by family R4, RZ, R7, and R12) and two RH domain-containing systems (RhoGEFs and GRKs). The published specificity of the RGS proteins for different Gα subunits is shown with the majority (listed above the Gα subunits) being nonselective among Gi/o and Gq alpha subunits. The more selective RGS proteins and RH-domain-containing proteins are illustrated below with an indication of their general selectivity.

There is, also, clear specificity at the level of RGS expression in different cell types and brain regions (20–22). Furthermore, substantial emerging literature addresses specificity driven by RGS complex formation with other signaling molecules in cells – including receptors and several scaffold proteins such as spinophilin, R7BP, and GIPC (23–26).

The main focus of the present chapter, however, is on the in vivo physiological functions of RGS proteins. First, we will review the existing literature on standard RGS knockout models and/or in vivo RNAi studies, then we will introduce an approach developed in our lab that uses knock-in mice with RGS-insensitive (RGSi) Gα subunits expressed from the genomic locus. These mice should have enhanced signaling from the mutant Gα subunits in physiological contexts where the Gα subunits are under the control of one or more RGS proteins. Furthermore, the RGSi mutants overcome the potential redundancy from the 20+ RGS proteins that act on the Gi and Gq family G proteins. An unexpected dividend of this approach is that it also has helped dissect the in vivo functions of closely related Gi/o family members. As will be discussed below, very different effects are seen from enhancing signaling by $G\alpha_{i2}$ and $G\alpha_o$. This also provides potentially important information about expected actions of biased GPCR agonists that can selectively activate one Gi/o subtype or another.

2. Studies with RGS Knockout Mice

In order to better understand the role played by RGS proteins, a number of labs have used traditional RGS knockout mice with disrupted RGS genes. So far, reports are available on nine RGS proteins that have been genetically knocked out.

2.1. RGS1

RGS1 was first identified as a gene upregulated in phorbol-ester-activated B lymphocytes (27). RGS1$^{-/-}$ mice were generated to understand the role of RGS1 in immune system and general physiology (28). They had no gross abnormalities and were viable through development. The expression level of other RGS proteins was not affected (28). B cells from RGS1$^{-/-}$ mice were shown to have an increased and prolonged elevation of intracellular Ca^{2+} in response to CXCL2 and enhanced motility in response to chemokines. They also had faster entry into peripheral lymph node and splenic B cell follicles. These effects were blocked by *pertussis* toxin treatment (28). This study showed that RGS1 alters CXCL2 mediated responses in immune cells most likely by increasing the rate of $G\alpha_i$ protein deactivation.

2.2. RGS2

RGS2 has major effects on the cardiovascular system – both on vasculature and in the heart. RGS2$^{-/-}$ animals have significantly higher mean arterial blood pressure compared to wild-type animals (29). The increased BP was attributed to increased vascular tone in response to endogenous angiotensin II stimulation. AT1 antagonist treatment decreases the blood pressure to baseline level in mutant mice (29). Loss of RGS2 also increases vascular smooth muscle cell Ca^{++} responses to vasopressin and reduces the effectiveness of NO-mediated vasodilation (30). Another group attributed some of the increased blood pressure to enhanced sympathetic tone (31). Loss of RGS2 also increased susceptibility to atrial fibrillation induced by burst pacing and programmed electrical stimulation, an effect attributed to increased M3 muscarinic receptor signaling (32). RGS2 knockout mice also experience worsened heart failure after aortic constriction (33). Interestingly, a number of rare nonsynonymous polymorphisms in RGS2 have been found in human hypertensive patients and two of these mutations have been shown to have functional effects in cell systems (34, 35).

RGS2 also appears to play an important role in T-cell function. Knockout mice show decreased interleukin-2 production and T-cell proliferation in response to various stimuli (36). RGS2$^{-/-}$ mice also have central nervous system alterations. They show increased anxiety-like behavior as indicated by increased time in the dark half of a light–dark box compared to wild-type mice and they have reduced spine density in CA1 hippocampal neurons (36).

2.3. RGS4

RGS4$^{-/-}$ mice have a relatively mild phenotype which is surprising given the broad expression of RGS4 in the brain (37). They perform poorly on the rotating rod test and have slightly lower than normal body weight. There are no differences in body temperature, grooming, posture, or righting reflex (37). One major question about RGS4 was its role in schizophrenia. A clinical study demonstrated significant downregulation of RGS4 in schizophrenic patients (38) but the RGS4$^{-/-}$ mice showed no change in prepulse inhibition (37), a behavioral test that differs in those suffering from schizophrenia compared to control individuals. This raises doubts about a potentially causal role of RGS4 alterations in schizophrenia, but it is also clear that animal models of neuropsychiatric diseases may or may not faithfully represent the human condition.

A recent study also showed a strong role for RGS4 in chronotropic control of the heart. A mouse expressing *lacZ* from the RGS4 locus (which also disrupts RGS4 expression) showed remarkably localized expression of RGS4 in the sino-atrial node (39). Homozygous RGS4$^{-/-}$ mice showed slowed recovery of G protein-coupled inwardly rectifying potassium (GIRK) currents after removal of a muscarinic agonist on SA node cells. In vivo, they showed enhanced carbachol-mediated bradycardia supporting a major role for RGS4 in chronotropic control of the heart.

2.4. RGS5

RGS5 is strongly expressed in vascular pericytes (40). All major arteries in the body have detectable expression levels (41). These findings led to the hypothesis that RGS5 plays a significant role in angiogenesis and vascular remodeling. RGS5$^{-/-}$ mice are viable, with no apparent developmental or behavioral defects. In the initial studies, no difference could be found in kidney or brain morphology, kidney function, angiogenesis during tumor growth, or pericyte abundance in retinal vasculature. The main difference between RGS5$^{-/-}$ and WT animals was that the knockouts had lower blood pressure (42, 43). Another study (44) demonstrated vascular normalization, improved blood supply, enhanced late-stage tumor growth, and worsened survival in a mouse model of induced pancreatic islet tumors. However, the normalized blood vessels permitted a much more efficient immune therapy in the same model (44).

2.5. RGS8

Another member of the RGS4 family, RGS8 is highly expressed in brain stem and in cerebral Purkinje cells (20). Like many other RGS knockout mice, RGS8$^{-/-}$ mice showed no gross abnormality. Due to the location of RGS8 expression, Purkinje cell morphology and cerebellar layers were examined without evidence of abnormalities (45). Consequently, no phenotype has yet been assigned to the RGS8$^{-/-}$ mice but only minimal testing has been reported to date.

2.6. RGS9

Of all the RGS proteins identified so far, RGS9 has the most limited expression pattern and the knockout has one of the most striking phenotypes. The two splice variants, RGS9-1 and RGS9-2, do not overlap in expression. RGS9-1 is highly expressed in retina and rod outer segments and RGS9-2 is highly expressed in striatum and other dopaminergic target tissues. RGS9-1 is the major determinant of GTP hydrolysis in the visual system. The recovery from a flash response is much slower in rod cells from RGS9$^{-/-}$ mice as compared to that seen in WT rod cells, highlighting the role of this protein in vision (46). RGS9-2 overexpression in rat nucleus accumbens causes decreased locomotion in response to cocaine and D2 agonists whereas RGS9$^{-/-}$ mice exhibit increased dopamine-induced locomotion and reward behavior (47). RGS9$^{-/-}$ mice also show enhanced analgesia as well as enhanced dependence in response to morphine (48). Thus RGS9 plays significant roles in vision and in behavioral responses to abused drugs.

2.7. RGS10

Homozygous RGS10$^{-/-}$ mice have growth retardation as seen by lower body weight as early as 9 days of age (49). The homozygous mice only survived up to 3 weeks and have severe symptoms of osteopetrosis. The RGS10$^{-/-}$ mice have impaired activation of NFκB by receptor activator of nuclear factor kappa B ligand (RANKL) and loss of RANKL-induced differentiation to osteoclasts. This appears to be due to a loss of calcium oscillation in the RGS10$^{-/-}$ osteoclast precursors but interestingly, the model proposed does not include a G protein in the mechanism. This novel hypothesis, however, will need to be directly tested. Regardless, RGS10 has an important role in skeletal development and bone remodeling.

2.8. RGS13

RGS13 is one of the major RGS proteins expressed in mast cells (50). Mast cells from RGS13$^{-/-}$ mice have increased degranulation in response to antigen. As expected for antigen which is not known to act through a GPCR, *pertussis* toxin treatment had no effect on the degranulation. Furthermore, the RGS13-mediated suppression of signaling was still observed upon transfection of a GAP-deficient mutant RGS13. Consequently, this effect of RGS13 is probably not related to loss of GAP activity (51). The precise mechanism of this non-GAP role of RGS13 in allergic responses will need to be established.

2.9. RGS14

An early report (52) found that RGS14$^{-/-}$ mutants were lethal at a very early developmental stage (prior to the first cell division). This was attributed to a general role of RGS14 in mitosis – perhaps through functions of the GoLoco motif. A report at a recent meeting (ASPET RGS Colloquium, San Diego, 2008), however, showed that RGS14 knockouts are viable and have behavioral changes. This discrepancy is probably due to either genetic background differences or to differences in the specific gene structure of the two knockout constructs.

2.10. What Have We Learned from RGS Knockout Animal Models?

The phenotypes of RGS knockout mice have been surprisingly modest given their strong effects on the fundamental process of G protein signaling. Two RGS mutant mice have been reported with strongly altered viability (RGS10 and RGS14 – though the latter is questionable at this point). Most RGS knockout mice appear grossly normal unless careful testing is done to elicit a phenotype, with effects on behavior being common (e.g., RGS 2, 4, 9, & 14). This is not surprising given the abundant expression of this protein family in the CNS (20, 22, 53). Several RGS proteins with specific tissue expression show prominent effects in knockouts that relate to their tissue locus (RGS1 in lymphocytes, RGS4 in SA node, RGS9 in eye and striatum, and RGS13 in mast cells).

The lack of prominent effects in several RGS knockout mouse models is probably due in part to the functional redundancy among RGS proteins (18). For example, in atrial myocytes there are 7 RGS proteins with abundant RNA expression (21) and 5 of them (RGS3, 4, 10, 17, and 19) have broad specificity for Gi and Gq proteins (Fig. 1). Another potential reason for the limited phenotypes is that RGS proteins may play a more prominent physiological role either under stress or in pathological situations, thus requiring analysis of these animals in specific disease models.

Given their functional redundancy and the large number of different RGS proteins, it will be virtually impossible to undertake combinatorial knockouts of all RGS combinations to understand the full contribution of RGS proteins to physiological processes. The RGS-insensitive Gα subunit approach outlined below, is one way to disrupt the action of multiple RGS proteins to begin to understand RGS function in vivo.

3. RGS-Insensitive Gα Subunits as Probes to Understand the Role of RGS Proteins

In 1998, Dohlman and colleagues undertook a genome-wide mutagenesis study in yeast to discover mutations that could phenocopy the enhanced pheromone sensitivity of the sst2 RGS knockout allele. The only mutation identified was a glycine-to-serine mutation at residue 302 in the Gα subunit Gpa1 (54). The mutation had no effect on nucleotide binding to or release from Gpa1. However, the RGS protein GAIP (a.k.a. RGS19) failed to increase GTP hydrolysis of the G302S mutant Gpa1 (54).

Subsequent biochemical studies showed that this mutation had general effects across multiple Gα and RGS proteins. Lan et al. (55) showed that the analogous G184S mutations in $G\alpha_o$ and $G\alpha_{i1}$ blocked GAP activity and binding of the mutant Gα subunits to RGS4 and 7 (55)[1]. There was a >100–1,000-fold reduction in the affinity of AlF_4-activated $G\alpha_o$ and $G\alpha_{i1}$ mutant G184S subunits for the RGS proteins and GAP activity was undetectable. This mutation in the critical switch 1 glycine of the Gα subunit is precisely in the contact interface between RGS4 and $G\alpha_{i1}$ (PDB: 1agr; (56)) as well as in other RGS proteins whose co-crystal structures with a Gα subunit have been solved.

A different situation exists for RGS homology (RH) domain-containing proteins that do not function primarily as GAPs for a Gα subunit (including the RH-domain-containing RhoGEFs such as LARG, PDZRhoGEF, and p115rhoGEF as well as the G protein-coupled receptor kinase or GRK RH domains). Their interaction mode with Gα is quite different (57). Distinct surfaces on those RH domains bind to the Gα subunit and they contact different surfaces on the Gα subunit as well. In the case of these RH domains, while the glycine is conserved, the G > S mutation in switch 1 of the Gα subunit does not prevent Gα/RH binding or function (57).

3.1. General Concepts Regarding the Use of RGS-Insensitive Mutant Gα Subunits to Understand RGS Function

For all examples to date in the "classical" RGS proteins (families R4, R7, R12, and RZ) the G > S mutation appears to abolish RGS binding to Gα which also prevents GAP function. Importantly, there has been no demonstrated effect on other functions of the Gα subunit such as: binding to Gβγ, activation by receptor, or coupling to effector (58). Furthermore, in vitro and in vivo, there is no apparent change in protein expression compared to wild-type Gα subunits (59, 60). Consequently, we will call these switch 1 G > S mutants RGS-insensitive (RGSi) Gα subunits.

An advantage of this approach compared to RGS knockouts or knockdowns is the elimination of the action of all RGS proteins upon the mutant Gα subunit, overcoming potential functional redundancy. In addition, the results differ from an RGS knockout in that only effects mediated through the RGS domain/Gα interaction would be affected. The role of other functional domains (e.g., GoLoco or RhoGEF) should not be altered in these mutants but would be lost in an RGS knockout. In that sense, the Gα

[1]The use of the terminology G183S for $G\alpha_{i1}$ and G184S for $G\alpha_o$ in the original Lan et al. paper (55) was due to the use of a protein-based, methione-deleted numbering for $G\alpha_{i1}$ but a gene-based nomenclature for $G\alpha_o$ which includes the initiator methionine. The latter is recommended for use to correlate with human genetic mutations (77) so we now use G184S for all such mutations in $G\alpha_i$ and $G\alpha_o$ proteins since they share the same number of residues to this position.

RGSi mutants give better insights into the actions of potential drugs that would block RGS binding to Gα subunits (17, 61, 62) than a knockout might. A primary disadvantage of the Gα G>S mutant is that it does not tell which RGS protein is involved in the process being studied. Also, for example, with the $G\alpha_{i2}$ G184S mutant, the effect would not mimic the effect of a drug-inhibiting RGS4 since RGS4 can act on all Gi/o subtypes as well as Gq subtypes. So further interpretation of results with RGSi mutants in many cases will require identification of the specific RGS involved and what other actions it might have beyond those on that one Gα subunit. Still, as noted below, the effects of RGSi mutant Gα subunits when either overexpressed or when knocked-in at the endogenous locus are frequently more dramatic and intense than those of individual RGS knockouts.

A final relatively unexpected benefit from studies of Gα RGSi mutant subunit knock-in mice (described below) is the evidence that they provide on differential signaling mediated by very closely related Gα subunits (e.g., $G\alpha_o$ vs. $G\alpha_{i2}$ and potentially others). Thus the "gain-of-function" nature of the mutation may provide insights that are not accessible via knockout studies given the high level of redundancy both at the G protein and RGS levels.

3.2. Cellular Studies

In the original paper defining the yeast G>S mutation, Dibello et al. (54) also showed a functional effect of the analogous mutation in mammalian $G\alpha_q$. In CHO cells transfected with the 5-HT_{2c} receptor, serotonin increased calcium mobilization through Gq activation. RGS7 co-expression along with WT Gq reduced this calcium mobilization but the effect of RGS7 was abolished when a G188S mutant Gαq subunit was co-transfected with the RGS (54).

Similarly strong effects of the G>S mutation have been defined on Gi/o functions in cellular systems. One commonly used tool to study inhibitory G proteins (i.e., the Gi/o family) is *pertussis* toxin which ADP-ribosylates a cysteine in the C-terminus of the Gi/o alpha subunits (with the exception of Gz) preventing coupling to GPCRs. Mutating that cysteine to a nonreactive amino acid (e.g., Ser, Gly, Ile, etc.) protects the Gα from modification, making it insensitive to *pertussis* toxin (PTXi). Once control experiments with adequate *pertussis* toxin pretreatment (generally 30–100 ng/ml overnight) have shown no residual signal, the PTXi mutants along with pertussis toxin pretreatment can be used effectively to ensure that only signals due to the transfected PTXi G protein are being measured. This has been used to probe the role of different Gi/o family members and to assess the function of mutant Gi proteins.

Clark et al. (59) used this approach to determine the effect of the G184S mutation in $G\alpha_o$ on opioid-induced inhibition of adenylyl cyclase in C6-mu cells – a rat glioma cell line stably

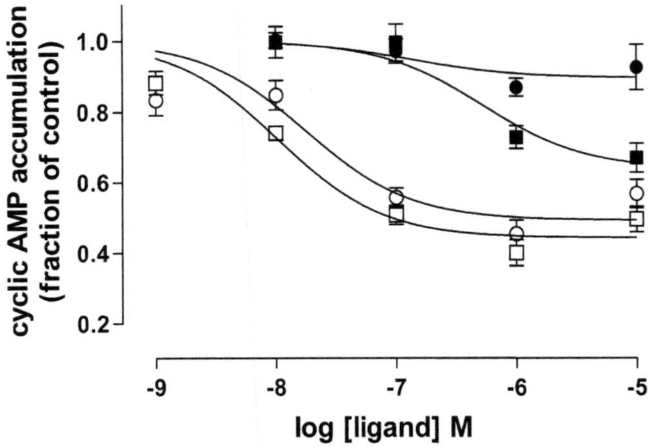

Fig. 2. Potentiation of opioid inhibition of cAMP production by the RGS-insensitive Gα_o G184S mutant. C6mu cells stably expressing either the PTXi Gα_o (*filled symbols*) or the PTXi/RGSi Gα_o (*open symbols*) subunit were pretreated with *pertussis* toxin then inhibition of cAMP production in whole cells was tested. All samples contained 30 μM forskolin and 1 mM 3-isobutyl-1-methylxanthine along with the indicated concentrations of DAMGO (*squares*) or morphine (*circles*). This research was originally published in *The Journal of Biological Chemistry*. Clark, M. J., Harrison, C., Zhong, H., Neubig, R. R., and Traynor, J. R. Endogenous RGS protein action modulates mu-opioid signaling through Galphao. Effects on adenylyl cyclase, extracellular signal-regulated kinases, and intracellular calcium pathways. *J Biol Chem* 2003; **278:** 9418–9425. © The American Society for Biochemistry and Molecular Biology.

expressing mu-opioid receptors. *Pertussis* toxin treatment abolished the opioid-dependent adenylyl cyclase inhibition which was restored by stable expression of the PTXi-Gα_o. Use of a PTXi/RGSi double mutant (59) showed a strikingly greater inhibition of AC with morphine being converted from a weak partial agonist to a full agonist. Also, the full agonist (D-Ala2, N-MePhe4, Gly-ol]-enkephalin (DAMGO) showed a nearly 50-fold left shift of the dose–response curve (Fig. 2). These data were interpreted to indicate that endogenous RGS proteins were strongly suppressing the opioid inhibition of adenylyl cyclase through Gα_o. Subsequent studies extended this finding to other Gi/o family members and substantially larger effects were seen on the maximum adenylyl cyclase inhibition for partial agonists while full agonists generally showed an increase in potency (decrease in EC_{50}) when RGSi Gα subunits were expressed (63).

The RGSi mutant Gα subunits also profoundly change ion channel regulation in primary neuron cultures. Using this approach, Jeong and Ikeda (64) showed that norepinephrine-induced inhibition of calcium currents in rat superior cervical ganglion (SCG) neurons is subject to regulation by RGS proteins. SCG neurons were transfected by intranuclear injections with PTXi Gα_o or PTXi/RGSi Gα_o. The kinetics of calcium current inhibition and recovery with a 60 s pulse of NE were similar for neurons with PTXi Gα_o

compared to control neurons. PTXi/RGSi $G\alpha_o$ mutants showed a similar maximum norepinephrine-induced inhibition of the calcium current but the rate of recovery from the inhibition was greatly slowed in PTXi/RGSi mutants as compared to that seen in PTXi-transfected or normal cells (from 10–30 s in controls to >1–3 min with RGSi $G\alpha_o$). In addition to the change in channel kinetics, the PTXi/RGSi mutations also resulted in an eightfold leftward shift in the dose–response curve for norepinephrine-induced current inhibition (64). This was one of the first demonstrations that elimination of endogenous RGS function could strongly potentiate agonist function in a mammalian system. These actions of RGS function on calcium channels also have implications for synaptic function. Chen and Lambert (65) showed that adenosine-mediated presynaptic inhibition in primary cultures of rat hippocampal neurons was restored to *pertussis* toxin-treated cells by viral transduction with PTXi G proteins ($G\alpha_o$ or $G\alpha_{i1}$). Furthermore, the recovery from adenosine-induced presynaptic inhibition was much slower in neurons expressing PTXi/RGSi Gα subunits as compared to those with only PTXi G protein. The time constant for recovery in the RGSi/PTXi $G\alpha_o$ mutant was increased to 40 s as compared to 3 s in the PTXi mutant $G\alpha_o$ (65). Consequently, RGS protein function, as detected by use of the RGSi Gα subunits is profound – especially in neural systems.

4. In Vivo Studies of RGSi Gα to Understand the Role of RGS Proteins

4.1. Developing In Vivo Models

Given the pronounced effects of the RGSi Gα subunit mutation in the cellular studies described above, we embarked on an effort to apply this system to in vivo studies in whole animals. One key consideration was how to maintain normal patterns and levels of expression of the mutant protein. To address this, we chose a knock-in strategy where the mutant gene replaces the wild-type gene at its normal genomic locus. The details of how this was accomplished are outlined in previous studies (58, 60, 66). In brief, the targeting construct contains the mutant codon (G184S) in exon 5 of the $G\alpha_o$ or $G\alpha_{i2}$ gene as well as a diagnostic restriction site (*Pvu*I) that is compatible with the coding sequence in the mutant protein. The *neo* selection marker for isolating targeted embryonic stem (ES) cells was placed in the intron between exons 5 and 6. Preliminary studies (58) showed that leaving the entire *neo* marker intact lead to markedly reduced expression of the mutant $G\alpha_o$ so we introduced loxP sites flanking the *neo* marker to permit its removal after the mice were generated. Introduction of cre recombinase by either transfection in ES cells or by breeding mutant strains with cre-expressing mice left only the small single loxP site in the intron which permitted normal levels of Gα subunit expression (58). Furthermore, we showed by Western blots that in homozygous $G\alpha_{i2}^{G184S/G184S}$ mice the $G\alpha_{i2}$ protein showed normal levels and tissue patterns of expression (60).

Other approaches to this problem could be taken. One option is the use of transgenic animals. However, this system appears to have some drawbacks as illustrated by a study of transgenic rats expressing the $G\alpha_q$ G188S mutant G protein. The G > S rats showed markedly enhanced 5HT2A signaling and lethality to the agonist (−) DOI (67). However, a significant increase in signaling was also seen in transgenic rats expressing the wild-type $G\alpha_q$. Therefore, overexpression of the Gα protein, regardless of RGS sensitivity, was contributing to the phenotype. Differences between WT and G188S $G\alpha_q$ transgenics were observed but the interpretation was complicated by the abnormal expression pattern in the transgenic animals. One other option that is worth considering is the use of BAC transgenics (68). These usually exhibit reasonably normal patterns and levels of expression. Introducing the G>S mutation into a BAC containing the appropriate Gα subunit and its endogenous promoter should be feasible. The BAC transgenic model could be further refined by breeding the transgenic with the corresponding KO mice for the given G protein thus allowing for the BAC-derived RGSi Gα to be the only endogenous source of that Gα subunit.

4.2. Gene Dosage Effects

The Gα G184S mutation, differing somewhat from RGS knockout mutants, has a dominant gain-of-function phenotype. To understand this, it is worth considering a scenario in which there is one G protein and one RGS protein in a system and the RGS protein suppresses the action of the G protein by 99% due to acceleration of turnoff. In Table 1, the predicted effects of a heterozygous mutation of $G\alpha_{i2}^{G184S/+}$ (or generically $G\alpha^{GS/+}$) compared to a heterozygous RGS knockout ($RGS^{+/-}$) is illustrated

Table 1
Predicted effects due to mutations in Gα subunit and RGS protein

	Signal strength		
	$RGS^{+/+}$	$RGS^{+/-}$	$RGS^{-/-}$
$G\alpha^{+/+}$	1	2	91
$G\alpha^{GS/+}$	46	46.5	91
$G\alpha^{GS/GS}$	91	91	91

This table illustrates predictions of a simple model of G protein activation and deactivation and compares results for loss-of-function mutations in the RGS vs. RGSi mutations in the Gα subunit. A simple equilibrium is assumed between an inactive G protein (G) and an active G protein G*. The total amount of G protein is 100. The rate of activation is constant for all situations ($1\ s^{-1}$). The rate of deactivation is equal to $0.1\ s^{-1}$ for G protein with no RGS present and is $100\ s^{-1}$ with the full amount of RGS present (1,000× stimulation). A heterozygote $RGS^{+/-}$ is presumed to have half as much RGS so would have half the rate of deactivation. A heterozygous Gα GS/+ is presumed to have half of its G protein behave like the $RGS^{+/+}$ situation and half like the $RGS^{-/-}$ situation. The signal strength is calculated to be equal to the amount of G* with 100 being full activation using a simple model based on rates of RGS-mediated Gα subunit deactivation (*see legend to* Table 1 *for parameters*).

It is striking that, in this model, a heterozygous RGS knockout ($RGS^{+/-}$) shows only a twofold increase in signaling while a heterozygous Gα mutant ($G\alpha^{GS/+}$) shows a 46-fold increase (Table 1). Homozygotes of both sorts show a strong 91-fold increase since both produce a complete loss of RGS function. Different parameters for the rates of activation and basal and stimulated deactivation would alter the magnitude of the effects but the qualitative result that the heterozygous RGSi Gα mutants give a larger effect than the heterozygous RGS knockout would always be the case. Furthermore, if there was more than one RGS in the system, the homozygous knockout of a single RGS might behave more like the heterozygote $RGS^{+/-}$ in this model since some RGS activity would remain from other RGS family members. Consequently, the Gα G>S mutants are expected to have much stronger phenotypes, with $G\alpha^{GS/+}$ heterozygotes showing clear effects perhaps approaching those of a homozygous $G\alpha^{GS/GS}$ mouse.

4.3. Role of Genetic Background

As in any mouse model study, the genetic background of the mice is very important! Thus it is critical to maintain accurate animal breeding records and to consistently use the exact same strain (e.g., C57BL/6J) as breeding partners for backcrossing. It is common practice to do such experiments on mice that have been backcrossed onto the desired strain at least five times (i.e., N5 animals or greater). Comparing results from animals with different genetic backgrounds can make interpretation difficult, so littermate controls with all animals at the same backcross generation are optimal. However, extensive backcrossing onto C57BL/6J (B6 for short) for the $G\alpha_{i2}$ G184S mutant mice has lead to reduced viability of mutants (Table 2). For example, the frequency of $G\alpha_{i2}{}^{GS/GS}$ mice surviving to weaning from het × het crosses is low after five backcross (N5) generations on the B6 background (34% of expected). It is even worse at N13 with only 25% of the expected numbers. Consequently, we have generally used N5 or N6 mice for our most recent studies. The reduced viability as the mice become congenic is probably due to B6 alleles that, when present in the homozygous state, are leading to interactions with the $G\alpha_{i2}{}^{G184S}$ mutation. An even more striking result is seen with $G\alpha_o$ mice in that homozygotes are virtually 100% embryonic/neonatal lethal (Table 2). Studies are underway to better understand this phenomenon. Heterozygote $G\alpha_{i2}{}^{GS/+}$ mice are generally born at or near the expected Mendelian ratios from N5 crosses.

As one option to obtain more homozygous RGSi mice, we have also generated F1 crosses. To do this, $G\alpha_{i2}{}^{GS/+}$ heterozygotes that are nearly congenic on the FVB or the C57Bl/6J backgrounds (N6 and N13, respectively) are crossed. This F1 hybrid approach significantly increased the yield of $G\alpha_{i2}{}^{GS/GS}$ mice from 25 to 33% of the expected numbers on the pure strains to 65% for the F1

Table 2
Frequency of different genotypes at weaning

	+/+	GS/+	GS/GS	Significantly different from Mendelian ratios	Reference
$G\alpha_{i2}$ – B6					
N5 (het × het)	145 (1.00)	193 (1.33)	49 (0.34)	*	(60)
N5 (het × wt)	37 (1.00)	33 (0.86)	NA	NS	(60)
N13(het × het)	103 (1.00)	150 (1.46)	26 (0.25)	*	
N13 (het × wt)	94 (1.00)	63 (0.65)	NA	*	
$G\alpha_{i2}$ – FVB					
N5 (het × het)	15 (1.00)	24 (1.60)	5 (0.33)	NS	
N7 (het × wt)	92 (1.00)	95 (1.03)	NA	NS	
$G\alpha_{i2}$ – B6-FVB F1					
N13/N6 (het × het)	69 (1.00)	118 (1.71)	45 (0.65)	NS	
$G\alpha_{o}$ – B6					
N4 (het × het)	34 (1.00)	21 (0.62)	0 (0.00)	*	
N5 (het × wt)	50 (1.00)	21 (0.42)	NA	*	
N8 (het × wt)	69 (1.00)	29 (0.42)	NA	*	

The number of pups alive at weaning for each genotype for $G\alpha_{i2}$ and $G\alpha_{o}$ G184S mutant crosses is shown for different degrees of backcrossing onto the C57BL/6J or FVB genetic background. N5 indicates the fifth backcross generation, etc. Values in the table are actual numbers of offspring while values in parentheses are the fraction of the number of WT (+/+) mice from the same litters. For het × het crosses the expected ratios are 1:2:1 while for het × WT crosses the expected ratio is 1:1. Ratios that differ from the expected Mendelian frequencies are marked with * in the fifth column. NA means not applicable. NS means not significantly different from the expected values

hybrids. Offspring from F1 crosses have one B6 and one FVB allele at each genomic locus so they represent a pure genetic strain, in contrast mice from mixed backgrounds or early backcross generations (<N5). This pure (though hybrid) genetic background should reduce experimental variability while increasing homozygote yield. It is necessary, however, to ensure that experimental phenotypes are also observable on this F1 strain as it is likely to differ in some respects from both of the parental strains.

5. Biological Results from RGSi Gα Subunit Knock-In Mice

5.1. Cellular Signaling in Gα RGSi Mutant Mice

Mouse embryo fibroblasts (MEFs) from $G\alpha_{i2}$ G184S mutant mice were isolated and showed alterations in lysophosphatidic acid (LPA)-induced inhibition of cAMP accumulation and stimulation of phospho-Akt and phosphor-ERK levels as compared to the WT cells (Fig. 3). It was surprising that the effect of the mutation on LPA inhibition of cAMP accumulation was quite modest while the enhancement of p-Akt was quite striking (60). Indeed, even heterozygous $G\alpha_{i2}{}^{G184S/+}$ MEFs showed a marked increase in

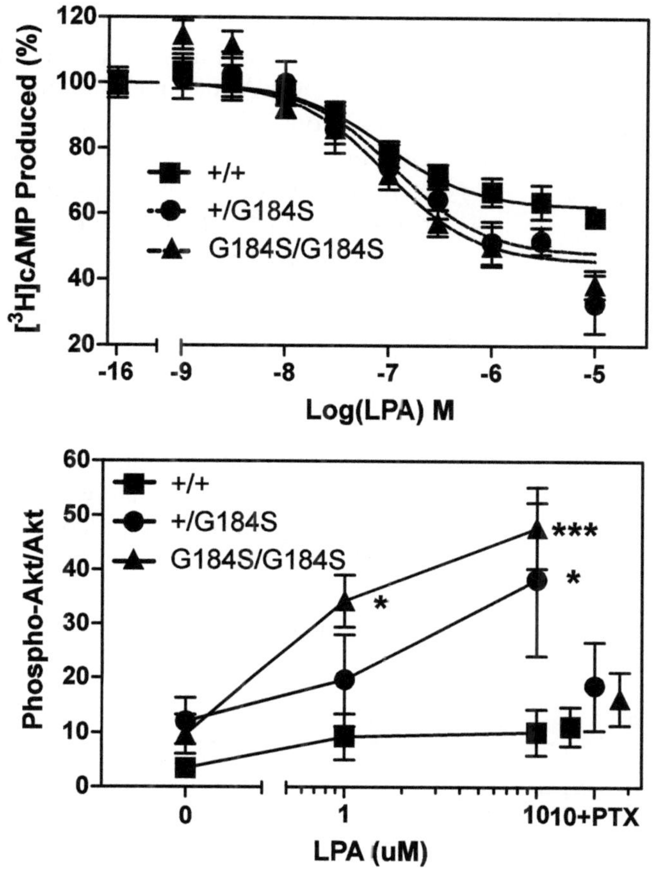

Fig. 3. Enhanced signaling in MEFs from knock-in $G\alpha_{i2}$ G184S mutant mice. Embryo fibroblasts from $G\alpha_{i2}$ G184S mutant mice were immortalized by serial passaging. Cells from $G\alpha_{i2}$ WT (*filled squares*, +/+) or heterozygous (*filled circles*, +/G184S) or homozygous (*filled triangles*, G184S/G184S) mutant embryos were tested for responses to LPA. *Top*: There was a small enhancement of LPA-mediated inhibition of cAMP production ($p<0.05$) for both +/G184S and G184S/G184S. *Bottom*: Signaling through the PI3K/Akt pathway as measured by Western blot analysis of p-Akt was markedly increased in both mutant cell lines. Previously published in: Huang, X. et al. *Mol Cell Biol* 2006; **26**: 6870–6879, DOI: 10.1128/MCB.00314-0. Copyright © American Society for Microbiology.

LPA-stimulated p-Akt that was fully blocked by *pertussis* toxin. Consequently, signaling mechanisms are sensitized but not to equivalent extents. This may be due to differential usage of Gα subunits (e.g., $G\alpha_{i1}$ or $G\alpha_{i3}$ rather than $G\alpha_{i2}$ for AC inhibition). It could also be due to differential dependence on RGS effects. One mechanism could be localization of RGS proteins leading to more or less effect on pools of Gα interacting with different effectors. Alternatively, interactions of the RGS with one effector (such as with RGS9 and PDE in the retina) could alter the RGS effect. Finally, it could simply be related to the observation of Traynor and colleagues (63) that low efficacy stimuli are more strongly affected by RGS actions.

5.2. Role of RGS Proteins in the Cardiovascular System: Selective Signaling Through $G\alpha_o$ vs. $G\alpha_{i2}$

Early studies with RGSi mutants were done before intact mice were available and used embryonic stem-cell-derived cardiomyocytes (ESDC). A key conclusion from these studies is that different "Gi/o" coupled receptors appear to differentially utilize $G\alpha_o$ and $G\alpha_{i2}$ in signaling. G184S mutant ES cells (both $G\alpha_o$ and $G\alpha_{i2}$) were converted to homozygosity by selection in high concentrations of G418 (58, 66). They were then differentiated into beating "atrial-nodal" like aggregates by forming embryoid bodies in hanging drops followed by withdrawal of leukemia inhibition factor. Beating rates were stimulated by isoproterenol then inhibition by either phenylisopropyl-adenosine (R-PIA) or carbachol was tested. $G\alpha_{i2}$ RGSi cells had a markedly increased response to carbachol (muscarinic M2 agonist) that was blocked by the GIRK channel inhibitor tertiapin Q (66). Surprisingly, the RGSi $G\alpha_{i2}$ mutant cells showed only a modest enhancement of the response to R-PIA (adenosine A1 agonist; approximately twofold decrease in IC_{50}). In contrast, $G\alpha_o$ RGSi ESDC had strongly increased sensitivity to R-PIA and this appeared to be independent of GIRK channel function (66) (Fig. 4).

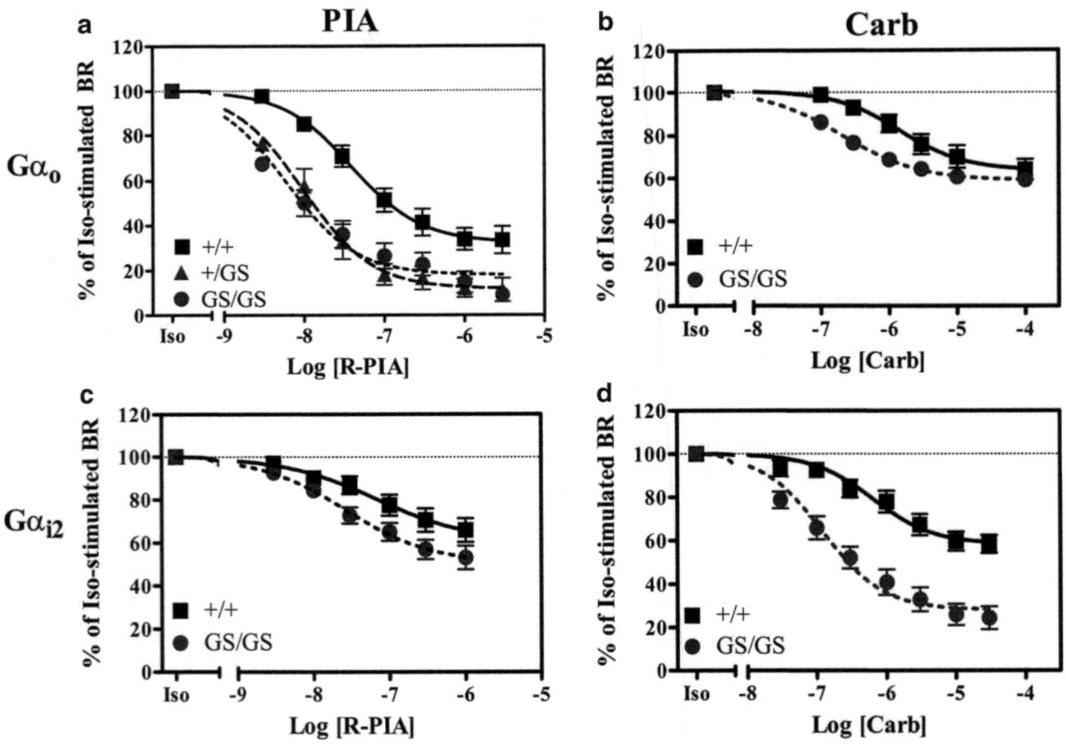

Fig. 4. Gα subunit-specific effects on muscarinic and adenosine-mediated slowing of atrial-nodal cell-beating rates. Embryonic stem-cell-derived cardiocytes were stimulated with 100 nM isoproterenol then the indicated concentrations of the M2 muscarinic agonist carbachol (*left panels*) and the A1 adenosine receptor agonist PIA (*right panels*) were added. Beating rates were measured every 5 min and are expressed as percent of the control value. ESDC that are wild type (*filled squares*) for either $G\alpha_o$ (*top panels*) or $G\alpha_{i2}$ (*bottom panels*) or that have either one copy (*filled triangles*) or two copies (*filled circles*) of the G184S mutant Gα subunit were tested. Previously published in: Fu, H. et al. *Circ Res* 2006; **98**:659–666.

The selective effects of the $G\alpha_{i2}$ G184S mutant on M2 receptor-mediated chronotropic control was also seen in vivo using telemetry in conscious, unrestrained $G\alpha_{i2}$ RGSi mice (66). This differential use of $G\alpha_o$ and $G\alpha_{i2}$ by A1 adenosine and M2 muscarinic receptors, respectively, was surprising. This key role for $G\alpha_{i2}$ in muscarinic heart rate control was further confirmed by Tinker and colleagues using $G\alpha_i$ subtype-selective knockout mice (69). These results raise important questions about the general practice of describing some GPCRs as "Gi/o" coupled receptors. While this may be true in overexpression studies in vitro, their functions in vivo appear to be much more refined.

The strong in vivo effect of $G\alpha_{i2}$ RGSi mutants on cholinergic chronotropic control begged the question of which RGS protein was involved. Recently, Heximer and colleagues reported that RGS4 is very strongly expressed in SA node, much more even than in atrium (39). They went on to show that $RGS4^{-/-}$ mice had profoundly slowed recovery of GIRK channel activity after removal of a muscarinic stimulus. The mice also showed strongly enhanced carbachol-mediated bradycardia – very similar in magnitude to the effect that we saw in the $G\alpha_{i2}^{G184S/G184S}$ mutant mice (66). The combination of the RGSi $G\alpha_{i2}$, RGS4 KO, and $G\alpha_{i2}$ KO data, as well as effects of tertiapin Q, provide a clear picture of the control of muscarinic bradycardia involving $G\alpha_{i2}$, RGS4, and Kir3.1/3.4 channels.

5.3. RGS-Regulated, Gα Subtype-Selective Signaling in the Central Nervous System

In another system, a similarly striking selectivity for Gi/o subtypes was found but in this case the response (via an α_{2a} adrenergic receptor) was mediated by $G\alpha_o$ and not $G\alpha_{i2}$. The CA3 region of the hippocampus has one of the lowest seizure thresholds in the CNS due to recurrent collaterals that can mediate synchronous bursting behavior. This is brought on by bicuculline-mediated blockade of the inhibitory $GABA_A$ receptors. This epileptiform activity is suppressed by α_2 adrenergic agonists such as epinephrine and UK14304, an effect mediated by α_{2a} adrenergic receptors (70). $G\alpha_o^{G184S/+}$ RGSi mice show an eightfold increase in epinephrine potency compared to wild-type mice (70). $G\alpha_{i2}^{G184S/+}$ RGSi mice show no change from WT mice (Fig. 5). The selective potentiation of the α_{2A} adrenergic receptor effect by the RGSi $G\alpha_o$ mutant suggests that this mechanism is primarily mediated by $G\alpha_o$ (at least compared to $G\alpha_{i2}$). While it is possible that $G\alpha_{i2}$ is involved but is not regulated by RGS proteins, the contrast of this result to those in the heart and for serotonin signaling (see below) suggests otherwise. The in vivo significance of this finding, as well as the generality to other receptors in the hippocampal CA3 region or to α_{2A} adrenergic signaling in other neural loci, will need to be evaluated.

Another exciting phenotype that further supports the role of differential Gi/o subtype function in the CNS relates to the actions of serotonin in murine models of antidepressant action.

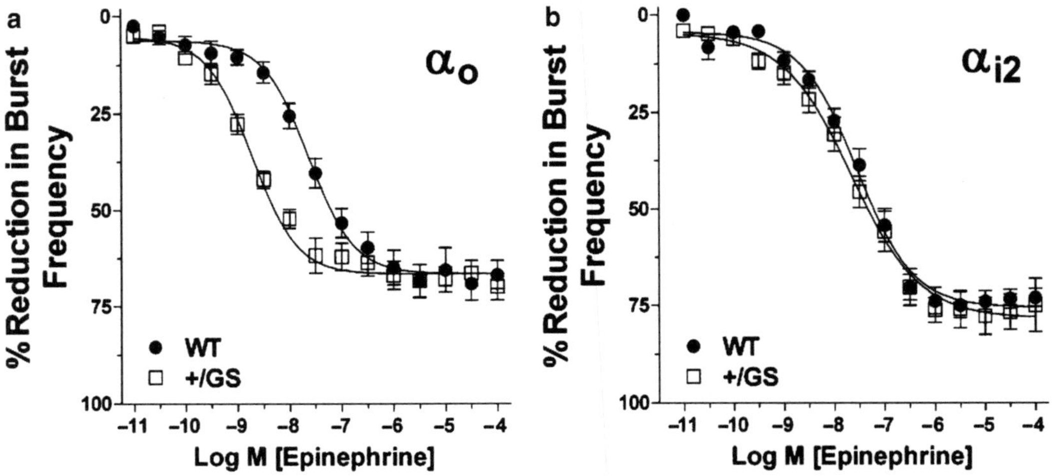

Fig. 5. Gα subunit-specific effects on α_{2a} adrenergic receptor-mediated anti-epileptiform activity in hippocampus. (**a**) Epinephrine is more potent on slices from $G\alpha o^{+/GS}$ mice (2.5 ± 0.9 nM) vs. control mice (19 ± 5 nM). (**b**) Epinephrine is equipotent on wild-type and $G\alpha i2^{+/GS}$ mice. Previously published in: Goldenstein, B. L., Nelson, B. W., Xu, K., Luger, E. J., Pribula, J. A., Wald, J. M., O'Shea, L. A., Weinshenker, D., Charbeneau, R. A., Huang, X., Neubig, R. R., and Doze, V. A. Regulator of G protein signaling protein suppression of Galphao protein-mediated alpha2A adrenergic receptor inhibition of mouse hippocampal CA3 epileptiform activity. *Mol Pharmacol* 2009; **75,** 1222–30.

In tail suspension tests, antidepressants reduce the immobility time of mice. The $G\alpha_{i2}^{G184S/G184S}$ mice show a spontaneous and nearly maximal reduction of immobility time that is reversed by a 5HT1A antagonist (71) suggesting that endogenous serotonin is signaling sufficiently strongly in the mutants to produce an antidepressant-like effect. The heterozygotes produce an intermediate reduction of immobility times and in those mice the dose–response for the 5HT1A agonist 8-OH-DPAT is "left-shifted" 14-fold while that for the selective-serotonin-reuptake inhibitor (SSRI), fluvoxamine is left-shifted sixfold. Consequently, an RGS protein action at $G\alpha_{i2}$ seems to be strongly suppressing the antidepressant-like actions of serotonin. Intriguing evidence of the specificity of this effect is seen in two observations. First, norepinephrine-dependent antidepressants (e.g., amitriptyline) are not potentiated in the $G\alpha_{i2}$ RGSi mutant mice. Second, there is no effect of the $G\alpha_{i2}^{G184S/+}$ mutation on the hypothermic effect of 8-OH-DPAT. The strong effect of the RGSi $G\alpha_{i2}$ mutation on 5HT-dependent antidepressant-like effects but no effect on hypothermia raises the possibility that modulating RGS actions could greatly enhance the specificity of drug action – enhancing beneficial effects while not altering side effects.

5.4. A Step-By-Step Approach to the Use of RGSi Gα Subunit Knock-In Models

1. For Gi/o family proteins, test to see if your system is sensitive to pertussis toxin in a cell-based or in vivo assay or find in the literature evidence that receptors involved in your system are coupled to Gi/o GPCRs.

2. Obtain $G\alpha_o$ and $G\alpha_{i2}$ RGSi mice on the C57BL/6 J background. Experiments should be performed on $G\alpha_o^{G184S/+}$ mice and their wild-type littermates as well as $G\alpha_{i2}^{G184S/+}$ mice and their wild-type littermates.
3. As an example, consider a receptor and agonist that prevents mice from sneezing. Furthermore, presume that there are other Gi/o coupled receptors that can cause this same response. Perform dose–response studies with both agonists.
4. If, in one of the RGSi Gα mutants, you found a shift in an agonist dose–response curve (presumably to a more potent effect) then proceed on to: (a) assessing the generality of this effect for other receptors or other responses to the same receptor, (b) mechanism of the enhancement or inhibition of signaling and (c) defining the RGS protein involved.
5. The follow-up steps from here would follow standard biomedical research approaches and would utilize all known information about the system. What other functions are known for the receptor that gives enhanced sneezing in your system? Are these other functions also potentiated by the Gα RGSi mutant that increased the inhibition of sneezing? What are known signal outputs from those G proteins and can they be measured easily in your system? If so, is one particular biochemical output selectively enhanced? Which RGS proteins are known to interact with the Gα subunit of interest (see Fig. 1)? What is the tissue distribution of the RGS protein candidates? Can you find RGS KO mice for that RGS or would an RNAi approach be usable in your system?
6. The conclusion from such a study would suggest that your biological system and agonist are controlled by an RGS-mediated effect on a specific Gα subunit. That opens the door to use of an agonist that either selectively activates that Gα subunit or to approaches to suppress the RGS protein involved in the system. With active work on RGS inhibitors (17, 61, 72), those tools may provide a novel way to modulate this function experimentally and ultimately therapeutically.

6. Advantages and Disadvantages of an RGSi Gα Subunit Knock-In Model System for Evaluating RGS Function In Vivo

As described above, the RGSi $G\alpha_{i2}$ and $G\alpha_o$ mutant mice provide useful information about how RGS proteins suppress signaling by particular agonists in specific physiological and pharmacological situations. Enhancement of a signaling response by one or another mutant Gα subunit is most simply interpreted to show that the endogenous receptor is able to activate that Gα subunit to lead to the observed response and that an endogenous RGS protein is able to suppress that signal. This provides insights into the most likely signal pathway (receptor and Gα subunit) that underlies that response in vivo. The striking specificity for "Gi/o" coupled receptors to utilize $G\alpha_{i2}$ and $G\alpha_o$ differentially was unexpected and further studies with alternate approaches to confirm conclusions of this sort will be very interesting, as would studies with additional RGSi Gα mutants. The information from this approach also suggests that a biased agonist that can direct a signal output from a receptor to a specific subtype of Gα subunit (e.g., 5HT1A agonist that selectively activates $G\alpha_{i2}$) might be quite useful as novel pharmacological agents.

The primary advantages of the use of RGSi Gα subunit mutants in evaluating RGS function include: (1) overcoming the functional redundancy of RGS proteins which often produces a strong phenotype (see also Table 1), (2) the gain-of-function mechanism brings out actions of specific subtypes of Gα subunits permitting an analysis of the roles of different but closely related Gα subuntis (i.e., $G\alpha_{i2}$ and $G\alpha_o$ in work to date), (3) the Gα RGSi mutation only eliminates RGS functions that relate to GAP activity or possibly to recruitment of the RGS to a Gα subunit; the functions of scaffolding functions and of other domains of the RGS proteins are left intact, and (4) the use of the knock-in approach in our studies avoids the pitfalls of overexpression that transfected cells and transgenic mice suffer.

Limitations to the conclusions from such studies are also apparent. Given that all RGS proteins are prevented from acting upon the mutated Gα subunit, this approach is unable to define which RGS is involved in the observed changes. Additional studies, such as RNAi or knockout methods would be needed to define the RGS protein mediating the effects. Also, if expression of a mutant Gα subunit does not lead to enhanced signaling, it is plausible that the Gα is involved but for whatever reason there is no effective RGS control of that signal (see above). Furthermore, using this system one can only investigate the G protein-mediated role of RGS proteins, while it is clear that RGS proteins have an expanding role outside of G protein GAP activity (15, 73). Specifically, RGS2 modulates protein translation (74) and RGS13 suppresses mast cell degranulation by a process that is not effected by *pertussis* toxin treatment implicating a non-GAP function for RGS13 (51). These phenomena would not be found in Gα RGSi mutant mice.

Note Added in Proof

Two very recent studies also demonstrate a role for RGS6 in control of carbachol-induced bradycardia (76, 77).

References

1. Sternweis, P. C., Northup, J. K., Smigel, M. D., and Gilman, A. G. (1981) The regulatory component of adenylate cyclase. Purification and properties. *J Biol Chem* **256,** 11517–26.
2. Ross, E. M., and Gilman, A. G. (1980) Biochemical properties of hormone-sensitive adenylate cyclase. *Annu Rev Biochem* **49,** 533–64.
3. Chabre, M., and Deterre, P. (1989) Molecular mechanism of visual transduction. *Eur J Biochem* **179,** 255–66.
4. Arshavsky, V., and Bownds, M. D. (1992) Regulation of deactivation of photoreceptor G protein by its target enzyme and cGMP. *Nature* **357,** 416–7.
5. Dratz, E. A., Lewis, J. W., Schaechter, L. E., Parker, K. R., and Kliger, D. S. (1987) Retinal rod GTPase turnover rate increases with concentration: a key to the control of visual excitation? *Biochem Biophys Res Commun* **146,** 379–86.
6. Adari, H., Lowy, D. R., Willumsen, B. M., Der, C. J., and McCormick, F. (1988) Guanosine triphosphatase activating protein (GAP) interacts with the p21 ras effector binding domain. *Science* **240,** 518–21.
7. Dohlman, H. G. (2002) G proteins and pheromone signaling. *Annu Rev Physiol* **64,** 129–52.
8. Chan, R. K., and Otte, C. A. (1982) Isolation and genetic analysis of Saccharomyces cerevisiae mutants supersensitive to G1 arrest by a factor and alpha factor pheromones. *Mol Cell Biol* **2,** 11–20.
9. Dietzel, C., and Kurjan, J. (1987) Pheromonal regulation and sequence of the Saccharomyces cerevisiae SST2 gene: a model for desensitization to pheromone. *Mol Cell Biol* **7,** 4169–77.
10. Dohlman, H. G., Apaniesk, D., Chen, Y., Song, J., and Nusskern, D. (1995) Inhibition of G-protein signaling by dominant gain-of-function mutations in Sst2p, a pheromone desensitization factor in Saccharomyces cerevisiae. *Mol Cell Biol* **15,** 3635–43.
11. Koelle, M. R., and Horvitz, H. R. (1996) EGL-10 regulates G protein signaling in the C. elegans nervous system and shares a conserved domain with many mammalian proteins. *Cell* **84,** 115–25.
12. Druey, K. M., Blumer, K. J., Kang, V. H., and Kehrl, J. H. (1996) Inhibition of G-protein-mediated MAP kinase activation by a new mammalian gene family. *Nature* **379,** 742–6.
13. Berman, D. M., Wilkie, T. M., and Gilman, A. G. (1996) GAIP and RGS4 are GTPase-activating proteins for the Gi subfamily of G protein alpha subunits. *Cell* **86,** 445–52.
14. Dohlman, H. G., and Thorner, J. (1997) RGS proteins and signaling by heterotrimeric G proteins. *J Biol Chem* **272,** 3871–4.
15. Hollinger, S., and Hepler, J. R. (2002) Cellular regulation of RGS proteins: modulators and integrators of G protein signaling. *Pharmacol Rev* **54,** 527–59.
16. Ross, E. M., and Wilkie, T. M. (2000) GTPase-activating proteins for heterotrimeric G proteins: regulators of G protein signaling (RGS) and RGS-like proteins. *Annu Rev Biochem* **69,** 795–827.
17. Neubig, R. R., and Siderovski, D. P. (2002) Regulators of G-protein signalling as new central nervous system drug targets. *Nat Rev Drug Discov* **1,** 187–97.
18. Zhong, H., and Neubig, R. R. (2001) Regulator of G protein signaling proteins: novel multifunctional drug targets. *J Pharmacol Exp Ther* **297,** 837–45.
19. Traynor, J. R., and Neubig, R. R. (2005) Regulators of G protein signaling & drugs of abuse. *Mol Interv* **5,** 30–41.
20. Gold, S. J., Ni, Y. G., Dohlman, H. G., and Nestler, E. J. (1997) Regulators of G-protein signaling (RGS) proteins: region-specific expression of nine subtypes in rat brain. *J Neurosci* **17,** 8024–37.
21. Doupnik, C. A., Xu, T., and Shinaman, J. M. (2001) Profile of RGS expression in single rat atrial myocytes. *Biochim Biophys Acta* **1522,** 97–107.
22. Grafstein-Dunn, E., Young, K. H., Cockett, M. I., and Khawaja, X. Z. (2001) Regional distribution of regulators of G-protein signaling (RGS) 1, 2, 13, 14, 16, and GAIP messenger ribonucleic acids by in situ hybridization in rat brain. *Brain Res Mol Brain Res* **88,** 113–23.
23. Liu, W., Yuen, E. Y., Allen, P. B., Feng, J., Greengard, P., and Yan, Z. (2006) Adrenergic modulation of NMDA receptors in prefrontal cortex is differentially regulated by RGS proteins and spinophilin. *Proc Natl Acad Sci U S A* **103,** 18338–43.
24. Anderson, G. R., Lujan, R., Semenov, A., Pravetoni, M., Posokhova, E. N., Song, J. H., Uversky, V., Chen, C. K., Wickman, K., and Martemyanov, K. A. (2007) Expression and localization of RGS9-2/G 5/R7BP complex in vivo is set by dynamic control of its constitutive degradation by cellular cysteine proteases. *J Neurosci* **27,** 14117–27.
25. Drenan, R. M., Doupnik, C. A., Jayaraman, M., Buchwalter, A. L., Kaltenbronn, K. M., Huettner, J. E., Linder, M. E., and Blumer, K. J. (2006) R7BP augments the function of RGS7*Gbeta5

complexes by a plasma membrane-targeting mechanism. *J Biol Chem* **281,** 28222–31.

26. De Vries, L., Lou, X., Zhao, G., Zheng, B., and Farquhar, M. G. (1998) GIPC, a PDZ domain containing protein, interacts specifically with the C terminus of RGS-GAIP. *Proc Natl Acad Sci U S A* **95,** 12340–5.
27. Murphy, J. J., and Norton, J. D. (1990) Cell-type-specific early response gene expression during plasmacytoid differentiation of human B lymphocytic leukemia cells. *Biochim Biophys Acta* **1049,** 261–71.
28. Moratz, C., Hayman, J. R., Gu, H., and Kehrl, J. H. (2004) Abnormal B-cell responses to chemokines, disturbed plasma cell localization, and distorted immune tissue architecture in Rgs1−/− mice. *Mol Cell Biol* **24,** 5767–75.
29. Heximer, S. P., Knutsen, R. H., Sun, X., Kaltenbronn, K. M., Rhee, M. H., Peng, N., Oliveira-dos-Santos, A., Penninger, J. M., Muslin, A. J., Steinberg, T. H., Wyss, J. M., Mecham, R. P., and Blumer, K. J. (2003) Hypertension and prolonged vasoconstrictor signaling in RGS2-deficient mice. *J Clin Invest* **111,** 445–52.
30. Sun, X., Kaltenbronn, K. M., Steinberg, T. H., and Blumer, K. J. (2005) RGS2 is a mediator of nitric oxide action on blood pressure and vasoconstrictor signaling. *Mol Pharmacol* **67,** 631–9.
31. Gross, V., Tank, J., Obst, M., Plehm, R., Blumer, K. J., Diedrich, A., Jordan, J., and Luft, F. C. (2005) Autonomic nervous system and blood pressure regulation in RGS2-deficient mice. *Am J Physiol Regul Integr Comp Physiol* **288,** R1134-42.
32. Tuomi, J. M., Chidiac, P., and Jones, D. L. (2009) Evidence for enhanced M3 muscarinic receptor function and sensitivity to atrial arrhythmia in the RGS2 deficient mouse. *Am J Physiol Heart Circ Physiol.* **298,**H554-61.
33. Takimoto, E., Koitabashi, N., Hsu, S., Ketner, E. A., Zhang, M., Nagayama, T., Bedja, D., Gabrielson, K. L., Blanton, R., Siderovski, D. P., Mendelsohn, M. E., and Kass, D. A. (2009) Regulator of G protein signaling 2 mediates cardiac compensation to pressure overload and antihypertrophic effects of PDE5 inhibition in mice. *J Clin Invest* **119,** 408–20.
34. Bodenstein, J., Sunahara, R. K., and Neubig, R. R. (2007) N-terminal residues control proteasomal degradation of RGS2, RGS4, and RGS5 in human embryonic kidney 293 cells. *Mol Pharmacol* **71,** 1040–50.
35. Gu, S., Tirgari, S., and Heximer, S. P. (2008) The RGS2 gene product from a candidate hypertension allele shows decreased plasma membrane association and inhibition of Gq. *Mol Pharmacol* **73,** 1037–43.
36. Oliveira-Dos-Santos, A. J., Matsumoto, G., Snow, B. E., Bai, D., Houston, F. P., Whishaw, I. Q., Mariathasan, S., Sasaki, T., Wakeham, A., Ohashi, P. S., Roder, J. C., Barnes, C. A., Siderovski, D. P., and Penninger, J. M. (2000) Regulation of T cell activation, anxiety, and male aggression by RGS2. *Proc Natl Acad Sci U S A* **97,** 12272–7.
37. Grillet, N., Pattyn, A., Contet, C., Kieffer, B. L., Goridis, C., and Brunet, J. F. (2005) Generation and characterization of Rgs4 mutant mice. *Mol Cell Biol* **25,** 4221–8.
38. Mirnics, K., Middleton, F. A., Stanwood, G. D., Lewis, D. A., and Levitt, P. (2001) Disease-specific changes in regulator of G-protein signaling 4 (RGS4) expression in schizophrenia. *Mol Psychiatry* **6,** 293–301.
39. Cifelli, C., Rose, R. A., Zhang, H., Voigtlaender-Bolz, J., Bolz, S. S., Backx, P. H., and Heximer, S. P. (2008) RGS4 regulates parasympathetic signaling and heart rate control in the sinoatrial node. *Circ Res* **103,** 527–35.
40. Bondjers, C., Kalen, M., Hellstrom, M., Scheidl, S. J., Abramsson, A., Renner, O., Lindahl, P., Cho, H., Kehrl, J., and Betsholtz, C. (2003) Transcription profiling of platelet-derived growth factor-B-deficient mouse embryos identifies RGS5 as a novel marker for pericytes and vascular smooth muscle cells. *Am J Pathol* **162,** 721–9.
41. Cho, H., Kozasa, T., Bondjers, C., Betsholtz, C., and Kehrl, J. H. (2003) Pericyte-specific expression of Rgs5: implications for PDGF and EDG receptor signaling during vascular maturation. *FASEB J* **17,** 440–2.
42. Nisancioglu, M. H., Mahoney, W. M., Jr., Kimmel, D. D., Schwartz, S. M., Betsholtz, C., and Genove, G. (2008) Generation and characterization of rgs5 mutant mice. *Mol Cell Biol* **28,** 2324–31.
43. Cho, H., Park, C., Hwang, I. Y., Han, S. B., Schimel, D., Despres, D., and Kehrl, J. H. (2008) Rgs5 targeting leads to chronic low blood pressure and a lean body habitus. *Mol Cell Biol* **28,** 2590–7.
44. Hamzah, J., Jugold, M., Kiessling, F., Rigby, P., Manzur, M., Marti, H. H., Rabie, T., Kaden, S., Grone, H. J., Hammerling, G. J., Arnold, B., and Ganss, R. (2008) Vascular normalization in Rgs5-deficient tumours promotes immune destruction. *Nature* **453,** 410–4.
45. Kuwata, H., Nakao, K., Harada, T., Matsuda, I., and Aiba, A. (2008) Generation of RGS8 null mutant mice by Cre/loxP system. *Kobe J Med Sci* **53,** 275–81.
46. Chen, C. K., Burns, M. E., He, W., Wensel, T. G., Baylor, D. A., and Simon, M. I. (2000)

Slowed recovery of rod photoresponse in mice lacking the GTPase accelerating protein RGS9-1. *Nature* **403,** 557–60.

47. Rahman, Z., Schwarz, J., Gold, S. J., Zachariou, V., Wein, M. N., Choi, K. H., Kovoor, A., Chen, C. K., DiLeone, R. J., Schwarz, S. C., Selley, D. E., Sim-Selley, L. J., Barrot, M., Luedtke, R. R., Self, D., Neve, R. L., Lester, H. A., Simon, M. I., and Nestler, E. J. (2003) RGS9 modulates dopamine signaling in the basal ganglia. *Neuron* **38,** 941–52.
48. Zachariou, V., Georgescu, D., Sanchez, N., Rahman, Z., DiLeone, R., Berton, O., Neve, R. L., Sim-Selley, L. J., Selley, D. E., Gold, S. J., and Nestler, E. J. (2003) Essential role for RGS9 in opiate action. *Proc Natl Acad Sci U S A* **100,** 13656–61.
49. Yang, S., and Li, Y. P. (2007) RGS10-null mutation impairs osteoclast differentiation resulting from the loss of [Ca2+]i oscillation regulation. *Genes Dev* **21,** 1803–16.
50. Hernandez-Hansen, V., Bard, J. D., Tarleton, C. A., Wilder, J. A., Lowell, C. A., Wilson, B. S., and Oliver, J. M. (2005) Increased expression of genes linked to FcepsilonRI Signaling and to cytokine and chemokine production in Lyn-deficient mast cells. *J Immunol* **175,** 7880–8.
51. Bansal, G., Xie, Z., Rao, S., Nocka, K. H., and Druey, K. M. (2008) Suppression of immunoglobulin E-mediated allergic responses by regulator of G protein signaling 13. *Nat Immunol* **9,** 73–80.
52. Martin-McCaffrey, L., Willard, F. S., Oliveira-dos-Santos, A. J., Natale, D. R., Snow, B. E., Kimple, R. J., Pajak, A., Watson, A. J., Dagnino, L., Penninger, J. M., Siderovski, D. P., and D'Souza, S. J. (2004) RGS14 is a mitotic spindle protein essential from the first division of the mammalian zygote. *Dev Cell* **7,** 763–9.
53. Larminie, C., Murdock, P., Walhin, J. P., Duckworth, M., Blumer, K. J., Scheideler, M. A., and Garnier, M. (2004) Selective expression of regulators of G-protein signaling (RGS) in the human central nervous system. *Brain Res Mol Brain Res* **122,** 24–34.
54. DiBello, P. R., Garrison, T. R., Apanovitch, D. M., Hoffman, G., Shuey, D. J., Mason, K., Cockett, M. I., and Dohlman, H. G. (1998) Selective uncoupling of RGS action by a single point mutation in the G protein alpha-subunit. *J Biol Chem* **273,** 5780–4.
55. Lan, K. L., Sarvazyan, N. A., Taussig, R., Mackenzie, R. G., DiBello, P. R., Dohlman, H. G., and Neubig, R. R. (1998) A point mutation in Galphao and Galphai1 blocks interaction with regulator of G protein signaling proteins. *J Biol Chem* **273,** 12794–7.
56. Tesmer, J. J., Berman, D. M., Gilman, A. G., and Sprang, S. R. (1997) Structure of RGS4 bound to AlF4-activated G(i alpha1): stabilization of the transition state for GTP hydrolysis. *Cell* **89,** 251–61.
57. Tesmer, J. J. Structure and Function of Regulator of G Protein Signaling Homology Domains. In: Fisher R., ed. Molecular Biology of RGS proteins. Amsterdam: Elsevier; 2010:75–113.
58. Fu, Y., Zhong, H., Nanamori, M., Mortensen, R. M., Huang, X., Lan, K., and Neubig, R. R. (2004) RGS-insensitive G-protein mutations to study the role of endogenous RGS proteins. *Methods Enzymol* **389,** 229–43.
59. Clark, M. J., Harrison, C., Zhong, H., Neubig, R. R., and Traynor, J. R. (2003) Endogenous RGS protein action modulates mu-opioid signaling through Galphao. Effects on adenylyl cyclase, extracellular signal-regulated kinases, and intracellular calcium pathways. *J Biol Chem* **278,** 9418–25.
60. Huang, X., Fu, Y., Charbeneau, R. A., Saunders, T. L., Taylor, D. K., Hankenson, K. D., Russell, M. W., D'Alecy, L. G., and Neubig, R. R. (2006) Pleiotropic phenotype of a genomic knock-in of an RGS-insensitive G184S Gnai2 allele. *Mol Cell Biol* **26,** 6870–9.
61. Roman, D. L., Talbot, J. N., Roof, R. A., Sunahara, R. K., Traynor, J. R., and Neubig, R. R. (2007) Identification of small-molecule inhibitors of RGS4 using a high-throughput flow cytometry protein interaction assay. *Mol Pharmacol* **71,** 169–75.
62. Jin, Y., Zhong, H., Omnaas, J. R., Neubig, R. R., and Mosberg, H. I. (2004) Structure-based design, synthesis, and pharmacologic evaluation of peptide RGS4 inhibitors. *J Pept Res* **63,** 141–6.
63. Clark, M. J., Linderman, J. J., and Traynor, J. R. (2008) Endogenous regulators of G protein signaling differentially modulate full and partial mu-opioid agonists at adenylyl cyclase as predicted by a collision coupling model. *Mol Pharmacol* **73,** 1538–48.
64. Jeong, S. W., and Ikeda, S. R. (2000) Endogenous regulator of G-protein signaling proteins modify N-type calcium channel modulation in rat sympathetic neurons. *J Neurosci* **20,** 4489–96.
65. Chen, H., and Lambert, N. A. (2000) Endogenous regulators of G protein signaling proteins regulate presynaptic inhibition at rat hippocampal synapses. *Proc Natl Acad Sci U S A* **97,** 12810–5.
66. Fu, Y., Huang, X., Zhong, H., Mortensen, R. M., D'Alecy, L. G., and Neubig, R. R. (2006) Endogenous RGS proteins and Galpha subtypes differentially control muscarinic and

adenosine-mediated chronotropic effects. *Circ Res* **98,** 659–66.

67. Shi, J., Damjanoska, K. J., Zemaitaitis, B., Garcia, F., Carrasco, G., Sullivan, N. R., She, Y., Young, K. H., Battaglia, G., Van De kar, L. D., Howland, D. S., and Muma, N. A. (2006) Alterations in 5-HT2A receptor signaling in male and female transgenic rats over-expressing either Gq or RGS-insensitive Gq protein. *Neuropharmacology* **51,** 524–35.
68. Heintz, N. (2001) BAC to the future: the use of bac transgenic mice for neuroscience research. *Nat Rev Neurosci* **2,** 861–70.
69. Zuberi, Z., Birnbaumer, L., and Tinker, A. (2008) The role of inhibitory heterotrimeric G proteins in the control of in vivo heart rate dynamics. *Am J Physiol Regul Integr Comp Physiol* **295,** R1822-30.
70. Goldenstein, B. L., Nelson, B. W., Xu, K., Luger, E. J., Pribula, J. A., Wald, J. M., O'Shea, L. A., Weinshenker, D., Charbeneau, R. A., Huang, X., Neubig, R. R., and Doze, V. A. (2009) Regulator of G protein signaling protein suppression of Galphao protein-mediated alpha2A adrenergic receptor inhibition of mouse hippocampal CA3 epileptiform activity. *Mol Pharmacol* **75,** 1222–30.
71. Talbot, J. N., Jutkiewicz, E. M., Graves, S. M., Clemans, C. F., Nicol, M. R., Mortensen, R. M., Huang, X., Neubig, R. R., and Traynor, J. R. (2010) RGS inhibition at G(alpha)i2 selectively potentiates 5-HT1A-mediated antidepressant effects. *Proc Natl Acad Sci U S A* **107,** 11086–91
72. Roof, R. A., Jin, Y., Roman, D. L., Sunahara, R. K., Ishii, M., Mosberg, H. I., and Neubig, R. R. (2006) Mechanism of action and structural requirements of constrained peptide inhibitors of RGS proteins. *Chem Biol Drug Des* **67,** 266–74.
73. Willars, G. B. (2006) Mammalian RGS proteins: multifunctional regulators of cellular signalling. *Semin Cell Dev Biol* **17,** 363–76.
74. Nguyen, C. H., Ming, H., Zhao, P., Hugendubler, L., Gros, R., Kimball, S. R., and Chidiac, P. (2009) Translational control by RGS2. *J Cell Biol* **186,** 755–65.
75. den Dunnen, J. T., and Antonarakis, S. E. (2001) Nomenclature for the description of human sequence variations. *Hum Genet* **109,** 121–4.
76. Posokhova, E., Wydeven, N., Allen, K. L., Wickman, K., Martemyanov, K. A. (2010) RGS6/Gß5 complex accelerates IKACh gating kinetics in atrial myocytes and modulates parasympathetic regulation of heart rate. *Circ Res* **107,** 1350–4.
77. Yang, J., Huang, J., Maity, B., Gao, Z., Lorca, R. A., Gudmundsson, H., Li, J., Stewart, A., Swaminathan, P. D., Ibeawuchi, S. R., Shepherd, A., Chen, C. K., Kutschke, W., Mohler, P. J., Mohapatra, D. P., Anderson, M. E., Fisher, R. A. (2010) RGS6, a modulator of parasympathetic activation in heart. *Circ Res* **107,** 1345–9.

Chapter 5

Bioinformatic Approaches to Metabolic Pathways Analysis

Stuart Maudsley, Wayne Chadwick, Liyun Wang, Yu Zhou, Bronwen Martin, and Sung-Soo Park

Abstract

The growth and development in the last decade of accurate and reliable mass data collection techniques has greatly enhanced our comprehension of cell signaling networks and pathways. At the same time however, these technological advances have also increased the difficulty of satisfactorily analyzing and interpreting these ever-expanding datasets. At the present time, multiple diverse scientific communities including molecular biological, genetic, proteomic, bioinformatic, and cell biological, are converging upon a common endpoint, that is, the measurement, interpretation, and potential prediction of signal transduction cascade activity from mass datasets. Our ever increasing appreciation of the complexity of cellular or receptor signaling output and the structural coordination of intracellular signaling cascades has to some extent necessitated the generation of a new branch of informatics that more closely associates functional signaling effects to biological actions and even whole-animal phenotypes. The ability to untangle and hopefully generate theoretical models of signal transduction information flow from transmembrane receptor systems to physiological and pharmacological actions may be one of the greatest advances in cell signaling science. In this overview, we shall attempt to assist the navigation into this new field of cell signaling and highlight several methodologies and technologies to appreciate this exciting new age of signal transduction.

Key words: Signaling, Network, Pathway, Phenotype, Receptor

1. Introduction

1.1. The Relentless Progression in Complexity

Many research scientists familiar with signal transduction research have in recent years realized that despite their enhanced output technologies, genomic, proteomic or metabolomic, they often consider themselves somewhat hampered by analytical techniques that do not seem able to adequately appreciate mass datasets. Our consideration of the nature of signal transduction systems has likely forever moved away from linear enzymatic cascades with near-Brownian modes of motion of individual signaling factors in intermediary metabolic systems. Current hypotheses, of at least

Louis M. Luttrell and Stephen S.G. Ferguson (eds.), *Signal Transduction Protocols*, Methods in Molecular Biology, vol. 756,
DOI 10.1007/978-1-61779-160-4_5,

receptor-mediated signal transduction pathways, include the presence of substate-specific isoforms of receptors coupled to preassembled signal transduction cascades consisting of subtype-specific, stable multiprotein signaling complexes that possess distinct subcellular targeting mechanisms (1, 2). Despite this conversion of thinking and the wider appreciation of the inherent increase in the complexity of signaling systems, the potential for hindrance of pharmacological research has not been seen, actually quite the reverse. The more subtle our appreciation of the intricate nature of receptor response mechanisms and their contextual variety, then the more selective and specific rationally designed pharmacotherapies may become (3, 4).

With the ability to rapidly and accurately measure multiple differences (genomic or proteomic) between either physiological or drug-induced states, our appreciation of the complex nature of biological processes has forced us to consider that often physiological disorders or drug responses are mediated by alterations in whole gene/protein networks, as opposed to simple activation or inhibition of a linear signal transduction pathway. A common phrase often used to describe this changing mindset in molecular biology is *"pathways no longer exist, there are only networks."* This statement however does not negate the many years of prior signal transduction research but suggests perhaps that the delineation of discrete signaling pathways is likely an abstraction of the true hyper-complex signaling network due to our previous deficiencies in analytical technology. There are a huge variety of efficient and sensitive techniques which an investigator can use to assess genomic or proteomic differences in distinct pathophysiological or pharmacological scenarios, including fluorometric gene array analysis, genome-wide association screening and massive parallel sequencing, ChIP(chromatin immunoprecipitation)-on chip, antibody arrays, protein-binding microarrays, differential in-gel electrophoresis and quantitative mass spectrometry (MS). These techniques have been thoroughly discussed in recent years and therefore will not be repeated here. These era-changing technologies, however, often leave experimenters feeling lost in a mass of data that may or may not contain the specific scientific answers they are seeking. The application of biologically relevant mathematical processes to divine the eventual physiological meaning of these datasets will be the primary subject of this overview. We intend to provide a simple primer that researchers can use as a reference for interpretation of their complex datasets. The analytical tools and processes described will be applicable to both genomic and proteomic data and will hopefully facilitate a more holistic understanding of the creation and eventual pharmacological targeting of signal transduction networks. The primary goal of these bioinformatic analytical tools is the rational and biologically relevant condensation of these mass data lists into outputs that may predict the functional activities of the genes/proteins modulated between the control and test datasets. The clustering of gene/protein factors into functional groups or even signaling pathways will help to categorize characteristic gene/protein sets for future diagnostic and therapeutic use. Therefore in the future patient diagnosis, drug development, testing, and design may all take place initially at the signaling network level rather than at the single gene/protein measurement index level.

We shall consider the most commonly used techniques to extract functionally relevant and experimentally actionable information from mass data lists and then describe the most apt future uses of these paradigms. Even before more complex functional analysis can begin we shall discuss several important considerations with respect to the initial generation of the dataset and the relative merits and detractions of genomic/proteomic techniques.

1.2. Textual Definitions

In this overview, we shall consider both gene and protein datasets and will describe both as the same, that is, "*dataset*." For most postexperimental analytical algorithms we find that the *Gene Symbol* nomenclature often provides the most reliable and flexible gene/protein annotation platform and therefore we shall primarily consider these in this overview. Individual genes or proteins will be individually and interchangeably described as "*factors*" in this overview.

2. Extracting Multiple Relevant Factors from Datasets

Since the advent of facile technologies that can generate large complex datasets, the primary goal of such experiments has been to identify many relevant *factors* (*gene* or *protein*) that may explain the pathophysiological outcome or drug response in the experimental paradigm. Typically, a single control and one or multiple test conditions are analyzed in a simple comparative manner. After the creation of the first readily available gene arrays, the primary data selection processes applied to these datasets were developed by classical statistical analysis (5). With respect to modern fluorometric gene arrays such as Illumina and also to quantitative proteomic techniques, the initial choices for data filtration are distinct due to the unique properties of either of the mass analytical techniques. Many of the analytical modes can be swapped between genomic or proteomic platforms but one must always take into account that often mass spectrometry is a discovery process while gene (and also antibody or protein) microarrays provide a standard reproducible platform for each experiment. The functional annotation of datasets provides an invaluable approach for divination of the physiological "meaning" of the output but specifically in the case of mass spectrometry proteomics provides a vital support for analysis of variability of function between experiments. This important aspect of functional annotation of proteomic data will be expanded upon in subsequent sections.

2.1. Fluorescent Microarrays

Using differential fluorescent dye attachment (typically Cy3 or Cy5) relative quantitative changes in mRNA expression are easily obtainable on a large scale (6, 7). As with most technologies based upon fluorescent dye usage, the presence of background residual signal can be problematical. Subtraction of such background intensity is achieved by statistically computing the average background intensity and using the standard deviation among this intensity to calculate a confidence interval, the upper limit of which is used for the subsequent background correction. To assist the comparison of multiple gene regulation profiles between microarray chips, normalization of the data is paramount. One of the most common methods employed for normalization of the respective gene fluorescent signal is the use of "housekeeping" genes. The valid employment of housekeeping genes to normalize biologically relevant fluctuating data on the array relies on the assumption that there is a set of standard genes whose expression does not change with experimental condition or ligand stimulation. However, with respect to our current thinking of physiological response/signal transduction networks, the concept of a nonchanging *factor* on the array unfortunately becomes less and less likely. Clearly, there will be a spectrum of perturbation of factors on the array and some genes may indeed be unperceivably altered and thus provide a *de facto* basis for normalization. It is likely though that in the next few years the reliance upon "housekeeping" *factors* will be an increasingly redundant concept even though it may be practically effective. Internal spotted standards of a control *factor*, for example, bovine serum albumin, can often provide an adequate control for the output from the assay chip instead of using an experimental sample. However, this merely controls for experimental detection process itself and not the differential *factor* data per se. An alternative approach though is the more reliable use of whole-array normalization. Typically, whole-array normalization is performed using linear or logarithmic regression techniques (8–12). The reliability of this process is likely to be affected by the network connectivity of the targets under study and the target selectivity of the experimental effect(s). This whole-array normalization also relies upon a potentially anachronistic assumption, that is, the majority of genes on the array are nondifferentially expressed between the experimental states, and that varying genes are not solely associated with one of the fluorescent labels. The latter assumption can be checked easily by dye-swapping paradigms in which fluorescent labels are reversed and experimental data obtained again. This can also be applied to quantitative proteomic technologies that we shall describe in later sections. As mentioned previously this assumption that there is only a minimal perturbation of genes on the array constructively reinforces our old concept of linear discrete signaling pathways. Practically, however, this technique may still yield the production of a de facto valid data set based on the "broadness" of the spectrum of variation in the response to the experimental actions (Fig. 1). To further prepare microarray data for functional analysis, it is typical to apply a log transformation to the fluorescent data to make numerical manipulation more acceptable.

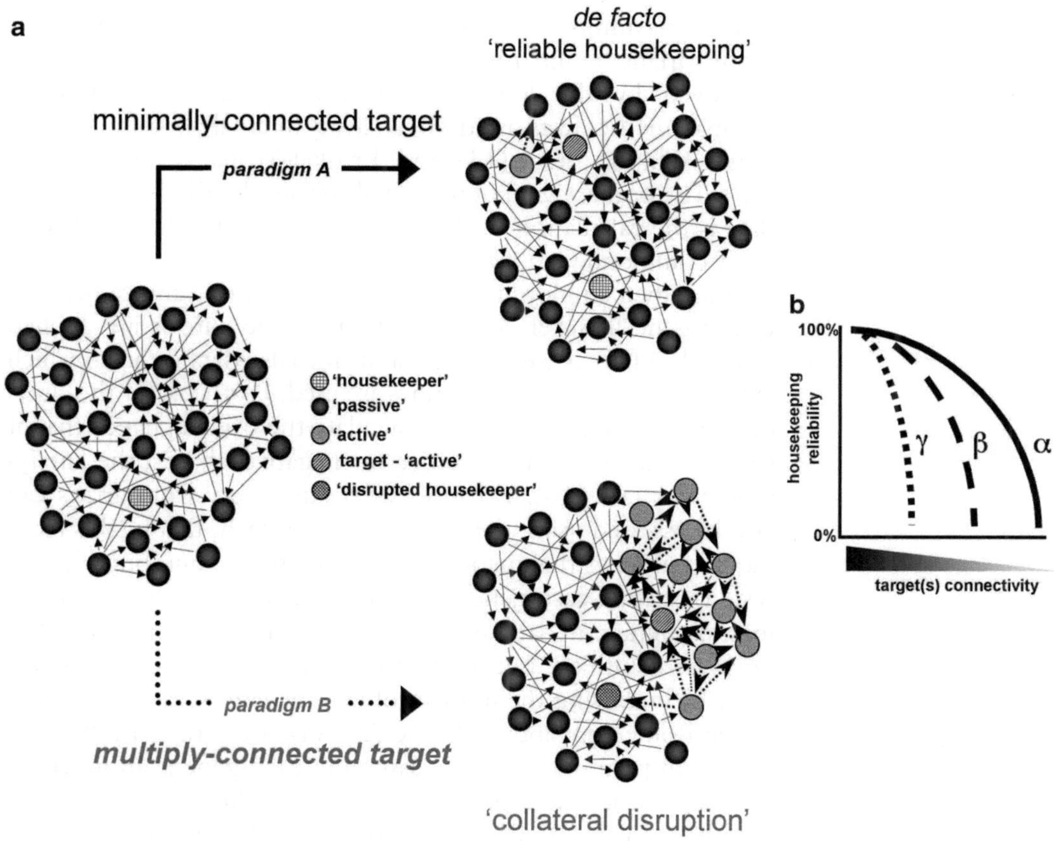

Fig. 1. Contextuality of dataset housekeeping reliability. Accepting a high level of connectivity of signaling factors introduces the likelihood of disruption of potential "housekeeping" factors. In paradigm A where a relatively selective activation of a target that possesses only minimal connectivity with the greater network of factors does not perceptibly disrupt the chosen housekeeper and therefore creates a de facto housekeeping factor. However, in paradigm B where the target factor is multiply connected to other factors in the network an increased likelihood of the loss of housekeeper reliability is seen (**a**). The potential effects of the connectivity in the network of the target factor and the target selectivity of a biological perturbing action (α, highly selective acting on minimal targets, β moderately selective acting on several targets, γ poorly selective acting on multiple targets). Highly connected targets possess a greater chance of disrupting housekeeping reliability and perturbations to the network that are nonselective are also likely to disrupt housekeeping reliability (**b**).

Parametric tests used for statistical analysis of the *factor* variation are the most commonly utilized, as these tests are much more sensitive and require the data to be normally distributed. This is usually achieved by using log transformation of the spot intensities to achieve a Gaussian distribution of the data. To extract the actual differential expression profile of genetic factors from microarray data, a ratio of intensity (as a measure of expression level: *z*-ratio) between two samples is used. As with all biological experiments, replicates of array data are required if a fold-change cutoff of *z*-ratios is used to primarily filter the data set. Several model-based techniques have been developed that facilitate the assumption of multiplicative noise, and eliminate statistically significant outliers from the data (13). The typical parametric analytical methods applied to primary gene array data management include maximum-likelihood analysis, F-statistic, ANOVA (analysis of variance), and *t*-tests. The results of these tests are often improved by the log transformation of the primary data. Nonparametric tests used to analyze microarray data include Mann–Whitney tests (14) and Kruskal–Williams rank analysis (15). The primary goal of the initial statistical analysis of the array data is the calculation of significance values for gene expression, most commonly as a "*p*-value." *P*-values, either fixed to 0.05 or 0.01 are then employed to reduce the dataset to significantly regulated gene lists before *z*-ratio/fold-change cutoffs are applied (typically ±1.5) as well as provisions for false data creation which are highly likely when large arrays are used. Protocols for the elucidation of random false results calculate the overall chance that at least one gene is a false-positive or -negative, that is, the family-wise error rate (16). Erroneous data discovery from arrays can also be assessed using the Bonferroni approach, that is, this technique multiplies the uncorrected *p*-value by the number of genes tested, treating each gene as an individual test. This protocol can increase significant data specificity by reducing the number of false-positives identified, but unfortunately attenuates the array sensitivity by increasing the number of false-negatives. A modification of the Bonferroni approach, the false-discovery rate (FDR), uses a random permutation while assuming each gene is an independent test. In addition, bootstrapping approaches can improve significantly on the Bonferroni approach, as they are less stringent (17). Resampling-based false discovery rate-controlling procedures can also be used (18). These array data extraction protocols can be applied to other array platforms, for example, antibody or protein arrays, as essentially the chip data can be easily analogized. However, one caveat is of course required, that is, the likelihood of high logarithmic increases in protein expression is highly unlikely as even a twofold change of protein expression may be sufficient to generate profound signaling actions, especially if the protein possesses enzymatic activity.

2.2. Quantitative Mass Spectrometry

The primary contrast between proteomic datasets and those from array experiments is the expectation of inclusion of certain datapoints, that is, proteins. Standard arrays provide a reproducible experimental platform while the recovery of the same protein between experiments is often unlikely. The use therefore of pathway bioinformatics, which can infer function from a variety of related proteins and not just based on individual identity, in such experiments may be paramount for the future use of proteomics. There are also recent advances in MS-based technologies that can be applied to mass spectrometers that can facilitate the accurate selection of protein species to be identified from a desired list (selective reaction monitoring, SRM; 19) in-part recreating the desired scanning pattern of an array. Such specific monitoring modes of MS may considerably slow down the rate of data retrieval and may only be suitable for experiments in which high levels of starting extract are available. In contrast to array technology though, the detection through SRM is still dependent on the ability of the MS to physically detect the specified peptides. This detection reliability is often more likely to demonstrate experiment to experiment variability than gene array platforms.

In this overview, our major focus is upon the functional interpretation of gene/protein datasets using bioinformatic approaches and therefore we shall focus upon the most commonly used current quantitative proteomic technique, that is, isobaric mass-tag labeling.

Mass-tag labeling (Fig. 2), for example, iTRAQ (isobaric tag for relative and absolute quantitation), SILAC (stable incorporation of labeled amino acids in culture) or SILAM (stable incorporation of labeled amino acids in mammals), allows the rapid ratiometric analysis of multiple peptides separated by multidimensional cation-exchange liquid chromatography (LC) identified with either time-of-flight (TOF) or linear ion-trap tandem mass spectrometry (LC-MS^2) with modified dissociation techniques such as PQD (20) and HCD (21). These instruments, and the diverse workflows they support, have in common that they both generate up to thousands of fragment ion spectra per hour of data acquisition. The assignment of these fragment ion spectra to peptide sequences, the inference of the proteins represented by the identified peptides and the determination of their abundances in the analyzed sample present complex computational and statistical challenges. It is important for the future use of MS and proteomics in metabolic signaling analysis to develop technological solutions to these issues that provide accurate and reproducible quantitative differential protein expression data. To this end, one of the major advances will be the application of accurate functional annotation and categorization into metabolic pathways of the protein sets created. As MS generally does not provide a *factor* identification process as reliable as microarrays, the physiological and rational

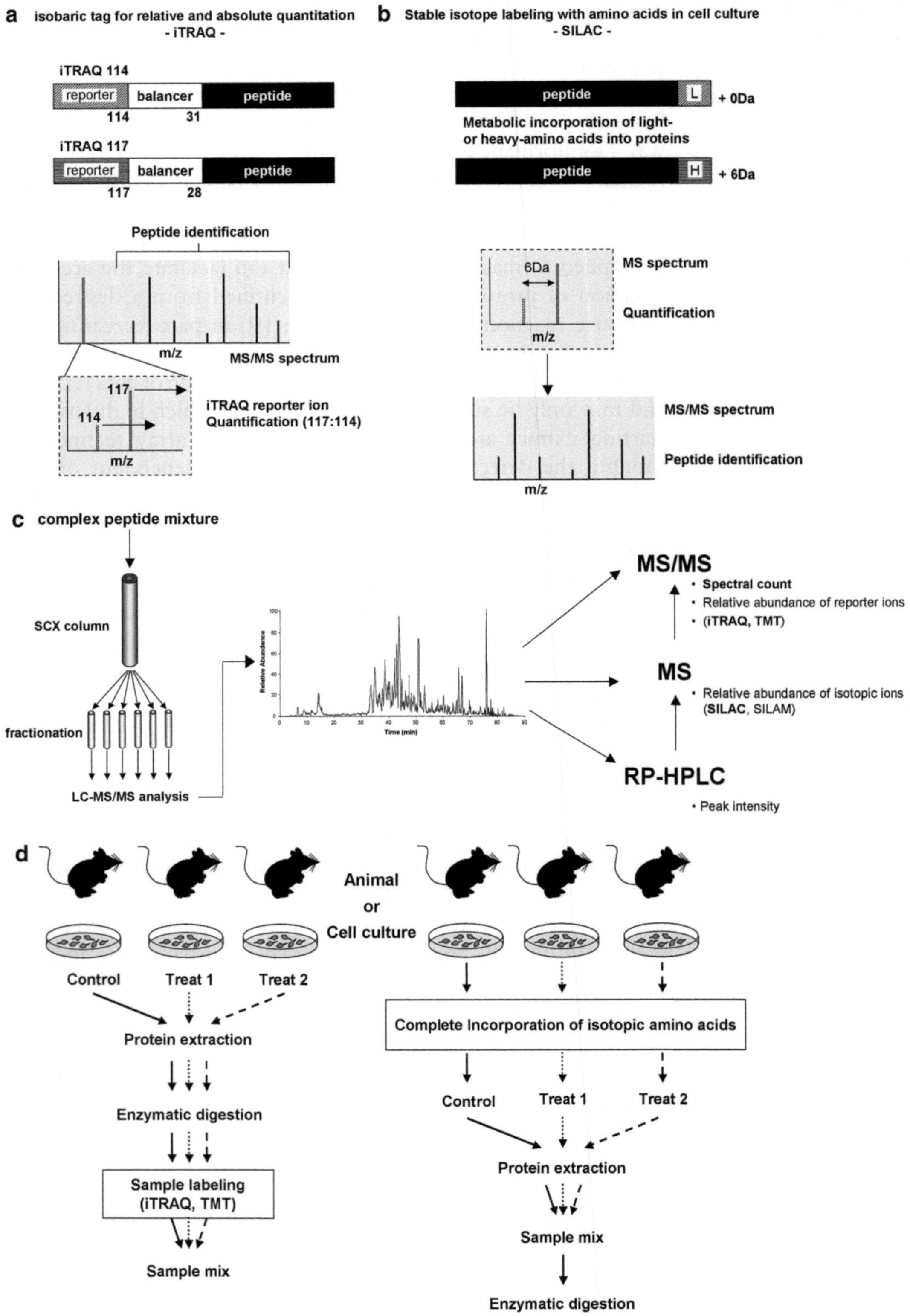
a isobaric tag for relative and absolute quantitation - iTRAQ -
iTRAQ 114
reporter balancer peptide
114 31
iTRAQ 117
reporter balancer peptide
117 28
Peptide identification
m/z
MS/MS spectrum
117
114
m/z
iTRAQ reporter ion Quantification (117:114)
b Stable isotope labeling with amino acids in cell culture - SILAC -
peptide L + 0Da
Metabolic incorporation of light- or heavy-amino acids into proteins
peptide H + 6Da
6Da
m/z
MS spectrum
Quantification
MS/MS spectrum
Peptide identification
m/z
c complex peptide mixture
SCX column
fractionation
LC-MS/MS analysis
Relative Abundance
Time (min)
MS/MS
• Spectral count
• Relative abundance of reporter ions
• (iTRAQ, TMT)
MS
• Relative abundance of isotopic ions (SILAC, SILAM)
RP-HPLC
• Peak intensity
d
Animal
or
Cell culture
Control
Treat 1
Treat 2
Protein extraction
Enzymatic digestion
Sample labeling (iTRAQ, TMT)
Sample mix
Complete Incorporation of isotopic amino acids
Control
Treat 1
Treat 2
Protein extraction
Sample mix
Enzymatic digestion

prediction of the signaling consequences of the protein streams will facilitate experiment to experiment comparison.

In contrast to array-based technologies, the primary concerns for MS-based dataset creation approaches involves the actual accurate identification of the proteins in the sample, for example, control versus test. For TOF and LC-MS^2 the identification of proteins in the sample is based upon fragmentation ion spectrum (MS^2-spectrum) of a specific peptide ion that is broken down into its constituent components in a gas-filled collision cell. Due to the enormous complexity of peptides composed of 20 amino acids, however, a large number of MS/MS spectra do not contain sufficient identity information to allow error-free peptide definition. In order to minimize false identification, a strict filtering criterion is required, which can be enforced, for example, by searching retrieved MS/MS spectra against a composite of both "target" and "decoy" (often reverse peptide alignments) sequence database (22). Much of the statistical manipulation used for protein datasets has focused upon the actual generation of the identified protein list rather than on the bioinformatic/pathway structure of the resultant data list itself. In recent years, however, with the advent of sophisticated automated identification software more attention is now paid to the physiological relevance of the mass datasets. The correct correlation and attribution of an MS^2-spectrum to its originating peptide sequence followed by eventual protein matching and identification is the first and central step in proteomic data processing. Numerous computational approaches and software tools have been developed to automatically assign candidate peptide sequences to fragment ion spectra, for example, SEQUEST, MASCOT, ProteinProspector, or

Fig. 2. Principle of isobaric mass-tags in quantitative mass spectrometry. (**a**) Several combinations of different-sized reporters of iTRAQ tags facilitate quantification of up to 8 different samples (masses from 113 to 121, excluding 120 as this corresponds to phenylalanine). Quantitative information is obtained from relative intensities of reporter ions in MS/MS spectrum. TMT (tandem mass tag: Thermo Electron Corporation) has the same property with iTRAQ but has different reporter and balancer chemistry. (**b**) In SILAC, isobaric amino acids are metabolically incorporated into all the cellular proteins. Animals can be fed and bred through multiple generations using feed with differential amino acid composition [SILAM: 60]. The equal amount of samples are combined and then applied to LC-MS/MS analysis. Quantitative information is obtained from relative intensities of light- and heavy-peptide ions in MS spectrum. (**c**) A representative analytical procedure of quantitative MS. In the bottom-up approach, complex peptide mixtures are fractionated through strong cation-exchange chromatography (SCX), which is essential for reducing sample complexity and increasing the number of identified peptides. Each fraction is analyzed through reverse-phase (RP) LC-MS/MS. For the nonisotopic study, quantitative information is obtained through peak intensity of specific peptides in ion chromatogram and more widely through counting finally matched MS/MS spectra and statistical manipulation. In case of using isobaric-tags, differentially labeled samples are combined before SCX chromatography. Quantitative information is obtained from MS or MS/MS spectrum, dependent on the property of isobaric tag. (**d**) Modes of sample preparation, labeling, and mixing for MS analysis. For mass-tag labeling procedures such as iTRAQ the individual extraction of proteins, then peptides from each sample is followed by individual mass-tag labeling and then mixing for single-run MS analysis. For stable isotope incorporation procedures, sufficient cell passages or animal generations in the presence of differential isotopes is required before mixing for single-run MS analysis.

ProbID (23–26). These computational approaches can involve database searching, where peptide sequences are identified by correlating acquired fragment ion spectra with theoretical spectra predicted for each peptide contained in a protein sequence database, or by correlating acquired fragment ion spectra with libraries of experimental MS^2 spectra identified in previous experiments. In addition de novo sequencing can also be used, where peptide sequences are explicitly read out directly from fragment ion spectra as well as hybrid computational approaches, such as those based on the extraction of short sequence tags of three to five residues in length, followed by "*error-tolerant*" database searching (27). For the majority of signal transduction laboratories, database searching remains the most frequently used peptide identification method. The use of MS-based techniques to identify quantitative protein profiles from animals/tissues has been excellently reviewed elsewhere (28–30) and therefore the focus of the rest of this overview is the predictive pathway analysis of mass datasets either from MS- or array-based experiments to appreciate *factor* expression at a network level.

While the accurate and unbiased collection of *factor* data is paramount, one extremely important caveat with respect to data retrieval and metabolic pathway analysis, is the need to physically retain both significant and nonsignificant *factor* data. The nature of the "nonsignificantly regulated" data may yet yield significance when the co-existence of related *factors* is analyzed using functional annotation-based bioinformatic strategies. Often subtle differences between experimental conditions may be missed as no individually dramatically modulated factors may present themselves. If, however, we consider our posit that metabolic and signaling functions are indeed composed of multiple interlaced network activities, the appreciation and functionally relevant correlation of these small changes with each other may illuminate a more realistic view of cellular physiology.

3. Bioinformatic Analysis of Quantitative Mass Analytical Datasets

With application of an initial data-filtering statistical analysis to each *factor* individually (compared to background), it is frequently the case that a large (100–1,000s) dataset of significantly regulated *factors* remains. In the first decade of mass biological data analysis only the highest and lowest regulated factors were often considered for further analyses. This approach, despite yielding some actionable data to describe the signaling function or physiological state under study, is often criticized for ignoring the correlated biological relevance of the multiple *factors* arranged in the large dataset that do not individually demonstrate significant differential regulation. Hence, we assume that genes and proteins function together and interact with each other in relevant groups and in specific microdomains but the analysis of the datasets often does not include this biologically vital information. However, if we consider that functional signaling responses or physiological states are the functional composite of multiple linked networks then an appreciation of the entire set in a mechanism analogous to signaling networks is needed. Gene-class, or pathway-level testing, integrates *factor* annotation and significance signaling pathway population tests (with geneset enrichment analysis) for coordinated changes at the system level. These approaches can both increase power for detecting differential *factor* expression and allow for a better understanding of the underlying biological processes associated with variations in signal transduction outcome. One of the earliest developed processes that allowed facile classification of *factor* function was Gene Ontology (http://www.geneontology.org/index.shtml) analysis.

3.1. Gene Ontology Classification

To create a rational and physiological/pharmacologically relevant appreciation of large datasets the first most reasonable goal is to look for methods in which to cluster the *factors* that are related to each other either by function, linkage in a metabolic process, or by subcellular localization. The number of these associations and the strength of observing multiple *factors* possessing the same associations within a large dataset provides the first level of "contextual" relevance of the mass dataset. An exemplar of the importance of elucidating common functional attributes for factors would be a protein such as actin, which conceivably may be directly involved in approximately 90% of all cellular processes either directly or distant by just one level from nearly all the *factors* in the dataset. To begin to appreciate what particular functional relevance the presence of actin has in one's dataset, the ability to look for functional groups in which to assign actin would start to narrow down the number of functional effects that the experimental changes in actin may be inducing. One of the primary levels of analysis of mass datasets to yield functional metabolic insights into its nature is the use of functional Gene Ontology (GO) analysis.

After many of the genomes of the major experimental eukaryotic organisms were fully sequenced, it became clear that a large majority of the genes controlling the fundamental biological processes and signaling pathways were common across multiple species. Therefore, an analytical method to allow inference and analogy of data between the diverse experimental organisms was required to potentially identify conserved signaling mechanisms. The GO project is an ongoing academic effort to address the need for consistent descriptions of gene products in different databases. The project began in 1998 as a collaboration between three model organism databases, FlyBase (*Drosophila:* http://flybase.org/), the *Saccharomyces* Genome Database (SGD: http://www.yeastgenome.org/) and the Mouse Genome Database (MGD: http://www.informatics.jax.org/). Since inception, the GO Consortium has grown to include many databases, including several of the world's major repositories for plant, animal, and microbial genomes. Functional biological knowledge is inherently complex and so cannot readily be integrated into existing databases of molecular (for example, sequence) data. An ontology is a formal way of representing knowledge in which concepts are described both by their meaning and their relationship to each other. Unique identifiers that are associated with each concept in biological ontologies (bio-ontologies) can be used for linking to and querying molecular databases.

The Gene Ontology Consortium (http://www.geneontology.org/GO.doc.shtml) was developed to provide a dynamic and controllable functional terminology syntax that can be used to accommodate the exponential increase in knowledge of *factor* connectivity in functional metabolic pathways. To initiate a mechanism by which *factors* (genes initially) could be associated with an expanding list of signaling functions, three major ontological databases were created, freely available on the internet (http://www.geneontology.org). These three databases would assist in assigning biologically relevant information to identified *factors* so that associations between functions and factors in a dataset can be ascertained and the relative significance of these within the dataset can be assessed. Biological Gene Ontology has two fundamental components: the ontologies themselves, which are the defined terms and the structured relationships between them (GO ontology), and the associations between gene products and the terms (GO annotations). GO provides both ontologies and annotations for three distinct areas of cell biology: molecular function, biological process, and cellular component or location.

3.2. Gene Ontology Categorization

The three main GO categories commonly used to cluster *factors* into related and biologically relevant groups are as follows: biological process (GObp), molecular function (GOmf), and cellular component (GOcc). Biological process, molecular function, and cellular component are all attributes of genes, gene products, or gene-product groups. Each of these may be assigned independently to *factors* in a dataset. The relationships between a given *factor* and biological process, molecular function, and cellular component are one-to-many, reflecting the biological reality that a particular protein may function in several processes, contain domains that carry out diverse molecular functions, and participate in multiple alternative interactions with other proteins, organelles, or locations in the cell. Within all of these three subgroups, there are hierarchies of GO terms ranging from extremely broad categories that can encompass hundreds of *factors* to GO terms that may only be associated with a handful of *factors*. An ontology comprises a set of well-defined terms with well-defined relationships. The ontological structure itself reflects the current representation of biological knowledge and therefore should be considered highly plastic and can act as a guide for organizing new data. Data can be annotated to varying levels depending on the amount and completeness of the available information. This flexibility also allows users to narrow or widen the focus of queries (31). The Gene Ontologies are formalized representations of current molecular and cellular biology knowledge. The GO ontology functional classification structure can be represented as a directed acyclic graph (DAG) in which the terms are nodes and the relationships among them are edges. Key characteristics of a DAG in the context of GO are that: parent–progeny relationships are defined, with parent terms representing more general biochemical functions than their progeny terms; and, unlike a simple tree (Fig. 3), a term in a DAG can have multiple parents. These characteristics of the GO structure facilitate facile grouping, searching, and analysis of multiple relevant *factors*.

GObp terms refer to biological objectives to which the *factor* contributes. The process is accomplished via one or more ordered assemblies of molecular functions. The specific functional processes often involve a chemical or physical transformation of a protein or a gene, for example, broad (high level) GObp terms are *"cell communication"* or *"negative regulation of cellular process."* Examples of more specific (lower level) process terms include, *"pyrimidine metabolism"* or *"cAMP biosynthesis"* and the most specific GObp terms include items such as "*cytoplasmic sequestering of transcription factor*" or "*protein import into mitochondrial matrix.*" GOmf terms are defined as a biochemical activity (including specific binding to ligands or structures) of an individual *factor*. This definition also applies to the capability that a *factor* carries as a potential. GOmf terms describe only what the *factor* can carry out without specifying where or when the biochemical event actually occurs. Examples of broad functional terms are "*enzyme*," *"transporter,"* or "*ligand*." Examples of narrower functional terms are *"Insulysin activity"* or *"Peptide YY receptor activity."* GOcc terms refer to the subcellular localization in the cell where the given *factor* is active. GOcc terms includes such terms as "*ribosome*" or "*proteasome*," *"nuclear membrane"* or *"Golgi apparatus"* specifying where multiple factors would be found. An important note, however, with respect to the usage of GO terms is the fact that due to the multispecies nature of their inception, GO terms may often not be fully transferable across species boundaries. Therefore, not all GO terms are applicable to all organisms; however, the full gamut of GO terminology is meant to be as inclusive as possible.

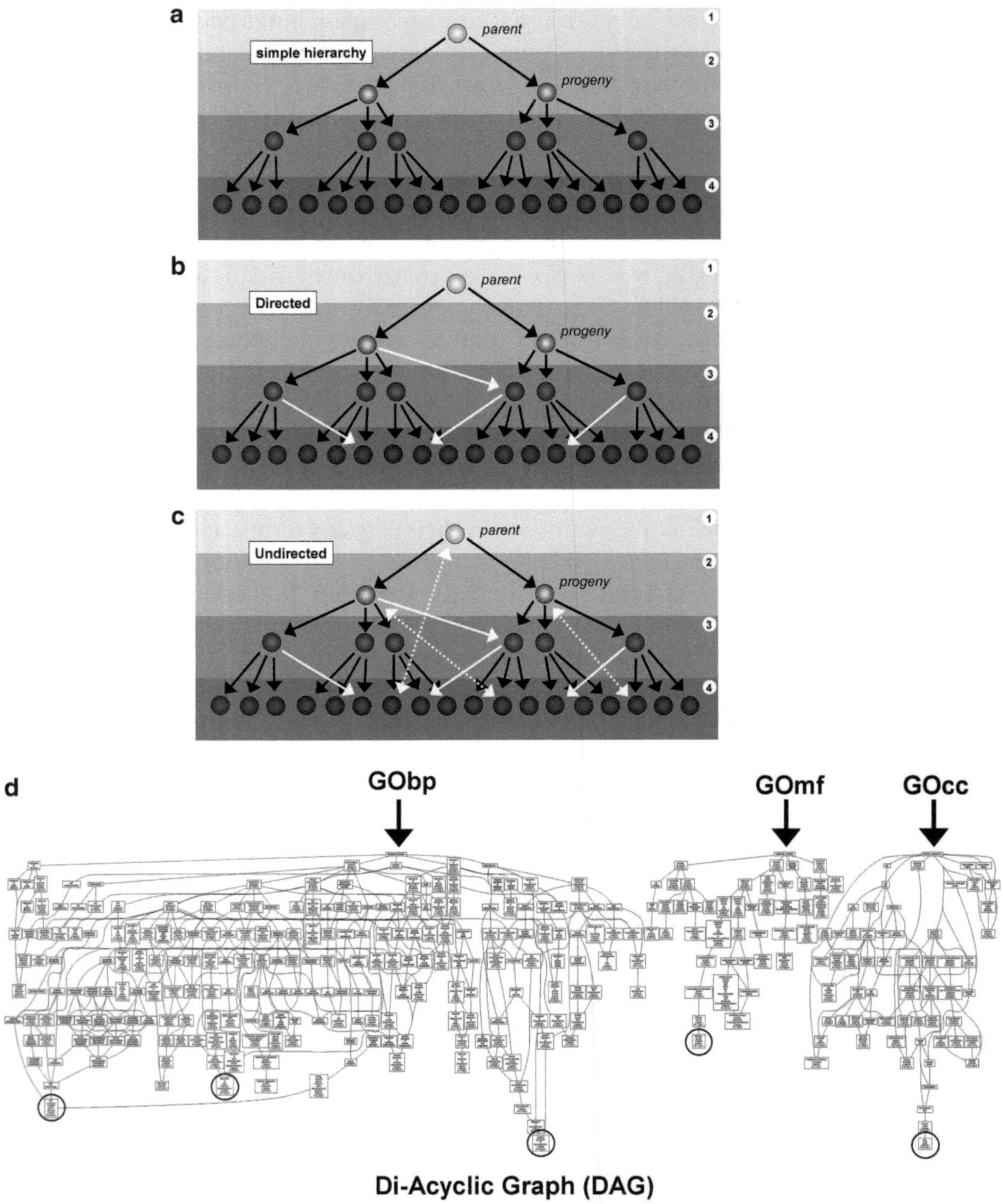

Fig. 3. Representation of ontological structures. Ontology of biologically relevant factors can be represented in a simple graphical structure in which parent Gene Ontology terms give rise to progeny terms (**a**). Parent terms are typically of a broad nature with their successive progeny possessing increasingly specific annotation (level **1** to **4**). This simple graphical ontology representation though can be governed by both directed and nondirected rules. Directed ontological relationships imply a classical hierarchical parent–progeny linking between the terms, that is, parent–progeny relationships are directed downward from less complex terms to more complex terms (*black arrows, panel A*). However, as broad-level parent terms may lead to multiple more specific ontological terms the simple one-parent one-progeny relationship may be less likely to reflect physiological systems than the one-parent multiple-progeny ontology (**b**). Undirected ontological representations, however, may allow nondirected progeny to parent relationships (**c**). Undirected representations may

3.3. Application of Gene Ontology Annotation

The GO project is currently one of the most widely used biological annotation databases for bioinformatic computational analyses. Upon interrogation of NCBI-Pubmed (http://www.ncbi.nlm.nih.gov/sites/entrez) there are currently over 2605 publications citing gene ontology as a crucial technique in functional signaling annotation, despite the first citation only occurring in 1997. GO annotation of datasets has been demonstrated to be vital for a variety of applications, for example, genome sequencing (32), network modeling (33), text data mining (34, 35), and for applied clinical situations (36). One of the first large-scale applications of GO term analysis of mass datasets was the creation of gene-GO term matrices, generating heatmap structures, to annotate sections of the *Drosophila Melanogaster* genome (37, 38). The ability to show increases in relevance (demonstrated by heatmap clusters) of certain GO terms ascribed to a subfamily of *factors* often represents the first level of revelation of the potential functional outputs of the experimental dataset (Fig. 4). The application of the appropriate GO terms to a dataset of significant factors is the first step in the process by which the statistical elucidation of the most likely clustering of the *factors* to a certain set of GO terms that can predict biologically relevant actions. There are now a plethora of excellent computational devices to achieve this first level of dataset functional analysis (Table 1). For the majority of the analytical tools indicated in Table 1, GO term annotation is used to analyze results from mass analytical techniques, primarily gene arrays but also more recently from quantitative proteomic studies. For these datasets, GO annotations are applied to greatly simplify and to determine which biological processes, functions and/or cellular locations are significantly over- or under-represented in the whole group of *factors*. This classification facilitates the determination of what new functions can be inferred on the basis of the data and how the given *factors* are distributed across a predefined set of biological GO term categories. As the primary goal of analysis of mass datasets is the revelation of physiologically/biologically relevant predictive functions that are distinct between the control and experimental scenarios, a quantitative assessment of the presence or absence of certain GO term groups is vital. The relative over- or under-representation of certain GO term groups can then be statistically assessed using various techniques.

Fig. 3. (continued) lead to cyclic closed relationship loops. If, however, all of the ontological relationships are directed then it is possible to represent biological linkages into a directed acyclic graph (DAG). (**d**) An example of an actual DAG from input signaling data. The three major classes of ontology (GObp, GOmf, GOcc) are shown. GO term specificity increases with descent into progeny branches of the DAG. Therefore, the most statistically significantly populated ontology terms are found in the lowest areas of the DAG diagram (e.g., circled GO term groups).

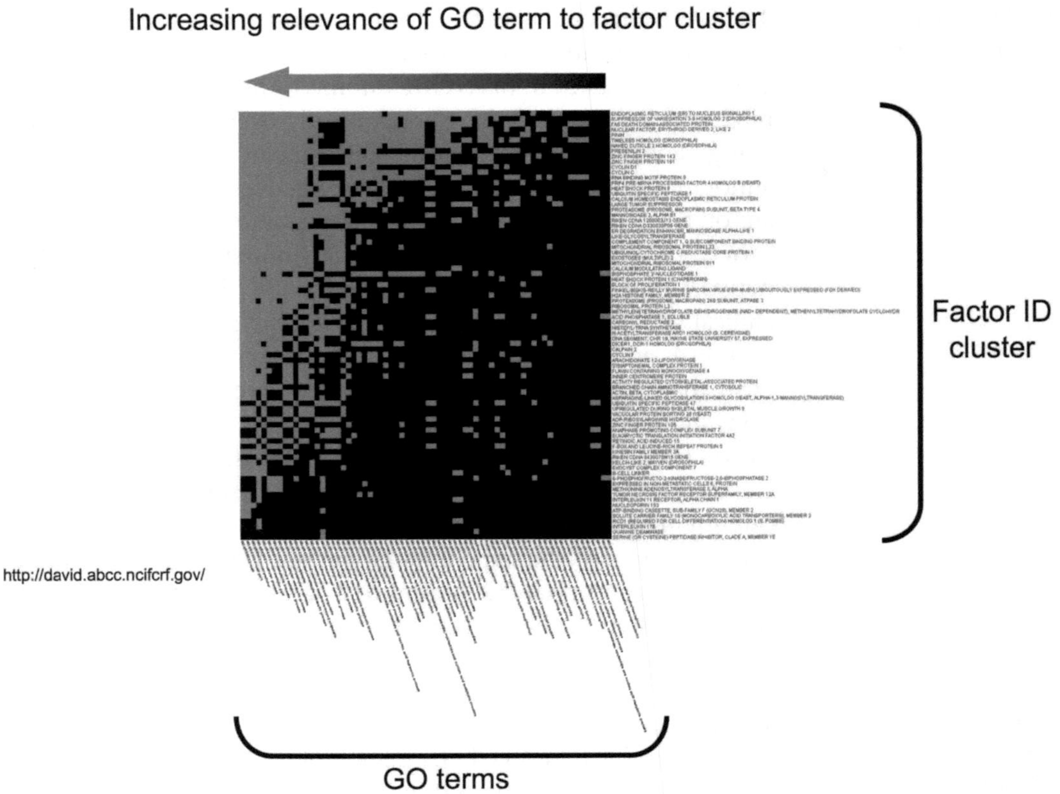

Fig. 4. Heatmap clustering for Gene Ontology annotation. Functional annotation of factor datasets using analytical tools such as DAVID (Database for Annotation, Visualization and Integrated Discovery: http://david.abcc.ncifcrf.gov/) allow the creation of visual factor heatmap clusters according to their most commonly descriptive GO terms. A large input dataset is broken down into smaller clusters that demonstrate commonality of related GO terms. The degree of correlation intensity between the input factors and the GO terms that most closely link the majority of the factors is demonstrated by the increased presence of correlating blocks (*grey*). Hence, in the figure depicted the GO terms (arranged *horizontally*) on the far left (end of *arrow*) are more likely to describe the functional output of the vertically arranged factor list.

3.4. Functional GO Term Enrichment and Categorization

Clustering of functionally correlated factors into common GO term groups can be used to infer which specific signaling functions the genes/proteins may be creating. The co-expression of these *factors* and the most common similarities in their functional common GO term annotation can demonstrate a potential predictive output of the dataset. The goal of mass analytical experimentation is the generation of differential datasets that, with variable isolation, can be linked to a biochemical function, physiological response, or even an organismal phenotype. This generation of a functional signaling "profile" of the dataset will allow correlation of *factor* expression to resultant function, with the most profoundly enriched *factor* clusters in the dataset being more reliably linked to the resultant output. Practically the "profile" of the dataset is often conducted by determining which GO terms are represented differently, in a significant fashion more or

Table 1
Computational programs for Gene Ontology term analysis of large datasets

Applications	URL
GO term retrieval	
AmiGO	http://amigo.geneontology.org/cgi-bin/amigo/go.cgi
CGAP GO browser	http://cgap.nci.nih.gov/
COBrA	http://www.xspan.org/
Comparative toxicogenomics database	http://www.mdibl.org/
DAVID	http://david.abcc.ncifcrf.gov/
DynGO	http://gauss.dbb.georgetown.edu/liblab/
Gene-class expression	http://gdm.fmrp.usp.br/
GeneInfoViz	http://www.utmem.edu/
GenNav	http://www.nlm.nih.gov/
GO consortium	http://geneontology.org
GOblet	http://www.molgen.mpg.de/
GoFish	http://llama.med.harvard.edu/
GONUTS	http://www.ecolicommunity.org/
MGI GO browser	http://www.informatics.jax.org/
Onto-express	http://vortex.cs.wayne.edu/projects.htm
Ontology evolution explorer (OnEX)	http://www.izbi.uni-leipzig.de/index.php
Ontology lookup service	http://www.ebi.ac.uk/
PANDORA	http://www.huji.ac.il/huji/eng/index_e.htm
QuickGO	http://www.ebi.ac.uk/QuickGO/
TAIR keyword browser	http://www.arabidopsis.org/
Tk-GO	http://www.illuminae.com/
GO term functional annotation	
Blast2GO	http://bioinfo.cipf.es/
g:Profiler	http://www.ut.ee/
GeneTools	http://www.microarray.no/index.php?section=1
GOanna	http://www.agbase.msstate.edu/
GoAnnotator	http://xldb.fc.ul.pt/
GOCat	http://eagl.unige.ch/GOCat/
GoPubMed	http://gopubmed.org/web/gopubmed/
GOtcha	http://www.compbio.dundee.ac.uk/Software/GOtcha/gotcha.html
InGOt (proprietary)	http://www.inpharmatica.co.uk/ingot/
InterProScan	http://www.ebi.ac.uk/Tools/InterProScan/
Manatee	http://manatee.sourceforge.net/
PubSearch	http://pubsearch.stanford.edu/
GO cluster analysis	
BiNGO	http://www.psb.ugent.be/cbd/papers/BiNGO/
CLASSIFI	http://pathcuric1.swmed.edu/pathdb/classifi.html
CLENCH	http://www.stanford.edu/~nigam/cgi-bin/doku-wiki/doku.php?id=clench
ClueGO	http://www.ici.upmc.fr/cluego/
DAVID	http://david.abcc.ncifcrf.gov/
EASE	http://david.abcc.ncifcrf.gov/content.jsp?file=/ease/ease1.htm&type=1

(continued)

Table 1 (continued)

Applications	URL
eGOn v2.0	http://www.genetools.microarray.ntnu.no/common/intro.php
ermineJ	http://bioinformatics.ubc.ca/ermineJ/
FIVA	http://bioinformatics.biol.rug.nl/standalone/fiva/
FuncAssociate	http://llama.med.harvard.edu/cgi/func/funcassociate
FuncExpression	http://www.plexdb.org/plex.php?database=Barley/funcexpression.php
FunCluster	http://corneliu.henegar.info/FunCluster.htm
FunNet	http://www.funnet.info/
G-SESAME	http://bioinformatics.clemson.edu/G-SESAME/
GENECODIS	http://genecodis.dacya.ucm.es/
GFINDer: genome function	http://www.medinfopoli.polimi.it/GFINDer/
GOALIE	http://bioinformatics.nyu.edu/Projects/GOALIE/
GOdist	http://basalganglia.huji.ac.il/links.htm
GOEAST	http://omicslab.genetics.ac.cn/GOEAST/
Gene ontology explorer (GOEx)	http://pcarvalho.com/patternlab/goex.shtml
GoMiner and MatchMiner	http://discover.nci.nih.gov/gominer/htgm.jsp
GOrilla	http://cbl-gorilla.cs.technion.ac.il/
GOstat	http://gostat.wehi.edu.au/
GoSurfer	http://bioinformatics.bioen.uiuc.edu/gosurfer/
GOTM (gene ontology tree machine)	http://bioinfo.vanderbilt.edu/gotm/
GOToolBox	http://burgundy.cmmt.ubc.ca/GOToolBox/
GraphWeb	http://biit.cs.ut.ee/graphweb/
L2L	http://depts.washington.edu/l2l/
MAPPFinder	http://www.genmapp.org/
MetaGP	http://metagp.ism.ac.jp/
MultiExperiment viewer	http://www.tm4.org/mev/
The ontologizer	http://compbio.charite.de/index.php/ontologizer2.html
Probe explorer	http://probeexplorer.cicancer.org/principal.php
ProfCom	http://webclu.bio.wzw.tum.de/profcom/
SeqExpress	http://www.seqexpress.com/
SerbGO	http://estbioinfo.stat.ub.es/apli/serbgov131/index.php
Source	http://smd.stanford.edu/cgi-bin/source/sourceSearch
STEM: short time-series expression miner	http://www.cs.cmu.edu/~jernst/stem/
T-Profiler	http://www.t-profiler.org/
THEA	http://thea.unice.fr/index-en.html

less often than expected by chance within the *factor* set compared to say their expression in a reference set (39–42). The most commonly applied approach for this is the calculation of "enrichment" for each GO term (i.e., a higher proportion of factors with certain common annotations among the differentially expressed factors than among all of the background factors in the study). The main problem here is that any enrichment value can occur just by chance. Therefore, enrichment alone should not be interpreted as unequivocal evidence implicating the GO term in the phenomenon studied without application of an appropriate statistical test. More sophisticated approaches calculate the probability of observing a particular enrichment value just by chance using a binomial model (43). This is a good approximation for large reference sets (e.g., whole-genome microarrays). However, it has been demonstrated that in many practical examples, better-suited models include the hypergeometric distribution or the Chi-squared (44) distribution, both of which take into consideration how the probabilities change when a *factor* is picked. More recent approaches perform the analysis while considering information about the relative position of the GO terms in the hierarchical tree (Fig. 3, 45–47).

Two types of questions can be addressed when performing functional GO term profiling: hypothesis-generating queries, for example, "which GO terms are significant in a particular set of factors?" or hypothesis-driven queries. An unbiased search for significant GO term associations can be performed with a standard "bottom-up" approach: for every progeny GO term, *p*-values for the *factors* are directly associated with it. If any term is significant, then analysis is not propagated to *factors* above it in the hierarchy. This would provide the most specific node that is significant in that particular DAG branch. If a term is not significant, the annotations are propagated to its parent and are recalculated with the parent term. The *factor* analysis will then propagate upward until a significant node is found or until the root is reached. To minimize false discovery rates, it may be more prudent in the future to precollapse many of the possible DAG branches to prevent "overtesting" of the dataset. To do this, a specific section of the tree organization may be reduced before any *p*-values are calculated, on the basis of the biological hypotheses tested. Unfortunately, most tools that are currently available are limited to performing analysis either at a fixed depth or with all nodes, thus preventing the customized collapsing of the GO that could improve significance in most circumstances. However, one of the more recently developed GO term analytical tools, QuickGO, was created to specifically facilitate this form of flexible analysis (31). QuickGO (http://www.ebi.ac.uk/QuickGO) allows users to individually tailor annotation sets using multiple filtering options as well as to construct specific and targeted subsets of the GO terms, called "*GO slims*" to "map-up" annotations allowing a general overview of the attributes of a set of *factors*. Collections of initial enriched GO terms primary dataset analysis can then be employed to construct a desired *GO slim* analytical subset. Broad "first pass" analysis annotations can then be "mapped up" or "slimmed" to these selected GO terms. Predetermined *GO slims* created by groups in the GO Consortium can also be used.

However, it is likely for anything other than primary discovery analysis that the majority of users in the future will be primarily interested in using their personal *GO slims* based on empirical data from other experimental sources.

Another common application of GO is to categorize genes on the basis of a relatively small set of heavily *factor*-populated high-level GO terms. Results of the functional categorization are frequently shown as pie charts or bar charts (48) based on the number or *p*-value of the *factors* present in that GO term group from the primary dataset. This involves the mapping of a set of annotations for the *factors* of interest to a specified subset of high-level GO terms. This is a typical way of providing an overview of the broad biology encoded by a differential expression patterns (49).

4. Geneset Enrichment and Pathway Analysis

While GO-based annotation techniques provide an excellent appreciation of the biologically relevant biases in a dataset there are additional, more in-depth, formats that can be applied to mass datasets. For example, analysis can be focused upon individual chemical molecular activity, promoter and regulatory network analysis, or by employing the vast-accumulated knowledge from the literature to carry out metabolic signaling pathway analysis. Signaling pathway analysis focuses on physical and functional interactions between factors within a preset signal transduction framework rather than taking the *factor*-centered view of GO-based database analyses (50). The simplest forms of pathway analysis analyze the distribution of *factors* within the dataset into precompiled functional signaling pathways in order to elucidate the most likely functional signaling relationships between the individual *factors* in the dataset. This is typically conducted using a process termed geneset enrichment analysis (GSEA). As this was primarily developed for genomics, the term GSEA has remained although this can be directly applied to proteomic data as well. GSEA typically employs predefined *factor* sets to identify significant biological changes in microarray/proteomic datasets. The *EcoCyc* database was perhaps one of the first computational attempts to methodically apply pathway analysis (51, 52). There are various efforts aimed toward the establishment of an accepted standard or ontology to represent functional pathway data. Defined signaling pathways usually include three major classes, (1) the molecules involved in the pathways, (2) the chemical reactions in which these molecules are involved, and (3) the location of the reactions. A pathway ontology should not only represent all these three classes of data, but also capture the intricate relationships among them. For example, a molecule can be related to a reaction as a reactant or a product. The transition from a reactant to a product can be affected by another molecule called a modifier. The modifier can exert various effects to the transition, such as catalysis, stimulation, inhibition, or modulation. Furthermore, the relationship between reactions and cellular components describes the location of these reactions. Such a higher level of functional correlation cannot be adequately captured using GObp as it does not capture all the dynamic inter-relationships in the pathways.

4.1. Statistical Analysis of Pathway Enrichment

Pathway enrichment analysis is a statistical approach used to discover a statistically significant representation of a functional pathway class within a selection of *factors* from a heterogeneous *factor* population. Enrichment analysis can be applied in any situation where important physiological/pharmacological activity is suspected in the choice of a subset of members from a reference dataset. Enrichment analysis requires calculations on thousands of sets against thousands of candidate classifiers, generating often large output datasets containing both significant and nonsignificant data. There are multiple freely available pathway databases and facile calculation programs now able to facilitate these computational issues for molecular biologists (Table 2). As with GO term analysis, there are several important issues to consider with respect to the enrichment analysis. The appropriate choice of the reference dataset with which the experimental dataset is compared is vital. Unlike many simple statistical algorithms for accurate enrichment analysis, the accommodation of nonindependent association of factors is required. This allows empirically known physiological interactions to be included into the enrichment inference. In addition, as with GO term analysis, multiple-testing errors need to be accounted for as lack of independence among *factor* classifiers (seen in many datasets), for example, the hierarchical organization of multiple ontologies, often complicates estimation of false discovery. A simple paradigm for the statistical elucidation of enrichment analysis for a given signaling pathway is depicted in Fig. 5. As with all technological applications subsequent iterations and developments can quickly surpass previous techniques. For example, in recent years the use of simple GSEA has been largely replaced by a parametric version of this process (PAGE, parametric geneset enrichment analysis; 53). GSEA employs a distribution-free, nonparametric approach to the analysis of the significance of population (normally at least two factors in each pathway are required for effective "population" of that pathway) of signaling pathways by the input dataset. PAGE and other parametric GSEA tools use a *Central Limit Theorem*, which states that "*when the sampling size is large*

Table 2
Computational programs for signaling and metabolic pathway analysis of large datasets

Applications	URL
Signaling pathway databases	
BBID	http://bbid.grc.nia.nih.gov/
BioCarta	http://www.biocarta.com/genes/index.asp
BioModels – biomodels database	http://www.ebi.ac.uk/biomodels-main/
DOQCS – database of quantitative cellular signaling	http://doqcs.ncbs.res.in/
DSM – dynamic signaling maps	http://www.hippron.com/hippron/index.html
eMIM – electronic molecular interaction map	http://discover.nci.nih.gov/mim/index.jsp
GeneNet – genetic networks	http://wwwmgs.bionet.nsc.ru/mgs/gnw/genenet/
GenMAPP – gene microarray pathway profiler	http://www.genmapp.org/
GON – genomic object net	http://genome.ib.sci.yamaguchi-u.ac.jp/~gon/index.html
HCPIN – human cancer protein interaction network	http://nesg.org:9090/HCPIN/
INOH – integrating network objects with hierarchies	http://www.inoh.org/
JWS online – online cellular systems modeling	http://jjj.biochem.sun.ac.za/
KEGG	http://www.genome.jp/kegg/pathway.html
Millipore pathways	http://www.millipore.com/pathways/pw/pathways
NetPath	http://www.netpath.org/
PANTHER – protein analysis through evolutionary relationships	http://www.pantherdb.org/
PC – pathway commons	http://www.pathwaycommons.org/pc/
PDS – pathways database system	http://nashua.case.edu/pathwaysweb/
PID – NCI-nature pathway interaction database	http://pid.nci.nih.gov/
pSTIING	http://pstiing.licr.org/
Reactome – reactome knowledgebase	http://www.reactome.org/
RGD – rat genome database pathway resource	http://rgd.mcw.edu/wg/pathway
ROSPath – reactive oxygen species related signaling pathway	http://rospath.ewha.ac.kr/
Signaling gateway – UCSD-nature signaling gateway	http://www.signaling-gateway.org/
SigPath – signaling pathway information system	http://icb.med.cornell.edu/crt/SigPath/index.xml
SMPDB – small molecule pathway database	http://www.smpdb.ca/
SPIKE – signaling pathway integrated knowledge engine	http://www.cs.tau.ac.il/~spike/
STCDB – signal transduction classification database	http://bibiserv.techfak.uni-bielefeld.de/stcdb/
TRMP – therapeutically relevant multiple pathways database	http://bidd.nus.edu.sg/group/trmp/trmp_ns.asp
TRRD – transcription regulatory regions database	http://wwwmgs.bionet.nsc.ru/mgs/gnw/trrd/
WikiPathways – WikiPathways	http://wikipathways.org/index.php/WikiPathways

(continued)

Table 2 (continued)

Applications	URL
Metabolic pathway databases	
aMAZE – protein function and biochemical pathways project	http://www.amaze.ulb.ac.be/
BioCyc – biocyc knowledge library	http://biocyc.org/
BioModels – biomodels database	http://www.ebi.ac.uk/biomodels-main/
Biopath – biochemical pathways database	http://www.molecular-networks.com/databases/biopath
BRENDA – braunschweig enzyme database	http://www.brenda-enzymes.info/
CellML repository – CellML model repository	http://models.cellml.org/
CPDB – ConsensusPathDB	http://cpdb.molgen.mpg.de/
ERGO – ERGO genome analysis and discovery system	http://www.ergo-light.com/
ExPASy biochemical pathways	http://www.expasy.org/tools/pathways/
GeneNet – genetic networks	http://wwwmgs.bionet.nsc.ru/mgs/gnw/genenet/
HMDB – human metabolome database	http://www.hmdb.ca/
HumanCyc – encyclopedia of homo sapiens genes and metabolism	http://humancyc.org/
IntEnz – integrated relational enzyme database	http://www.ebi.ac.uk/intenz/index.jsp
LIGAND – database of chemical compounds and reactions	http://www.genome.jp/ligand/
MetaCyc – metabolic pathway database	http://metacyc.org/
MetNetDB – metabolic network exchange	http://www.metnetdb.org/MetNet_db.htm
MouseCyc – mouse pathway database	http://mousecyc.jax.org/
NetBiochem – medical biochemistry resource	http://library.med.utah.edu/NetBiochem/NetWelco.htm
PathCase – CASE pathways database system	http://nashua.cwru.edu/PathwaysWeb/
PATRIC – PathoSystems resource integration center	http://patric.vbi.vt.edu/
PharmGKB	http://www.pharmgkb.org/
Pathway analytical applications	
Ariadne genomics: pathway studio	http://www.ariadnegenomics.com/pathway-studio/
ArrayXPath	http://www.snubi.org/software/ArrayXPath/
Biochip core laboratory – CRSD	http://140.120.213.10:8080/crsd/
Cpath	http://cbio.mskcc.org/software/cpath/
D-GEM (disease-to-gene expression mapper)	http://dgem.cs.iupui.edu/
ErmineJ	http://www.bioinformatics.ubc.ca/ermineJ/
Gene set enrichment analysis – molecular signatures database	http://www.broadinstitute.org/gsea/
GeneTrial	http://genetrail.bioinf.uni-sb.de/

(continued)

Table 2 (continued)

Applications	URL
GenMAPP	http://www.genmapp.org/
Genome expression pathway analysis tool	http://gepat.sourceforge.net/
Ingenuity pathway analysis	www.ingenuity.com/
KEGG pathway database	http://www.genome.jp/kegg/pathway.html
KOBAS: KO-based annotation system	http://kobas.cbi.pku.edu.cn/
Onto-express – intelligent systems and bioinformatics laboratory	http://vortex.cs.wayne.edu/ontoexpress/
PathExpress	http://bioinfoserver.rsbs.anu.edu.au/utils/PathExpress/
PathJam – biological pathway integration tool	http://www.pathjam.org/
Pathway miner – genes and their pathways	http://www.biorag.org/index.php
PROPA: probabilistic pathway annotation	http://www.stat.duke.edu/research/software/west/propa/
VisANT: an integrative platform for network/pathway analysis	http://visant.bu.edu/
WebGestalt: Web-based gene set analysis toolkit	http://bioinfo.vanderbilt.edu/webgestalt

enough, distribution of an average of sampled observations is normal regardless of the nature of parent distribution." Statistical PAGE analysis intentionally directs the analysis of predefined signaling pathways in datasets rather than of individual *factors*. To generate easy to appreciate data with respect to differential metabolic/physiological states, PAGE uses the fold change between the control and experimental groups to calculate Z-scores of the predefined gene sets (various database sources can be used) and normal distribution to assign statistical significance to the gene sets (53). The list of all of the factors used in the dataset and their Z-scores are put into the analysis and Z-scores are assigned to the functional signaling sets within each experimental group. Traditional large dataset analysis requires that individual genes have significantly different expression levels in order for them to be considered differentially regulated. PAGE specifically takes into account that *factors* are both co-regulated and co-present, to help populate discrete signaling pathways. Therefore, it is possible that factors individually may not be significantly regulated above or below baseline, but significant regulation of pathways can be generated by such *factors* by grouping them significantly into the predefined signaling sets. By looking at groups of *factors* involved in a specific function, significant differences between their relative population may represent a biologically meaningful result. The

a

$$k_e = \left(\frac{n}{m}\right) \times j$$

b

$$r = \frac{k}{k_e}$$

c

$$P = \sum_{i=k}^{n} \frac{\binom{m-j}{n-i}\binom{j}{i}}{\binom{m}{n}}$$

d

$$P = \sum_{i=k}^{n} \frac{\binom{n}{i}\binom{m}{j+k-i}}{\binom{m+n}{j+k}}$$

e

$$c > \left(\frac{d}{b}\right) \times a$$

f

$$P = \sum_{i=c}^{d} \frac{\binom{b-a}{d-i}\binom{a}{i}}{\binom{b}{d}}$$

g

$$c < \left(\frac{d}{b}\right) \times a$$

h

$$P = \sum_{i=0}^{c} \frac{\binom{b-a}{d-i}\binom{a}{i}}{\binom{b}{d}}$$

Fig. 5. Functional *factor* enrichment. To identify functional categories with significantly enriched *factor* numbers within the input experimental dataset comparison is needed between the input dataset with a reference dataset. The input dataset needs in this case to be a subset of the reference dataset. For a theoretical scenario we may have *n* factors in the experimental dataset (**a**) and *m* factors in the reference dataset (**b**). For a given functional category of interest (e.g., a KEGG signaling pathway, **c**) there may be *k* number of factors from A and *j* number of factors from B. Based on the reference dataset (**b**) the expected value of *k* (k_e) is depicted in *panel A*. If *k* exceeds k_e then the specific category C is said to be enriched. Derivation of the index of the degree of pathway C enrichment (*r*) in the experimental dataset A is depicted in *panel B*. Analysis of the significance of the enrichment of pathway C in dataset B compared to dataset A, using a hypergeometric test is demonstrated in *panel C*. If, however, datasets A and B are independent, a Fisher's exact test may be more appropriate (**d**). Advanced pathway analysis software such as WebGestalt also allow the user to reduce their scope of pathway analysis in a similar manner to *GO slims*, for example, inspecting tissue-specific enrichment. For another *factor*, there may be *d* examples of a selected *factor* in all tissues and *b* examples for all factors in all tissues. In addition, if there are *c* number of a selected *factor* in a selected tissue and *a* number of all factors in that tissue, the over-representation of the specific *factor* in that tissue can be calculated as depicted (**e**). Calculation of the significance of over-representation in the specific tissue is depicted in *panel* (**f**). Mathematical under-representation of the specific *factor* in the selected tissue is described by the equation in *panel* (**g**) with the significance of the under-representation denoted in *panel* (**h**).

polarity (up or downregulated) of the respective PAGE signaling pathway is determined by the sum of the Z-scores of the *factors* present in the experimental dataset that then fall into the set of *factors* used to describe the predetermined signaling pathway.

4.2. Pathway Analysis Applications

GSEA is especially powerful for the largest datasets that will have an increased likelihood of retrieved *factor* identity variation between experiments (especially the case for MS-based proteomics) or when there are subtle differences between control and experimental paradigms. With respect to the latter issue, a specific example of the power of GSEA techniques was the successful demonstration of prediction of significant metabolic pathway activation (oxidative phosphorylation) from a human dataset in which no one single gene out of 20,000 tested yielded an individually significant perturbation between control and diabetic patient muscle tissue (54). Thus the ability to apply significance of predicted functional output no longer rests upon individual *factors* but on co-expression and coherent regulation of these factors, reflecting the coordinated, interconnected nature of metabolic pathways themselves. Therefore, across diverse samples the signaling functionality can be correlated even if the identity of the regulated *factors* are not identical but still fall within the same functional preset pathway. Such flexibility is crucial for the analysis of MS-based quantitative proteomic data as the detection of exactly the same stream of proteins is highly unlikely over what can be long term experiments (10–20 h of run time).

In complex biological systems, coordinated metabolic functions are created by the summation of multiple interconnected pathways forming networks of varying sizes and relative importance. Using statistical processes to specifically search for these may greatly expand our understanding of the subtleties of disease processes or drug responses. Not only can these techniques be used for the investigation of dynamic experimental responses but may also illuminate how cells/tissues/animals react in response to spontaneous disease or genetically implied pathophysiology (48). Hence, not only may "disease-causing" networks of *factors* exist; "disease-management" *factor* networks are also likely as flexible and reactive biological systems attempt to ameliorate perturbations and achieve homeostasis.

With respect to the practical implementation of pathway analysis for large datasets there are multiple excellent databases of precompiled pathways available for pathway analysis as well as freely accessible software applications to perform the analysis (Table 2). However, not all signaling pathways are equally suitable for various experimental paradigms. For example, metabolic signaling pathways are controlled to a large extent by protein-based events that are not observable on microarrays as only steady-state levels of mRNAs are monitored. Kinase-based signaling cascades also do not necessarily involve changes in mRNA levels. The best case for microarray-based pathway analysis is transcriptional-signaling pathways that are directly coupled to de novo transcription. One of earliest developed tools for pathway analysis is the GenMAPP tool (55) that allots *factors* to preset pathways, as well as allowing user-based pathway generation. There are many excellent Web-based Pathway analysis tools such as Pathway Miner that provides ranking of the gene/pathway groups via a Fisher's exact test on top of the gene–pathway association analysis (56) and WebGestalt, that can generate GO DAG diagrams as well as KEGG and BioCarta pathway enrichment analysis (57). An example of the practical workflow and functioning of pathway analysis tools (e.g., WebGestalt) is depicted in Fig. 6. An extensive list of available programs is listed in Table 2. These tools often share similar lists of signaling pathways consisting of the relative

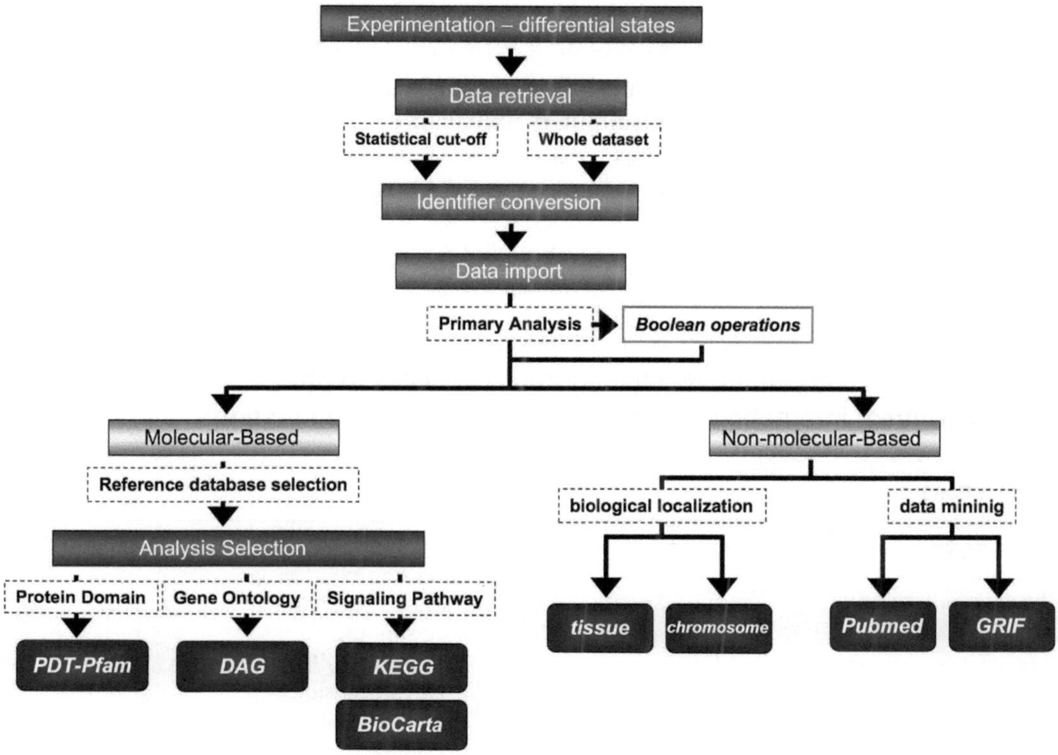

Fig. 6. Archetypical Pathway Analysis workflow. A typical flow of information processing to create a metabolic signaling output pathway using the WebGestalt analytic process is demonstrated in a series of logical steps. After data retrieval from mass analytical techniques primary statistical analysis can be employed using empirically derived cutoffs or whole-dataset data may be used instead. After uploading, the data can be converted to various identifiers, for example, Locus Links, Uniprot, or Unigene symbols. The software allows simple dataset Boolean operations as well before the two major forms of dataset analysis, that is, molecular or non-molecular-based. Non-molecular-based analyses include the investigation of enriched *tissue* or *chromosome*-specific expression of factors in the dataset. In addition, *Pubmed* (Gene-Association publication database) or *GRIF* (Gene expression into Function: http://generifs_basic.gz) Tables demonstrate co-expression of various factors in the dataset within the same publications. Multiple forms of biological signaling information can also be generated in parallel to these outputs. With selection of appropriate comparative base datasets (built-in) statistical enrichment of factors in the primary dataset into protein domain tables (Pfam: http://pfam.sanger.ac.uk/), directed acyclic gene ontologies (*DAG*) or discrete *KEGG*/*BioCarta* signaling pathways is determined.

factors allotted to them based on meta-literature searches. Again, as with the analytical tools themselves there are multiple sources of rationally created signaling and metabolic pathways. Some of the most commonly employed are the KEGG database (http://www.genome.jp/kegg/pathway.html) of metabolic and signaling pathways (58), the BioCarta database (http://www.biocarta.com/genes/index.asp) and the excellent and authoritative MIT/Harvard Broad Institute Molecular Signatures Database (MsigDB: http://www.broadinstitute.org/gsea/msigdb/). All of these databases provide easy open access to the pathways and associated diagrams for use with geneset enrichment software. In addition to these excellent resources for metabolic pathway analysis, correlated investigational technologies employing similar methodologies of functional inference are now widely used for transcription promoter analysis, protein–protein interaction and resultant mammalian phenotype prediction (Table 3). These analysis modules can often be used to supplement and support findings derived from GO and signaling pathway analysis.

Table 3
Databases and computational tools for mass analysis of promoter activity, protein–protein interaction and mammalian phenotype annotation

Applications	URL
Transcriptional promoter databases/tools	
DBTBS	http://dbtbs.hgc.jp/
TRED	http://rulai.cshl.edu/cgi-bin/TRED/tred.cgi?process=home
Mammalian promoter database	http://rulai.cshl.edu/CSHLmpd2/
Eukaryotic promoter database	http://www.epd.isb-sib.ch/
DBTSS	http://dbtss.hgc.jp/index.html
Jaspar	http://jaspar.cgb.ki.se/cgi-bin/jaspar_db.pl
TRRD	http://www.mgs.bionet.nsc.ru/mgs/gnw/trrd/
cisRED	http://www.cisred.org/
Protein interaction databases/tools	
STRING	http://string.embl.de/
MIPS	http://mips.helmholtz-muenchen.de/proj/ppi/
HPID	http://165.246.44.48/hpid/webforms/intro.aspx
EMBL-EBI-IntAct	http://www.ebi.ac.uk/intact/main.xhtml
BioGrid	http://www.thebiogrid.org/
DIP	http://dip.doe-mbi.ucla.edu/dip/Main.cgi
HUGE ppi	http://www.kazusa.or.jp/huge/ppi/
KEGG BRITE	http://www.genome.jp/brite/brite.html
MINT	http://mint.bio.uniroma2.it/mint/Welcome.do
PRIME	http://prime.ontology.ims.u-tokyo.ac.jp:8081/
SNAPPIView	http://www.compbio.dundee.ac.uk/SNAPPI/downloads.jsp
PPID	http://www.anc.ed.ac.uk/mscs/PPID/
Reactome	http://www.reactome.org/
Mammalian phenotype databases/tools	
Jackson labs mouse genome database	http://www.informatics.jax.org/
Phenomics	http://www.phenomicDB.de
Polydoms	http://polydoms.cchmc.org/polydoms/
Rat genome database	http://rgd.mcw.edu/
Phenotype and trait ontology (PATO)	http://www.obofoundry.org/
HUGE navigator	http://www.hugenavigator.net/
GenomeWeb	http://www.biologie.uni-hamburg.de/b-online/library/genomeweb/comp-gen-db.html
IKMC	http://www.knockoutmouse.org/

5. Future Aspects for Signaling Pathway Analysis

The combined employment of mass data collection and signaling pathway analytical tools is likely to revolutionize signal transduction research in the next several decades. The ability to accurately appreciate and perhaps predict a global cellular impact of physiological or pharmacological perturbations may facilitate an understanding of disease etiology and eventual drug control of disease at the level of the *factor* network rather than the linear signaling pathway level. The appreciation of a network hypothesis for biological activity presents many important new avenues for signal transduction and pharmacological research. For example, the ability to identify "keystone" *factors* within a network that exert the most profound actions upon the state of a given pathological network may facilitate the creation of indirect pharmacological strategies. Such agents may be able to ensure a profound regulation of the keystone factors via modulation of multiple parts of the signaling network that have subsequent synergistic actions upon the keystones. These agents may be therefore more efficacious in smaller doses as their effects are amplified greatly by the reinforced network before hitting the keystone itself. In addition, as they may be inducing regulation of the network keystone through multiple mechanisms, such therapeutics may be more resistant to the development of desensitization, tolerance, or resistance. Hence these agents may present a polypharmacological network profile, but through careful knowledge-based design may effectively result in a more discrete resultant phenotypic action.

One important consideration of signaling pathway analysis that is often overlooked is the huge potential for temporal plasticity in signaling networks. The majority of mass analytical datasets are usually "snapshots" in time, as the expense of gaining multiple, temporally distinct, datasets is currently prohibitive. However, as the cost of mass analysis is likely to be reduced, our conversion of signaling pathways from rigid to plastic will undoubtedly assist in the greater appreciation of how signaling systems are integrated to form the basis of complicated physiological states and also drug responses. An understanding of the therapeutic at effective temporal windows may increase the potentiation of drug efficacy, again allowing a potential reduction in applied dose, thus minimizing side-effects or contra-indications. At a very crude level we are already demonstrating such a temporal drug response concept by the use of "chronotherapeutics" for anti-cancer drugs (59).

In conclusion, it is clear that the relentless increase in the intricacy of our understanding of molecular signaling has presented many challenges both in technological methodology and in computational analysis. Our ability to combine these two approaches for diagnostic and predictive capacities will only serve to improve our appreciation of disease pathophysiology and the mechanism of action of pharmacological agents. Appreciating these two coordinated factors at a systemic network level may allow the generation of far more efficacious and better-tolerated drug treatments for a wide variety of diseases and pathophysiological states.

Acknowledgments

This work was supported entirely by the Intramural Research Program of the NIH, National Institute on Aging.

References

1. Luttrell, L. M. (2003) "Location, location, location": activation and targeting of MAP kinases by G protein-coupled receptors. *J Mol Endocrinol* **30**, 117–26.
2. Maudsley, S., Martin, B. and Luttrell, L. M. (2005) The origins of diversity and specificity in G protein-coupled receptor signaling. *J Pharmacol Exp Ther* **314**, 485–494.
3. Schadt, E. E., Lamb, J., Yang, X., Zhu, J., Edwards, S., Guhathakurta, D., Sieberts, S. K., Monks, S., Reitman, M., Zhang, C., Lum, P. Y., Leonardson, A., Thieringer, R., Metzger, J. M., Yang, L., Castle, J., Zhu, H., Kash, S. F., Drake, T. A., Sachs, A. and Lusis, A. J. (2005) An integrative genomics approach to infer causal associations between gene expression and disease. *Nat Genet* **37**, 710–7.
4. Chen, Y., Zhu, J., Lum, P. Y., Yang, X., Pinto, S., MacNeil, D. J., Zhang, C., Lamb, J., Edwards, S., Sieberts, S. K., Leonardson, A., Castellini, L. W., Wang, S., Champy, M. F., Zhang, B., Emilsson, V., Doss, S., Ghazalpour, A., Horvath, S., Drake, T. A., Lusis, A. J. and Schadt, E. E. (2008) Variations in DNA elucidate molecular networks that cause disease. *Nature* **45**, 429–35.
5. Hilsenbeck, S. G., Friedrichs, W. E., Schiff, R., O'Connell, P., Hansen, R. K., Osborne, C. K. and Fuqua, S. A. (1999) Statistical analysis of array expression data as applied to the problem of tamoxifen resistance. *J Natl Cancer Inst* **91**, 453–9.
6. Martin, B., Pearson, M., Brenneman, R., Golden, E., Wood, W., Prabhu, V., Becker, K. G., Mattson, M. P. and Maudsley, S. (2009) Gonadal transcriptome alterations in response to dietary energy intake: sensing the reproductive environment. *PLoS One* **4**, e4146.
7. Martin, B., Brenneman, R., Golden, E., Walent, T., Becker, K. G., Prabhu, V. V., Wood, W. 3 rd, Ladenheim, B., Cadet, J. L. and Maudsley, S. (2009) Growth factor signals in neural cells: coherent patterns of interaction control multiple levels of molecular and phenotypic responses. *J Biol Chem* **284**, 2493–511.
8. Quackenbush, J. (2002) Microarray data normalization and transformation. *Nat Genet* **32**, 496–501.
9. Zhao, Y., Li, M. C. and Simon, R. (2005) An adaptive method for cDNA microarray normalization. *BMC Bioinformatics* **6**, 28.
10. Kepler, T. B., Crosby, L. and Morgan, K. T. (2002) Normalization and analysis of DNA microarray data by self-consistency and local regression. *Genome Biol* **3**, RESEARCH0037.
11. Yang, Y. H., Dudoit, S., Luu, P., Lin, D. M., Peng, V., Ngai, J. and Speed, T. P. (2002) Normalization for cDNA microarray data: a robust composite method addressing single and multiple slide systematic variation. *Nucl Acids Res* **30**, e15.
12. Zien, A., Aigner, T., Zimmer, R. and Lengauer, T. (2001) Centralization: a new method for the normalization of gene expression data. *Bioinformatics* **17**, S323–31.
13. Sasik, R., Calvo, E. and Corbeil, J. (2002) Statistical analysis of high-density oligonucleotide arrays: a multiplicative noise model. *Bioinformatics* **18,** 1633–40.
14. Troyanskaya, O. G., Garber, M. E., Brown, P. O., Botstein, D. and Altman, R. B. (2002) Nonparametric methods for identifying differentially expressed genes in microarray data. *Bioinformatics* **18**, 1454–61.

15. Lee, M. L., Whitmore, G. A., Björkbacka, H. and Freeman, M. W. (2005) Nonparametric methods for microarray data based on exchangeability and borrowed power. *J Biopharm Stat* **15**, 783–97.
16. Li, H., Wood, C. L., Getchell, T. V., Getchell, M. L. and Stromberg, A. J. (2004) Analysis of oligonucleotide array experiments with repeated measures using mixed models. *BMC Bioinformatics* **5**, 209.
17. Meuwissen, T. H. and Goddard, M. E. (2004). Bootstrapping of gene-expression data improves and controls the false discovery rate of differentially expressed genes. *Genet Sel Evol* **36**, 191–205.
18. Reiner, A., Yekutieli, D. and Benjamini, Y. (2003) Identifying differentially expressed genes using false discovery rate controlling procedures. *Bioinformatics* 19, 368–75.
19. Lange, V., Picotti, P., Domon, B., Aebersold, R. (2008) Selected reaction monitoring for quantitative proteomics. *Mol Systems Biol* 4, 222.
20. Griffin, T. J., Xie, H., Bandhakavi, S., Popko, J., Mohan, A., Carlis, J. V. and Higgins, L. (2007) iTRAQ reagent-based quantitative proteomic analysis on a linear ion trap mass spectrometer. *J Proteome Res* **6**, 4200–4209.
21. Dayon, L., Pasquarello, C., Hoogland, C., Sanchez, J. C. and Scherl, A. (2010) Combining low- and high-energy tandem mass spectra for optimized peptide quantification with isobaric tags. *J Proteomics* **73**, 769–77.
22. Elias, J. E. and Gygi, S. P. (2007) Target-decoy search strategy for increased confidence in large-scale protein identifications by mass spectrometry. *Nat Methods* **4**, 207–14.
23. Eng, J. K., McCormack, A. L. and Yates, J. R. (1994) An approach to correlate tandem massspectral data of peptides with amino-acid-sequences in a protein database. *J Am Soc Mass Spectrom* **5**, 976–89.
24. Perkins, D. N., Pappin, D. J. C., Creasy, D. M. and Cottrell, J. S. (1999) Probability-based protein identification by searching sequence databases using mass spectrometry data. *Electrophoresis* **20**, 3551–67.
25. Clauser, K. R., Baker, P. and Burlingame, A. L. (1999) Role of accurate mass measurement (+/− 10 ppm) in protein identification strategies employing MS or MS/MS and database searching. *Anal Chem* **71**, 2871–2882.
26. Zhang, N., Aebersold, R. and Schwilkowski, B. (2002) ProbID: a probabilistic algorithm to identify peptides through sequence database searching using tandem mass spectral data. *Proteomics* **2**, 1406–12.
27. Creasy, D. M. and Cottrell, J. S. (2002) Error tolerant searching of uninterpreted tandem mass spectrometry data. *Proteomics* **2**, 1426–34.
28. Nesvizhskii, A. I., Vitek, O. and Aebersold, R. (2007) Analysis and validation of proteomic data generated by tandem mass spectrometry. *Nat Methods* **4**, 787–97.
29. Gstaiger, M. and Aebersold, R. (2009). Applying mass spectrometry-based proteomics to genetics, genomics and network biology. *Nat Rev Genet* **10**, 617–27.
30. Mueller, L. N., Brusniak, M-Y., Mani, D. R. and Aebersol, R. (2008). An assessment of software solutions for the analysis of mass spectrometry based quantitative proteomic data. *J Proteome Res* 7, 51–61.
31. Binns, D., Dimmer, E., Huntley, R., Barrell, D., O'Donovan, C. and Apweiler, R. (2009) QuickGO: a web-based tool for Gene Ontology searching. *Bioinformatics* **25**, 3045–6.
32. Adams, M. D., Celniker, S. E., Holt, R. A., Evans, C. A., Gocayne, J. D., Amanatides, P. G., Scherer, S. E., Li, P. W., Hoskins, R. A., Galle, R. F., *et al.* (2000) The genome sequence of *Drosophila melanogaster*. *Science* **287**, 2185–95.
33. Liu, M., Liberzon, A., Kong, S. W., Lai, W. R., Park, P. J., Kohane, I. S. and Kasif, S. (2007) Network-based analysis of affected biological processes in type 2 diabetes models. *PLoS Genetics* **3**, e96.
34. Hirschman, L., Yeh, A., Blaschke, C. and Valencia, A. (2005) Overview of BioCreAtIvE: critical assessment of information extraction for biology. *BMC Bioinformatics* **6**, S1.
35. Camon, E. B., Barrell, D. G., Dimmer, E. C., Lee, V., Magrane, M., Maslen, J., Binns, D. and Apweiler, R. (2005) An evaluation of GO annotation retrieval for BioCreAtIvE and GOA. *BMC Bioinformatics* **6**, S1-S17.
36. Dressman, H. K., Muramoto, G. G., Chao, N. J., Meadows, S., Marshall, D., Ginsburg, G. S., Nevins, J. R. and Chute, J. P. (2007) Gene expression signatures that predict radiation exposure in mice and humans. *PLoS Medicine* **4**, e106.
37. Eisen, M., Spellman, P. T., Brown, P. O. and Botstein, D. (1998) Cluster analysis and display of genome-wide expression patterns. *Proc Natl Acad Sci USA* **95**, 14863–8.
38. Spellman, P. T., Sherlock, G., Zhang, M. Q., Iyer, V. R., Anders, K., Eisen, M. B., Brown, P. O., Botstein, D. and Futcher, B. (1998) Comprehensive identification of cell cycle-regulated genes of the yeast Saccharomyces cerevisiae by microarray hybridization. *Mol Biol Cell* 9, 3273–3297.
39. Faustino, R. S., Behfar, A., Perez-Terzic, C. and Terzic, A. (2008) Genomic chart guiding embryonic stem cell cardiopoiesis. *Genome Biol* 9, R6.
40. Martin, B., Pearson, M., Brenneman, R., Golden, E., Keselman, A., Iyun, T., Carlson, O. D.,

Egan, J. M., Becker, K. G., Wood, W. 3 rd, Prabhu, V., de Cabo, R., Maudsley, S., Mattson, M. P. (2008) Conserved and differential effects of dietary energy intake on the hippocampal transcriptomes of females and males. *PLoS One* **3**, e2398.

41. Stranahan, A. M., Lee, K., Becker, K. G., Zhang, Y., Maudsley, S., Martin, B., Cutler, R. G. and Mattson, M. P. (2008) Hippocampal gene expression patterns underlying the enhancement of memory by running in aged mice. *Neurobiol Aging* [Epub ahead of print]
42. Ginos, M. A., Page, G. P., Michalowicz, B. S., Patel, K. J., Volker, S. E., Pambuccian, S. E., Ondrey, F. G., Adams, G. L. and Gaffney, P. M. (2004) Identification of a gene expression signature associated with recurrent disease in squamous cell carcinoma of the head and neck. *Cancer Res* **64**, 55–63.
43. Draghici, S., Kulaeva, O., Hoff, B., Petrov, A., Shams, S. and Tainsky, M. A. (2003) Noise sampling method: an ANOVA approach allowing robust selection of differentially regulated genes measured by DNA microarrays. *Bioinformatics* **19**, 1348–59.
44. Man, M. Z., Wang, X. and Wang, Y. (2000) POWER_SAGE: comparing statistical tests for SAGE experiments. *Bioinformatics* **16**, 953–9.
45. Alexa, A., Rahnenfuhrer, J. and Lengauer, T. (2006) Improved scoring of functional groups from gene expression data by decorrelating GO graph structure. *Bioinformatics* **22**,1600–7.
46. Grossmann, S., Bauer, S., Robinson, P. N. and Vingron, M. (2007) Improved detection of overrepresentation of Gene Ontology annotations with parent child analysis. *Bioinformatics* **23**, 3024–31.
47. Schlicker, A., Rahnenfuhrer, J., Albrecht, M., Lengauer, T. and Domingues, F. S. (2007) GOTax: investigating biological processes and biochemical activities along the taxonomic tree. *Genome Biol* **8**, R33.
48. Martin, B., Brenneman, R., Becker, K. G., Gucek, M., Cole, R. N. and Maudsley, S. (2008) iTRAQ analysis of complex proteome alterations in 3xTgAD Alzheimer's mice: understanding the interface between physiology and disease. *PLoS One* **3**, e2750.
49. Qin, X., Ahn, S., Speed, T. P. and Rubin, G. M. (2007) Global analyses of mRNA translational control during early *Drosophila* embryogenesis. *Genome Biol* 8, R63.
50. Thomas, P. D., Mi, H. and Lewis, S. (2007) Ontology annotation: mapping genomic regions to biological function. *Curr Opin Chem Biol* **11**, 4–11.
51. Karp, P. D., Riley, M., Paley, S. M. and Pelligrini-Toole, A. (1996) EcoCyc: an encyclopedia of Escherichia coli genes and metabolism. *Nucl Acids Res* 24, 32–9.
52. Keseler, I. M., Collado-Vides, J., Gama-Castro, S., Ingraham, J., Paley, S., Paulsen, I. T., Peralta-Gil, M. and Karp, P. D. (2005) EcoCyc: a comprehensive database resource for *Escherichia coli*. *Nucl Acids Res* **33**, D334–7.
53. Kim, S-Y. and Volsky, D. J. (2005) PAGE: Parametric analysis of geneset enrichment. *BMC Bioinformatics* 6, 144–156.
54. Mootha, V. K., Lindgren, C. M., Eriksson, K. F., Subramanian, A., Sihag, S., Lehar, J., Puigserver, P., Carlsson, E., Ridderstråle, M., Laurila, E., Houstis, N., Daly, M. J., Patterson, N., Mesirov, J. P., Golub, T. R., Tamayo, P., Spiegelman, B., Lander, E. S., Hirschhorn, J. N., Altshuler, D. and Groop, L. C. (2003). PGC-1alpha-responsive genes involved in oxidative phosphorylation are coordinately downregulated in human diabetes. *Nat Genet* 34, 267–73.
55. Dahlquist, K. D., Salomonis, N., Vranizan, K., Lawlor, S. C. and Conklin, B. R. (2002) GenMAPP, a new tool for viewing and analyzing microarray data on biological pathways, *Nat Genet* **31**, 19–20.
56. Pandey, R., Guru, R. K. and Mount, D. W. (2004) Pathway Miner: extracting gene association networks from molecular pathways for predicting the biological significance of gene expression microarray data. *Bioinformatics* **20**, 2156–8.
57. Zhang, B., Kirrov, S. and Snoddy, J. (2005) WebGestalt: an integrated system for exploring gene sets in various biological contexts. *Nucl Acids Res* **33**, W741–8.
58. Kanehisa, M., Goto, S., Kawashima, S. and Nakaya, A. (2002) The KEGG databases at GenomeNet. *Nucl Acids Res* **30**, 42–6.
59. Bouchahda, M., Adam, R., Giacchetti, S., Castaing, D., Brezault-Bonnet, C., Hauteville, D., Innominato, P. F., Focan, C., Machover, D. and Lévi, F. (2009) Rescue chemotherapy using multidrug chronomodulated hepatic arterial infusion for patients with heavily pretreated metastatic colorectal cancer. *Cancer* **115**, 4990–9.
60. McClatchy, D. B., Liao, L., Park, S. K., Venable, J. D. and Yates, J. R. (2007) Quantification of the synaptosomal proteome of the rat cerebellum during post-natal development. *Genome Res* **17**, 1378–8.

Part II

Receptor–Ligand Interactions

Chapter 6

Studying Ligand Efficacy at G Protein-Coupled Receptors Using FRET

Jean-Pierre Vilardaga

Abstract

Drug "ligands" that bind G protein-coupled receptors (GPCRs) can either stimulate, fully (full agonists) or partially (partial agonists), or reduce (inverse agonists) basal receptor activity, by stabilizing different receptor conformations. The term "intrinsic efficacy" was introduced as a parameter to express the ability of a ligand to activate its receptor and to differentiate the varying signaling capacity of diverse ligands when they occupy the same fraction of a single receptor. Most methods use downstream biochemical and physiological responses as proxies of "intrinsic efficacy" but cannot measure it directly at the level of the receptor. Here I describe the development of a Förster resonance energy transfer (FRET) approach that permits the rigorous measurement of the intrinsic efficacy of a ligand directly at the level of a GPCR and independent from variation in experimental conditions. This approach also allows intrinsic efficacies of ligands to be linked with the effects of receptor polymorphisms or receptor heterodimerization.

Key words: G protein-coupled receptor, α_{2A}-Adrenergic receptor, β_1-Adrenergic receptor, GPCR heterodimer, GPCR polymorphisms, Heterotrimeric G proteins, Förster resonance energy transfer, Ligand efficacy, Conformational changes, FlAsH labeling

1. Introduction

G protein-coupled receptors (GPCRs) serve as cell surface switches to transmit extracellular sensory (e.g., light, taste, and odorant molecules), chemical (e.g., drugs), and physiological (e.g., hormones, ions) signals into cells. These receptors regulate cell functions primarily by activating heterotrimeric Gαβγ-proteins located on the cytoplasmic face of the cell membrane. Activated G proteins then regulate the activity of effector molecules such as adenylyl cyclases and phospholipase C, which control the production of intracellular second messengers such as cAMP and inositol 1,4,5-trisphosphate, respectively. These responses are usually

Louis M. Luttrell and Stephen S.G. Ferguson (eds.), *Signal Transduction Protocols*, Methods in Molecular Biology, vol. 756, DOI 10.1007/978-1-61779-160-4_6, © Springer Science+Business Media, LLC 2011

terminated by the binding of the receptor to cytosolic arrestins (β-arrestin-1 and -2) that coordinate in space and time receptor and G protein uncoupling, as well as receptor internalization (1). This latter event may also promote various distal signaling responses such as those mediated by mitogen-activated protein (MAP) kinase cascades. Key regulators of a large palette of physiological functions such as heart rate, bone formation, and mineral metabolism among others, GPCRs are also involved in many pathological processes, and are the targets of a majority of clinical drugs.

A fundamental property of drugs acting at receptors is their capacity to produce either a maximal (full agonists) or submaximal (partial agonists) receptor-mediated response even upon total occupancy of receptors by the drug, or reduce (inverse agonists) basal levels of receptor signaling. The term "intrinsic efficacy" was introduced as a parameter to express the ability of a drug "ligand" to activate its receptor and produce a functional response (2). The intrinsic efficacy of a ligand can thus be used to differentiate the varying signaling capacity of diverse ligands when they occupy the same fraction of a single receptor. Most methods use downstream biochemical and physiological responses as proxies of "intrinsic efficacy" but cannot measure it directly. These methods rely traditionally on measurements of second messenger production, protein phosphorylation, level of gene expression, or even smooth muscle cell relaxation. Since the varying number of receptors, G proteins, and effector molecules encountered in diverse cell types and tissues unavoidably affects these measurements, the efficacy of a ligand at the level of its receptor has remained inaccessible. One of the difficulties in measuring rigorously the intrinsic efficacy of a compound is illustrated in Fig. 1. In this example, VIP_{2-28}, a fragment of the vasoactive intestinal peptide (VIP), increased cAMP levels in cells expressing a high

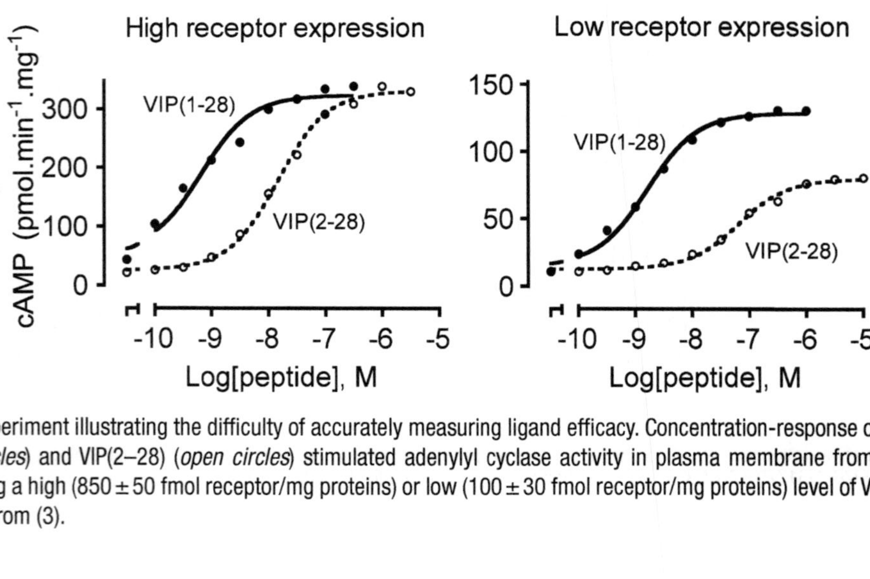

Fig. 1. Experiment illustrating the difficulty of accurately measuring ligand efficacy. Concentration-response of VIP(1–28) (*filled circles*) and VIP(2–28) (*open circles*) stimulated adenylyl cyclase activity in plasma membrane from CHO cells expressing a high (850 ± 50 fmol receptor/mg proteins) or low (100 ± 30 fmol receptor/mg proteins) level of VIP receptor. Adapted from (3).

amount of VIP type 1 receptors (VIP_1-R) as efficiently as the native hormone, but displayed partial agonism in cell expressing ≈eightfold less VIP_1-R (3). The measurement of ligand efficacy is thus sensitive to experimental conditions. These difficulties can now be circumvented by an optical approach based on the FRET principle, which directly measures changes in receptor conformation upon ligand binding (see below). This strategy permits the rigorous measurement of the intrinsic efficacy of a ligand directly at the level of the receptor and independent from variation in either receptor number or cell conditions.

1.1. The FRET Principle

Initially described by the French physicist Jean Perrin in 1930 (4) and later elucidated by the German scientist Theodor Förster in 1948 (5), *F*örster *r*esonance *e*nergy *t*ransfer (FRET) is a process by which an excited fluorophore molecule (the donor) transfers its energy nonradiatively to an acceptor fluorophore partner. This energy transfer occurs when both donor and acceptor molecules are <100 Å of each other, and share a spectral overlap between the donor's emission and acceptor's absorption spectra. The efficiency of the energy transfer (ET) depends on dipole orientation and distances between donor and acceptor molecules – ET falls off with the sixth power of the distance between the two fluorophore moieties. FRET can thus report interactions between different proteins by the measurement of intermolecular FRET (in this case, the two fluorophores are in two separate proteins), as well as conformational changes within a single protein by the measurement of intramolecular FRET (the fluorophores are within a single protein) (6).

The cyan fluorescent protein (CFP) and yellow fluorescent protein (YFP) form a popular donor/acceptor FRET pair for studies of protein–protein interactions or conformational changes in a given protein (7). This pair has a strong spectral overlap and permits excellent resolution for recording FRET with both dual emission photometry and imaging systems and can provide powerful insights into kinetics and mechanisms of protein associations, and protein conformational changes in live cells (8, 9). These two GFP variants can be easily incorporated into proteins by genetic engineering, and can be used to monitor the complex formation between two proteins, and quantify kinetics of the initial steps involved in GPCR signaling such as in ligand binding, receptor activation, receptor and G protein coupling, G protein activation, and second messenger propagation in live cells (10).

A limitation is the requirement for external excitation of the CFP to initiate the fluorescence transfer, which leads to a slight direct excitation of the YFP (called cross-talk). In addition to this direct excitation of the YFP by light at 436 nm, another critical source of consideration is the partial overlap of the CFP emission spectrum that is detectable in the channels used to detect YFP (called bleed-through). These two sources of non-FRET signal, however, can be easily controlled and corrected (see Subheading 6.3).

1.2. Design of GPCR Biosensors

Both the kinetics and magnitude of GPCR activation can be determined with high temporal resolution in live cells using GPCR biosensors that use FRET to measure intramolecular conformational changes. These biosensors are designed by inserting one fluorescent moiety into the third intracellular loop of the receptor and the other fluorescent moiety into the receptor carboxy terminus. For example, GPCR biosensors can be made with either CFP/YFP, or with CFP/FlAsH (*Fl*uorescein *Ars*enical *H*airpin binder) as FRET pairs (8, 9). In this latter case, CFP is generally fused (or inserted) to the carboxy terminus of the receptor and the tetracysteine motif CCPGCC, which can specifically bind the membrane-permeable dye FlAsH, is inserted in the third intracellular loop of the same receptor. These receptor constructs (referred to as $GPCR^{CFP/YFP}$ or $GPCR^{CFP/FlAsH}$) are usually well expressed and functional, although the size of the CFP/YFP protein tags can cause reduced signaling responses (9). Because CFP and YFP (or FlAsH) are in close proximity in the inactive receptor, energy is efficiently transferred and both the cyan and yellow light emitted by CFP and YFP are detected upon CFP illumination. When agonist binding induces a conformational change in the receptor, the relative distance and/or dipole–dipole orientation between the fluorescent partners in the FRET pair change, resulting in rapid loss of FRET. This FRET change can be monitored as the change in the ratio of emission intensities of yellow and cyan fluorescence, F_{CFP}/F_{YFP} (Fig. 2). Formation of the active receptor state in response to ligand binding is monitored as a fast decrease of the FRET ratio, which usually follows a monoexponential time course. This approach has been successfully applied to several different GPCRs, and allows detailed analysis of the kinetics of ligand-mediated receptor activation as well as its direct visualization in live cells (6).

1.3. Recording Ligand Efficacy

Experiments are usually performed under an inverted microscope, where light at 436 nm selectively excites a single cell expressing a GPCR biosensor ($GPCR^{Flash/CFP}$ or $GPCR^{CFP/YFP}$) to induce donor

Fig. 2. Comparing the CFP/YFP and FlAsh/CFP FRET pair constructs of the α_{2A}-adrenergic receptor. (**a**) Design of a GPCR biosensor (*blue circles* CFP, *greenish yellow circles* YFP/FlAsH). GPCR activation is monitored by a decrease in FRET between CFP and YFP (or FlAsH) inserted in the third intracellular loop and C-termini of the receptor, respectively. (**b**) *Upper panels*; confocal pictures show the cellular expression of the α_{2A}-$AR^{Flash/CFP}$ and α_{2A}-$AR^{YFP/CFP}$ in transiently transfected HEK-293 cells. Note that both receptor constructs exhibit predominant localization to the plasma membrane. *Lower panels*; changes in the fluorescence of CFP and Flash (*left*) or CFP and YFP (*right*), and corresponding FRET ratio F_{YFP}/F_{CFP} in response to a saturating concentration of norepinephrine (NE; 100 μM) recorded from a single HEK-293 cell expressing α_{2A}-$AR^{Flash/CFP}$ or α_{2A}-$AR^{YFP/CFP}$. Initial values of relative fluorescence (*cyan* traces for CFP, and *yellow* traces for YFP or Flash) and the FRET ratio (*red* traces) were set to one. The fractional decrease of FRET in response to NE reflects the intramolecular conformational switch accompanying receptor activation. Measurements were performed in single cells continuously perfused with buffer or with ligand for the time indicated by the horizontal bar. (**c**) FRET imaging of receptor activation in HEK293 cells transiently tranfected with α_{2A}-$AR^{CFP/YFP}$. The *left panel* shows the epifluorescence image and the next two panels present the pseudo-colored FRET ratio before and after stimulation by NE. Adapted from (9).

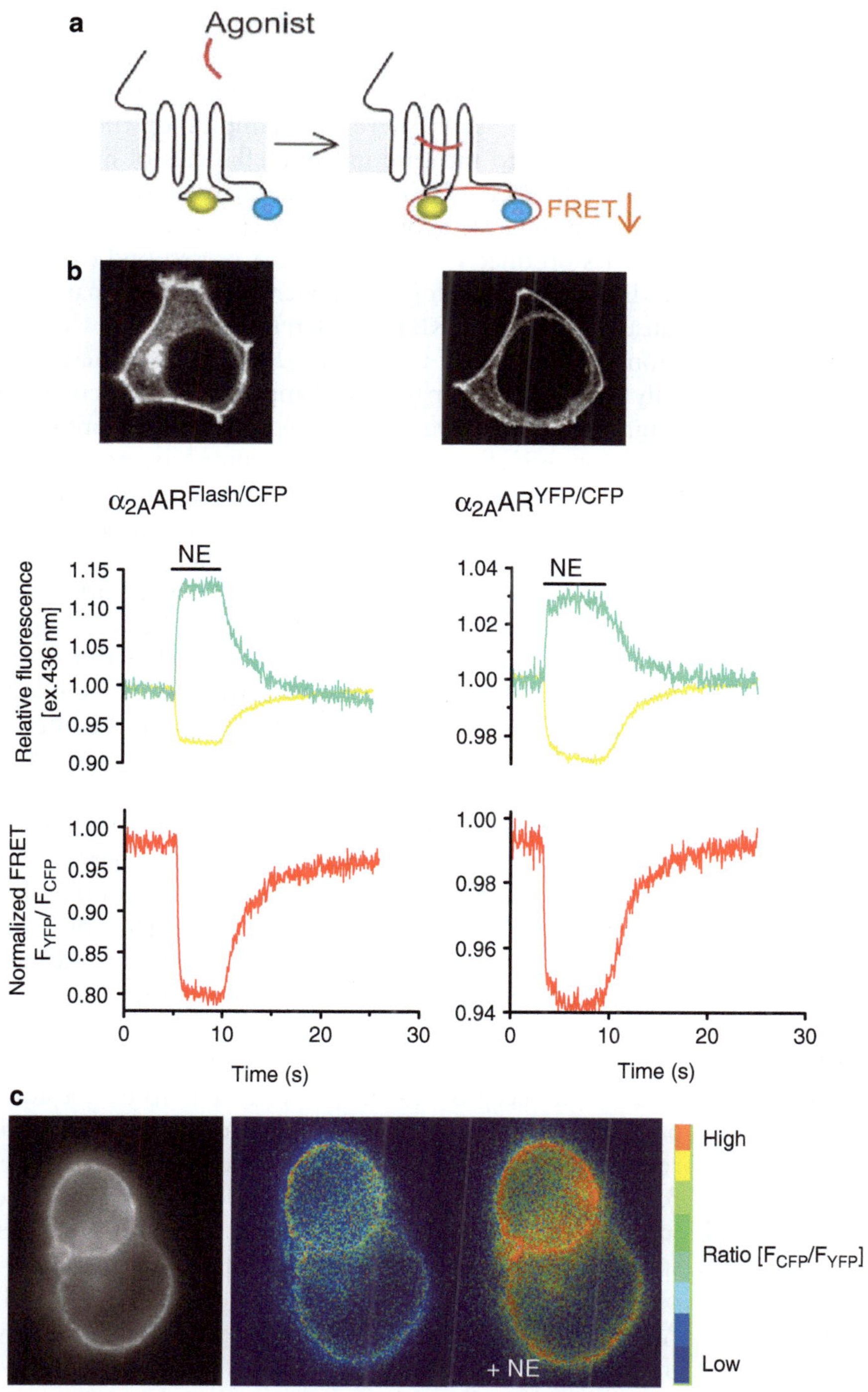
a
Agonist
FRET
b
α2AAR Flash/CFP
α2AAR YFP/CFP
NE
Relative fluorescence [ex.436 nm]
1.15
1.10
1.05
1.00
0.95
0.90
1.04
1.02
1.00
0.98
Normalized FRET F YFP/ F CFP
1.00
0.95
0.90
0.85
0.80
0.98
0.96
0.94
0
10
20
30
Time (s)
c
+ NE
High
Ratio [F CFP/F YFP]
Low

(CFP) and acceptor (FlAsH or YFP) emission fluorescence which is simultaneously recorded over time. Studies in live cells expressing a α_{2A}-adrenergic receptor (α_{2A}-AR) biosensor, α_{2A}-AR$^{CFP/YFP}$ or α_{2A}-AR$^{CFP/FlAsH}$, serve here as examples of the experimental strategy used to measure intrinsic efficacies of ligands acting at GPCRs (8, 9, 11–13). These studies have demonstrated that full, partial, and inverse agonists of different chemical structures and efficacies produce FRET signals that correspond exactly to their predicted pharmacological properties: full and partial agonists cause a decrease in FRET of different magnitudes, whereas inverse agonists cause FRET to increase (Fig. 3) (11). These ligands not only induce conformational changes of a different nature and magnitude, but divergence also appears in the kinetics of receptor activation: FRET changes of the receptor biosensor have revealed fast conformational changes for full agonists (rate constant $\tau \approx 50$ ms), smaller and slower changes for partial agonists ($\tau < 1$ s), and even slower changes ($\tau \approx 1$ s) in the opposite direction in response to inverse agonists (11, 12). These different kinetics depend on neither structural differences between ligands nor their binding affinities but correlate very well with the functional efficacy of ligand, as measured by activation of the inhibitory G protein G_i (Fig. 4) (12). These differences can be interpreted as evidence that GPCRs in live cells not only switch between an inactive "off" receptor state and a ligand-bound "on" active receptor state but also adopt distinct conformations in response to compounds of different intrinsic efficacies. Therefore, both the amplitudes and the kinetics of change in intramolecular FRET can be used to classify compounds as full, partial, or inverse agonists, and can be used to directly monitor intrinsic efficacy of compounds acting directly at a given receptor.

1.4. Ligand Efficacy Modulated by GPCR Polymorphisms

The capacity to record the effects of ligand efficacy directly at the level of the receptor also makes it possible to detect how GPCR polymorphisms might modify ligand efficacies. The results of a recent FRET study done with a β_1-adrenergic receptor biosensor, β_1-AR$^{CFP/YFP}$ (14), showed that carvedilol differentiated itself from metoprolol and bisoprolol, two other powerful β-blockers, by a specific and marked inverse agonist effects at the Arg389-variant of the receptor (Fig. 5). The specific effect of carvedilol on the Arg389-variant of the β_1AR revealed by this FRET study was further confirmed by a physiological assay that measured the beating frequency in cardiac myocytes expressing the two receptor variants (14). This FRET study thus opens a strategy to the determination and differentiation of the intrinsic efficacies of clinical drugs acting at variants of the same receptor.

1.5. Ligand Efficacy Modulated by GPCR Heterodimerization

Recent studies revealed that allosteric interactions between GPCR heterodimers, mediated by direct cross-conformational changes between receptors, can modulate the intrinsic efficacy of a ligand (13).

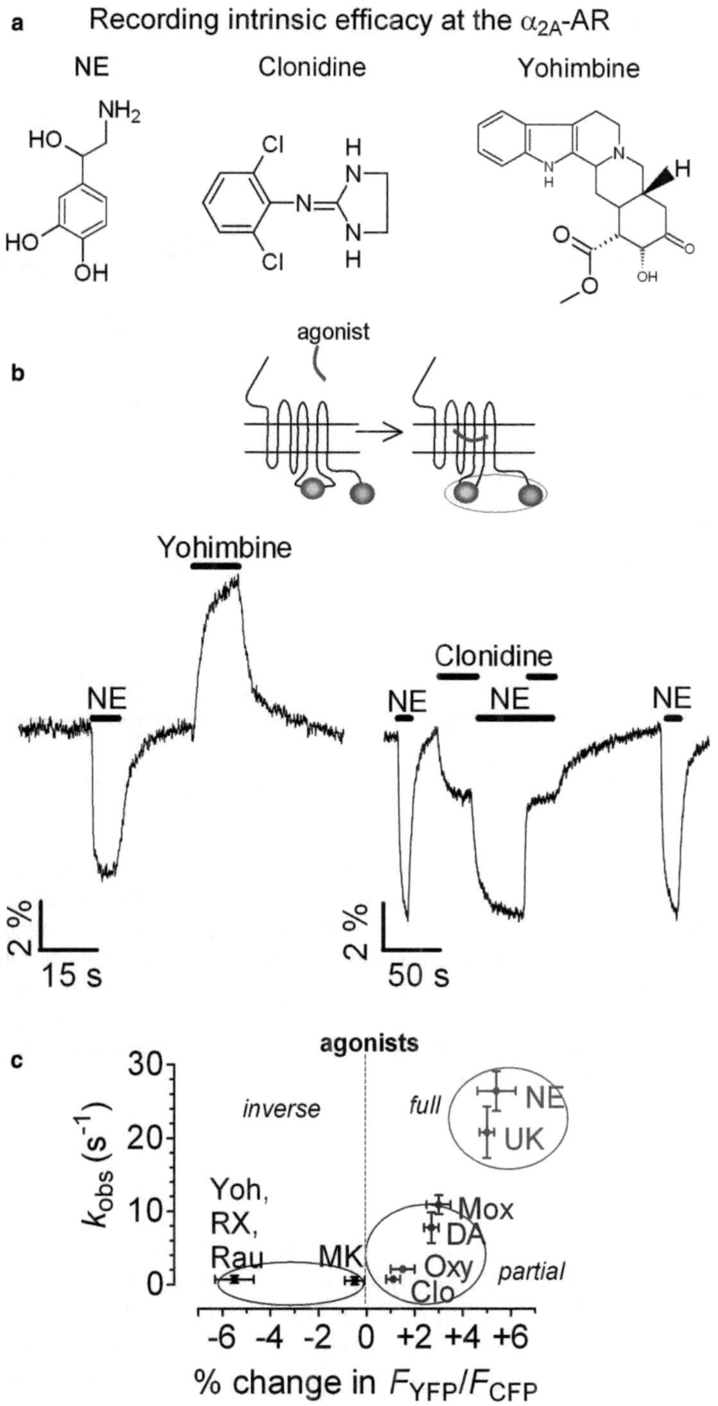

Fig. 3. Direct recording of intrinsic efficacy at a GPCR. (**a**) Chemical structure of norepinephrine (NE), clonidine (Clo), and yohimbine (Yoh) as examples of full, partial and inverse α_{2A}-AR agonists, respectively. (**b**) Example of FRET signals seen during sequential application of ligands of distinct efficacies to a single HEK-293 cell expressing $\alpha_{2A}AR^{YFP/CFP}$. The *left* trace represents the FRET signals mediated by NE, and yohimbine. The *right trace* represents the action of saturating concentrations of the NE or clonidine added alone or together. Note that the simultaneous application of NE and clonidine restored the partial response seen with clonidine alone. This corresponds to the predicted properties of a high-affinity partial agonist. (**c**) The correlation between the rate constant (k^{-1}) of receptor activation, and respective extent of FRET amplitude seen with ligands of different efficacies. Adapted from (11, 12).

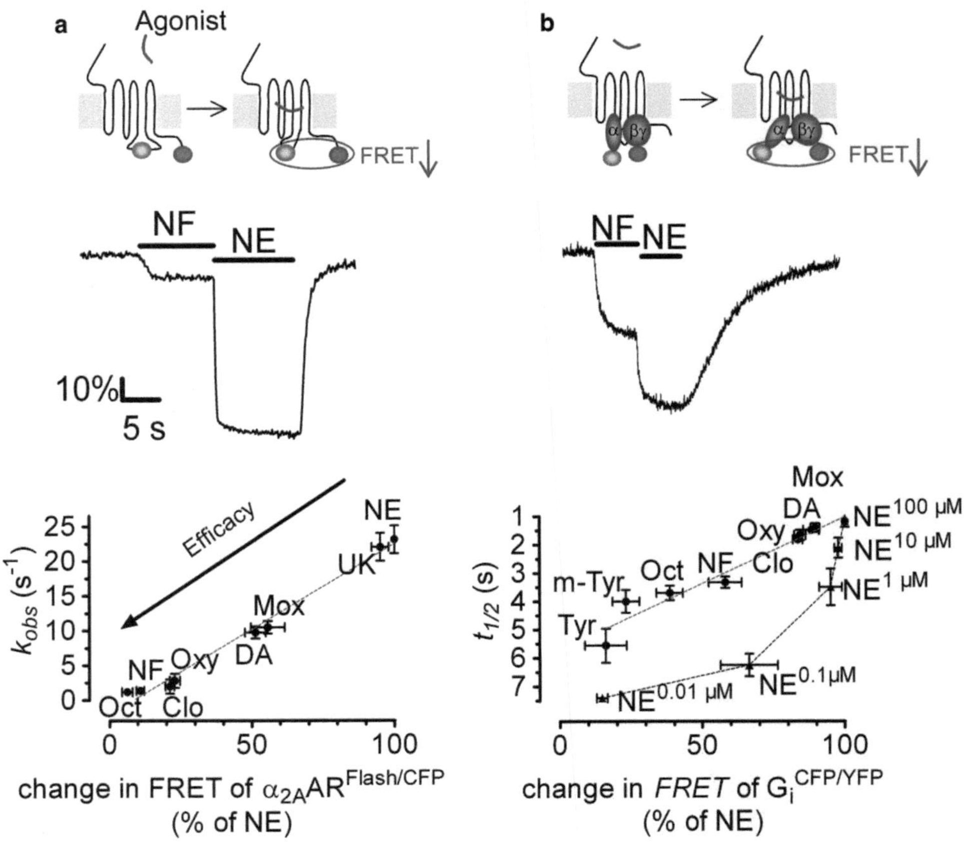

Fig. 4. Relation between ligand efficacy and kinetics of receptor activation and G protein activation. (**a**, **b**) Traces represent recording of the FRET ratio, F_{YFP}/F_{CFP}, as examples of action of agonists on HEK-293 cells expressing $\alpha_{2A}AR^{Flash/CFP}$ (**a**), or the wild-type α_{2A}AR together with $G\alpha_i^{YFP}\beta_1\gamma_2^{CFP}$ (**b**). Plots show the rate constant of receptor activation (**a**) or half-time of G protein activation (**b**) when the efficacy of the agonist varies. Values were obtained from fitting the kinetic recording with a monoexponential equation. Note that decreasing NE concentrations decrease the degree and rate of G_i activation but did not mimic the action of partial agonists. *α_{2A}AR full agonists*: norepinephrine (NE), UK-14,304 (UK); *α_{2A}AR partial agonists*: dopamine (DA), octopamine (Oct), norphenylephrine (NF), tyramine (Tyr), m-tyramine (m-Tyr), moxonidine (Mox), clonidine (Clo), oxymetazoline (Oxy). Adapted from (12).

For example, in the heterodimer formed by the two inhibitory G protein (G_i)-coupled α_{2A}-adrenergic/μ-opioid receptors, morphine binding to the μ-opioid receptor triggers a conformational change in the norepinephrine (NE)-bound α_{2A}-adrenergic receptor that decreases the efficacy of NE to activate its receptor. This allosteric regulation by morphine in turn decreases both NE-mediated inhibitory G protein (G_i) activation and NE-mediated MAP kinase activation (Fig. 6) (13). Thus, conformational cross-talk between receptors is able to modulate the intrinsic efficacy and physiological response to two different ligands acting simultaneously at a receptor heteromer.

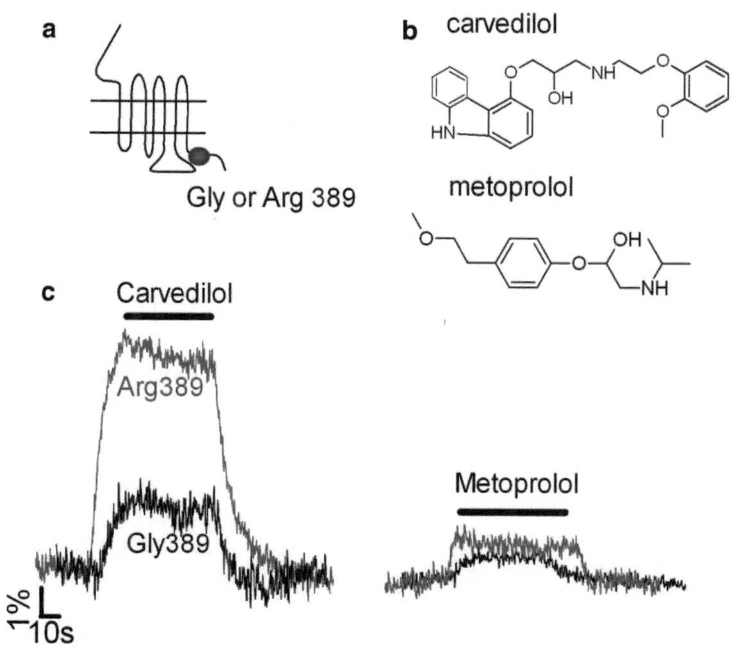

Fig. 5. Linking GPCR polymorphisms and ligand efficacy. (**a**) Schematic representation of β_1-adrenergic receptor variants. (**b**) Chemical structure of two β-blockers. (**c**) Effect of β_1-adrenergic receptor polymorphisms on beta-blocker responses. Examples of the FRET ratio F_{YFP}/F_{CFP} signals recorded from single cells expressing the Gly389–β_1AR$^{CFP/YFP}$ sensor (*black* traces) or the Arg389–β_1AR$^{CFP/YFP}$ sensor (*gray* traces) after application the beta-blockers carvedilol or metoprolol. Note the marked inverse agonist effect of carvedilol on the Arg389-β_1AR variant. Adapted from (14).

2. Materials

2.1. Cell Culture

1. Glass coverslips 24 mm.
2. Six-well and 10 cm tissue culture dishes.
3. Adherent cells such as human embryonic kidney (HEK-293) or rat osteosarcoma (ROS 17/2.8) cells among others (see Note 1).
4. Growth medium for HEK-293 cells: DMEM supplemented with 10% FCS (Fetal Calf Serum), 1% L-Glutamine (200 mM), 1% penicillin (10,000 units/ml)/Streptomycin (10 mg/ml) solution.
5. Transfection reagents such as Effectene (Qiagen), FuGene (Roche Pharmaceuticals), among others (see Note 2).
6. Phosphate-buffered saline (PBS).
7. Poly-L-lysine 0.01% w/v in phosphate-buffered saline (PBS) stored at 4°C.

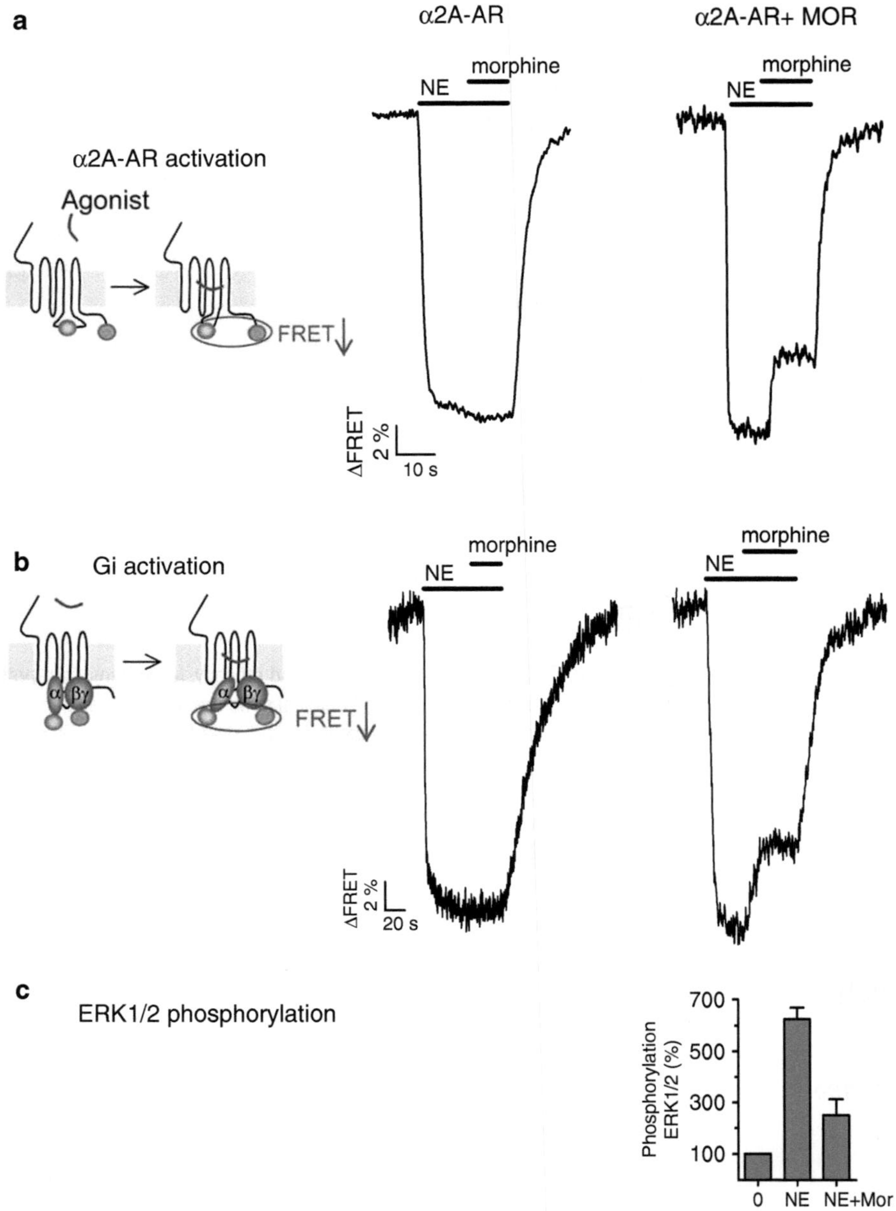

Fig. 6. Cross-conformational switching between GPCR heteromers modulate ligand efficacy. (**a**, **b**) *Left panels* illustrate the design of FRET sensors (*filled circles* CFP, *open circles* YFP/FlAsH) for receptor activation (**a**) and G protein activation (**b**), which are monitored by recording changes in FRET between CFP and FlAsH (or YFP); *center panels* show examples of activation of the α_{2A}-AR, and the inhibitory G protein, G_i, which were monitored in a single HEK-293 cell expressing α_{2A}-AR$^{FlAsH/CFP}$ (**a**) or the wild-type α_{2A}-AR and $G\alpha_i^{YFP}\beta_1\gamma_2^{CFP}$ (**b**); *right panels* show similar experiments in cell coexpressing the μ-opioid receptor MOR. Horizontal bars indicate the application of NE or morphine to the cell. (**c**) Maximal NE effect on phosphorylation of ERK1/2 in cells coexpressing α_{2A}-AR$^{FlAsH/CFP}$ and MOR. Note that NE acts as a partial agonist when the MOR is activated by morphine. Adapted from (13).

2.2. FlAsH Labeling

1. Hank's balanced salt solution (HBSS): 150 mM NaCl, 10 mM HEPES pH 7.4, 25 mM KCl, 4 mM $CaCl_2$, 2 mM $MgCl_2$, 10 mM glucose (add freshly).
2. 1,2-Ethanedithiol (EDT) stock (Fluka).
3. Dimethylsulfoxide (DMSO) stock.
4. TC-FlAsH™ stock solution: 1 mg/ml (Invitrogen).

2.3. FRET Assay

1. HEPES/BSA buffer: 137 mM NaCl, 5 mM KCl, 1 mM $CaCl_2$,1 mM $MgCl_2$, 20 mM HEPES, 0.1% (w/v) bovine serum albumin (BSA), pH 7.4.
2. Photometric detection system (see Fig. 7): inverted fluorescent microscope (e.g., Zeiss Axiovert 200) equipped with an oil immersion 60× or 100× objective and a dual emission

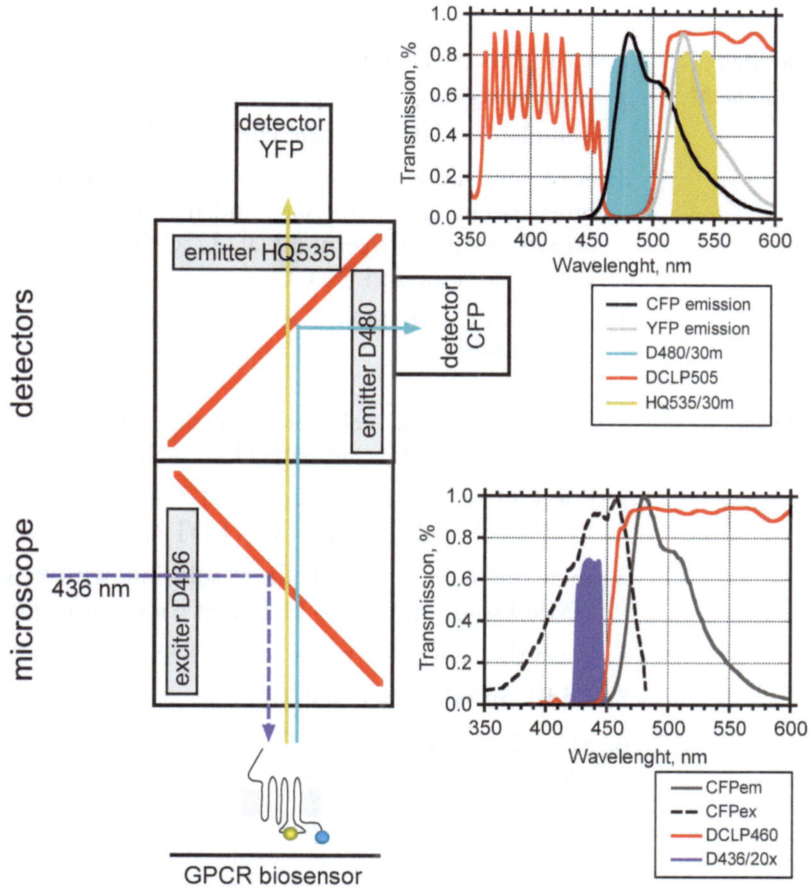

Fig. 7. Fluorescent microscope system for CFP/YFP FRET detection. Experiments are performed under an inverted fluorescent microscope equipped with a "FRET cube" containing an excitation filter D436/20 and beam splitter DCLP460 for CFP excitation. Light at 436 nm selectively excites cells expressing a GPCR biosensor to induce CFP and YFP emission fluorescence, which is detected by using a "beam splitter cube" with the following filter combination: 535/30 M (YFP emission), 480/30 M (CFP emission), DCLP505 (beam splitter). Plots represent excitation and emission characteristics of CFP and YFP, as well as spectral profiles of filters and mirrors used.

photometric system (TILL Photonics). Filter sets for CFP/YFP FRET (Chroma): excitation filter 436/20, beam splitter dichroic long-pass DCLP460 (CFP excitation); emission intensities are recorded by using 535/30 M (YFP emission), 480/30 M (CFP emission), DCLP505 (beam splitter).

3. Perfusion system: computer-assisted solenoid valve rapid superfusion device that permits rapid solution exchanges within 5–10 ms (ALA-VM8, ALA Scientific Instruments).
4. Imaging chamber: Attofluor cell chamber (Molecular Probes).
5. Analytical software: Origin 7.1 (Microcal); Clampex; GraphPAd Prism.

3. Methods

3.1. Cell Preparation for FRET Measurement

1. Soak glass coverslips in poly-l-lysine solution for 20 min.
2. Wash the coated coverslips with PBS and place them into six-well plates or 10 cm tissue culture dishes.
3. Seed cells stably or transiently expressing a GPCR biosensor onto coated glass coverslips and maintain the cell culture in growth medium overnight at 37°C in a humidified atmosphere (95% air and 5% CO_2).
4. Label cells with FlAsH if you are using cells expressing a $GPCR^{FlAsH/CFP}$.

3.2. FlAsH Labeling

1. For convenience, cells on coverslips are placed in six-well plates and then labeled with FlAsH–EDT.
2. For 1 plate, freshly prepare EDT/DMSO by mixing 2.1 μl EDT stock with 1 ml DMSO (EDT/DMSO solution).
3. Incubate 6.3 μl FlAsH stock solution with 6.3 μl EDT/DMSO solution for 10–15 min at room temperature.
4. Wash the cells three times with HBSS.
5. Add 1 ml HBSS per well.
6. Mix 12.6 μl FlAsH/EDT with 300 μl HBSS, then add an additional 6 ml of HBSS.
7. Incubate the cells with 1 ml of FlAsH/HBSS per well for 1 h at 37°C.
8. Freshly prepare the wash solution by mixing 42 μl EDT with 1 ml DMSO, and then adding 25 μl of this mix to 50 ml HBSS buffer.
9. Wash cells three times with 3 ml wash solution at 37°C; 10 min each wash.
10. Incubate the cells with 1 ml medium or HBSS/glucose until FRET recording.

3.3. Recording Ligand-Induced Changes in FRET

1. Mount a cover slip in an imaging chamber and carefully wash cells with HEPES/BSA buffer; place the chamber on a fluorescence inverted microscope equipped with an oil immersion 60× or 100× objective and a dual emission photometric system.
2. Turn on the monochromator, and under the binocular select a single cell expressing the $GPCR^{CFP/YFP}$ or $GPCR^{CFP/FlAsH}$ by exciting cells at 436 nm (CFP excitation). CFP emission is selectively observed using the DCLP460 dichroic and D480/30 m emission filter; YFP emission can be selectively observed using the DCLP505 dichroic and the HQ535/30 m emission filter (Fig. 7). For experiments involving G protein biosensors see Notes 3–6.
3. Perfuse the selected cell with HEPES/BSA buffer using a computer-assisted solenoid valve rapid superfusion device.
4. At the beginning of each experiment, record the emission intensity of YFP upon direct excitation at 500 nm (using HQ500/20 excitation and HQ535/30 m emission filters, and the DCLP505 dichroic). These data will be used in the data analysis (see Subheading 3.4).
5. Start FRET recording with excitation at 436 nm to induce cyan (CFP) and, via FRET, yellow (FlAsH or YFP) fluorescence, using a FRET cube containing the 436/10 excitation filter nm and the DCLP460 nm dichroic (Fig. 7). The illumination time is typically set to 5–10 ms with a frequency between 5 and 200 Hz. See Note 7.
6. Record the baseline for ≈20 s and use the perfusion system to exchange buffer and ligands as appropriate for the experiment. Collect data.

3.4. Data Analysis

1. The FRET ratio is corrected according to Eq. 1:

$$\text{Ratio}\left(\frac{F_{\text{YFP}}}{F_{\text{CFP}}}\right) = \frac{F_{\text{YFP}}^{\text{ex436/em535}} - a' \times F_{\text{CFP}}^{\text{ex436/em480}} - b' \times F_{\text{YFP}}^{\text{ex500/em535}}}{F_{\text{CFP}}^{\text{ex436/em480}}}, \quad (1)$$

where $F_{\text{YFP}}^{\text{ex436/em535}}$ and $F_{\text{CFP}}^{\text{ex436/em480}}$ represent, respectively, the emission intensities of YFP (recorded at 535 nm) and CFP (recorded at 480 nm) upon excitation at 436 nm; a and b represent correction factors for the bleed-through of CFP into the 535 nm channel (a) and the cross-talk due to the direct YFP excitation by light at 436 nm (b). $F_{\text{YFP}}^{\text{ex500/em535}}$ represents the emission intensity of YFP (recorded at 535 nm) upon direct excitation at 500 nm, and was recorded at the beginning of each experiment. Note that the bleed-through of YFP into the 480 nm channel usually is negligible. To ensure that CFP- and YFP-labeled molecule expression is similar in examined cells, experiments should be performed in cells displaying comparable fluorescence levels.

2. The means of FRET experiments is calculated according to Eq. 2, which normalizes for different expression levels of CFP and YFP molecules (15):

$$N_{\text{FRET}} = \frac{F_{\text{YFP}}^{\text{ex436/em535}} - a \times F_{\text{CFP}}^{\text{ex436/em480}} - b \times F_{\text{YFP}}^{\text{ex500/em535}}}{\sqrt{F_{\text{CFP}}^{\text{ex436/em480}} \times F_{\text{YFP}}^{\text{ex500/em535}}}} \quad (2)$$

3. The changes in the FRET ratio are usually analyzed using nonlinear regression to one- or two-exponential models:

$$F(t) = A_0 + A_1 \times e^{-k_1 \times t} + A_2 \times e^{-k_2 \times t},$$

where A_1 and A_2 are the amplitudes of the two phases, A_0 is the baseline at $t = \infty$, k_1 and k_2 are the rate constants (s^{-1}) for the two phases, and t is the time. The best fit can be judged by the analysis of residuals (i.e., differences between the experimental data and the calculated curve fits). Changes in emission due to photobleaching are systematically subtracted, and data are analyzed using Origin 7.1 software or Prism.

4. Notes

1. Transiently or stably transfected cells work equally well in the FRET assay described here. The following cell types have been successfully used in this FRET assay: Chinese hamster ovary cells (CHO), human embryonic kidney cells (HEK 293); Osteoblastic cell lines such as rat osteosarcoma cells (ROS 17/2.8), UMR 106, mouse MC-3T3; neuron-like rat PC12 cells; primary culture of rat hippocampal neurons.
2. Cells can be easily transfected using diverse transfection reagents such as Effectene™ (Qiagen), FuGENE™ (Roche), X-tremeGENE™ (Roche), or protocols based on the DEAE-dextran or calcium-phosphate methods (16). Stably expressing cells can be generally selected after 3–4 weeks of drug treatment (e.g., hygromycin).
3. Controls and conditions for studies involving G proteins. The physiological relevance of new information obtained by FRET approaches may be questioned by the use of recombinant fluorescent fusion proteins expressed at high levels in cultured cells. Therefore, a number of controls and conditions should be performed to test the effect of molecular crowding (17).
4. Design of G protein biosensors. For FRET assays involving the measurement of G proteins activation (Fig. 4b), GFP variants are attached to the G protein subunits at various positions. For example, CFP or YFP can be inserted at position 91 of

$G\alpha_i$, whereas $G\beta_1$- and $G\gamma_2$-subunits are fused to GFP variants at their C- or N-termini. These constructs are functional with respect to receptor coupling as well as effector activation. C-terminal labeling of the γ-subunit, however, resulted in loss of the lipid modification site and, thus, in reduced membrane targeting. Based on the classical model that receptor-mediated activation of G protein results in the dissociation between α and βγ subunits, activation of FRET-based G protein sensors ($G\alpha^{YFP}.G\beta\gamma^{CFP}$) should presumably drive a loss in FRET. Unexpectedly, an increase in FRET was observed in some cases, thus revealing that activation of G protein in live cells proceeds via conformational or dissociational events.

5. Optimize expression conditions to ensure low to moderate expression levels G protein constructs based on Western blot analysis of endogenous and expressed CFP or YFP-labeled G proteins using antibodies against the Gα subunit or fluorescent proteins.
6. For experiments involving the coexpression a GPCR and Gα-YFP plus Gβγ-CFP subunits for the diverse G proteins, expression conditions should be optimized to ensure a ≈1:1 molar expression ratio of CFP and YFP constructs at moderate expression levels. A CFP–YFP dimer construct in which the ratio is constrained to 1:1 could be used as a control. To ensure that G proteins expression (i.e., coexpression of $G\alpha^{YFP}$ and $G\beta\gamma^{CFP}$) is similar in examined cells, experiments should be performed in cells displaying comparable fluorescence levels.
7. CFP and YFP emissions are simultaneously recorded over time. FRET is monitored as a YFP/CFP emission intensity ratio upon excitation at 436 nm. The emission fluorescence intensities are determined at 535 ± 15 nm (YFP) and 480 ± 20 nm (CFP) with a beam splitter DCLP of 505 nm, and are detected by avalanche photodiodes and are digitalized using an analog/digital converter and stored on a personal computer using Clampex 10.0 software (Axon Instruments).

Acknowledgments

This work was supported by start-up funds from the Department of Pharmacology & Chemical Biology, University of Pittsburgh School of Medicine and by National Institutes of Health (NIH) grant DK087688. I thank Tim Feinstein for careful comments on this manuscript.

References

1. Kendall, R.T., and Luttrell, L.M. (2009) Diversity in arrestin function. *Cell Mol Life Sci* **66**, 2953–73.
2. Neubig, R.R., Spedding, M., Kenakin, T., and Christopoulos, A. (2003) International union of pharmacology nomemclature and rug classification. Upadate on tgerms and symbols in quantitative pharmacology. *Pharmcol Rev* **55**, 597–606.
3. Ciccarelli, E., Vilardaga, J.-P., De Neef, P., DiPaolo, E., Waelbroeck, M., Bollen, A., and Robberecht, P. (1994) Properties of the VIP-PACAP type II receptor stably expressed in CHO cells. *Regulatory Peptides* **54**, 397–407.
4. Perrin, J. Théorie quantique des transferts d'activation entre molécules de même espèce. (1932) Cas des solutions fluorescentes. *Ann Chim Phys* **17**, 283–313.
5. Förster, T. (1948) Zwischenmolekulare Energiewanderung und Fluoreszenz. *Ann Physik* (Leipzig) **2**, 55–75.
6. Vilardaga, J.-P., Bünemann, M., Feinstein, T.N., Lambert, N., Nikolaev, V.O., Engelhardt, S., Lohse, M.J., and Hoffmann, C. (2009) GPCR and G proteins: drug efficacy and activation in live cells. *Mol Endocrinol* **23**, 590–99.
7. Balla, T. (2009) Green light to illuminate signal transduction events. *Trends Cell Biol* **19**, 575–86
8. Vilardaga, J.-P., Bünemann, M., Krasel, C., Castro, M., and Lohse, M.J. (2003) A millisecond activation switch for G protein-coupled receptors in living cells. *Nat Biotechnology* **21**, 807–12.
9. Hoffmann, C., Gaietta,. G., Bünemann, M., Adams, R.S., Oberforff-Maas, S., Behr, B., Vilardaga. J.-P., Tsien, R.Y., Ellisman, M.H., and Lohse, M.J. (2005) A FLASH-based approach to determine G protein-coupled receptor activation in living cells. *Nature Methods* **2**, 171–76.
10. Ferrandon, S., Feinstein, T.N., Castro, C., Bouley, R., Potts, J.T., Gardella, T.J., and Vilardaga, J.-P. (2009) Sustained cyclic AMP production by parathyroid hormone receptor endocytosis. *Nat Chem Biol* **5**, 734–42.
11. Vilardaga, J.-P., , Steinmeyer, R., Harms, G., and Lohse M.J. (2005) Molecular basis of inverse agonism in a G protein-coupled receptor. *Nat Chem Biol* **1**, 25–8.
12. Nikolaev, O.V., Hoffmann, C., Bünemann, M., Lohse, M.J., and Vilardaga, J.-P., (2006) Molecular basis of partial agonism at the neurotransmitter α_{2A}-adrenergic receptor and G_i-protein heterotrimer. *J. Biol. Chem.* **281**, 24506–11.
13. Vilardaga, J.-P., Nikolaev, O.V., Lorentz, K., Ferrandon, S., Zhuang, Z., and Lohse, M.J., (2009) Direct inhibition of G protein signaling by cross-conformational switches between α_{2A}-adrenergic and m-opioid receptors. *Nat Chem Biol* **4**, 126–31.
14. Rochais, F., Vilardaga, J.-P., Bünemann, M., Lohse, M.J., and Engelhardt, S,. Real-time optical recording of β_1-adrenergic receptor activation reveals supersensitivity of the Arg389 variant to carvedilol. (2007) *J Clin Invest* **117**, 229–35.
15. Xia, Z., and Liu, Y. (2001) Reliable and global measurement of fluorescence resonance energy transfer using fluorescence microscopes. (2001) *Biophys. J.* **81**, 2395–02.
16. Ausubel, F. M., Brent, R., Kingston, R. E., Moore, D. D., Seidman, J. G., Smith, J. A., and Struhl, K. (1997) *Current Protocols in Molecular Biology*. Wiley, New York
17. Hein, P., Frank, M., Hoffmann, C., Lohse, M.J, and Bünemann, M. (2005) Dynamic of receptor/G protein coupling in living cells. *EMBO J* **24**, 4106–14.

Chapter 7

Using BRET to Detect Ligand-Specific Conformational Changes in Preformed Signalling Complexes

Nicolas Audet and Graciela Piñeyro

Abstract

Bioluminescence energy transfer (BRET) has become a powerful tool to study protein–protein interactions and conformational changes among interacting proteins. In particular, BRET assays performed in living cells have revealed that heptahelical receptors (7TMRs), heterotrimeric G proteins and their proximal effectors form constitutive signalling complexes. BRET technology has also allowed us to demonstrate that these multimeric protein arrays remain intact throughout initial stages of receptor signalling, thus providing a platform for direct transmission of conformational information from activated receptors to downstream signalling partners. A clear example of the latter are the distinct intermolecular re-arrangements undergone by 7TMRs and G protein subunits following activation of the receptor by different ligands. Here we present protocols describing the type of BRET assay that has been used to reveal the existence of constitutive signalling arrays formed by 7TMRs and proximal signalling partners as well as the ability of complex components to undergo ligand-specific conformational changes.

Key words: G protein-coupled receptor, Heterotrimeric guanine nucleotide-binding protein, Bioluminescence resonance energy transfer, Signal transduction, Protein complex

1. Introduction

The ability to sense and adapt to the environment is essential for cell survival. Heptahelical receptors (7TMRs), heterotrimeric guanine nucleotide-binding proteins (G proteins) and their proximal effectors contribute to environmental adaptation by translating a large variety of external stimuli (hormones, neurotransmitters, ions) into signals that can be decoded by the cell. The decoding process starts when an extracellular ligand binds to the receptor and triggers a series of intramolecular re-arrangements that result in its activation (1, 2). Conformational changes associated with

Louis M. Luttrell and Stephen S.G. Ferguson (eds.), *Signal Transduction Protocols*, Methods in Molecular Biology, vol. 756
DOI 10.1007/978-1-61779-160-4_7, © Springer Science+Business Media, LLC 2011

activation are then transmitted downstream, as the receptor interacts with heterotrimeric G proteins inducing exchange of GDP for GTP (3, 4). In turn, nucleotide exchange allows G protein subunits to modulate effectors which are capable of influencing vital cellular processes (3).

Although the order in which these transduction steps take place is fairly well agreed upon, the way in which information flows from one level to the next has been matter of much debate (5, 6). For quite some time the predominant model of signal transduction has been one in which 7TMRs stabilized in their active conformation were thought to shuttle in the membrane until randomly colliding with and activating heterotrimeric G proteins. The latter would in turn dissociate, and shuttle to collide with and activate their specific effectors. This model, known as *collision coupling*, is supported by evidence gathered for the most part using purification-reconstitution strategies, in vitro interaction assays, and kinetic modelling (5–8). More recently, the techniques of bioluminescence (BRET) and fluorescence (FRET) resonance energy transfer have allowed to monitor the interaction of signalling proteins within living cells (8–10). Results obtained applying these approaches have lent support to an increasingly accepted alternative model (*precoupling model*) in which receptors, G proteins and their effectors are thought to constitutively associate, forming multimeric signalling complexes (11–13). Here we will describe the type of BRET protocol that has allowed us to characterize constitutive complexes formed by 7TMRs, G proteins and their effectors, maintaining the description of the procedure general enough so that it may also be applied to other complexes of interest.

The BRET technique is based on a naturally occurring phenomenon that results from the non-radiative transfer of energy between a luminescent donor (*Renilla* luciferase) and a fluorescent acceptor (GFP or YFP) (14). There are two main pre-requisites for energy transfer to take place that (a) the emission spectrum of the donor overlaps the excitation spectrum of the acceptor and (b) the donor and acceptor exist within a distance of no more than 100 Å from each other. The latter property is the basis for monitoring in vivo interaction between different types of cellular proteins previously tagged with donor/acceptor BRET pairs (15, 16). The particular goal of a BRET experiment assessing whether two (or more) proteins associate to form a constitutive complex is to determine the existence of a specific spontaneous BRET signal between donor/acceptor constructs of the proteins of interest (11–13). As for other techniques that rely on the overexpression of tagged proteins, specificity controls are necessary for conclusions to be valid (17, 18). The way to carry out these controls is also described below.

Concomitant with the new views on the physical bases of signal transduction, our conception of how extracellular ligands modulate 7TMR signalling has also evolved. Until quite recently differences in drug efficacy had been exclusively considered in quantitative terms; i.e. the more efficacious the drug the greater its ability to stabilize a larger amount of a single active state of the receptor (19, 20). However, as technological development has allowed us to monitor receptor activation using a growing number of readouts, the validity of this model has been challenged. In particular, numerous reports have shown that ligand efficacy depends on the signalling pathway in which drugs are tested (21–25), calling for an amendment of the quantitative model since these observations could not be explained by the accumulation of a single active state of the receptor (26) (see also Ehlert (27)). Since then, the revised model has incorporated the notion that 7TMRs may adopt multiple active conformations (28, 29), implying the possibility of different ligands stabilizing different receptor states (26, 28, 29).

The ability of a receptor to adopt multiple signalling conformations is the basis for "*functional selectivity.*" This term describes ligand ability to stabilize a receptor conformation which triggers only a subset of the responses controlled by the receptor (26, 28, 29). Such a property raises the possibility of pharmacologically specifying the type of signal elicited by the activation of any given 7TMR via the development of ligands that induce/stabilize a conformation which only produces a desired set of responses. The prospect of producing such ligands could provide the basis for rational design of therapeutic agents with reduced side effects. In the following paragraphs we will describe the type of BRET experiments that have allowed us to unveil ligand-specific conformations for the delta opioid receptor (DOR; (13)). From a general perspective, the goal is to determine whether spontaneous BRET signals generated by proteins interacting within a multimeric complex are distinctively modified by different ligands which specifically bind one of the complex components. To be able to attain this goal it is crucial to produce BRET pairs that allow monitoring of the same interaction from different vantage points. This is usually achieved by labelling one of the interaction partners at a fixed position while the tag on the other is placed at alternate locations (see Fig. 1). By comparing the way in which different ligands change the signal generated by these BRET pairs it may be possible to determine (a) whether ligand binding produces a conformational re-arrangement among proteins interacting within a complex and (b) if the conformational changes are ligand specific (13).

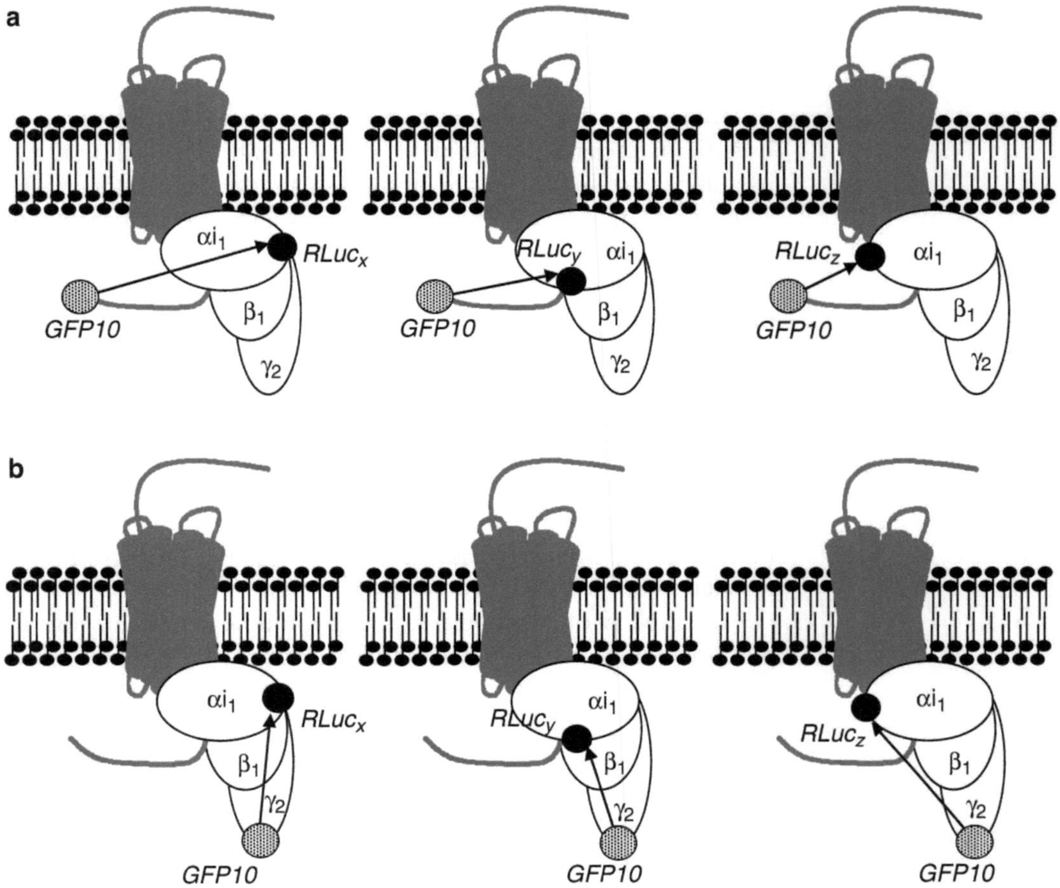

Fig. 1. Schematic representation of BRET constructs for monitoring conformational changes within multimeric complexes containing 7TMRs and heterotrimeric G protein. (**a**) Shows three BRET pairs which allow monitoring of the interaction between a 7TMR and a heterotrimeric Gα subunit from different vantage points. The distinct perspectives are given by the localization of the donor RLuc at different positions on Gα while the acceptor GFP remains at a constant position at the receptor C terminus. (**b**) Shows how the same Gα constructs may be used to evaluate conformational changes that take place between Gα and Gγ, simply by tagging the latter with the acceptor GFP at a constant position (N terminus).

2. Materials

2.1. Cell Culture and Transfection

1. Cells suitable for transfection, e.g. human embryonic kidney 293 (HEK293) cells.
2. 60 mm tissue culture dishes.
3. Trypsin/EDTA solution: 0.05% trypsin and 0.53 mM EDTA.
4. Complete culture medium: Dulbecco's modification of Eagle's medium (DMEM) with 4.5 g/L glucose and sodium pyruvate, supplemented with 10% fetal bovine serum (FBS), 2 mM L-glutamine, and 100 units/ml penicillin/streptomycin.

5. Plain culture medium: DMEM with 4.5 g/L glucose and sodium pyruvate.
6. Sterile, de-ionized water.
7. NaCl solution: 150 mM NaCl prepared with sterile de-ionized water.
8. Polyethyleneimine (L-PEI-25; Polyscience Inc.) dissolved in de-ionized water (1 mg/ml; pH 6.5–7.5) (see Note 1).
9. Validated cDNA constructs for BRET2; i.e. proteins of interest tagged with codon humanized Renilla Luciferase (hRLuc) and GFP10. Constructs are stored in sterile 50 mM Tris/HCl, pH 8.0 (see Notes 2 and 3).

2.2. BRET Assays

1. Sterile phosphate-buffered saline (PBS): 137 mM NaCl, 10 mM Na_2HPO_4, 2.68 mM KCl, 1.76 mM KH_2PO_4, pH 7.4.
2. Coelenterazine 400A for BRET2 assays (Biotium) and coelenterazine h (Nanolight Technology) needed for assessing *total* luminescence in samples (see Note 4).
3. 96-well plates: Isoplate-96 (White Frame Clear Well; Perkin Elmer) for measuring total fluorescence and luminescence in samples. Isoplate-96 (Black Frame White Well; Perkin Elmer) for BRET readings (see Note 5).
4. Mithras LB 940 plate reader (Berthold Technology; see Subheading 3.2 for specifications concerning filters required for different readings).
5. Bio-Rad DC Protein assay kit (Bio-Rad).

3. Methods

3.1. Cell Culture and Transfection

1. On day one of the experimental procedure, passage stock HEK293 cell cultures using trypsin/EDTA, and plate 1.2×10^6 cells in complete culture medium onto 60 mm culture dishes, such that they reach 40–60% confluence on the next day. Cells are grown at 37°C with humidified, 5% CO_2 atmosphere.
2. On day two of the experimental procedure, transfect cells with cDNA encoding the desired BRET constructs (see Notes 6 and 7). Transfection steps:
   (a) Replace culture medium with 2.5 ml of plain DMEM per 60-mm dish.
   (b) cDNAs should be diluted in 150 mM NaCl solution to a total volume of 250 μl followed by gentle mixing.
   (c) PEI in the amount of 3 μl/μg of DNA to be transfected should be diluted in sterile de-ionized water to a total volume of 250 μl, followed by gentle mixing.

(d) PEI solution should then be added to the cDNA solution, mixing gently. The mixture should then be left to incubate for 15 min at room temperature.

(e) After incubation the PEI/cDNA mixture is gently added to each culture dish, placing the pipette on the side of the dish and not directly on the cells.

(f) Culture dishes are returned to the incubator (37°C; 5% CO_2) where they will remain for 4 h. At the end of this incubation period, the transfection medium should be exchanged for 5 ml of complete culture medium supplemented with 10% FBS and the dishes returned to the incubator (see Note 8).

3. All transfection steps should be carried out in a biological hood under sterile conditions.

3.2. Generating Titration Curves to Assess the Existence of a Constitutive, Specific BRET Signal Between Two Proteins of Interest

1. BRET measurements should be carried out on the fourth day of the experiment. None of the steps involved in obtaining BRET measurements require a sterile environment.
2. The plate reader should be turned on before proceeding with steps described below.
3. Cells for this experiment should have been previously transfected with a fixed amount of donor construct (e.g. Protein A-Rluc; initial test amount: 0.2 μg DNA/60 mm petri dish) and increasing amounts of the acceptor (e.g. Protein B-GFP10: 0.001, 0.005, 0.01, 0.05, 0.1. 0.25, 0.5, 1, 3, 5 μg), each combination in separate petri dishes. Since total cDNA transfected must remain constant across all conditions, the empty vector in which the acceptor is expressed should be co-transfected in amounts that will allow to attain the same total amount of cDNA at varying levels of the GFP10 construct (see Note 9). BRET experiments also require a condition in which the donor is transfected at similar levels as the rest of samples but in the absence of acceptor (include empty vector to keep the total amount of cDNA constant). This sample will allow one to calculate net BRET values (see Step 9). Titration curves for negative controls should be prepared in the same manner as for proteins of interest, exchanging one of the BRET constructs above for a BRET-tagged protein which is known not to form part of the complex under study. A condition with a positive control to monitor whether the experiment was correctly run should also be included at cDNA levels known to produce a saturated BRET signal for that particular pair (see Note 10).
4. Cultured cells should be mechanically detached, washed, and then re-suspended in PBS at room temperature (see Note 11). Protein concentration should then be assessed in order to adjust volumes to obtain same concentrations across all samples (see Note 12).

5. 90 μl of each sample should then be added to Isoplate-96 (White Frame Clear Well) and total fluorescence measured (without adding coelenterazine) at settings that allow adequate excitation/emission by GFP10. For the Mithras plate reader, settings are as follows: reading time of 1 s, lamp energy 7,000, excitation filter of 390/20 nm. The emission reading is taken with a 515/30 nm filter. These steps will allow quantification of the total amount of fluorescence produced by the acceptor in each sample. These values will be necessary for plotting a titration curve from the data (see Step 9).
6. The same plate should then be used (covering the bottom with white paper; see Note 5) to measure total luminescence generated by coelenterazine h. At this time the stock solution for coelenterazine h should be diluted to a concentration of 10 μg/ml in PBS kept at room temperature. 10 μl of this solution are then added to each well (to a final concentration of 1 μg/ml), the plate is gently stirred, incubated 10 min in the dark at room temperature to stabilize coelanterazine h signal, then introduced into the Mithras reader using the following settings: reading time of 1 s, luminescence reading without filter. These steps will allow quantification of the total amount of luminescence produced by the donor in each sample. Luminescence values are required for plotting a titration curve from the data (see Step 9).
7. An Isoplate-96 (Black Frame White Well) plate containing 90 μl of sample/well should be prepared in order to obtain BRET2 readings.
8. Immediately after completing Step 7, the stock solution of coelantrazine 400A should be diluted to 50 μM in PBS. 10 μl should then be added to each well (to a final concentration of 5 μM), the plate gently stirred and BRET2 readings rapidly taken using the following settings for the Mithras plate reader: reading time of 0.5 s/reading, with emission filters of 410/80 nm and 515/30 nm for RLuc and GFP10, respectively.
9. Data obtained in Step 8 should be expressed as the ratio of fluorescence counts over luminescence counts to yield total BRET ratios. Subtracting total BRET ratios obtained from cells that were not transfected with the acceptor from the total BRET values obtained in cells expressing this construct, allows removal of the background from the measurement, yielding a net BRET signal. Net BRET values are then plotted vs. the ratio of corresponding total fluorescence/total luminescence readings, respectively, obtained in Steps 6 and 7 of this section. Typically, a specific interaction among proteins of interest yields a saturated curve indicating that all

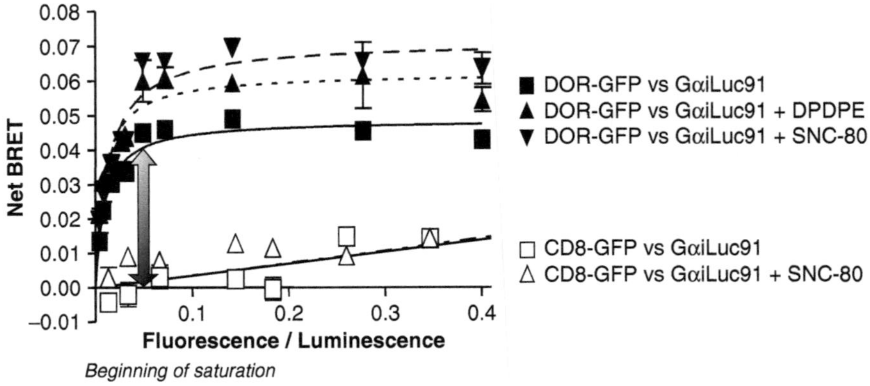

Fig. 2. Titration curves for specific and non-specific interactions between different BRET pairs. HEK293 cells were transfected with a fixed amount of Gαi1 tagged with RLuc at amino acid 91 (Gαi1-Luc91) (0.3 μg), and increasing amounts (0–3 μg) of the human delta opioid receptor (DOR) bearing GFP at its C terminus (DOR-GFP). Fixed amounts of untagged, complementary heterotrimeric Gβ1 (1.33 μg) and Gγ2 (1.33 μg) subunits were co-transfected with the Gαi1-Luc91/DOR-GFP pair. The spontaneous, hyperbolic BRET signal produced by these transfections, is typical of specific, constitutive interactions between two proteins. Modulation of the BRET signal by DOR agonists (SNC-80 and DPDPE; 10 μM, 2 min), is also typical of a specific interaction between a receptor and heterotrimeric G protein subunits. In contrast, HEK cells similarly transfected except for the substitution of DOR-GFP by an acceptor construct coding for a membrane protein that does not interact with G proteins (i.e. CD8-GFP), display minimal spontaneous transfer of energy, that does not follow saturation kinetics and is not modulated by DOR agonist SNC-80 (10 μM, 2 min). The *gray arrow* shows the ratio of donor/acceptor constructs that yield initial saturation points.

donor proteins are in the proximity of an acceptor. In contrast, non-specific interactions evaluated by the negative control produce marginal transfer of energy that does not follow saturation kinetics (Fig. 2).

3.3. Additional Controls for Establishing Specificity of Association Between BRET Constructs

BRET signals generated by specific interactions, but not those produced by negative controls, are expected to be modulated by a ligand that specifically binds one of the interacting partners in the complex of interest. Hence, the objective is to determine whether a ligand for one of the proteins in the complex modifies the spontaneous signal generated by BRET partners of interest, leaving the signal generated by negative controls unchanged.

3.3.1. Modulation by Ligands

1. To accomplish this goal, the steps described in Subheading 3.2 should be carried out using parallel sets of samples, one exposed to vehicle and the other to the ligand.
2. 10 μl of vehicle or ligand should be introduced in each well before coelenterazine (see Notes 13 and 14). In order to compensate for this injection volume, plates should be initially loaded with 80 μl of sample. Calculation of net BRET values and plotting of results should proceed as in Step 9 of the previous section (see Fig. 2).
3. Ligands should be used at concentrations known to produce a maximal functional response. Treatment duration for assessing conformational changes is usually short and generally does not require more than 2 min treatment.

3.3.2. Competition Assays

Typically, a signal arising from the specific interaction of two BRET constructs may be competed by keeping their expression constant and progressively increasing that of an untagged version of one of the interacting BRET partners. For this purpose, HEK293 cells should be transfected with a fixed amount of donor/acceptor constructs that will allow displacement at low expression levels of the competitor. This is usually achieved by using cDNA amounts that yield initial saturation points in the titration curve (indicated with arrow in Fig. 2).

1. The amounts of competitor cDNA to be transfected should be determined empirically. (We usually start with 0.000, 0.003, 0.017, 0.033, 0.167, 0.333, 0.833, 1.667, 3.333 μg/60 mm petri dish). Since total cDNA transfected must remain constant for all conditions, the empty vector in which the competitor is expressed should be co-transfected such that the same total amount of cDNA is maintained in cells expressing different levels of the competing construct.
2. Measurements should be taken on the fourth day of the experimental protocol repeating Steps 4–9 of Subheading 3.2. Total fluorescence (Step 5) and total luminescence (Step 6) permit control for constant expression of donor and acceptor BRET constructs at increasing levels of the competitor. BRET readings (Step 8) should be expressed as net BRET (Step 9). Net BRET values should be plotted as a function of the ratio of μg of competitor/μg of the corresponding BRET construct. The plot should yield a competition curve of the type shown in Fig. 3.

3.4. Ligand-Induced Changes in the Spontaneous BRET Signal Generated by Constitutively Interacting Proteins

1. cDNA for BRET constructs that permit monitoring the same interaction from different vantage points (e.g. Protein A-Luc$_x$ vs. Protein B-GFP10; Protein A-Luc$_y$ vs. Protein B-GFP10; Protein A-Luc$_z$ vs. Protein B-GFP10; Fig. 1b) should be separately transfected in sufficient amounts to achieve a saturated net BRET signal for each pair. cDNA amounts that are necessary to achieve saturation are inferred from the corresponding titration curves. Transfections should be performed as in Subheading 3.1.
2. On day four of the experimental protocol, BRET measures should be obtained as described in Steps 7 and 8 of Subheading 3.2, and injection of vehicle or ligand as recommended in Note 13.
3. Data for each pair should be first expressed as net BRET values obtained in presence or absence of different ligands. Ligand-induced changes in net BRET should then be calculated by subtracting the net BRET signal observed in the absence of ligand from similar value obtained in its presence. Figure 4 shows examples of ligand-induced changes in net

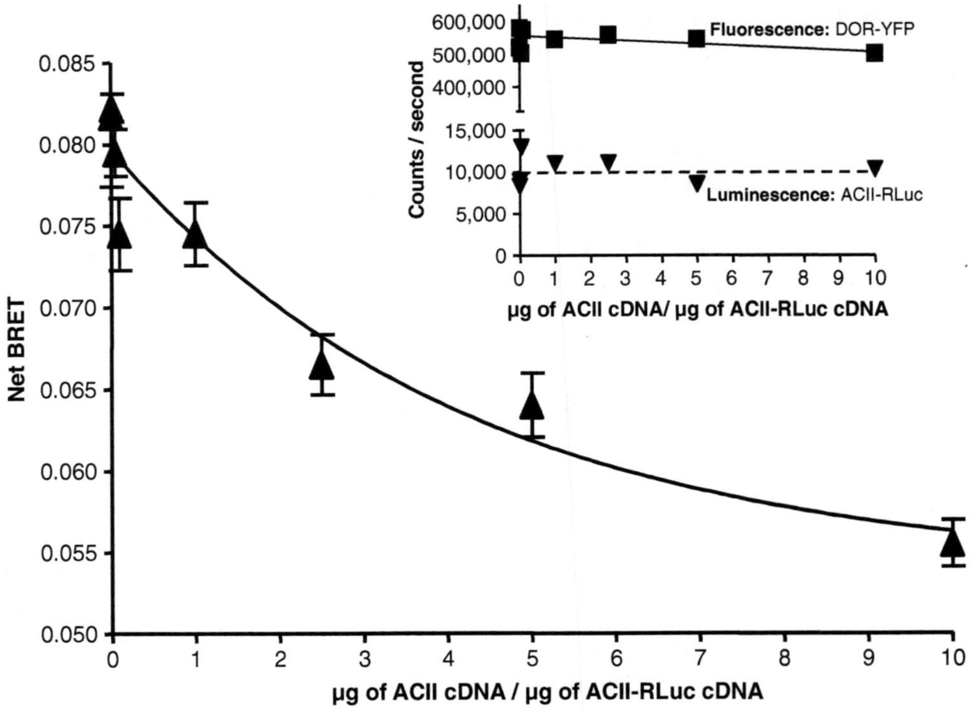

Fig. 3. BRET competition assays. Fixed amounts of adenylate cyclase II-Luc (ACII–Luc) and DOR-YFP yielding one of the initial saturation points in the titration curve for this pair (ACII-Luc: 0.33 μg; DOR-YFP: 0.33 μg) were co-transfected into HEK 293 cells together with increasing amounts of untagged ACII (0–6.33 μg). The reduction in energy transfer caused by progressively increasing ACII over its Luc-tagged counterpart is considered as evidence of the specific interaction between BRET constructs. *Inset* shows that the reduction in energy transfer ("competition") took place at constant levels of expression of donor and acceptor constructs.

BRET for three different BRET pairs monitoring the same interaction (Gα vs. Gγ; see legend to Fig. 4 for interpretation of results).

4. Notes

1. To prepare a stock solution (1 mg/ml) of PEI, 25 mg should be mixed with 20 ml of de-ioinized H_2O using 1 N HCl to reduce the pH to ≅3 to facilitate dissolution. The mixture should then be vortexed and once solution is attained (15–30 min), the pH must be adjusted to 6.5–7.5 using 1 N NaOH before adding de-ionized water to a final volume of 25 ml. We have found that preparing PEI in this manner allows it to be stored at 4°C for more than 1 month.
2. BRET assays may be classified into BRET1 or BRET2 depending on the kind of donor-acceptor pair used to reveal interactions between proteins of interest. In BRET1 RLuc oxidizes

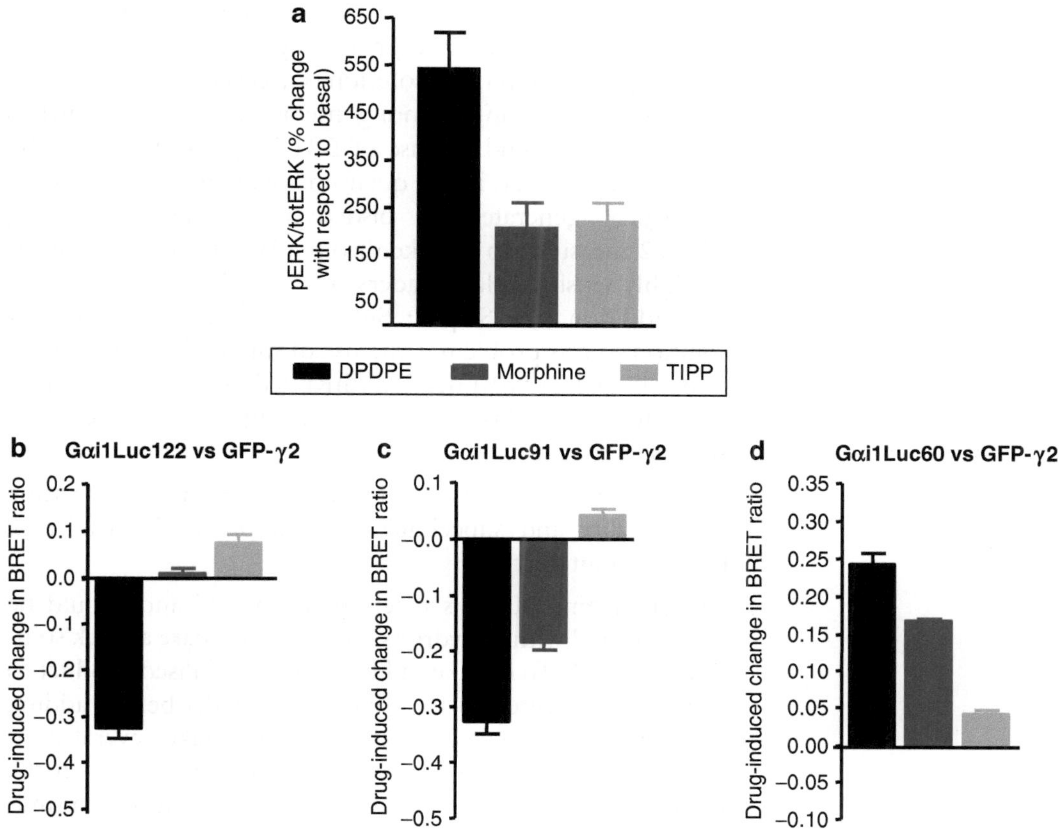

Fig. 4. Binding of DOR agonists to the receptor induces a conformational re-arrangement among heterotrimeric G protein subunits: Evidence for ligand-specific conformational changes. Panel (**a**) shows the relative efficacy of different DOR ligands in the MAPK pathway. Panels (**b–d**) show ligand-induced BRET changes at the Gα–Gγ interface as assessed by BRET pairs containing the acceptor at a fixed position (N terminus of the Gγ2 subunit) and the donor at different locations within the Gα subunit. Results show that: (1) the direction of ligand-dependent BRET changes is determined by tag position within Gα (e.g. compare **c** and **d**), (2) drugs of similar signalling efficacy like morphine and TIPP induce opposing BRET changes at the Gαi1Luc91/GFP-γ2 BRET pair, and (3) BRET changes produced by DPDPE and morphine differ only in magnitude. Intuitively, the first observation is better explained by a conformational re-arrangement of G protein subunits than by a change in the total number of heterotrimeric complexes containing Gα–Gγ interaction. The interpretation of the second observation may also be done in an intuitive manner, i.e. while morphine causes the acceptor on the N terminus of Gγ2 to separate from the donor at position 91 of the Gα subunit (decrease in BRET) TIPP induces an approach of both tags (increase in BRET). These ligand-specific changes are consistent with the notion that despite similar ability to stimulate the MAPK pathway, DOR activation by morphine and TIPP produces distinct conformational re-arrangements within the G protein heterotrimer. Similarly, opposing BRET changes produced by TIPP and DPDPE (**b**, **c**) indicate that these two agonists also produce ligand-specific conformational changes. In contrast, differences in the magnitude of BRET changes, as those observed between DPDPE and morphine do not exclude the possibility that the two agonists stabilize different amounts of the same active state of the receptor. To be able to confidently conclude as to whether DPDPE and morphine induce ligand-specific changes it would be necessary to compare the effect of both ligands at yet another set of BRET pairs (figure modified from ref. **13**).

coelenterazine h to excite YFP (14, 30). BRET2 assays rely on oxidation of bisdeoxycoelenterazine [coelenterazine 400A (Biotium) or DeepBlueC (Perkin Elmer)] to produce the excitation of GFP10 or GFP2 (11, 30). The practical difference between the two assays is a greater separation between donor and acceptor spectra in BRET2, resulting in reduced background and therefore greater signal resolution than in BRET1. A high level of resolution is necessary when studying conformational changes among interacting proteins, which is why we recommend the use of BRET2 pairs (11, 13). The drawback, however, is low quantum yield and rapid decay of the signal generated by bisdeoxycoelenterazine, forcing BRET2 measures to be taken from a larger number of cells, in highly sensitive plate readers, and rapidly after coelenterazine addition (see Step 8; Subheading 3.2). An emerging solution to this problem is the use of mutated variants of the donor known as "hRLuc2" or "hRLuc8" whose oxidation of coelenterazine yields increased light output and more stable signals (31, 32).

3. A validated BRET construct is one which maintains the same functionality and subcellular distribution as the wild-type protein of interest.
4. Coelenterazine 400A is received lyophilised and should be reconstituted with anhydrous ethanol to prepare a stock solution of 1 mM. To do so equilibrate the lyophilised product to room temperature (to be done in the dark) before adding ethanol. Vortex to resuspend (this might take several minutes) and aliquot to store away from light at −20°C. Lyophilised coelenterazine h should be reconstituted in similar manner to obtain a stock solution of 1 mg/ml. It is extremely important to use anhydrous ethanol because even small traces of water will oxidize coelenterazine.
5. In readers in which the excitation beam for fluorescence readings comes from below the plate, the use of clear wells ensures the excitation needed to measure total fluorescence. If the excitation beam comes from above, the use of plates with black wells may allow to minimize reflection and reduce background from fluorescence measures. On the other hand, when taking BRET measures in which fluorophore excitation is achieved via luminescence emitted by the donor, we prefer to use plates with white wells to increase reflection and signal intensity. This is also the reason for covering the plate's bottom when measuring total luminescence counts present in the sample (Step 6; Subheading 3.2).
6. Most frequently, constitutive complexes include the association of several proteins apart from those tagged with donor/acceptor BRET pairs. If this were the case for the complex under study, untagged, complementary proteins should be co-transfected with the BRET constructs.
7. We have observed that a 16 h delay between plating and transfection allows optimal incorporation of large amounts of DNA. Nonetheless, it should be kept in mind that more than

8 μg cDNA/60 mm dish may result in excessive cell death and insufficient material to carry out the experiment.

8. FBS interferes with the efficiency of transfection so it is preferable to remove all serum while PEI is present. On the other hand, serum deprivation may contribute to cytotoxicity. Hence, once the procedure is over, cells should be changed to medium supplemented with 10% FBS which contains survival/growth factors.
9. The amount of donor–acceptor constructs to be transfected needs to be established empirically for each pair. We have found that the amounts given in Step 3, Subheading 3.2 usually result in a good approximation to a saturated curve. If saturation is not achieved the first step should be to reduce the amount of donor transfected and test whether a saturated energy transfer is attained using the same amounts of acceptor as before. On the contrary, if the luciferase counts obtained are initially too low to produce stable transfer of energy, then the amount of cDNA for the Rluc construct should be increased.
10. Proteins acting as negative controls should be expressed at similar levels and have the same subcellular distribution as the protein they replace. Positive controls are proteins that have been previously shown to associate, producing a BRET signal (e.g. 7TMR dimerization) or alternatively, a construct containing Rluc and GFP10 joined by a linker.
11. Cells should be detached gently. If this is not possible, an option is to use PBS with 50 mM EDTA.
12. Theoretically, since BRET is a ratiometric measure, small differences in the total amount of protein present in each well should not influence readings. However, for titration curves, in which we take related readings from different plates (see Steps 5–8; Subheading 3.2) we prefer to adjust protein contents so as to reduce variability to a minimum. The minimal concentration of proteins to be used is determined by the limit of resolution of the plate reader. This limit may be determined by progressively diluting a positive control sample and monitoring the level of luminescence/fluorescence counts at which the BRET signal becomes unstable. All BRET measures should be taken in samples in which luminescence/fluorescence counts are above this limit.
13. As mentioned in Note 1, BRET2 measures should be rapidly taken after coelenterazine addition, implying that the ligand must be added immediately before introduction of the RLuc substrate. This can be easily achieved in readers that have an injector, but it is also feasible manually, as long as the time-lag between introduction of coelenterazine and BRET read-

ings remains constant across samples. If working manually, we recommend series of not more than four consecutive readings when using BRET2 and Rluc. The number of readings should not be a problem when using BRET2 in combination with RLuc2, since the luminescence signal is quite stable over time.

14. Ligand addition should not modify readings for total fluorescence and luminescence described in Subheading 3.2.

Acknowledgements

This work was supported by Discovery Grant from The National Sciences and Engineering Council of Canada (NSERC) to GP.

References

1. Yao, X., Parnot, C., Deupi, X., Ratnala, V. R., Swaminath, G., Farrens, D., and Kobilka, B. (2006) Coupling ligand structure to specific conformational switches in the beta2-adrenoceptor. *Nat Chem Biol* **2**, 417–22.
2. Granier, S., Kim, S., Shafer, A. M., Ratnala, V. R., Fung, J. J., Zare, R. N., and Kobilka, B. (2007) Structure and conformational changes in the C-terminal domain of the beta2-adrenoceptor: insights from fluorescence resonance energy transfer studies. *J Biol Chem* **282**, 13895–905.
3. Oldham, W. M., and Hamm, H. E. (2008) Heterotrimeric G protein activation by G-protein-coupled receptors. *Nat Rev Mol Cell Biol* **9**, 60–71.
4. Kapoor, N., Menon, S. T., Chauhan, R., Sachdev, P., and Sakmar, T. P. (2009) Structural evidence for a sequential release mechanism for activation of heterotrimeric G proteins. *J Mol Biol* **393**, 882–97.
5. Brinkerhoff, C. J., Traynor, J. R., and Linderman, J. J. (2008) Collision coupling, crosstalk, and compartmentalization in G-protein coupled receptor systems: can a single model explain disparate results? *J Theor Biol* **255**, 278–86.
6. Hein, P., and Bunemann, M. (2009) Coupling mode of receptors and G proteins. *Naunyn Schmiedebergs Arch Pharmacol* **379**, 435–43.
7. Citri, Y., and Schramm, M. (1980) Resolution, reconstitution and kinetics of the primary action of a hormone receptor. *Nature* **287**, 297–300.
8. Hebert, T. E., Gales, C., and Rebois, R. V. (2006) Detecting and imaging protein-protein interactions during G protein-mediated signal transduction in vivo and in situ by using fluorescence-based techniques. *Cell Biochem Biophys* **45**, 85–109.
9. Lohse, M. J., Nikolaev, V. O., Hein, P., Hoffmann, C., Vilardaga, J. P., and Bünemann, M. (2008) Optical techniques to analyze real-time activation and signaling of G-protein-coupled receptors. *Trends Pharmacol Sci* **29**, 159–65.
10. Pineyro, G. (2009) Membrane signalling complexes: implications for development of functionally selective ligands modulating heptahelical receptor signalling. *Cell Signal* **21**, 179–85.
11. Galés, C., Rebois, R. V., Hogue, M., Trieu, P., Breit, A., Hébert, T. E., and Bouvier, M. (2005) Real-time monitoring of receptor and G-protein interactions in living cells. *Nat Methods* **2**, 177–84.
12. Rebois, R. V., Robitaille, M., Galés, C., Dupré, D. J., Baragli, A., Trieu, P., Ethier, N., Bouvier, M., and Hébert, T. E. (2006) Heterotrimeric G proteins form stable complexes with adenylyl cyclase and Kir3.1 channels in living cells. *J Cell Sci* **119**, 2807–18.
13. Audet, N., Galés, C., Archer-Lahlou, E., Vallières, M., Schiller, P. W., Bouvier, M., and Pineyro, G. (2008) Bioluminescence resonance energy transfer assays reveal ligand-specific conformational changes within preformed signaling complexes containing

delta-opioid receptors and heterotrimeric G proteins. *J Biol Chem* **283**, 15078–88.

14. Angers, S., Salahpour, A., Joly, E., Hilairet, S., Chelsky, D., Dennis, M., and Bouvier, M. (2000) Detection of beta 2-adrenergic receptor dimerization in living cells using bioluminescence resonance energy transfer (BRET). *Proc Natl Acad Sci U S A* **97**, 3684–9.
15. Milligan, G., and Bouvier, M. (2005) Methods to monitor the quaternary structure of G protein-coupled receptors. *Febs J* **272**, 2914–25.
16. Marullo, S., and Bouvier, M. (2007) Resonance energy transfer approaches in molecular pharmacology and beyond. *Trends Pharmacol Sci* **28**, 362–5.
17. Bouvier, M., Heveker, N., Jockers, R., Marullo, S., and Milligan, G. (2007) BRET analysis of GPCR oligomerization: newer does not mean better. *Nat Methods* **4**, 3–4; author reply 4.
18. Salahpour, A., and Masri, B. (2007) Experimental challenge to a 'rigorous' BRET analysis of GPCR oligomerization. *Nat Methods* **4**, 599–600; author reply 601.
19. Maehle, A. H. (2005) The quantification and differentiation of the drug receptor theory, c. 1910–1960. *Ann Sci* **62**, 479–500.
20. Colquhoun, D. (2006) The quantitative analysis of drug-receptor interactions: a short history. *Trends Pharmacol Sci* **27**, 149–57.
21. Roettger, B. F., Ghanekar, D., Rao, R., Toledo, C., Yingling, J., Pinon, D., and Miller, L. J. (1997) Antagonist-stimulated internalization of the G protein-coupled cholecystokinin receptor. *Mol Pharmacol* **51**, 357–62.
22. Willins, D. L., and Meltzer, H. Y. (1998) Serotonin 5-HT2C agonists selectively inhibit morphine-induced dopamine efflux in the nucleus accumbens. *Brain Res* **781**, 291–9.
23. Azzi, M., Charest, P. G., Angers, S., Rousseau, G., Kohout, T., Bouvier, M., and Piñeyro, G. (2003) Beta-arrestin-mediated activation of MAPK by inverse agonists reveals distinct active conformations for G protein-coupled receptors. *Proc Natl Acad Sci U S A* **100**, 11406–11.
24. Audet, N., Paquin-Gobeil, M., Landry-Paquet, O., Schiller, P. W., and Pineyro, G. (2005) Internalization and Src activity regulate the time course of ERK activation by delta opioid receptor ligands. *J Biol Chem* **280**, 7808–16.
25. Groer, C. E., Tidgewell, K., Moyer, R. A., Harding, W. W., Rothman, R. B., Prisinzano, T. E., and Bohn, L. M. (2007) An opioid agonist that does not induce micro-opioid receptor--arrestin interactions or receptor internalization. *Mol Pharmacol* **71**, 549–57.
26. Urban, J. D., Clarke, W. P., von Zastrow, M., Nichols, D. E., Kobilka, B., Weinstein, H., Javitch, J. A., Roth, B. L., Christopoulos, A., Sexton, P. M., Miller, K. J., Spedding, M., and Mailman, R. B. (2007) Functional selectivity and classical concepts of quantitative pharmacology. *J Pharmacol Exp Ther* **320**, 1–13.
27. Ehlert, F. J. (2008) On the analysis of ligand-directed signaling at G protein-coupled receptors. *Naunyn Schmiedebergs Arch Pharmacol* **377**, 549–77.
28. Kenakin, T. (2005) New concepts in drug discovery: collateral efficacy and permissive antagonism. *Nat Rev Drug Discov* **4**, 919–27.
29. Kenakin, T. (2007) Functional selectivity through protean and biased agonism: who steers the ship? *Mol Pharmacol* **72**, 1393–401.
30. Pfleger, K. D., Seeber, R. M., and Eidne, K. A. (2006) Bioluminescence resonance energy transfer (BRET) for the real-time detection of protein-protein interactions. *Nat Protoc* **1**, 337–45.
31. Loening, A. M., Fenn, T. D., Wu, A. M., and Gambhir, S. S. (2006) Consensus guided mutagenesis of Renilla luciferase yields enhanced stability and light output. *Protein Eng Des Sel* **19**, 391–400.
32. Kocan, M., See, H. B., Seeber, R. M., Eidne, K. A., and Pfleger, K. D. (2008) Demonstration of improvements to the bioluminescence resonance energy transfer (BRET) technology for the monitoring of G protein-coupled receptors in live cells. *J Biomol Screen* **13**, 888–98.

Part III

Receptor–Receptor Interactions

Chapter 8

Reconstitution of G Protein-Coupled Receptors into a Model Bilayer System: Reconstituted High-Density Lipoprotein Particles

Gisselle A. Vélez-Ruiz and Roger K. Sunahara

Abstract

Reconstituted high-density lipoprotein particles (rHDL) are powerful platforms used as a model phospholipid bilayer system to study membrane proteins. They consist of a discoidal-shaped planar bilayer of phospholipids that is surrounded by a dimer of apolipoprotein A-I (apoA-I). The amphipathic nature of apoA-1 shields the hydrophobic acyl chains of the lipids from solvent and keeps the particles soluble in aqueous environments. These monodispersed, nanoscale discoidal HDL particles are approximately 10–11 nm in diameter with a thickness that is dependent on the length of the phospholipid acyl chain. Reconstituted HDL particles can be assembled in vitro using purified apoA-1 and purified lipids. Investigators have utilized this model bilayer system to co-reconstitute membrane proteins, and take advantage of the small size and its monodispersion. Our laboratory and others have utilized the rHDL approach to study the behavior of G protein-coupled receptors. In this chapter, we describe strategies for the preparation of rHDL particles containing GPCRs in their monomeric form and discuss various methodologies used to analyze the reconstituted receptor function.

Key words: Apolipoprotein A-I, High-density lipoprotein particles, Receptor, 1-Palmitoyl-2-oleoyl-*sn*-glycero-3-phosphocholine, 1-Palmitoyl-2-oleoyl-*sn*-glycero-3-[phosphor-*rac*-(1-glycerol)], Monomer, Oligomer

1. Introduction

G protein-coupled receptors (GPCRs) are the largest family of integral membrane proteins. Their vast diversity and their importance in cellular signaling make them prime therapeutic targets constituting nearly 50% of available drugs. Although their significance in biomedical research is undeniable, their study has been limited by the lack of experimental systems that resemble their

Louis M. Luttrell and Stephen S.G. Ferguson (eds.), *Signal Transduction Protocols*, Methods in Molecular Biology, vol. 756,
DOI 10.1007/978-1-61779-160-4_8,

natural environment. Nevertheless, over the past decades membrane protein research has escalated due to a diverse array of membrane modeling systems such as detergent micelles, bicelles, and liposomes (1–10). These have allowed the functional and structural characterization of a great number of membrane proteins, in particular GPCRs. However, in some instances it is unclear how these systems mimic the natural milieu. In most cases, the orientation and oligomerization state of the reconstituted proteins cannot be determined.

Recently, a new class of model membranes has been developed to study the function of isolated membrane proteins especially GPCRs (4). Now several integral membrane proteins have been reconstituted into rHDL particles such as cytochrome 450 (3), bacteriorhodopsin (2), bacterial chemoreceptor (5), voltage dependent anion channel (VDAC-1) (6), and EGF receptor (7) to name a few. In this approach, membrane proteins in detergent micelles such as GPCRs may be reconstituted into the phospholipids bilayers of a discoidal high-density lipoprotein (HDL) particle (Fig. 1). The reconstituted HDL (rHDL) particles are monodispersed, homogenous and in the case of GPCRs, preferentially incorporate monomeric receptors (1). Several forms of rHDL particles have been described and have successfully been used to reconstitute membrane proteins, e.g., nanodiscs and NABBs (nanoscale apolipoprotein-bound bilayers) (8).

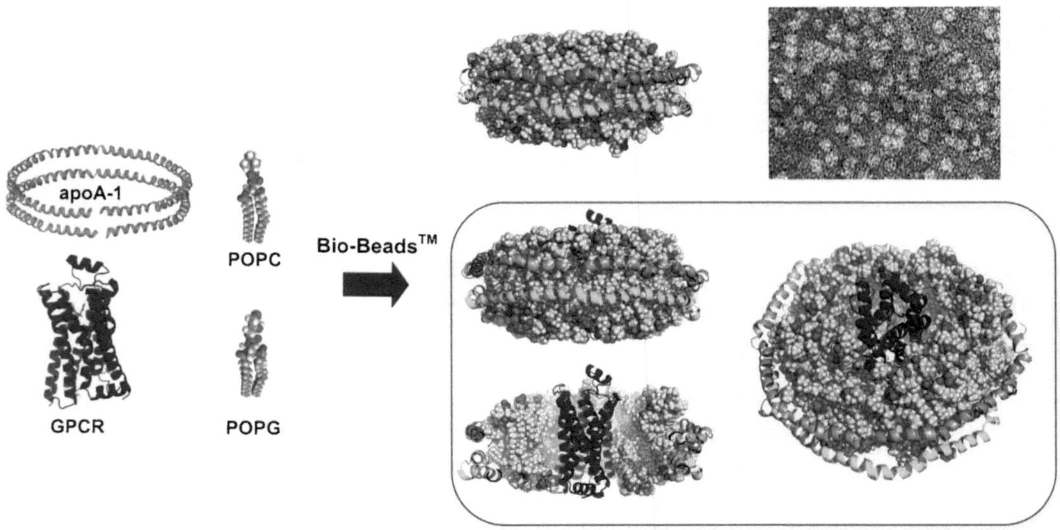

Fig. 1. Reconstitution of a prototypical GPCR into rHDL particle. Illustration of the procedure for the reconstitution of the β_2AR (PDB: 2RH1) into rHDL particles. Detergent-solubilized purified lipids and purified β_2AR are incubated with purified apolipoprotein A1 (apo-A1) as described in the text. Bilayer formation and self-assembly of the rHDL particle accompanies the detergent removal step through the addition of BioBeads™. Both empty and β_2AR-containg rHDL particles are illustrated (the latter are illustrated from multiple perspectives). Also illustrated is the electron micrograph of a typical rHDL preparation (Whorton and Sunahara, unpublished). The coordinates for the rHDL particle were based on the model reported by Segrest et al. (37) and used with the permission of Dr. Stephan Harvey (Georgia Institute of Technology).

For the remainder of this chapter we will simply refer to these apolipoprotein particles as rHDL particles. A variety of GPCRs have now been reconstituted into rHDL particles: rhodopsin (8–10), β_2-adrenergic receptor (1) and the μ-opioid receptor (11), each fully capable of activating its G protein when reconstituted as monomers in rHDL particles. Furthermore, reconstituted receptors display strong allosteric modulation by G proteins as well as arrestin (9).

In this chapter, we discuss the methodology behind the formation and incorporation of GPCRs in rHDL particles. We will discuss in detail the isolation, purification of apolipoprotein A-I, as well as the incorporation of GPCRs into rHDL particles.

2. Materials

2.1. Purification of Wild-Type ApolipoproteinA-I (WT apoA-I)

1. Human Serum stored at –20°C in 10 mM $CaCl_2$.
2. Cibacron blue F3GA-agarose resin (Sigma).
3. Q Sepharose column (Amersham Pharmacia).
4. SP Sepharose column (Amersham Pharmacia).
5. Superdex 200 size exclusion column (Amersham Pharmacia).
6. Amicon Centricon 10,000 MWCO (Millipore).
7. Buffer A: 50 mM Tris–HCl, pH 8.0, 1 mM $CaCl_2$, 3 M NaCl, and 5 mM EDTA.
8. Buffer B: 50 mM Tris–HCl, pH 8.0, 1 mM $CaCl_2$, and 5 mM EDTA.
9. Dilution Buffer: 25 mM Tris–HCl, pH 8.0, 1 mM $CaCl_2$, 5 mM EDTA, 0.2% Triton X-100.
10. Buffer C: 20 mM Tris–HCl, pH 8.0, 1 mM $CaCl_2$, 5 mM EDTA, and 0.1% Triton X-100.
11. Exchange buffer: 100 mM K-acetate, pH 5.0, 1 mM EDTA, 0.1% Triton X-100.
12. Buffer D: 25 mM K-acetate, pH 5.0, 1 mM EDTA, 0.1% Triton X-100.
13. Buffer E: 20 mM Hepes, pH 8.0, 100 mM NaCl, 1 mM EDTA.

2.2. Purification of Recombinant ApolipoproteinA-I (apoA-I)

1. Luria Broth (LB) medium containing 50 μg/ml of carbenicillin.
2. Isopropyl-β-D-thiogalactopyranoside (IPTG).
3. Nickel-nitrilotriacetic acid column (Ni-NTA; Qiagen).
4. Superdex 75 column (Amersham Pharmacia).

5. Amicon Centricon 10,000 MWCO (Millipore).
6. Buffer A: 10 mM Tris–HCl, pH 8.0, 100 mM NaH_2PO_4, 6 M Guanidine hydrochloride, and 1% Triton X-100.
7. Buffer B: 10 mM Tris–HCl, pH 7.0, 100 mM NaH_2PO_4, 6 M Guanidine hydrochloride, and 1% Triton X-100.
8. Buffer C: 50 mM NaH_2PO_4, pH 8.0, 300 mM NaCl, and 1% Triton X-100.
9. Buffer D: 50 mM NaH_2PO_4, pH 8.0, 300 mM NaCl, 250 mM Imidazole, and 1% Triton X-100.
10. Buffer E: 20 mM Hepes, pH 8, 100 mM NaCl, 1 mM EDTA, 20 mM sodium cholate.

2.3. β_2 Adrenergic Receptor Purification

1. Sf9 and Hi5™ cells.
2. Sf900 (Gibco) supplemented with 1% FBS or Insect Xpress (Lonza) depending on the cell line being used.
3. Transfer vector: pFastBac (Invitrogen).
4. Human β_2 Adrenergic Receptor DNA can be requested from Dr. Brian Kobilka (Stanford University).
5. *n*-dodecyl-β-D-maltoside (DDM; Dojindo Molecular Technologies).
6. 50 mM Tris–HCl, pH 7.4, and 150 mM NaCl (TBS).
7. Protease inhibitors (PI): 3.2 μg/ml leupeptin, 3.2 μg/ml ovomucoid trypsin inhibitor, 17.5 μg/ml phenylmethanesulfonyl fluoride, 16 μg/ml tosyl-L-lysine-chloromethylketone (TLCK), 16 μg/ml tosyl-L-phenylalanine chloromethyl ketone (TPCK).
8. Talon® metal-chelate affinity column (Clontech).
9. Source Q anion exchange column (GE Healthcare).
10. Superdex 200 size exclusion column (GE Healthcare).
11. M1-Flag affinity chromatography (Sigma).
12. Buffer A: 50 mM Tris–HCl, pH 8.0, 50 mM NaCl and PIs.
13. Buffer B: 50 mM Tris–HCl, pH 8.0, 300 mM NaCl, 0.1% DDM and PIs.
14. Buffer C: 50 mM Tris–HCl, pH 8.0, 50 mM NaCl, 0.1% DDM, 2.5 mM Imidazole and PIs.
15. Buffer D: 50 mM Tris–HCl, pH 8.0, 50 mM NaCl, 0.1% DDM, 100 mM Imidazole and PIs.
16. Buffer E: 20 mM Hepes pH 8.0, 1 mM EDTA, 0.1% DDM and PIs.
17. Buffer F: 20 mM Hepes pH 7.5, 100 mM NaCl, 1 mM $CaCl_2$, 0.1% DDM.
18. Buffer G: 20 mM Hepes pH 7.5, 100 mM NaCl, 1 mM EDTA, 0.1% DDM, and 200 μg/mL Flag peptide (Invitrogen).

2.4. In Vitro Reconstitution of rHDL

1. 1-palmitoyl-2-oleoyl-*sn*-glycero-3-phosphocholine (POPC) (Avanti Polar Lipids, Albaster, AL). Light sensitive, store at −20°C.
2. 1-palmitoyl-2-oleoyl-*sn*-glycero-3-[phosphor-*rac*-(1-glycerol)] (POPG) (Avanti Polar Lipids, Albaster, AL). Light sensitive, store at −20°C.
3. Sodium Cholate (Sigma, St. Louis, MO).
4. HNE: 20 mM Hepes, pH 8.0, 100 mM NaCl, 1 mM EDTA.
5. BioBeads™ (Bio Rad, Hercules, CA) at 0.5 mg/mL.
6. Superdex 200 size exclusion column (GE Healthcare).

3. Methods

3.1. High-Density Lipoprotein: Apolipoprotein A-I

Apolipoprotein A-I (apoA-I) is the major component of high-density lipoprotein particles (HDL). It contains a globular domain in the N terminus and a lipid-binding domain in the C-terminal (residues 44–243). Two models have been proposed to explain its geometry; "double belt" (12–14), and the "picket fence" model (15). The "picket fence" model proposes that apoA-I forms a series of amphipathic short, 22 residue, α-helices that span the lipid bilayer (perpendicular to the bilayer) surrounding the particle and keeping the acyl moieties of the lipids protected from solvent (15). However, recent evidence from mutagenesis studies (16), cross-linking (17, 18) FRET (19), fluorescence spectroscopy (20), and infrared spectroscopy (21) studies suggest that the two molecules of apoA-I form continuous but antiparallel amphipatic α-helices that are aligned with the plane of the bilayer. ApoA-I therefore serves as a molecular belt surrounding an island of phospholipids (Fig. 1).

Several native and recombinant forms of apoA-I have been used to support the bilayer such as apoA-I from human and zebrafish (zap1) as well as membrane scaffolding proteins (MSPs). For the incorporation of membrane proteins such as GPCRs into HDL particles it is possible to use both human and recombinant apoA-I interchangeably. Both can be purified to >90% homogeneity and have been shown to form stable and homogenous particles (1, 11). For recombinant proteins we can take advantage of bacterial expression as well as engineered affinity tags (6xHis) and metal-chelate affinity chromatography. ApoA-I, however, is highly abundant in serum (>1 g per liter of serum) making it a rich and inexpensive source of protein. The following paragraphs summarize the purification of native and recombinant forms of apoA-I.

3.1.1. Wild-Type Apolipoprotein AI

1. WT human apoA-I is purified from serum by a protocol adapted from Gan et al. (22) by Whorton et al. (1). Frozen serum (200–1,000 mL at −20°C in 10 mM $CaCl_2$) is thawed at 37°C, filtered through cheesecloth, and centrifuged to pellet any debris (5,000 × *g* for 10 min).
2. The clarified serum is diluted to a final concentration of 50 mM Tris–HCl, pH 8.0, 1 mM $CaCl_2$, 3 M NaCl, and 5 mM EDTA (Buffer A). The solution is mixed with equal serum volume of Cibacron blue F3GA-agarose resin equilibrated in Buffer A, and stirred for 30 min at room temperature (RT).
3. The resin is filtered through a Whatman #1 filter in a Büchner funnel. The resin cake is resuspended in 3X resin volume of Buffer A and then re-filtered as before. The resin is washed until absorbance at 280 nm of the filtrate is less than 0.025, then washed two more times with Buffer B. The cake is resuspended in an equal volume of Buffer B and loaded onto an empty column. The remaining apoA-I protein is eluted with Buffer B containing 5 mM Cholate. After this wash, apoA-I is ~80–90% pure.
4. ApoA-I containing fractions are pooled and concentrated using an Amicon stirred ultrafiltration cell affixex with a 10,000 MWCO filter and then diluted (1:1) in Dilution Buffer.
5. The solubilized material is loaded into a 70 ml Q Sepharose column equilibrated in Buffer C.
6. The protein is eluted with a shallow linear gradient with Buffer C with 1 M NaCl. ApoA-I usually elutes around 100–150 mM NaCl. Peak fractions are exchanged into 100 mM K-acetate, pH 5.0, 1 mM EDTA, 0.1% Triton X-100 and applied to a SP Sepharose column equilibrated in Buffer D and eluted with a linear gradient of Buffer D with 1 M NaCl.
7. Triton X-100 is exchanged for cholate by applying SP Sepharose fractions to a Superdex 200 size exclusion column in Buffer E with 20 mM cholate at 4°C.
8. ApoA-I fractions are pooled and concentrated to at least 10 mg/ml, dialyzed overnight against Buffer E with 5 mM cholate and stored in −80°C until further use.

3.1.2. Recombinant Apolipoprotein A-I

Secondary structure predictions, biochemical and crystallographic evidence suggest that the carboxy-terminal of apoA-I (residues 44–243) performs the predominant role of maintaining the discoidal HDL structure (23). Deletion of the globular N-terminal region (aa 1–43) does not alter the HDL structure but does impair the capacity of apoA-I to interact with the ABC transporter (ABCA1). ABCA1 serves as the putative cholesterol transporter to load cellular cholesterol onto HDL in vivo (24).

The C-terminal half of apoA-I (44–243) can be expressed as a recombinant protein in *Escherichia coli* as a hexahistidine-tagged (6xHis) version (11). Analysis of the secondary structure of apoA-I suggests that the C-terminal region is composed of repeating spans of 22 residues separated by proline residues. Positioning of the proline residues, reputable α-helix disrupters, fueled the original "picket fence" model but are now thought to serve as kinks within the "helical belt" that surrounds the island of phospholipids. Sligar and colleagues have taken advantage of these discrete repeating spans to engineer larger rHDL particles by inserting additional 22-residue cassettes within apoA-I (25). The modified forms of the apolipoprotein, termed membrane scaffolding proteins (MSPs) can produce particles with a Stokes radius of 10–13 nm, depending on the MSP subtype and the lipid:MSP ratio used during the reconstitution. The discoidal HDL formed by these MSPs are called nanodiscs. Apolipoproteins with 3 cassettes (MSP1E3) were successfully used to produce particles to accommodate the insertion of two rhodopsin molecules (9, 26). Wild-type apoA-I will comfortably generate rHDL particles of approximately 10 nm in diameter. However, increasing the lipid to protein ratios during the reconstitution can produce larger particles up to ~17 nm in diameter, albeit in a heterogeneous mixture with smaller particles (8, 23). While most reconstitution studies have been performed using sequences derived from human apoA-I, apolipoproteins from other species (*zap1*, from zebrafish, *Danio rerio*) (8) have been successfully used to support a phospholipid bilayer.

The method below describes the purification of a modified version of apoA-I with an N-terminal 43 amino acid deletion and a 6xHis tag (Δ1-43-His_6-TEV-apoA-I) expressed using a pET15b vector to transform *E. coli* cells (BL21) (11).

1. A starter culture is prepared by inoculating 20 ml of Luria Broth (LB) medium containing 50 μg/ml of carbenicillin with a single colony overnight for no more than 10–12 h at 37°C.
2. The starter culture is diluted 1:200 into the final culture. It is important that the cells are spun down and resuspended in fresh media before inoculating the final culture. Grow cells until OD_{600} reaches 0.6 (2–3 h) and induce with 1 mM isopropyl-β-D-thiogalactopyranoside (IPTG) for 3–4 h.
3. The cells are harvested by centrifugation at 4,200 ×*g* for 10 min. The cell pellets may be flash frozen in liquid nitrogen and stored at −80°C.
4. The cells are resuspended and lysed by gently vortexing in Buffer A. The lysate is fractionated by centrifugation at 10,000 ×*g* for 20 min at room temperature (RT).
5. The supernatant is loaded onto a Ni-NTA column by gravity flow. The column is washed with 10 column volumes of Buffer B and then with Buffer C.

6. Bound Δ1-43-His$_6$-TEV-apoA-I is eluted with Buffer D.
7. The fractions containing the protein of interested are further purified on a Superdex 75 column equilibrated with Buffer E.
8. Pooled Δ1-43-His$_6$-apoA-I is dialyzed against Buffer E containing 5 mM cholate. Purified protein is concentrated in an Amicon Centricon 10,000 MWCO to ~10 mg/ml, flash frozen in liquid nitrogen and stored at −80°C until use.

This tagged version of Δ1-43-His$_6$-TEV-apoA-I can be used for reconstitution experiments. HDL particles formed with the hexahistidine tag proteins are homogenous in size based on size exclusion chromatography and the incorporation efficiency is not affected. However, removal of the tag allows the separation of empty particles from those that contain hexahistidine-tagged proteins. To remove the tag the purified protein can be incubated with TEV protease.

TEV is the common name for the catalytic domain of the Nuclear Inclusion a (NIa) protein encoded by the tobacco etch virus (TEV) (27). This enzyme is commercially available; it contains a polyhistidine tag on the N terminus and a polyarginine tag on the C terminus. For digestion, we incubate TEV protease and Δ1-43-His$_6$-TEV-apoA-I overnight at a ratio of 1:7.5. After cleavage, the proteins are separated using Ni-NTA (IMAC), the TEV protease will bind to the column and the un-tagged apoA-1 will be present in the flow through.

3.2. β_2 Adrenergic Receptor (β_2AR) Purification

The β_2AR has several roles in the body, including the regulation of smooth muscle relaxation as well as other processes such as glycogenolysis and lipolysis. The β_2AR is primarily activated by endogenous epinephrine in the body. Receptor activation promotes the activation of the stimulatory G protein, G_s. GTP-bound G_s binds to and directly activates adenylyl cyclase causing an increase in levels of cAMP and activation of protein kinase A (PKA). PKA mediates various downstream effects. In recent years, work has shown that the β_2AR can signal through G protein-independent pathways, specifically arrestins (28).

To study the function of β_2AR we express either WT of CFP-fused receptor in Sf9 cells and solubilize using methods previously described (29, 30). The modified receptor is expressed using recombinant baculoviruses that were created using transfer vectors (pFastBac™) that encode a fusion protein of an N-terminal cleavable hemagglutinin signal sequence (MKTIIALSYIFCLVF), a FLAG epitope (DYKDDDD), a decahistidine tag, the monomeric and enhanced cyan fluorescent protein (Clontech), and the human β_2AR.

3.2.1. Preparation and Solubilization of Membranes

1. High titer viruses (10^7–10^8 plaque-forming units/ml) are used to infect Sf9 or High-Five™ ($2–3 \times 10^6$ cells/mL) suspension cultures at a multiplicity of infection of 0.5–1.
2. FLAG-His10-mECFP-B_2AR (CBAR) is expressed for 48–60 h.
3. After 60 h of infection, spin the cultures for 10 min at 500 × *g* to pellet cells.
4. Resuspend the cells in 1/10 the original culture volume of TBS plus PIs (see Note 1).
5. Lyse the cells by nitrogen cavitation (see Note 2).
6. Spin down debris at 500 × *g* for 10 min. Keep the supernatant and discard the pellet.
7. Spin the supernatant from the previous step for 35 min at 100,000 × *g* to pellet the membranes.
8. Resuspend the membrane containing fraction in 1/20th the culture volume with Buffer A and either store at −80°C or further process to purify receptor (see Note 3).
9. Membrane preparations are diluted to 5 mg/ml in Buffer A plus 1% DDM (w/v) final concentration and stir for 30–45 min in ice.
10. Spin down solubilized material for 35 min at 100, 000 × *g*.

3.2.2. Purification of Soluble CFP-β_2AR

1. For CFP-β_2AR: The DDM-solublized extract is applied to a Talon metal-chelate affinity column equilibrated by running 10× column volume of Buffer A with 0.1% DDM.
2. Wash the column with 10× column volume of Buffer B.
3. Wash the column with 5× column volume of Buffer C.
4. Elute eight half-column volume fractions with Buffer D.
5. Peak fractions are applied to a 1 mL Source Q anion exchange column in Buffer E.
6. CFP-β_2AR is eluted with a 15 mL 0–40% linear gradient of Buffer E with 1 M NaCl.
7. Peak fractions are pooled and resolved on a Superdex 200 size exclusion column in Buffer E with 50 mM NaCl to resolve the CFP-β_2AR from the clipped CFP.
8. The resultant CFP-β_2AR is greater than 95% pure and stored with 10% glycerol until use.

3.2.3. Purification of Soluble WT-β_2AR

1. For WT-β_2AR: The solubilized extract can be purified with a metal-chelate affinity column following the procedure described above.
2. $CaCl_2$ is added to the peak fractions from the previous step to a final concentration of 1 mM.

3. The β_2AR is purified by M1-Flag affinity chromatography (Sigma) equilibrated with Buffer F.
4. Wash the column with 10× column volume of Buffer F.
5. Elute eight half-column volume fractions with Buffer G (see Note 4).
6. Determine the concentration of functional, purified receptor using a saturating concentration of [^{3}H]-dihydroalprenolol as previously described (29) (see Note 5).

3.3. In Vitro Reconstitution of GPCRs in rHDL

3.3.1. Empty rHDL

To date a number of membrane proteins have been functionally incorporated into rHDL particles, nanodiscs or NABBs, including GPCRs (1, 2, 10, 11, 26), cytochrome P450 (3), the protein pump bacteriorhodopsin (2), SecYEG heterotrimer (31), and bacterial chemoreceptors (32), among others. Reconstituted HDL particles have been used in a great number of biochemical and biophysical assays, demonstrating their utility in membrane protein research.

HDL particles are formed by adding apoA-I to detergent-solubilized phospholipids in a specific ratio followed by removal of the detergent. Detergent removal is achieved by addition of hydrophobic adsorbent BioBeads™. For the formation of empty HDL particles, lipids are solubilized in sodium cholate in a three-fold molar excess to the lipid. The selection of lipids is critical for particle formation and efficient protein incorporation; it must be optimized for each protein of interest. A typical reconstitution has ratios of 1:50 to 1:100 apo:lipid ratio and a final concentration of 24 mM cholate.

1. POPC and POPG are combined at a 3:2 molar ratio, a mixture that mimics the zwitterionic environment of the cell membrane.
2. Dry lipids under argon (or nitrogen) from a chloroform solution and place in a vacuum dessicator for 30–60 min to remove residual traces of chloroform.
3. Solubilize lipids in HNE + 50 mM cholate, again flush with argon and reseal. For some lipids (e.g., POPC and POPG), it is hard to solubilize them with the working concentration of cholate. It helps to first solubilize in a higher concentration of cholate, then dilute this down to the working concentration before adding the receptor and apoA-I.
4. Allow the detergent buffer to solubilize the lipids for about 10–15 min; protect from light. The solution should become clear or translucent, but with no particles. Make sure to solubilize all lipids, especially on the sides of the tube.
5. Further dilute with HNE to achieve the desired concentration of cholate.

6. Finally, a concentrated stock of apoA-I is added such that the final concentrations of the components are 24 mM detergent, 8 mM lipids, and 100 μM apoA-I.
7. Incubate the solution for 1–2 h at 4°C. ($Tm_{POPC/POPG} = -2°C$).
8. The mixture is added to BioBeads (0.05 mg/ml reconstitution volume) and incubated for a minimum of 3 h to remove the detergent.
9. Samples can be store at 4°C until further use (see Note 6).

3.3.2. β_2 Adrenergic Receptor Reconstitution into rHDL

1. For the β_2 Adrenergic Receptor, zwitteronic environment is achieved by mixing POPC and POPG at a molar ratio of 3:2. The lipids are prepared the same way as for empty rHDL (see Note 7).
2. Add purified β_2AR (DDM solubilized and affinity purified as described in Subheading 3.2. (33)) and 100 μM apoA-I (see Note 8).
3. Following incubation on ice (1–2 h), BioBeads™ are added to the mix for detergent removal and spontaneous formation of homogenous particles that contain monomeric receptor.
4. The reconstituted sample can be centrifuged briefly to sediment the BioBeads. Recover the supernatant without disturbing the beads.
5. Subsequent size exclusion chromatography of the reconstituted sample in a Superdex 200 column will allow determination of size and homogeneity of the rHDL particles; size variation in the peak fractions indicate improper particle formation. In addition, empty particles can be separated from those containing membrane proteins by affinity chromatography if a tag is present (see Fig. 2 and Note 9).

3.4. GPCR Oligomers in rHDL

Clearly one of the major advantages of the rHDL technology is the isolation of homogeneous populations of reconstituted particles where the oligomeric state of the GPCRs may be controlled. The physical size limitations of the rHDL particle and the conditions under which the receptors are reconstituted can dictate the stoichiometry of receptors reconstituted within each particle. However, a technically challenging step is determining how many receptors are reconstituted within each particle. Equally important is the assessment of the GPCR:rHDL ratio based on biochemical properties that are not related to receptor function or activity. Since cooperatively factors, such as potential consequences of oligomerization, may influence receptor activity (e.g., radio-ligand binding or photoactivation), it is critical that the receptor:rHDL be quantified using biochemical properties of the receptor rather than activity alone.

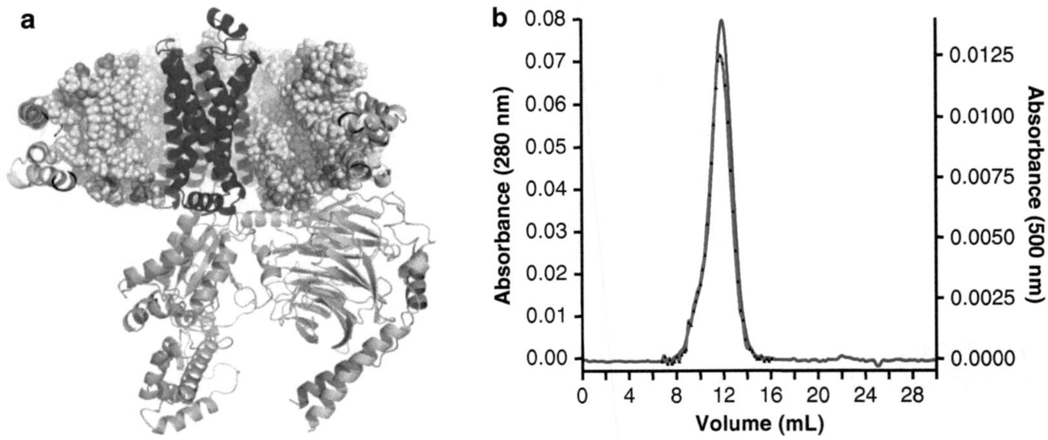

Fig. 2. Reconstitution of a prototypical GPCR with G proteins in rHDL particles. (**a**) Illustration of the β_2AR (PDB:2RH1) and heterotrimeric G protein (Gi$\alpha\beta\gamma$, PDB:1GP2) reconstituted into rHDL particles. (**b**) Size exclusion chromatography of rhodopsin-rHDL particles. Indicated are the UV absorption of protein (A_{280}) and dark-state rhodopsin (A_{500}). Adapted from Whorton et al. (1).

While it has been clearly demonstrated that monomeric GPCRs are fully functional in apolipoprotein particles, the contributions of oligomeric forms to signaling are far more complicated. Reconstituted particles containing two or more rhodopsin molecules (9, 26) have been reported to display activity (i.e., interaction with G protein or arrestin) that is lower than expected for two receptors. The assumption, of course, is that both reconstituted receptors share the same topology within the lipid bilayer, i.e., the N-termini are both on the same side of the membrane. However, as acknowledged by each group and as demonstrated by Banerjee et al. (8), this assumption should be taken with caution. Single molecule studies by electron microscopy (EM) analysis indicate that insertion of two or more rhodopsins into rHDLs leads to a random orientation where dimers equally distributed between parallel and antiparallel forms. Thus only ~50% of the particles containing two receptors have inserted with the correct topology, i.e., with their N-termini on the same side of the phospholipid bilayer. With this caveat in mind, the data suggests that dimeric forms couple to transducin half as efficiently as monomeric forms (8, 26). Similar observations are noted for arrestin–dimeric rhodopsin interactions (9). Additionally, in many cases receptor oligomerization has been implicated in the complex binding characteristics of various ligands to hormone receptors in cellular systems. Although it would seem likely that the rHDL system could be a logical preparation to address these properties, such emergent properties of oligomeric receptors reconstituted in rHDL particles have yet to be reported.

These examples illustrate the usefulness of rHDL particles as a powerful approach to study membrane proteins such as GPCRs. They provide the means to maintain membrane proteins embedded in a lipid environment as a soluble and monodispersed particle, mimicking that of the plasma membrane. This approach has allowed investigators to address a central question regarding the role of receptor oligomerization: what is the minimal functional unit required to activate a G protein? The physical properties of rHDL particles and their accessibility to G proteins on both sides of the bilayer are qualities that make this approach an adequate system to investigate this question. Further studies will undoubtedly reveal the functions of oligomerization and the contributions it makes toward the recruitment of signaling partners.

As with other model membrane preparations, rHDL particles have been limited to the use of purified-functional proteins (especially GPCRs), thus limiting the number of targets that can be studied. This has recently been overcome with a plant cytochrome P450 where it was expressed, solubilized, and directly incorporated in particles (34). In addition, reconstituted HDL technology has recently been commercialized by Invitrogen and sold as an in vitro membrane protein expression kit (MembraneMax™). Reconstituted HDL approaches have garnered subtle criticism as it is considered an ultra reductionist and simplistic method allowing direct characterization of GPCRs and its cognate partners in vivo, but removing the plethora of signaling partners that could normally have either direct or indirect effects on their activity in the cell. That being said, as new membrane protein interactors are identified we predict that the utilization of rHDL methodologies will continue to provide valuable information about membrane protein biophysics complementing other known membrane model systems.

4. Notes

1. To help maintain functional and stable receptor, 10 μM alprenolol can be added to the buffers used in the membrane solubilization. It should not be present in the buffers used in the consecutive steps as it can remain bound and interfere with subsequent functional assays.
2. The nitrogen cavitation chamber should be pre-chilled and filled with Nitrogen up to 600 psi. Allow it to equilibrate in the cold room for 30–45 min to allow optimal cell rupture.
3. All receptor purification procedures are performed at 4°C unless noted.
4. Add half a column volume at a time; let incubate 3–5 min on the column before adding the next elution.
5. WT-β_2AR can be further purified to separate active vs. inactive receptor. Flag-purified receptor can be purified by

alprenolol-Sepharose chromatography as previously described (29, 30). The peak fractions from this column can be loaded onto M1-Flag resin in order to remove free alprenolol. Two liters of Sf9 cells typically yield 500 μL of a 5 μM solution of β_2AR.

6. Reconstituted HDL particles are composed of two molecules of apoA-I, for particles with a Stokes diameter of ~10.8 nm the final preparation consists of 1 molecule of apoA-I per 80 lipid molecules.

7. In some cases (e.g., μ-opioid receptor) the lipid component may be modified with porcine polar brain lipid extract (Avanti Polar Lipids) in addition to POPC and POPG for a final concentration of 7 mM lipids and at a molar ratio of 1.07:1.5:1 brain lipid:POPC:POPG.

8. The final concentration of the GPCR varies from 100 nM to 1 μM, but must comprise no more than 20% of the total reconstitution volume in order to minimize detergent concentration.

9. It is critical that the particles are stored at 4°C in the absence of divalent cations, as we have observed (unpublished results) that the presence of <5 mM $MnCl_2$, $CaCl_2$, and/or $MgCl_2$ promotes rHDL aggregation. HDL particles have been shown to form "rouleau" or stacks of particles resembling stacks of coins (35). The interaction between HDL particles is thought to be facilitated through the bridging action of divalent cations between the phospholipid headgroups of each HDL particle. HDL aggregation leads to the loss of protein yield, homogeneity, and unknown stoichiometries. In addition, HDL particles are sensitive to detergents where exposure may lead to instantaneous re-solubilization of both membrane proteins and the lipids. To verify the homogeneity of the particles it is useful to analyze the samples through a size exclusion chromatography column, although dynamic light scattering (36) and ultracentrifugation may also be used (31).

Acknowledgments

This work is supported through funding of the National Institutes of Health (GM-068603 and GM-083118), the University of Michigan Biological Sciences Scholars Program and the Cellular and Molecular Biology Training Grant and the University of Michigan Rackham Merit Program.

References

1. Whorton, M. R., Bokoch, M. P., Rasmussen, S. G., Huang, B., Zare, R. N., Kobilka, B., and Sunahara, R. K. (2007) A monomeric G protein-coupled receptor isolated in a high-density lipoprotein particle efficiently activates its G protein. *Proc Natl Acad Sci U S A* 104, 7682–7. Copyright 2007 National Academy of Sciences USA.
2. Bayburt, T. H., and Sligar, S. G. (2003) Self-assembly of single integral membrane proteins into soluble nanoscale phospholipid bilayers. *Protein Sci* 12, 2476–81.
3. Baas, B. J., Denisov, I. G., and Sligar, S. G. (2004) Homotropic cooperativity of monomeric cytochrome P450 3A4 in a nanoscale native bilayer environment. *Arch Biochem Biophys* 430, 218–28.
4. Leitz, A. J., Bayburt, T. H., Barnakov, A. N., Springer, B. A., and Sligar, S. G. (2006) Functional reconstitution of Beta2-adrenergic receptors utilizing self-assembling Nanodisc technology. *Biotechniques* 40, 601–602, 604, 606.
5. Amin, D. N., and Hazelbauer, G. L. (2010) The chemoreceptor dimer is the unit of conformational coupling and transmembrane signaling. *J Bacteriol* 192, 1193–200.
6. Raschle, T., Hiller, S., Yu, T. Y., Rice, A. J., Walz, T., and Wagner, G. (2009) Structural and functional characterization of the integral membrane protein VDAC-1 in lipid bilayer nanodiscs. *J Am Chem Soc* 131, 17777–9.
7. Mi, L. Z., Grey, M. J., Nishida, N., Walz, T., Lu, C., and Springer, T. A. (2008) Functional and structural stability of the epidermal growth factor receptor in detergent micelles and phospholipid nanodiscs. *Biochemistry* 47, 10314–23.
8. Banerjee, S., Huber, T., and Sakmar, T. P. (2008) Rapid incorporation of functional rhodopsin into nanoscale apolipoprotein bound bilayer (NABB) particles. *J Mol Biol* 377, 1067–81.
9. Tsukamoto, H., Sinha, A., Dewitt, M., and Farrens, D. L. (2010) Monomeric rhodopsin is the minimal functional unit required for arrestin binding. *J Mol Biol* 399, 501–11.
10. Whorton, M. R., Jastrzebska, B., Park, P. S., Fotiadis, D., Engel, A., Palczewski, K., and Sunahara, R. K. (2008) Efficient coupling of transducin to monomeric rhodopsin in a phospholipid bilayer. *J Biol Chem* 283, 4387–94.
11. Kuszak, A. J., Pitchiaya, S., Anand, J. P., Mosberg, H. I., Walter, N. G., and Sunahara, R. K. (2009) Purification and functional reconstitution of monomeric mu-opioid receptors: allosteric modulation of agonist binding by Gi2. *J Biol Chem* 284, 26732–41.
12. Rogers, D. P., Roberts, L. M., Lebowitz, J., Datta, G., Anantharamaiah, G. M., Engler, J. A., and Brouillette, C. G. (1998) The lipid-free structure of apolipoprotein A-I: effects of amino-terminal deletions. *Biochemistry* 37, 11714–25.
13. Rogers, D. P., Roberts, L. M., Lebowitz, J., Engler, J. A., and Brouillette, C. G. (1998) Structural analysis of apolipoprotein A-I: effects of amino- and carboxy-terminal deletions on the lipid-free structure. *Biochemistry* 37, 945–55.
14. Segrest, J. P. (1977) Amphipathic helixes and plasma lipoproteins: thermodynamic and geometric considerations. *Chem Phys Lipids* 18, 7–22.
15. Nolte, R. T., and Atkinson, D. (1992) Conformational analysis of apolipoprotein A-I and E-3 based on primary sequence and circular dichroism. *Biophys J* 63, 1221–39.
16. Gorshkova, I. N., Liu, T., Kan, H. Y., Chroni, A., Zannis, V. I., and Atkinson, D. (2006) Structure and stability of apolipoprotein a-I in solution and in discoidal high-density lipoprotein probed by double charge ablation and deletion mutation. *Biochemistry* 45, 1242–54.
17. Bhat, S., Sorci-Thomas, M. G., Alexander, E. T., Samuel, M. P., and Thomas, M. J. (2005) Intermolecular contact between globular N-terminal fold and C-terminal domain of ApoA-I stabilizes its lipid-bound conformation: studies employing chemical cross-linking and mass spectrometry. *J Biol Chem* 280, 33015–25.
18. Thomas, M. J., Bhat, S., and Sorci-Thomas, M. G. (2006) The use of chemical cross-linking and mass spectrometry to elucidate the tertiary conformation of lipid-bound apolipoprotein A-I. *Curr Opin Lipidol* 17, 214–20.
19. Li, H., Lyles, D. S., Thomas, M. J., Pan, W., and Sorci-Thomas, M. G. (2000) Structural determination of lipid-bound ApoA-I using fluorescence resonance energy transfer. *J Biol Chem* 275, 37048–54.
20. Panagotopulos, S. E., Horace, E. M., Maiorano, J. N., and Davidson, W. S (2001) Apolipoprotein A-I adopts a belt-like orientation in reconstituted high density lipoproteins. *J Biol Chem* 276, 42965–70.
21. Koppaka, V., Silvestro, L., Engler, J. A., Brouillette, C. G., and Axelsen, P. H. (1999)

The structure of human lipoprotein A-I. Evidence for the "belt" model. *J Biol Chem* 274, 14541–4.

22. Gan, K. N., Smolen, A., Eckerson, H. W., and La Du, B. N. (1991) Purification of human serum paraoxonase/arylesterase. Evidence for one esterase catalyzing both activities. *Drug Metab Dispos* 19, 100–6.
23. Rogers, D. P., Brouillette, C. G., Engler, J. A., Tendian, S. W., Roberts, L., Mishra, V. K., Anantharamaiah, G. M., Lund-Katz, S., Phillips, M. C., and Ray, M. J. (1997) Truncation of the amino terminus of human apolipoprotein A-I substantially alters only the lipid-free conformation. *Biochemistry* 36, 288–300.
24. Attie, A. D., Kastelein, J. P., and Hayden, M. R. (2001) Pivotal role of ABCA1 in reverse cholesterol transport influencing HDL levels and susceptibility to atherosclerosis. *J Lipid Res* 42, 1717–26.
25. Denisov, I. G., Grinkova, Y. V., Lazarides, A. A., and Sligar, S. G. (2004) Directed self-assembly of monodisperse phospholipid bilayer Nanodiscs with controlled size. *J Am Chem Soc* 126, 3477–87.
26. Bayburt, T. H., Leitz, A. J., Xie, G., Oprian, D. D., and Sligar, S. G. (2007) Transducin activation by nanoscale lipid bilayers containing one and two rhodopsins. *J Biol Chem* 282, 14875–81.
27. Lucast, L. J., Batey, R. T., and Doudna, J. A. (2001) Large-scale purification of a stable form of recombinant tobacco etch virus protease. *Biotechniques* 30, 544–546, 548, 550.
28. Lefkowitz, R. J., and Shenoy, S. K. (2005) Transduction of receptor signals by beta-arrestins. *Science* 308, 512–7.
29. Kobilka, B. K. (1995) Amino and carboxyl terminal modifications to facilitate the production and purification of a G protein-coupled receptor. *Anal Biochem* 231, 269–71.
30. Swaminath, G., Deupi, X., Lee, T. W., Zhu, W., Thian, F. S., Kobilka, T. S., and Kobilka, B. (2005) Probing the beta2 adrenoceptor binding site with catechol reveals differences in binding and activation by agonists and partial agonists. *J Biol Chem* 280, 22165–71.
31. Alami, M., Dalal, K., Lelj-Garolla, B., Sligar, S. G., and Duong, F. (2007) Nanodiscs unravel the interaction between the SecYEG channel and its cytosolic partner SecA. *EMBO J* 26, 1995–2004.
32. Boldog, T., Grimme, S., Li, M., Sligar, S. G., and Hazelbauer, G. L. (2006) Nanodiscs separate chemoreceptor oligomeric states and reveal their signaling properties. *Proc Natl Acad Sci U S A* 103, 11509–14.
33. Devanathan, S., Yao, Z., Salamon, Z., Kobilka, B., and Tollin, G. (2004) Plasmon-waveguide resonance studies of ligand binding to the human beta 2-adrenergic receptor. *Biochemistry* 43, 3280–8.
34. Civjan, N. R., Bayburt, T. H., Schuler, M. A., and Sligar, S. G. (2003) Direct solubilization of heterologously expressed membrane proteins by incorporation into nanoscale lipid bilayers. *Biotechniques* 35, 556–60; 562–3.
35. Forte, T., Norum, K. R., Glomset, J. A., and Nichols, A. V. (1971) Plasma lipoproteins in familial lecithin: cholesterol acyltransferase deficiency: structure of low and high density lipoproteins as revealed by elctron microscopy. *J Clin Invest* 50, 1141–8.
36. Lima, E. S., and Maranhao, R. C. (2004) Rapid, simple laser-light-scattering method for HDL particle sizing in whole plasma. *Clin Chem* 50, 1086–8.
37. Segrest, J. P., Jones, M. K., Klon, A. E., Sheldahl, C. J., Hellinger, M., De Loof, H., and Harvey, S. C. (1999) A detailed molecular belt model for apolipoprotein A-I in discoidal high density lipoprotein. J Biol Chem 274, 31755–8.

Chapter 9

Using Quantitative BRET to Assess G Protein-Coupled Receptor Homo- and Heterodimerization

Lamia Achour, Maud Kamal, Ralf Jockers, and Stefano Marullo

Abstract

Over a period of 15 years the concept of G protein-coupled receptor (GPCR) dimerization moved from a challenging hypothesis to a scientific fact, which is now accepted by the vast majority of the scientists working in the field. However, several important issues remain debated such as the biological function of dimerization, or the actual complexity of the oligomeric organization. Because of its major potential implications in physiology and pharmacology, the question of GPCR heterodimerization (or hetero-oligomerization) is currently one of the most central. Several complementary experimental approaches are used to investigate these novel important aspects of GPCR biology. In this context, Bioluminescence Resonance Energy Transfer-based techniques are extremely powerful, provided that they are conducted with the appropriate (numerous) controls and correctly interpreted.

Key words: G protein-coupled receptor, Resonance energy transfer, Bioluminescence resonance energy transfer, Oligomerization, Endoplasmic reticulum, Quality control, Biosynthetic pathway, Allostery, Conformational change

1. Introduction

Studying interactions among proteins is one of the most important and challenging tasks of post-genomic biology. Among the available approaches to study these phenomena in living cells, Resonance Energy Transfer (RET)-based techniques have become increasingly popular over the past few years, in particular when investigating G protein-coupled receptor (GPCR) signaling and oligomerization. Indeed, these approaches do not necessitate any separation or purification, and are sensitive enough to allow studies at physiological concentrations of proteins. RET consists of the non-irradiative energy transfer between a donor and an acceptor.

Louis M. Luttrell and Stephen S.G. Ferguson (eds.), *Signal Transduction Protocols*, Methods in Molecular Biology, vol. 756, DOI 10.1007/978-1-61779-160-4_9,

Because the efficacy of the energy transfer varies inversely with the sixth power of the distance between the donor and acceptor molecules, a signal is obtained only if the two molecules are in close proximity (1–10 nm). Thus, the detection of an energy transfer between two proteins fused, respectively, to an energy donor and an energy acceptor, often reflects the existence of molecular interaction between the proteins of interest. In contrast, an absence of signal does not exclude the possibility that two proteins interact. It may be due to the particular conformations of the interacting partners that maintain the acceptor too distant from the donor or cause inappropriate relative orientation of the donor to acceptor molecules.

Bioluminescence RET (BRET) has been inspired by a natural phenomenon observed in glowing marine organisms. In the presence of its substrate, coelenterazine, *Renilla* luciferase (Rluc, the luminescent energy donor) transfers some energy to a GFP variant (the energy acceptor) (1). No excitation of the donor is required and the substrate, which is membrane permeable, can be added to the supernatant of cultured cells. BRET-based protocols have been designed to monitor and quantify both regulated and constitutive molecular interactions in intact cells, such as GPCR oligomerization (2). In this context, performing experiments in intact cells avoids possible artifacts due to receptor solubilization, an obligate step for other biochemical assays such as coimmunoprecipitation.

Most (if not all) GPCRs may exist as either homo- or heterodimers or as higher-order oligomers (3). Oligomerization is not an absolute prerequisite for proper signaling, as shown by functional studies on purified monomers (4, 5). Dimerization, instead, could have an important role during biosynthesis for the quality control of newly synthesized receptors (6, 7). Moreover, ligand-driven transactivation or inhibition, between protomers within a receptor dimer or between adjacent dimers in larger complexes, has been reported for an increasing number of receptors representing additional functions for GPCR oligomerization (8).

Except in few specific cases, such as the extensively investigated example of $GABA_B$ receptors (9), the issue of GPCR heterodimerization is a complex phenomenon to analyze experimentally and to interpret functionally, even with well-controlled BRET experiments. The hydrophobic properties of GPCR transmembrane domains may lead to false-positive interactions between different protomers in reconstituted systems. In this context, quantitative issues are critical to consider, since nonspecific interactions tend to increase with rising concentrations of the protein of interest. Also, because reconstitution systems used for BRET analysis artificially drive the synthesis of the receptors to be studied in the same cell at the same time, specific interactions may not be representative of a physiological situation simply because in real life the receptors are not synthesized in the same cell and/or at the same moment. The literature reports numerous examples of functional cooperativity between different GPCRs (reviewed in (10) and other articles of the same issue), which are attributed to GPCR heteromerization. However, at least in some cases, this functional cooperativity is more likely due to the interaction of distinct receptor homodimers within higher-order oligomers (11, 12). The protocols detailed below will allow the reader to quantitatively investigate the oligomerization of their favorite GPCRs using BRET and will provide some guidelines for the appropriate interpretation of their results.

2. Materials

2.1. Cell Culture and Transfection

1. Dulbecco's Modified Eagle's Medium (DMEM) supplemented with 10% (v/v) fetal bovine serum, 4.5 g/L glucose, 100 U/mL penicillin, 0.1 mg/mL streptomycin, and 1 mM glutamine (Invitrogen).
2. Solution of trypsin-EDTA (0.05%).
3. PBS-EDTA solution: Phosphate-buffered saline (PBS; 1×) without $CaCl_2$ and $MgCl_2$ (Invitrogen) plus 2 mM ethlenediaminetetraacetic acid (EDTA).
4. 6- and 12-well plates for cell culture, White 96-well plates with clear well sterile and tissue culture treated (Perkin Elmer).
5. Poly-L-lysine used to coat 96-white well plates for adherent cell experiments.
6. Transfection reagents: JetPEI (Polyplus transfection), GeneJuice (Novagen), FuGENE6 transfection reagent (Roche Applied Science).

2.2. Chemiluminescence, Fluorescence, and BRET-Ratio Measurements

1. Coelenterazine-h powder (Uptima; Interchim) or Deepblue C (Coelenterazine 400a; Uptima, Interchim) are dissolved in 100% ethanol at 1 mM (stock solution) and stored at −20°C in opaque microcentrifuge tubes (see Note 1).
2. Hank's balanced salt solution (HBSS; 1×) containing $CaCl_2$ and $MgCl_2$ (Invitrogen).
3. PBS (1×) containing $CaCl_2$ and $MgCl_2$ (Invitrogen).
4. Multi-mode microplate reader: Mithras (LB 940) (Berthold) or equivalent.
5. White 96-well plates (Perkin Elmer Optiplate™-96HB or equivalent) for BRET measurements.
6. Black 96-well plates (Perkin Elmer Optiplate™-96HB #6005279 or equivalent) for fluorescence measurements.

2.3. SDS-Polyacrylamide Gel Electrophoresis

1. Separating buffer (4×): 1.5 M Tris–HCl, pH 8.8, 0.4% SDS. Store at room temperature (RT).
2. Stacking buffer (4×): 0.5 M Tris–HCl, pH 6.8, 0.4% SDS. Store at RT.
3. Forty percent (w/v) acrylamide/bisacrylamide solution (Sigma). Acrylamide is a neurotoxin when unpolymerized and so care should be taken not to receive exposure.
4. *N,N,N,N′*-Tetramethyl-ethylenediamine (TEMED).
5. Ammonium persulfate solution: Prepare 10% solution in water and immediately freeze in single use (200 μL) aliquots at −20°C.
6. Water-saturated isobutanol: Shake equal volumes of water and isobutanol in a glass bottle and allow separation. Use the top layer. Store at RT.
7. Running buffer (5×): 125 mM Tris, 960 mM glycine, 0.5% (w/v) SDS. Store at RT.
8. Prestained molecular weight markers: e.g., BenchMark Prestained Protein Ladder (Invitrogen).

2.4. Western Blotting for Quantitative Assessment of GPCR Expression in BRET Experiments

1. BCA protein assay reagent.
2. Lysis Buffer: 50 mM HEPES (*N*-2-hydroxyethylpiperazine-*N′*-2ethanesulfonic acid), pH 7.4, 250 mM NaCl, 2 mM EDTA, 0.5% NP40, 10% glycerol, and complete protease inhibitors.
3. Laemmli sample buffer (5×): 312.5 mM Tris–HCl, pH 6.8, 50% glycerol (v/v), 10% SDS (w/v), 8% Dithiothreitol DTT (w/v), 1% Bromophenol blue (w/v).
4. Setup buffer: 25 mM Tris–HCl (do not adjust pH), 190 mM glycine, 20% (v/v) methanol.
5. Transfer buffer: Setup buffer plus 0.05% (w/v) SDS. Store in the transfer apparatus at RT.
6. Nitrocellulose Transfer Membrane: e.g., PROTRAN (Whatman).
7. PBS-Tween 0.2%: 500 ml PBS with 1 ml Tween-20.
8. Blocking buffer: 5% (w/v) nonfat dry milk in PBS-Tween 0.2%.
9. Secondary antibodies: Peroxydase-conjugated Affinipure antibodies (Jackson Immuno-research).
10. ECL™ western blotting detection reagent (Amersham™, GE Healthcare).
11. High performance chemiluminescence film: e.g., Amersham Hyperfilm™ ECL from GE Healthcare.

2.5. Radioligand Binding Assays for Quantitative Assessment of GPCRs in BRET Experiments

1. TEM lysis buffer: 25 mM Tris pH 7.4, 2 mM EDTA, 10 mM $MgCl_2$, containing protease inhibitors (2 μg/ml, benzamidine 1 mM AEBSF, 1 μg/ml pepstatin A and 1 μg/ml leupeptin).
2. Ultra Turrax® tissue disperser (Janke and Kunkel).

3. Methods

3.1. Preliminary Notions

To monitor the homo- or heterodimerization of GPCRs, one GPCR protomer is fused to Rluc (BRET-donor) and the other to a fluorescent BRET-acceptor (YFP for BRET1 or GFP2 for BRET2; see Note 2). The energy transfer requires that both the donor and the acceptor be on the same side of the cellular membrane. In theory, it should be possible to fuse Rluc or fluorescent proteins to the extracellular (or intraluminal) extremity of a GPCR, but in this case signal peptides should be added N-terminally to facilitate receptor export (these type of constructs have been used for FRET experiments). In general, for BRET experiments, the donor and the acceptor are fused at the C-terminal tail of a GPCR and are thus located in the cytosol. Depending on the type of BRET assay to be used (BRET1 or BRET2), the appropriate fusion protein constructs are generated using classical cloning methods (see Note 3). BRET measurements can be performed in various types of intact mammalian cell lines; either adherent (e.g., HEK-293 T, CHO, COS, Hela cells) or in suspension (e.g., THP-1, Jurkat). Optimization of the conditions of transfection of BRET constructs is necessary (see Note 4). BRET measurements consist of calculated ratios of the light emitted by the luciferase (donor) over the fluorescence emitted by the YFP (acceptor). They require BRET readers that can measure rapidly, simultaneously (at both wavelengths) and repetitively luminescence (produced by the BRET-donor) and fluorescence (emitted by the BRET-acceptor) values in the same well. These readers are generally controlled by software, programmed by the user to accomplish the measurements, and also calculate the BRET ratios. Both raw data and calculations are saved on a spreadsheet compatible with Microsoft Excel or similar.

3.2. Setting Up a BRET Donor Saturation Experiment

The BRET-donor saturation assay has been developed from original basic BRET experiments for a more quantitative and precise interpretation of BRET signals (13, 14). The level of expression of the BRET-donor used in saturation experiments (GPCR-Rluc, in the present case) should correspond to the lowest amount of protein required to obtain a detectable and robust BRET signal.

Cells are transfected with a constant amount of plasmid DNA coding for the BRET-donor fusion protein in the presence or absence of increasing concentrations of the plasmid for the BRET-acceptor fusion protein. The BRET signal will increase with the concentration of the acceptor up to a maximum that is achieved when all BRET-donor molecules are in proximity of BRET-acceptor molecules. In case of a specific interaction, the BRET signal increases hyperbolically and reaches an asymptote. In contrast, in case of nonspecific interaction resulting from random proximity, the "bystander BRET" signal increases almost linearly and eventually saturates at very high expression levels of the BRET-acceptor protein (15) (see Note 5).

3.2.1. Detailed Protocols to Detect GPCR Dimerization in HEK-293T Cells

We will detail below typical BRET1 assays for detection and analysis of GPCR dimers using either adherent cells or cells in suspension.

3.2.2. BRET Measurements on Adherent Cells

1. HEK-293T cells are seeded 24 h before transfection in 12-well plates ($250–500 \times 10^3$ cells per well in 2 mL of complete DMEM medium).
2. Cells are transfected with 1 μg of total DNA per well using one of the transfection reagents indicated in Subheading 2.1, according to the manufacturer's instructions. This DNA quantity comprises a fixed amount of plasmid encoding the BRET-donor (GPCR1-Rluc: 10–100 ng, depending on the obtained expression level); increasing concentrations of the energy acceptor (GPCR2-YFP: 0, 25, 50, 100, 200, 300 … 990 ng) and sufficient "empty" vector (such as pCDNA3.1 or any other cloning vector) to bring the total amount of DNA in the transfection to 1 μg.
3. White 96-well plates are incubated with poly-L-lysine (50 μL per well) for 10 min at 37°C. The poly-L-lysine solution is then removed and the wells washed once with complete medium. Note that this step can be skipped when using cell lines that adhere tightly to the plastic.
4. Twenty-four hours after transfection, the cells of each well of the 12-well plates are washed with PBS, detached in 200 μL PBS-EDTA (5 min at 37°C) and resuspended in 2 ml of complete medium. Cells in suspension are distributed into white 96-well plates: 200 μl per well (corresponding to about 50,000 cells) and incubated at 37°C for 24 h before BRET measurements.
5. The next day, cells are washed with PBS containing $MgCl_2$ and $CaCl_2$ (see Note 6). The BRET signal is measured after adding 50 μl of a mixture containing 40 μL of PBS-$CaCl_2$/$MgCl_2$ and 10 μL of a freshly prepared solution of 25 μM coelenterazine-h diluted in HBSS or PBS-$CaCl_2$/$MgCl_2$ to each well.

6. After 10 min incubation at RT, BRET readings are started using a lumino/fluorometer that allows sequential integration of luminescence signals detected with two filter settings (Rluc Filter: 485 ± 10 nm; YFP filter: 530 ± 12.5 nm) (see below and Note 7 for the detailed procedure).
7. Fluorescence measurements to quantify the amount of expressed acceptor are performed using the same equipment. Cells are detached from spare wells of the white 96-well plate using 100 μL PBS-EDTA as described, washed twice with PBS and collected by centrifugation for 5 min at 300 × *g* at RT. The pellet is resuspended in 200 μL PBS and transferred to a black 96-well plate for fluorescence measurement at 530 ± 12.5 nm.

3.2.3. BRET Measurements on Cells in Suspension

1. Forty-eight hours after transfection (performed as described above), cells are detached from 12-well plates with PBS-EDTA (5 min at 37°C), transferred to a microcentrifuge tube, and collected by centrifugation for 5 min at 300 × *g* at RT.
2. Pellets are resuspended in 250 μl HBSS (or PBS-$CaCl_2$/$MgCl_2$) and 40 μl of the cell suspension are distributed into each well of a white 96-well plate.
3. BRET measurements are performed directly after adding 10 μl of 25 μM coelenterazine-h solution.
4. For YFP fluorescence measurements, 50 μl of the cell suspension is distributed into black 96-well plates and readings are performed as described above.

3.3. Calculating the BRET Ratio and Data Analysis

1. The BRET ratio is the fluorescence signal (filter 530 ± 12.5 nm) over the Rluc signal (filter 485 ± 10 nm) measured simultaneously. This ratio (automatically calculated by the software of BRET readers) is obtained by measuring each well for 1 s. The readings are repeated three to six times to obtain average values and all data are saved on a spreadsheet. The specific BRET ratio is calculated by subtracting from the mean BRET ratio value above the background BRET ratio, which corresponds to the signal obtained with cells expressing the BRET-donor alone (not expressing the BRET-acceptor). Results are expressed in milli-BRET units (mB) by multiplying the values × 1,000 to avoid the need to manipulate decimal numbers.
2. To quantify the amount of BRET-donor in each well, the average luminescence values at 485 ± 10 nm are calculated (see Note 8). It is important that BRET-donor levels are relatively constant throughout the experiment. In case of significant variation (difference of 30% or more from the average value) the corresponding points should be excluded from the final plot.

3. To quantify the amount of BRET-acceptor in each well, the fluorescence is measured at 530 ± 12.5 nm after external excitation at 480 nm. Background fluorescence measured in cells not expressing the BRET-acceptor (YFP_0) is subtracted from fluorescence values measured in cells expressing the increasing amounts of BRET-acceptor (YFP) to obtain YFP-YFP_0 values. Depending on the application, it may be necessary to convert luminescence and fluorescence values into actual protein expression levels using standard curves correlating luminescence and fluorescence signals with protein amounts (see Note 9).
4. From the values of *Points 2* and *3*, YFP-YFP_0/Rluc values are calculated. BRET ratios from *Point 1* are then plotted as a function of YFP-YFP_0/Rluc values. A slightly different calculation method can be used (see Note 10), which is of interest when comparing experiments conducted at different times. Data are fit using a nonlinear regression equation assuming a single binding site (GraphPad Prism) to estimate BRETmax and BRET50 values (see Note 11 and Fig. 1).

3.4. BRET Displacement Assays

A hyperbolic BRET-saturation curve using two different GPCRs, as BRET-acceptor and BRET-donor, does not have an unequivocal interpretation. It may reflect true heterodimerization or proximity of distinct homodimers. BRET competition experiments can be used to assess GPCR heterodimerization. If two GPCRs do form heterodimers, an excess of one of them should be capable of displacing the homodimerization of the other (Fig. 2).

1. A typical BRET-saturation experiment is conducted, as described in Subheading 3.2, using the GPCR1 as both BRET-donor and BRET-acceptor.

Fig. 1. Prototypical BRET donor saturation experiment to study GPCR heterodimerization. (**a**) BRET data acquisition and calculations. In the presented experiment, cells have been transfected with a constant amount of GPCR1-Rluc : (10 ng; column 1) and increasing amounts of GPCR2-YFP (25, 50, 100, 250, 500, 1000; columns 2–7). Cells from each transfection have been distributed into three wells (triplicate) of a white 96-well plate and BRET measurements have been repeated six times for each well. Because of the limited space, the values obtained for only one well are shown in the figure, although the values in lines 23–32 have been calculated from the average values of triplicates. The six measurements of the Rluc signal are in lanes 3–8, those for the YFP signal in lanes 10–15. The mean BRET ratio calculated from values of lanes 17–22 is shown in lane 23. The background ratio, which corresponds to the BRET signal obtained in samples expressing the GPCR1-Rluc alone (column 1), is shown in lane 24. Specific BRET ratios (lane 25) are calculated by subtracting from each mean BRET ratio (lane 23) the background BRET ratio (lane 24); values are then multiplied × 1,000 to obtain mBU (lane 26). The average of the specific BRET ratio (lane 27) represents the average of specific BRET ratios from the triplicate multiplied by 1,000. The average of luminescence values from triplicates is used to quantify the BRET donor (GPCR1-Rluc) in the transfection (lane 28). The quantity of BRET acceptor (GPCR2-YFP) expressed in each transfection (lane 31), is then calculated by subtracting the background fluorescence (YFP_0, lane 30) measured in cells expressing the BRET-donor alone (column 1), from fluorescence values of lane 29. YFP-YFP_0/Rluc are then calculated and multiplied ×100 (lane 32) to avoid the manipulations of decimal numbers. (**b**) A BRET-saturation curve is obtained by plotting BRET ratios from A (lane 27) as a function of YFP-YFP_0/Rluc (lane 32) and fitting the data with a hyperbolic equation. Two important parameters are defined from the curve: the $BRET_{max}$, which represents the maximal signal reached at saturation, and the $BRET_{50}$, which corresponds to the BRET ratio giving 50% of the maximal BRET signal.

a

1	Transfection	1	2	3	4	5	6	7
2	Rluc Signal							
3	00:00:000	106030	93920	98140	95920	102750	91520	73370
4	01:56:330	117770	118040	114230	121720	118970	92560	91520
5	03:52:690	112280	119730	112120	121790	115300	87300	92560
6	05:48:990	102790	112360	102770	114430	104200	84910	87300
7	07:45:270	93260	103570	94120	105140	94650	87850	79640
8	09:41:630	86790	95140	86310	96950	86610	83220	72770
9	YFP signal							
10	00:00:000	68460	62530	65910	65328	72620	64594	52639
11	01:56:330	75290	78840	76290	82419	82468	65385	65152
12	03:52:690	71310	79780	75340	82474	80722	61937	66311
13	05:48:990	65760	75060	68830	76536	72711	60024	61936
14	07:45:270	59830	68180	62740	70300	65954	61702	57655
15	09:41:630	55720	64030	57520	65761	60146	59024	49998
16	BRET ratio							
17	00:00:000	0,6457	0,6658	0,6716	0,6811	0,7068	0,7058	0,7174
18	01:56:330	0,6393	0,6679	0,6679	0,6771	0,6932	0,7064	0,7119
19	03:52:690	0,6351	0,6663	0,6720	0,6772	0,7001	0,7095	0,7164
20	05:48:990	0,6398	0,6680	0,6697	0,6688	0,6978	0,7069	0,7095
21	07:45:270	0,6415	0,6583	0,6666	0,6686	0,6968	0,7024	0,7239
22	09:41:630	0,6420	0,6730	0,6664	0,6783	0,6944	0,7093	0,6871
23	Mean BRET ratio	0,6406	0,6653	0,6696	0,6746	0,6989	0,7062	0,7158
24	Background BRET ratio	0,6416						
25	Specific BRET ratio	0	0,0247	0,0280	0,0330	0,0573	0,0646	0,0742
26	Specific BRET ratiox1000	0,00	24,71	27,96	32,98	57,34	64,60	74,24
27	Average specific BRET ratio	0,00	25,03	28,71	34,36	56,84	64,47	70,62
28	Average luminescence (Rluc)	106217	106803	102238	100998	106536	95820	79932
29	Fluorescence (YFP)	692	1250	1570	1900	2650	3362	4130
30	YFP0	692						
31	YFP-YFP0	0	558	878	1208	1958	2670	3438
32	(YFP-YFP0/Rluc)x100	0,00	0,52	0,86	1,88	2,49	3,51	5,17

b

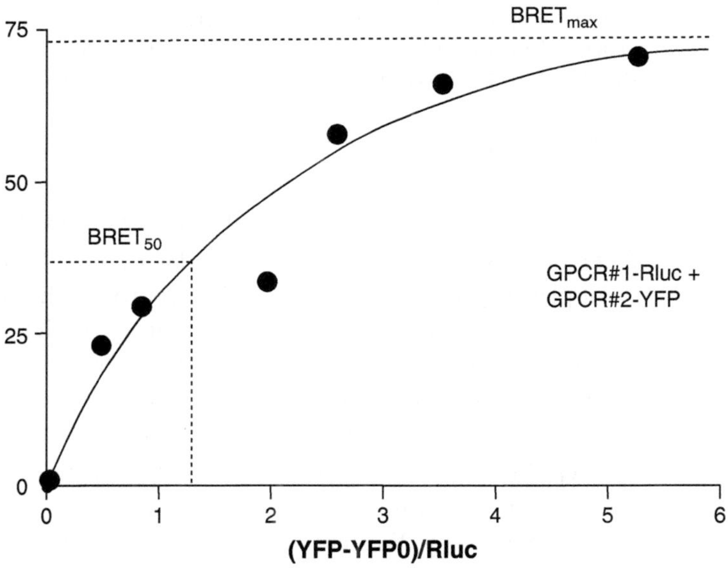

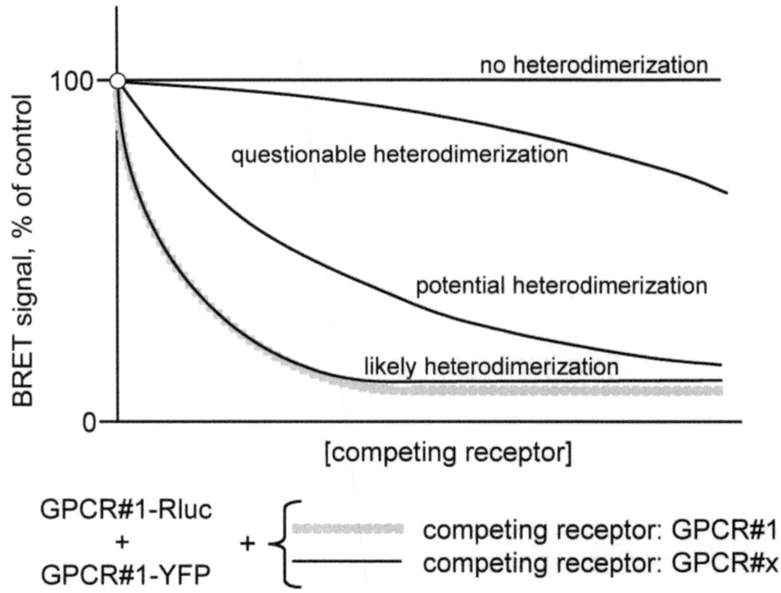

Fig. 2. BRET competition experiments to assess GPCR heterodimerization. The possibility of true heterodimerization between 2 GPCRs, GPCR1 and GPCRX, which interact in a BRET-donor saturation experiment, is investigated in this example by studying the displacement of the interaction between the protomers of one receptor (GPCR1-Rluc and GPCR1-YFP) in the presence of excess untagged GPCR1 or GPCRX. Cells are co-transfected with a constant amounts of GPCR1-Rluc and GPCR1-YFP (giving BRET signals falling in the ascending portion of the saturation curve, just before the plateau) in the presence of increasing amounts of plasmid coding for untagged receptor 1 or receptor X. BRET ratios are calculated and expressed as a function of the concentration of the competing receptor, determined by immunoblot or using biding experiments. Depending on the aspect of the competition curve, the indicated conclusions can be drawn.

2. The actual amount of donor and acceptor plasmids giving an YFP-YFP$_0$/Rluc value in the ascending portion of the curve before the plateau is selected for the displacement assay.
3. HEK-293T cells are then transfected with the selected amount of plasmids for BRET-donor and acceptor in the presence of increasing quantities of plasmid coding for native GPCR2 or GPCR1. As in classical saturation experiments, the total amount of transfected DNA is maintained constant using appropriate amounts of "empty" cloning vector.
4. BRET ratios are calculated as described in Subheading 3.3 and plotted as a function of the expression of native receptor determined by Western blot or binding experiments (as detailed in Subheading 3.6). The interpretation of the data is described in Note 12.

3.5. Measuring the Kinetics of BRET Changes Upon Ligand Binding to GPCRs

For this type of experiment, cells are transfected with a plasmid coding for the BRET-donor in the presence of a single concentration of the plasmid encoding the BRET-acceptor and BRET measurements are performed over time (up to 20–30 min).

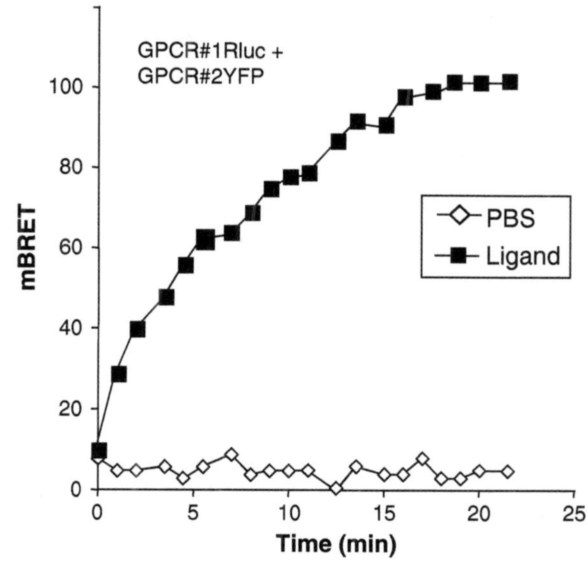

Fig. 3. BRET changes induced by ligand binding. Cells are transfected with a constant amounts of GPCR1-Rluc (10 ng) and GPCR2-YFP (500 ng). BRET measurements are carried out at the indicated times after addition of the appropriate ligand or buffer alone (PBS).

Detailed below is an example of a ligand-dependent BRET signal modulation for a GPCR heterodimer in adherent HEK-293T cells.

1. HEK-293T cells are prepared and transfected as in Subheading 3.2.
2. The BRET signal is directly measured in the same plate. Forty microliter of a mixture composed of 30 μL of PBS-$CaCl_2$/$MgCl_2$ and 10 μL of 25 μM coelenterazine-h solution is distributed in each well, then 10 μL of ligand (at saturating concentration) in PBS or of PBS alone (basal control condition) is added to each well.
3. BRET measurements are started and repeated for the desired duration in order to measure the evolution of the BRET ratio with time. BRET ratios are then plotted over time (Fig. 3).

3.6. Assessing GPCR Expression Levels in BRET Experiments

Determination of GPCR expression levels in BRET experiments can be relevant to ascertain that the expression level of fusion proteins falls within the physiological range. It can also be useful to determine the true acceptor/donor ratio in BRET-donor saturation experiments. Relative expression levels can be obtained by western blotting using specific antibodies, whereas actual receptor levels can be determined using radioligand binding assays.

3.6.1. Preparation of Samples for Western Blotting

1. Cells from BRET experiments are collected in a microcentrifuge tube by centrifugation at 300 × *g* for 5 min at RT.

2. The supernatant is aspirated and the pellet resuspended in 300 μl of lysis buffer for 15–30 min in ice.
3. Cell lysates are then centrifuged at 12,000 × *g* for 15 min at 4°C to remove debris.
4. Supernatants are transferred in clean microcentrifuge tubes and 5 μl of each sample is used to determine its protein concentration (with the BCA protein Assay).
5. 50–100 μg of protein in 40 μl is mixed with 10 μl of Laemmli 5× buffer and denatured for 1–6 h at RT. Samples can then be stored at −20°C or used immediately for separation by SDS-polyacrylamide gel electrophoresis (SDS-PAGE) and western blotting using appropriate antibodies.

3.6.2. SDS-PAGE and Western Blotting

1. The glass plates for the gels are cleaned with ethanol and rinsed extensively with distilled water.
2. The separating gel is prepared. For example, to prepare a 10% gel; mix 7.5 ml of 4× separating buffer, with 10 ml acrylamide/bis solution 40% (w/v), 12.5 ml water, 100 μl ammonium persulfate solution and 20 μl TEMED. Pour the gel, leave space for a stacking gel, and overlay with water-saturated isobutanol. The gel should polymerize in about 20–30 min. The isobutanol is then poured off and the gel is rinsed twice with water.
3. The stacking gel is prepared by mixing 2.5 ml of 4× stacking buffer with 1.3 ml acrylamide/bis solution, 6.1 ml water, 50 μl ammonium persulfate solution, and 10 μl TEMED. The stacking gel is then poured and the comb is inserted.
4. Once the stacking gel is polymerized, the comb is removed carefully and the gel is rinsed with water.
5. The gel is then ready and 50 μl of each sample are loaded into each well (10 μl of prestained molecular weight markers is added in one well).
6. After separation by SDS-PAGE, samples are transferred to nitrocellulose membranes electrophoretically (a gel of 15 × 6 cm requires a transfer at 100 mA for 2 h).
7. Once the transfer is completed, the nitrocellulose membrane is incubated in 10 ml of blocking buffer for 1 h at RT on a rocking platform.
8. The nitrocellulose membrane is then incubated with the primary antibody needed to detect the specific GPCR, washed three times in blocking buffer and then incubated with the appropriate secondary antibody.
9. The membrane is next washed three to four times for 30 min with PBS-Tween 0.2% before incubation in the ECL detection reagent.

10. The membrane is then removed from ECL and placed in an X-ray film cassette with film for the suitable exposure time (typically from 30 s to a few minutes).

3.6.3. Establishing Correlation Curves Between Fluorescence or Luminescence Values and Receptor Expression by Radioligand Binding

As an alternative to western blotting, true GPCR expression levels can be determined for each BRET experiment by radioligand binding when appropriate radioligands are available. If BRET-donor and acceptor receptors both bind the same radioligand, determination of the expression level of each fusion protein can be challenging. We therefore recommend the generation of correlation curves between luminescence (Rluc fusion protein) or fluorescence (YFP fusion proteins) values and radioligand binding sites in independent sets of experiments (16).

1. Cells are transfected with different quantities of either the BRET-donor or the acceptor fusion protein plasmid. At least 10 different independent transfections should be performed for each fusion protein to obtain a sufficient number of data points for correlation curves.
2. Luciferase activity (for the BRET-donor GPCR) and fluorescence values (for the BRET-acceptor GPCR) are determined as described above.
3. Radioligand binding experiments using saturating ligand concentrations are performed in parallel using appropriate assay conditions for each GPCR (see Note 13).
4. The number of radioligand binding sites is then plotted against fluorescence or luminescence values measured in the same sample; a linear correlation is expected. Provided that the settings for fluorescence and luminescence measurements are identical to those used in BRET experiments, these curves can be used to convert fluorescence and luminescence values measured in BRET experiments into actual receptor amounts (see Note 14).

4. Notes

1. Coelenterazine-h is sensitive to degradation by light and oxygen. Working solutions are freshly prepared, typically at 25 μM in HBSS or PBS-$CaCl_2$+$MgCl_2$. Coelenterazine-h solutions should be protected from light during long incubation periods. Whereas coelenterazine-h is used for routine BRET experiments, other *Renilla* luciferase substrates can be used for more specific applications. For BRET2 assays (see Note 2) use of Deepblue C (Coelenterazine 400a, UPBB8392; Uptima, Interchim) is recommended. The recently developed ViviRen™ substrate (Promega) generates a rapid light burst

(three- to fivefold increase of the luminescence peak) followed by accelerated decline. On the other hand, EnduRen™ (Promega) generates long-lasting luminescence signals (up to 24 h) but with 10–25 times lower amplitude. All three substrates are used at working concentrations of 60 μM. ViviRen™ and EnduRen™ can only be used for BRET measurements in intact cells because they are pro-substrates necessitating transformation by cellular esterases to become effective *Renilla* substrates.

2. BRET1 is the most widely used BRET assay. The energy donor in this assay is the luciferase of *Renilla reniformis* (Rluc) and the energy acceptor, the enhanced yellow fluorescent protein (YFP), a codon-humanized enhanced and yellow-shifted mutant of the *Aequorea victoria* GFP (green fluorescent protein). The substrate for BRET1 assays is coelenterazine-h. Rluc variants, such as Rluc8, and the YFP variant YPet have been obtained by random mutagenesis (17, 18). Superior BRET signals have recently been reported with this optimized Rluc8/YPet BRET pair, compared to the "classical" Rluc/YFP pair, due to increased assay sensitivity (19). In the BRET2 assay, the YFP is replaced by GFP2, a GFP mutant that can be excited at 400 nm, and the Rluc substrate is the Deepblue C (also known as coelenterazine 400a) (see Note 1). The advantage of BRET2 is a superior separation of donor and acceptor peaks. Indeed, in the presence of Deepblue C, Rluc emits light at 400 nm, a wavelength that excites GFP2, which, in turn, emits light at 510 nm. The recommended filter sets for BRET2 are therefore 400 ± 10 nm and 515 ± 10 nm. The disadvantage of BRET2, compared to BRET1 is the 100–300 times lower intensity of emitted light. To compensate this loss of amplitude, fusion proteins are often overexpressed, which may cause problems for interpreting the results, in particular when GPCR oligomerization is the object of the study.

3. For BRET experiments, C-terminal fusion proteins are used. To generate fusion constructs, the entire GPCR coding sequence is typically amplified by PCR without its stop-codon using sense and antisense primers harboring unique restriction sites. The sequence is then subcloned upstream and in frame with YFP (or GFP2) and Rluc, resulting in GPCR-YFP and GPCR-Rluc, respectively. Fusion of Rluc and YFP at the N-terminal end of GPCRs is also possible but interferes in many cases with proper export of receptors to the cell surface and requires addition of signal peptide sequences.

4. Depending on each construct and on the quality of the plasmid preparation, the actual amount of expressed fusion protein after transfection of a given amount of DNA may change. It is recommended to establish experimentally the amount of

BRET-donor DNA leading to a suitable luminescence signal. This signal should correspond to GPCR levels compatible with physiological conditions and to luminescence counts that are sufficiently high over background. Once the conditions for the BRET-donor are established, a range of different BRET-acceptor protein levels are coexpressed in order to determine the conditions that correspond to the highest specific BRET signals.

5. Negative control fusion proteins should be included in BRET-donor saturation assays to verify the specificity of the generated BRET. Ideally, negative control proteins should have a similar topology as the GPCR of interest (i.e., another GPCR) and should be localized in the same subcellular compartment. As many GPCRs have a natural tendency to associate (see Subheading 1), finding a true negative control may be challenging. Therefore, membrane proteins with different topologies (the single membrane-spanning protein CD4, for example) can be used as negative control protein.

6. To perform measurements on attached cells, the culture medium has to be removed by washing once with PBS (containing $CaCl_2/MgCl_2$), as the phenol red, present in most culture mediums, quenches the luciferase signal measured at 485 nm.

7. Improved BRET1 filter sets (480 ± 10 nm and 540 ± 20 nm) have been described recently. In our hands, BRET values determined with optimized filter sets were increased by approximately 50% for various BRET pairs (link to Berthold Application Note: http://www.berthold.com/ww/en/pub/bioanalytik/overview/notes.cfm). These filters are now included in the filter set package of some microplate BRET readers (e.g., Mithras, Berthold).

8. Rluc light emission is transient. The typical profile of light emission in the presence of coelenterazine-h is composed of a rapid raise, a transient stabilization at maximal values and a slow decline in light output. The duration of each phase may vary with assay conditions. Whereas maximal values are reached rapidly (less than 3 min) when working with cell extracts and purified proteins, approximately 5 and 15 min are necessary to complete the rise when experiments are performed with cell suspensions and adherent cells, respectively. According with these considerations, to estimate the amount of expressed donor in BRET donor saturation experiments, we determine the maximal luciferase value obtained by measuring luminescence values repeatedly at 485 nm in the appropriate time window.

9. Conversion of luminescence and fluorescence values into actual receptor amounts depends on the availability of appropriate

tools (antibodies, radioligands) and on the specific question to be answered. Whereas radioligand binding assays provide absolute values, western blot detection using anti-receptor antibodies provides only relative values (i.e., comparison with receptor expression levels in tissue samples examined in parallel). A common question in the context of GPCR homo- and heterodimerization is the relative propensity of formation of such complexes. The most straightforward approach to investigate this issue is based on the parallel study of the coexpression of a given GPCR-Rluc fusion protein with different GPCR-YFP fusion proteins. The $BRET_{50}$ values obtained in BRET-donor saturation assays for each pair of GPCRs provide an estimate for the relative propensity of the corresponding interaction. To obtain meaningful results, the amount of expressed Rluc fusion protein must be equivalent in all assays (a maximal variation of 30% is tolerated). This approach is based on the assumption that the YFP-associated fluorescence of all GPCR-YFP fusion proteins is comparable (which appears to be the case for most, but not all, fusion proteins). If a more precise quantification of fusion proteins is necessary, fluorescence and luminescence values should be converted into absolute expression levels. To determine the physiological relevance of the interactions, conversion into absolute receptor expression levels (or comparison with endogenous expression levels in western blot experiments with tissue lysates) are mandatory. Conversion into absolute expression levels is obligate when comparing GPCR interactions with different BRET-donors. Semi-quantification based on luminescence and fluorescence values are sufficient, however, to analyze BRET changes induced by ligands.

10. We noticed that background fluorescence may vary, not only according to the cell type in which BRET experiments are conducted, but also between two experiments conducted at different times in the same cell type. Moreover, for some constructs, the transfection of the same amount of BRET-donor plasmid can result in interday fluctuation in luminescence signals. In these cases, when data obtained from experiments performed at different times are plotted together, it may be convenient to normalize the background fluorescence and luminescence values. BRET values are thus plotted as a function of $[(YFP\text{-}YFP_0)/YFP_0]/[Rluc/Rluc_0]$ where YFP_0 corresponds to background fluorescence measured in cells not expressing the BRET-acceptor and $Rluc_0$ to the average luminescence value in cells expressing the BRET-donor alone.

11. In the case of specific interactions, hyperbolic BRET-donor saturation curves are expected. Maximal BRET values, which correspond to the saturation of all BRET donors by BRET acceptors, are defined as the $BRET_{max}$. Half-maximal BRET values,

corresponding to the saturation of 50% of BRET donors by BRET acceptors, are defined as the $BRET_{50}$. $BRET_{50}$ values give an approximation of the relative affinity of receptors for each other. When studying heterodimers, $BRET_{50}$ values are indicative of the likelihood of the examined interaction in natural cells. The closer the measured $BRET_{50}$ is to the $BRET_{50}$ values measured between homodimers, the greater the possibility that the heterodimer forms in native cells.

12. In case of true heterodimerization, native GPCR2 is expected to displace the BRET signal. If the BRET signal remains unmodified by increasing concentrations of GPCR2, heterodimerization can be ruled out. If the displacement curve is close to that obtained with untagged GPCR1, one can assume that the heterodimerization can occur in natural cells provided that they express a corresponding level of endogenous receptors synthesised at the same moment. In case of partial displacement by GPCR2 or displacement requiring high concentrations of GPCR2, heterodimerization in native cells is unlikely, but final conclusions will require additional investigation using other approaches.
13. Radioligand binding assays for GPCRs can be performed on crude membranes or intact cells. Assay conditions can vary depending on the specific GPCR and available radioligands, but should always be designed to detect all receptor binding sites, whether they are located at the plasma membrane or on intracellular membranes. Indeed BRET values reflect the entire receptor population expressed in the cell. Fluorescence-labeled ligands can be used to estimate receptor binding sites, provided that interference with the excitation spectrum of BRET-acceptor are avoided.
14. Independent determination of donor (luminescence) and acceptor (fluorescence) quantities in cells coexpressing both fusion proteins, represents a key feature of the BRET assay and is a clear advantage compared to FRET assays based on two GFP variants.

References

1. Xu, Y., Piston, D. W., and Johnson, C. H. (1999) A bioluminescence resonance energy transfer (BRET) system: application to interacting circadian clock proteins. *Proc Natl Acad Sci U S A* **96,** 151–6.
2. Issad, T., and Jockers, R. (2006) Bioluminescence resonance energy transfer to monitor protein-protein interactions. *Methods Mol Biol* **332,** 195–209.
3. Fung, J. J., Deupi, X., Pardo, L., Yao, X. J., Velez-Ruiz, G. A., Devree, B. T., Sunahara, R. K., and Kobilka, B. K. (2009) Ligand-regulated oligomerization of beta(2)-adrenoceptors in a model lipid bilayer. *Embo J* **28,** 3315–28.
4. Chabre, M., and le Maire, M. (2005) Monomeric G-protein-coupled receptor as a functional unit. *Biochemistry* **44,** 9395–403.

5. Kuszak, A. J., Pitchiaya, S., Anand, J. P., Mosberg, H. I., Walter, N. G., and Sunahara, R. K. (2009) Purification and functional reconstitution of monomeric mu-opioid receptors: allosteric modulation of agonist binding by Gi2. *J Biol Chem* **284,** 26732–41.
6. Bulenger, S., Marullo, S., and Bouvier, M. (2005) Emerging role of homo- and heterodimerization in G-protein-coupled receptor biosynthesis and maturation. *Trends Pharmacol Sci* **26,** 131–7.
7. Achour, L., Labbe-Juillie, C., Scott, M. G. H., and Marullo, S. (2008) An escort for G Protein Coupled Receptors to find their path: implication for regulation of receptor density at the cell surface. *Trends Pharmacol Sci* **29,** 528–35.
8. Fuxe, K., Marcellino, D., Leo, G., and Agnati, L. F. (2010) Molecular integration via allosteric interactions in receptor heteromers. A working hypothesis. *Curr Opin Pharmacol* **10,** 14–22.
9. Bouvier, M. (2001) Oligomerization of G-protein-coupled transmitter receptors. *Nat Rev Neurosci* **2,** 274–86.
10. Ayoub, M. A., and Pfleger, K. D. G. (2010) Recent advances in bioluminescence resonance energy transfer technologies to study GPCR heteromerization. *Curr Opin Pharmacol* **10,** in press.
11. Contento, R. L., Molon, B., Boularan, C., Pozzan, T., Manes, S., Marullo S, and Viola, A. (2008) CXCR4-CCR5: a couple modulating T cell functions. *Proc Natl Acad Sci U S A* **105,** 10101–6.
12. Sohy, D., Yano, H., de Nadai, P., Urizar, E., Guillabert, A., Javitch, J. A., Parmentier, M., and Springael, J. Y. (2009) Hetero-oligomerization of CCR2, CCR5, and CXCR4 and the protean effects of "selective" antagonists. *J Biol Chem* **284,** 31270–9.
13. Mercier, J. F., Salahpour, A., Angers, S., Breit, A., and Bouvier, M. (2002) Quantitative assessment of beta 1 and beta 2-adrenergic receptor homo and hetero-dimerization by bioluminescence resonance energy transfer. *J Biol Chem* **277,** 44925–31.
14. Couturier, C., and Jockers, R. (2003) Activation of the leptin receptor by a ligand-induced conformational change of constitutive receptor dimers. *J Biol Chem* **278,** 26604–11.
15. Marullo, S., and Bouvier, M. (2007) Resonance Energy Transfer approaches in Molecular Pharmacology and beyond. *Trends Pharmacol Sci* **28,** 362–5.
16. Ayoub, M. A., Levoye, A., Delagrange, P., and Jockers, R. (2004) Preferential formation of MT1/MT2 melatonin receptor heterodimers with distinct ligand interaction properties compared to MT2 homodimers. *Mol Pharmacol* **66,** 312–21.
17. Loening, A. M., Fenn, T. D., Wu, A. M., and Gambhir, S. S. (2006) Consensus guided mutagenesis of Renilla luciferase yields enhanced stability and light output. *Protein Eng Des Sel* **19,** 391–400.
18. Nguyen, A. W., and Daugherty, P. S. (2005) Evolutionary optimization of fluorescent proteins for intracellular FRET. *Nat Biotechnol* **23,** 355–60.
19. Kamal, M., Marquez, M., Vauthier, V., Leloire, A., Froguel, P., Jockers, R., and Couturier, C. (2009) Improved donor/acceptor BRET couples for monitoring β-arrestin recruitment to G protein-coupled receptors. *Biotechnol J* **4,** 1337–44.

Chapter 10

Cell-Surface Protein–Protein Interaction Analysis with Time-Resolved FRET and Snap-Tag Technologies: Application to G Protein-Coupled Receptor Oligomerization

Laëtitia Comps-Agrar, Damien Maurel, Philippe Rondard, Jean-Philippe Pin, Eric Trinquet, and Laurent Prézeau

Abstract

G protein-coupled receptors (GPCRs) are key players in cell–cell communication, the dysregulation of which has often deleterious effects leading to pathologies such as psychiatric and neurological diseases. Consequently, GPCRs represent excellent drug targets, and as such are the object of intense research in drug discovery for therapeutic application. Recently, the GPCR field has been revolutionized by the demonstration that GPCRs are part of large protein complexes that control their pharmacology, activity, and signaling. Moreover, in these complexes, one GPCR can either associate with itself, forming homodimers or homooligomers, or with other receptor types, forming heterodimeric or heterooligomeric receptor entities that display new receptor features. These features include alterations in ligand cooperativity and selectivity, the activation of novel signaling pathways, and novel processes of desensitization. Thus, it has become necessary to identify GPCR-associated protein complexes of interest at the cell surface, and to determine the state of oligomerization of these receptors and their interactions with their partner proteins. This is essential to understand the function of GPCRs in their native environment, as well as ways to either modulate or control receptor activity with appropriate pharmacological tools, and to develop new therapeutic strategies. This requires the development of technologies to precisely address protein–protein interactions between oligomers at the cell surface. In collaboration with Cisbio Bioassay, we have developed such a technology, which combines TR-FRET detection with a new labeling method called SnapTag. This technology has allowed us to address the oligomeric state of many GPCRs.

Key words: Fluorescence resonance energy transfer, G protein-coupled receptor, Dimerization, SnapTag, $GABA_B$ receptor

1. Introduction

An increasing number of G protein-coupled receptors (GPCRs) have been shown to form oligomers (1–3). The first well-characterized functional homodimeric receptor complex was demonstrated for

Louis M. Luttrell and Stephen S.G. Ferguson (eds.), *Signal Transduction Protocols*, Methods in Molecular Biology, vol. 756, DOI 10.1007/978-1-61779-160-4_10, © Springer Science+Business Media, LLC 2011

the metabotropic glutamate (mGlu) receptors, in which both protomers are linked by a disulfide bond in their extracellular domains (4). The $GABA_B$ receptor, also a Class C GPCR, was the first to be described as a functional obligatory heterodimer (5). The two homologous subunits, $GABA_{B1}$ which binds the endogenous ligand GABA, and $GABA_{B2}$ that couples to G proteins, are required for the formation of an active receptor at the cell surface. Thus, these GPCRs have been models of choice for developing technological tools to study in detail their dimerization states and their roles in receptor function.

Recently, several biophysical methods based on resonance energy transfer (RET), like Bioluminescence Resonance Energy Transfer (BRET) and Förster Resonance Energy Transfer (FRET) have been developed for the analysis of protein–protein interactions in living cells, notably GPCR interactions (6–8). However, these techniques suffer a number of limitations, especially when studying receptors at the cell surface. First, the receptors of interest needs to be fused to fluorescent proteins [such as the green fluorescent protein (GFP) family proteins], meaning that all receptors contribute to the recorded fluorescent signal. This includes receptors that are expressed at the cell surface as well as those localized to intracellular compartments, e.g., within the synthetic, endocytic, and degradation pathways. Second, because absorption and emission spectra of the donor and acceptor fluorophores are not well separated, the excitation of the donor fluorophore leads to a contaminating excitation of the acceptor fluorophore, and the recorded emission spectra of the acceptor fluorophore is contaminated by the emission of the donor fluorophore.

In order to avoid these two major problems, a new technology called Homogeneous Time Resolved FRET (HTRF®) has been developed based on an energy transfer between an europium cryptate (or a terbium cryptate, see Note 1) as the donor fluorophore and the Cy5-like dye (d2) as acceptor (9–11). Thanks to the peculiar properties of these rare earth cryptates, which display a long fluorescence emission lifetime in the millisecond range, compared to the nanosecond range for standard fluorophores, it is possible to record the FRET signal in a time resolved manner (Time Resolved-FRET or TR-FRET). The application of a 50 μs delay between the excitation of the donor and the TR-FRET measurement (emission of the acceptor) allows the investigator to remove the background that arises from either the autofluorescence of the cells or the free acceptor, as their fluorescence is quickly switched off. Moreover, europium cryptate and d2 present optimal spectral properties, such that d2 emits in a spectral range where the emission of the europium cryptate is barely detectable. Each of these features allow a large increase in the signal to noise ratio when compared to the classical RET methods.

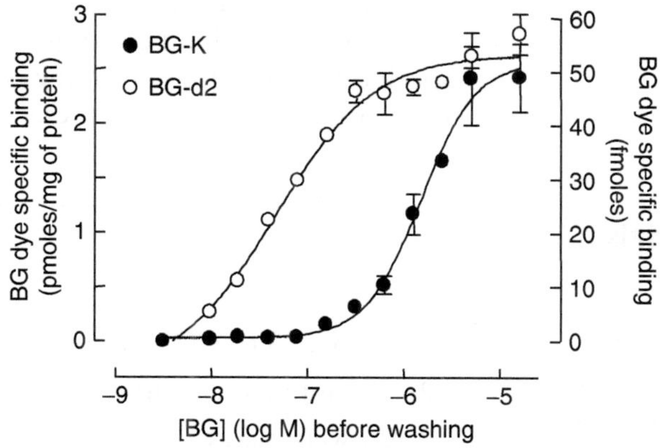

Fig. 1. Cell surface-specific binding with increasing concentrations of either BG-K or BG-d2 for cells expressing constant amount of ST-GB1 and GB2. Data are represented as the number of mol of either BG-K or BG-d2 fixed to the GB1 subunit per well (*right scale*) or per milligram of protein (*left scale*). Reproduced from (13) with permission from Nature Publishing Group NPG.

In our laboratory, a recent improvement of this technology has been the coupling of the TR-FRET approach to the Snap-tag labeling method (12), which allows the covalent labeling of one protein with one fluorophore at the cell surface (see Fig. 1 and Note 1) (13). The Snap-tag (ST) 20 kDa derives from the DNA repair protein O^6-alkylguanine-DNA alkyltransferase (AGT), that recognizes O^6-alkylated guanines in DNA and irreversibly transfers the alkyl group to one of its reactive cysteine residues. The Snap-tag corresponds to a modified AGT that displays faster reaction kinetics with O^6 benzylguanine (BG) substrates, and no longer interacts with DNA. Practically, the Snap-tag protein is fused to the extracellular extremity of the protein of interest and expressed in cells. Cells are then incubated with BG labeled with either cryptate donor or d2 acceptor fluorophores. Thus, it is possible to perform FRET experiments between adjacent Snap-tagged proteins covalently labeled with a fluorophore (11). Incubation with a precise ratio of donor and acceptor BG-fluorophores leads to several possible dimer combinations: 25% of dimers containing both receptors labeled with the donor, 25% of the dimers containing both receptors labeled with the acceptors, and 50% of the dimers containing one receptor labeled with the donor and one receptor labeled with the acceptor, which will provide the FRET signal. Methodology for performing TR-FRET, as well as the analysis and the interpretation of the data is described below (13). The newly commercial version of the technology, using reagents based on Terbium crytates and known as the Tag-Lite substrates (Cisbio Bioassay; Bagnols/Cèze, France) are referenced (see Note 1). An alternative method to the Snap-tag is the use of antibodies conjugated to HTRF fluorophores to specifically label receptors at the cell surface. This approach is also described at the end of the present article (11).

2. Materials

2.1. Cell Culture and Transfection with Electroporation

1. HEK 293 or COS7 cells.
2. 100 mm tissue culture dishes.
3. 96-well plates (black plate with black bottom) (CellStar®; Greiner).
4. Phosphate-Buffered Saline (PBS): 1× prepared from PBS 10× (Lonza).
5. Trypsin/EDTA solution: 0.05% trypsin and 0.53 mM EDTA (Gibco/BRL-Life Technologies).
6. Complete Dulbecco's Modified Eagle's Medium (DMEM) medium: DMEM supplemented with 10% fetal bovine serum, 1% penicillin/streptomycin and 1% nonessential amino acids. All the products used for cell culture are purchased from Gibco/BRL-Life Technologies.
7. Braun water (Braun).
8. 5× Electroporation buffer (EB): 250 mM KH_2PO_4, 100 mM CH_3COOK, 100 mM KOH. Prepare a concentrated buffer stock (5× EB) that will be diluted with water for the buffer (1× EB).
9. 1 M $MgSO_4$ solution.
10. Electroporator (Gene pulser®; Biorad) and adapted electroporation cuvettes.
11. 2.53 mM polyornithine solution diluted in PBS and kept at 4°C.

2.2. Snap-Tag and Antibody Labeling of Receptors

1. Benzylguanine conjugated with fluorophores: europium cryptate for donor (BG-K) and d2 for acceptor (BG-d2) calibrated at 1 mM each (see Note 1 for the commercially available version of these reagents (Cisbio Bioassay)).
2. Tris KREBS buffer (TK buffer): 20 mM Tris, 118 mM NaCl, 5.6 mM Glucose, 1.2 mM KH_2PO_4, 1.2 mM $MgSO_4$, 4.7 mM KCl, 1.8 mM $CaCl_2$, pH 7.4.
3. Monoclonal anti-HA (12CA5) and anti-Flag (M2) antibodies conjugated with europium cryptate and d2 provided (Cisbio Bioassay). See manufacturer recommendations for preparation and storage of the different compounds.

2.3. FRET Measurements

1. Time-resolved fluorimeter, e.g., RUBYStar® plate reader (BMG Labtechnologies).
2. An adapted plate reader is required when reading the specific d2 fluorophore signal at 680 nm, e.g., Analyst® plate reader (Molecular Devices).

3. Methods

The analysis of receptor oligomerization relies on the capacity of the methods to detect proximity between proteins at the cell surface. Thus, it is crucial to know the percentage of proteins labeled with the fluoropohores. Usually, it is necessary to perform experiment with near 100% of the proteins of interest labeled by the fluorophores you intend to use.

3.1. Preparation of Expression Plasmids

The plasmid used for mammalian cell expression should contain several specific restriction sites upstream of the receptor cDNA sequence to allow extracellular localization of labeling proteins at the amino-terminus of the GPCRs.

1. Subclone the Snap-tag cDNA sequence (obtained from the pSST26m plasmid from Covalys) upstream of the cDNA sequence encoding the protein of interest.
2. Upstream of the Snap-tag, a tag (e.g., HA, Flag or myc) is added in order to perform ELISA experiments to allow the determination of cell surface expression of labeled receptors.
3. The HA, Flag, and myc tags can also be useful to perform HTRF technology with commercially available fluorophores conjugated antibodies against these tags. This will also permit orthogonal labeling between a Snap-tag and an epitope-tag to measure TR-FRET.

3.2. Preparation of the Cells and Transfection

The method below describes the transfection of 1×10^7 cells by electroporation (HEK 293 or COS-7 cell lines). To electroporate 5×10^6 cells, all the quantities mentioned bellow should be divided by two.

1. HEK 293 or COS7 cells are cultured in complete DMEM incubated at 37°C, 5% CO_2. Cells are split into 100-mm dishes when approaching confluence to provide new cell cultures.
2. Greiner CellStar® 96-well plates should be treated with 50 μl per well of 2.53 mM polyornithine and incubated at 37°C in 5% CO_2 for at least 30 min.
3. Prepare the plasmid cDNA mix encoding the proteins of interest and completed to a total amount of 10 μg plasmid cDNA

with pRK5 empty vector. For example, to obtain maximal expression of the $GABA_B$ receptor, use 1 μg of pRK-$GABA_{B1}$ (GB1 subunit) and 1 μg of pRK-$GABA_{B2}$ (GB2 subunit) supplemented with 8 μg of pRK5. For FRET experiments, a mock with only pRK5 empty vector must be included.

4. To the cDNA mix, add 142 μl of Braun water, 40 μl of 5× EB, and 8 μl of 1 M $MgSO_4$.
5. Take a 100-mm dish of cells from the incubator. Wash the cells once with PBS solution (10 ml/dish) and then dissociate the cells using prewarmed trypsin/EDTA solution (5 ml/dish) for 10 min at 37°C in 5% CO_2.
6. Neutralize the trypsin by adding 5 ml of prewarmed complete DMEM to each dish. Cells are then transferred to a 50 ml conical polypropylene tissue culture tube and gently triturated. A small volume is removed and used to count the cells with a Malassez cell/hemocytometer.
7. Centrifuge the cells for 5 min at 167×*g* and remove the supernatant.
8. Resuspend the cell pellet in 1× EB. The volume of 1× EB to add should be previously determined in order to obtain a concentration of 1×10^7 cells per 100 μl.
9. Add 100 μl of cell suspension to each cDNA mix (total volume for electroporation: 300 μl) and gently resuspend cells using a pipette. Incubate the cDNA/cell mix for 10 min at room temperature.
10. During this time, select the electroporation parameters for the pulse mode: for 1×10^7 HEK 293 or COS7 cells, the electroporation parameters are 250 V–1,000 μF and 280 V–1,000 μF, respectively.
11. Transfer each eletroporation mix to an electroporation cuvette and place the cuvette in the electroporator. Deliver the electric shock for about 40 ms in the pulse mode.
12. Remove the cells from the cuvette and resuspend in 10 ml of fresh complete DMEM.
13. Take out the Greiner CellStar® 96-well plates from the incubator, remove the polyornythine, seed wells with 100 μL of cell suspension (1×10^5 cells per well) and incubate for 24 h at 37°C in 5% CO_2.

3.3. Optimization of Snap-Tag Labeling

First, it is necessary to determine the optimal concentration of fluorophore to use in order to achieve 100% protein labeling, which is required for reliable interpretation.

1. Cells are transiently transfected with plasmids encoding the protein of interest fused in N-terminal end with the Snap-tag, such as ST-GB1 and ST-GB2 as described in Subheading 3.2.

2. Prepare BG-fluorophores (both K and d2) in prewarmed complete DMEM at concentrations ranging between 10 nM and 10 μM. Each concentration of fluorophore is distributed to three wells in a 100 μL volume.
3. Wash the cells once with prewarmed complete DMEM.
4. Add each concentration of BG-fluorophores in triplicate wells of the 96-well plate which is subsequently incubated for 1 h at 37°C in 5% CO_2 (see Note 2).
5. Following the incubation step, each well of the 96-well plate is washed four times with 100 μL of TK buffer, which can be removed by aspiration, but care should be taken not to aspirate the cells (see Notes 3–5).
6. Add 100 μl per well of TK buffer and measure the specific fluorescence emission of the BG-K and BG-d2 at 620 and 682 nm, with excitation at 337 and 640 nm, respectively (total fluorescence minus that measured with mock transfected cells) (Fig. 1) on a fluorimeter. The fluorescence signal should become saturated with increasing fluorophore concentration.
7. To ensure that 100% of the proteins of interest are labeled, an additional experiment is performed. For cells expressing increasing amounts of ST-receptors, the specific labeling of the ST-receptor with the BG-dye is plotted as a function of the receptor density at the cell surface. Receptor density can be determined by radioligand binding. The slope of the line provides the labeling efficiency, which is defined as the ratio between the number of labeled ST-receptors to the total population of ST-receptors. If the slope of the straight line is equal to one, it is concluded that 100% of the ST-receptors expressed at the cell surface are labeled each with one fluorophore.

After the Snap-tag labeling optimization step, two options are possible to perform FRET assay: a Snap-tag/Snap-tag labeling (Subheading 3.4) or Snap-tag/Antibody labeling (Subheading 3.5).

3.4. Snap-Tag/Snap-Tag TR-FRET Assay After Optimization of the Snap-Tag Labeling

To perform TR-FRET experiments between two ST-GPCRs, like ST-$GABA_{B1}$ and ST-$GABA_{B2}$, it is necessary to determine the labeling conditions that will ensure equivalent labeling of the Snap-tags with either fluorophore (50% of each protein labeled with the donor and 50% with the acceptor). This is necessary as the BG-K and the BG-d2 have slightly different labeling kinetics depending on their chemical environment.

3.4.1. Determination of Optimal Labeling Conditions

1. Transfect either HEK 293 or COS7 cells with plasmid cDNAs encoding the Snap-tagged receptors of interest as described in Subheading 3.1.
2. Twenty-four hours following transfection, dilute BG-K in complete DMEM at a concentration corresponding to the maximal occupancy of the Snap-tag sites as outlined in Subheading 3.3.

3. Using the BG-K solution, prepare BG-d2 and O^6-BG ("cold"-BG, if both BG-d2 and O^6-BG display the same reactivity; otherwise see Note 6) at concentrations ranging from 10 nM to 10 μM (three wells per concentration tested, 100 μl per well).
4. Wash the cells once with prewarmed complete DMEM.
5. 100 μL volumes of either BG-K/BG-d2 or BG-K/O^6-BG mix are added to triplicate wells (for each label) of a 96-well plate at each of the concentrations tested. The 96-well plate is then incubated for 1 h at 37°C, 5% CO_2 (see Note 2).
6. After the labeling, carefully wash the cells four times with the TK buffer (100 μl per well) (see Notes 3–5).
7. Add 100 μL of TK buffer to each well and record the signal at 665 nm using the time-resolved fluorimeter. The specific FRET signal is determined by substracting the signal recorded at 665 nm for cells labeled with BG-K/BG-d2 from the signal recorded at 665 nm for cells labeled with BG-K/O^6-BG.
8. FRET signal is represented as a function of the increasing concentrations of BG-d2 to obtain a bell curve. The ratio BG-K/BG-d2 allowing the equivalent labeling of the Snap-tagged receptors with either fluorophore will correspond to maximal point of the curve (Fig. 2).

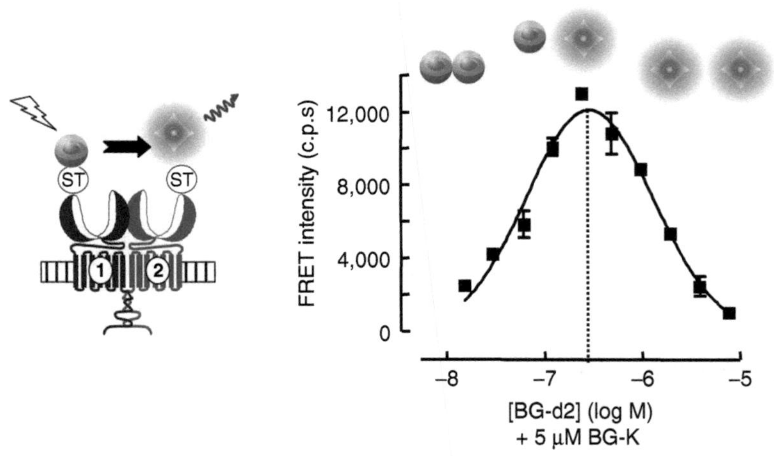

Fig. 2. FRET intensity depending on the ratio BG-K/BG-d2 applied to cells expressing ST-GB1 and ST-GB2. A constant concentration of BG-K is used combined with increasing concentrations range of BG-d2. The ratio given rise to the maximal FRET signal corresponds to the peak of the bell curve. Reproduced from (13) with permission from Nature Publishing Group NPG.

3.4.2. FRET Determination Using the Defined Conditions

1. Transfect either HEK 293 or COS7 cells with increasing amounts of plasmid cDNAs encoding the Snap-tagged receptors of interest and empty pRK5 vector as a control, as described in Subheading 3.1. Distribute the electroporated cells in a Greiner CellStar® 96-well plate as described in Subheading 3.2: six wells per transfection. Plate the extra cells on another plate to determine the cell surface expression of the receptors for each transfection (perform either an ELISA assay or radioligand binding).
2. Twenty-four hours after transfection, prepare two mixtures in complete DMEM: either BG-K/BG-d2 or BG-K/cold-BG. The optimal ratio of BG-dyes to use was determined previously (see Subheading 3.4.1). The cold-BG is diluted at the same concentration than the BG-d2.
3. Wash the cells once with prewarmed complete DMEM.
4. Add the BG-dyes to the cells: three wells with BG-K/cold-BG and three other wells with BG-K/BG-d2 per transfection. Incubate the 96-well plate for 1 h at 37°C at 5% CO_2 (see Note 2).
5. Wash the cells four times with TK buffer (see Notes 3–5).
6. Add 100 μl per well of TK buffer and read the fluorescence signal on the time-resolved fluorimeter. The specific FRET signal is determined by the calculation of the Δ665, as detailed below (Subheading 3.6.1).
7. FRET signal is plotted against the cell surface expression of the proteins of interest (Fig. 3).

3.5. Snap-Antibody Labeling Associated with FRET Assay

To perform an orthogonal labeling of receptor, it is possible to combine the Snap-tag and antibody labeling. To do this, cells should be transfected with plasmid cDNA encoding a Snap-tagged receptor and with plasmid cDNA encoding an amino-terminal epitope-tagged (e.g., HA, Flag, or myc) receptor.

1. Transfect either HEK 293 or COS7 cells with increasing amounts of a plasmid cDNA encoding a Snap-tagged receptor and another plasmid cDNA encoding an amino-terminal epitope-tagged receptor as outlined in Subheading 3.1 and plate cells in a Greiner CellStar® 96-well plate, six wells per transfection as outlined in Subheading 3.2 in order to perform TR-FRET analysis. The extra cells are plated in parallel in another plate to determine the cell surface expression of the receptors.
2. Twenty-four hours after transfection, label the Snap-tagged protein with the BG-d2 (for optimal FRET pairs, see Note 7). Note that the concentration of the BG-fluorophore required

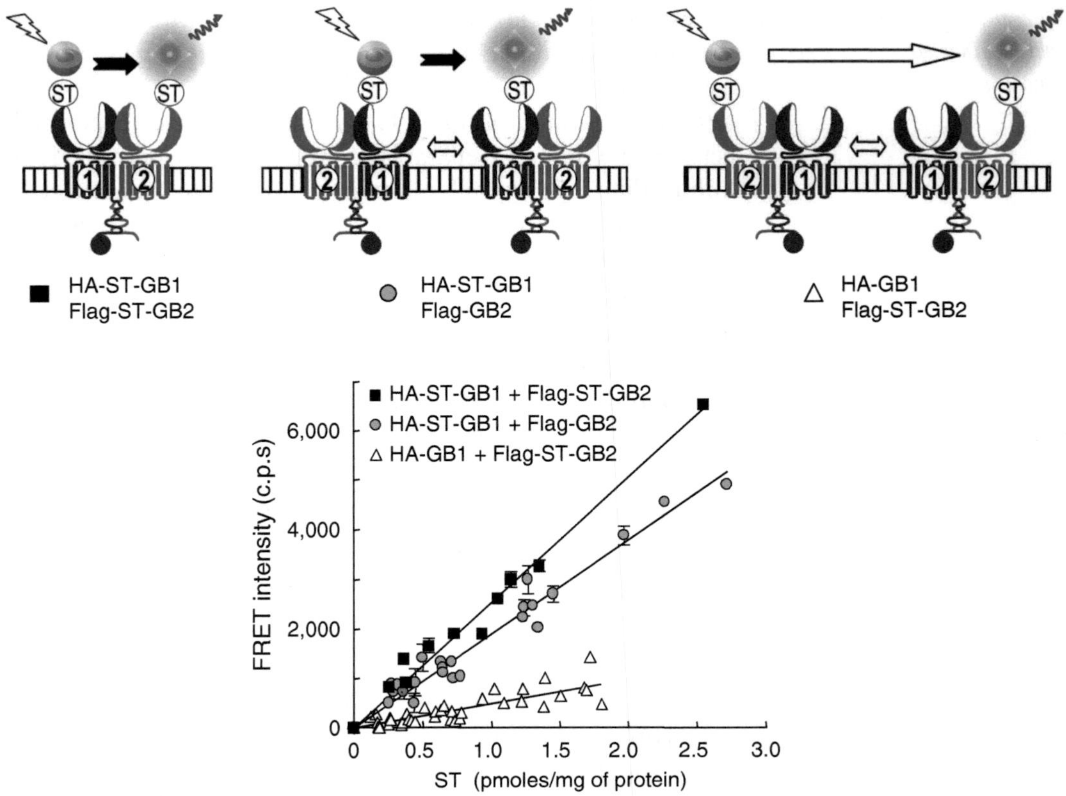

Fig. 3. Detection of higher-order multimers of $GABA_B$ dimers at the cell surface using Snap-tag labeling associated with TR-FRET assay. FRET intensity is recorded on cells expressing increasing amounts of different combinations of ST-$GABA_B$ receptors (*see scheme*): (1) both subunits GB1 and GB2 carry a Snap-tag, (2) only GB1 carries a Snap-tag, and (3) only GB2 carries a Snap-tag. FRET intensity is plotted as a function of the amount of snap-tags at the cell surface. These results support a model of higher-ordered oligomer, whereby $GABA_B$ heterodimers interact each other via the GB1 subunit. Reproduced from (13) with permission from Nature Publishing Group NPG.

for 100% labeling of the Snap-tagged protein is determined as described above in Subheading 3.3.

3. Wash cells four times with TK buffer, add 100 μL of anti-Tag-K antibody diluted in TK buffer to a final concentration of 2 nM, and incubate the 96-well plate overnight at 4°C (see Note 8). We have observed that the FRET signal is higher when the FRET donor is carried by the antibody and the FRET acceptor is carried by the BG rather than the inverse.
4. Read the signal on the time-resolved fluorimeter. The specific FRET signal is determined by the calculation of the Δ665 as detailed below.
5. FRET signal is plotted against the cell surface expression of the proteins of interest (Fig. 3).

3.6. Data Analysis and Presentation

3.6.1. Determination of the Specific Δ665 FRET Signal

The calculation of the Δ665 FRET signal allows the determination of the specific FRET signal and thus removes nonspecific FRET due to random collisions and the weak signal due to the contamination of the donor emission at 665 nm (see Note 9).

1. Δ665 represents the signal at 665 nm measured on cells co-labeled with the donor and the acceptor (positive) from which the signal recorded on cells labeled with the donor in absence of acceptor (negative) is subtracted: Δ665 = (signal at 665 nm from the positive) – (signal at 665 nm from the negative) (Table 1).
   (a) In the case of the Snap-tag/Snap-tag FRET signal, the Δ665 is obtained by subtraction of the signal recorded at 665 nm from cells labeled with BG-K/cold-BG from the one measured from cells labeled with BG-K/BG-D2.
   (b) In the case of the Snap-tag/Antibody FRET signal, the Δ665 is obtained by subtraction of the signal recorded at 665 nm from control cells expressing two proteins that do not interact (one Snap-tagged and one epitope-tagged), labeled with BG-acceptor and donor anti-epitope antibodies from the one measured from cells expressing the proteins of interest labeled with BG-acceptor and the donor anti-epitope antibody. If the control cells are not available, they can be replaced by mock cells treated the same way.

3.6.2. Data Interpretation

1. FRET signal is plotted against the amount of receptor protein of interest that is expressed at the cell surface (Fig. 3). If the receptors interact with each other, the curve should fit with a linear regression analysis.

Table 1
Determination of the positive and negative parameters for calculation of FRET analysis in various conditions

	Positive		Negative	
Fusion proteins	**Cells**	**Fluorophores**	**Cells**	**Fluorophores**
Snap-tag + Snap-tag	Snap-tag + Snap-tag	BG-K + BG-d2	Snap-tag + Snap-tag	BG-K + cold-BG
Snap-tag + Epitope-tag	Snap-tag + Epitope-tag	BG-d2 + anti-Tag-K	Snap-tag + Epitope-tag BUT non interacting proteins (by default use mock cells)	BG-d2 + anti-Tag-K

2. As TR-FRET intensity is a relative unit, the analyzed curve has to be compared with the curves obtained with at least two other conditions, named "Positive control" using two proteins already known to interact with each other, and "Negative control" using two proteins known to not interact. The Positive control cells and Negative Control cells have to be treated the same way and at the same time as the cells expressing the proteins of interest.
3. If the curve does not fit with a linear regression analysis, but rather with a hyperbolic curve analysis, the signal can reflect the clustering of the proteins rather than their oligomerization.

4. Notes

1. Three Tag-lite substrates are commercially available from Cisbio Bioassays to perform GPCR oligomerization assays. The SNAP-Lumi4 Tb bearing a terbium cryptate (5 moles; Reference SSNTBE) can be either associated with the SNAP-red substrate (20 nmoles; Ref SSNPREDE) or with the SNAP-green substrate (5 nmoles; Reference SSNGRNE). These new benzylguanine derivatives conjugated with different fluorophores are significantly more reactive than the reagents used in the present chapter. However, the protocols to be used to carry out experiments with these reagents are very similar to those described in this chapter. Additional information can be found on http://www.htrf.com.
2. If dyes sensitive to the light are used (e.g., fluorescein), protect your 96-well plate from the light with an aluminum foil to avoid the bleaching of dyes.
3. The presence of BSA in TK buffer during the wash steps does not increase or decrease the nonspecific signal.
4. Washes in TK buffer rather than DMEM are more efficient to obtain a higher signal to noise ratio.
5. HEK 293 cells are less adherent than COS7 cells to the 96-well plate even if treated with polyornithine 1×, so the washes should be done carefully. Usually, plates were seeded with a higher number of HEK cells (1.5×10^5 cells) compared to COS7 cells (1×10^5 cells) to obtain comparable results.
6. If the BG-d2 and the O^6-BG do not have the same reactivity, an additional experiment will be required to determine the concentration allowing 50% of the Snap-tag sites to be occupied by the O^6-BG in the presence of the BG-K. On cells expressing the Snap-tagged receptors of interest, increasing concentrations of O^6-BG are added in combination with a

constant amount of BG-K (see Subheading 3.3). The O^6-BG concentration to use is defined as the concentration for which the BG-K emission is half of the maximal emission (as measured in the absence of O^6-BG).

7. Of note, the FRET signal is significantly higher when using BG fused with an acceptor and antibody fused with a donor rather than the inverse. In the current protocol this is the only approach described.
8. Due to the low antibody concentration (2 nM) and the low temperature used, the labeling kinetic is quite slow, hence a long incubation time is required. However, incubating cells with antibodies at low temperature prevents receptors from clustering and internalizing and limits the nonspecific antibody binding.
9. In screening assays, the emission at 620 nm is used as an internal reference to correct the signal at 665 nm from possible interfering artifacts like absorption by the assay medium at the excitation wavelength. By dividing the signal at 665 nm by the signal at 620 nm it is possible to introduce a correction of the FRET signal, which is now independent of the media optical properties. Here, we chose not to represent the TR-FRET signal with a ratiometric representation but with the Δ665. It is more convenient to use this representation as TR-FRET measurement cannot be performed in homogeneous format due to the labeling steps needed to wash out the excess of free benzylguanine derivatives (see Subheading 3.6). In this heterogeneous format the emission of the europium cryptate is not constant and cannot be used as internal reference. It is important to mention here that Δ665 can be used as a TR-FRET representation if experiments have been performed in the same conditions (i.e., same buffers, same plate reader).

Acknowledgments

We thank Eric Trinquet and his team at Cisbio Bioassay, for the technological and scientific collaboration, the fluorophores, and the development of new fluorescent tools dedicated to TR-FRET analysis. We thank K. Johnsson (Ecole Polytechnique Fédérale de Lausanne) for his support to this project, and for providing some Snap-tag tools. The work described was made possible thanks to the screening facilities (Plate-forme de Pharmacologie Criblage Interactome) of the Institut Fédératif de Recherche 3. This work has been supported by CNRS, INSERM, Cisbio Bioassay, and by Grants from the French Ministry of Research, Action Concertée

Incitative "Biologie Cellulaire Moléculaire et Structurale" (ACI-BCM 328), the Agence Nationale de la Recherche (ANR-05-PRIB-02502, ANR-BLAN06-3_135092 and ANR-05-NEUR-035) and by an unrestricted grant from Senomyx.

References

1. Bouvier, M. (2001) Oligomerization of G-protein-coupled transmitter receptors. *Nature Rev* **2**, 274–86.
2. Milligan, G. (2006) G-protein-coupled receptor heterodimers: pharmacology, function and relevance to drug discovery. *Drug Discov Today* **11**, 541–9.
3. Pin, J. P., Neubig, R., Bouvier, M., Devi, L., Filizola, M., Javitch, J. A., Lohse, M. J., Milligan, G., Palczewski, K., Parmentier, M., and Spedding, M. (2007) International Union of Basic and Clinical Pharmacology. LXVII. Recommendations for the recognition and nomenclature of G protein-coupled receptor heteromultimers. *Pharmacol Rev* **59**, 5–13.
4. Romano, C., Yang, W. L., and O'Malley, K. L. (1996) Metabotropic glutamate receptor 5 is a disulfide-linked dimer. *J Biol Chem* **271**, 28612–6.
5. White, J. H., Wise, A., Main, M. J., Green, A., Fraser, N. J., Disney, G. H., Barnes, A. A., Emson, P., Foord, S. M., and Marshall, F. H. (1998) Heterodimerization is required for the formation of a functional GABA(B) receptor. *Nature* **396**, 679–82.
6. Angers, S., Salahpour, A., Joly, E., Hilairet, S., Chelsky, D., Dennis, M., and Bouvier, M. (2000) Detection of β_2-adrenergic receptor dimerization in living cells using bioluminescence resonance energy transfer (BRET). *Proc Natl Acad Sci U S A* **97**, 3684–9.
7. Overton, M. C., and Blumer, K. J. (2000) G-protein-coupled receptors function as oligomers in vivo. *Curr Biol* **10**, 341–4.
8. Rocheville, M., Lange, D. C., Kumar, U., Patel, S. C., Patel, R. C., and Patel, Y. C. (2000) Receptors for dopamine and somatostatin: formation of hetero-oligomers with enhanced functional activity. *Science* **288**, 154–7.
9. Bazin, H., Trinquet, E., and Mathis, G. (2002) Time resolved amplification of cryptate emission: a versatile technology to trace biomolecular interactions. *Rev Mol Biotech* **82**, 233–50.
10. Mathis, G. (1995) Probing molecular interactions with homogeneous techniques based on rare earth cryptates and Xuorescence energy transfer. *Clin Chem* **41**, 1391–7.
11. Maurel, D., Kniazeff, J., Mathis, G., Trinquet, E., Pin, J. P., and Ansanay, H. (2004) Cell surface detection of membrane protein interaction with homogeneous time-resolved fluorescence resonance energy transfer technology. *Anal Biochem* **329**, 253–62.
12. Keppler, A., Gendreizig, S., Gronemeyer, T., Pick, H., Vogel, H., and Johnsson, K. A (2003) General method for the covalent labeling of fusion proteins with small molecules in vivo. *Nat Biotechnol* **21**, 86–9.
13. Maurel, D., Comps-Agrar, L., Brock, C., Rives, M. L., Bourrier, E., Ayoub, M. A., Bazin, H., Tinel, N., Durroux, T., Prézeau, L., Trinquet, E., and Pin, J. P. (2008) Cell-surface protein–protein interaction analysis with time-resolved FRET and snap-tag technologies: application to GPCR oligomerization. *Nat Methods* **5**, 561–7.

Chapter 11

Analysis of GPCR/Ion Channel Interactions

Christophe Altier and Gerald W. Zamponi

Abstract

Voltage-gated calcium channels are key regulators of calcium homeostasis in excitable cells. A number of cellular signaling pathways serve to fine tune calcium channel activity, including the activation of G protein-coupled receptors. Besides regulating channel activity via second messengers, GPCRs can also physically associate with calcium channels to directly regulate their functions, as well as their trafficking to and from the plasma membrane. Here we provide some methods that can be used to examine channel–receptor interactions and co-trafficking. While we focus on voltage-gated calcium channels, the techniques described herein are broadly applicable to other types of channels.

Key words: Voltage-gated calcium channel, G protein-coupled receptor, Luminometry, Confocal microscopy

1. Introduction

Calcium entry via voltage-gated calcium channels (VGCCs) mediates a number of downstream effects that include activation of calcium-dependent enzymes, the release of neurotransmitters, cardiac muscle contraction, and gene transcription (1, 2). Therefore, intracellular calcium levels must be precisely controlled. One such mechanism is the regulation of calcium channel activity by G protein-coupled receptors (GPCRs), which upon agonist binding, activate G protein cascades and initiate a plethora of intracellular signaling pathways that have the propensity to regulate channel function (3). Our laboratory was the first to show that certain types of GPCRs, including members of the opioid receptor and dopamine receptor families, can also physically associate with N-type channels to not only regulate channel function, but also to regulate the density of channels in the plasma membrane. The latter is predicted to affect

Louis M. Luttrell and Stephen S.G. Ferguson (eds.), *Signal Transduction Protocols*, Methods in Molecular Biology, vol. 756, DOI 10.1007/978-1-61779-160-4_11, © Springer Science+Business Media, LLC 2011

the amount of calcium entering into a given cell (4–7). We found that both dopamine and nociceptin receptors can increase cell surface expression by enhancing channel trafficking to the plasma membrane, and furthermore, to specific subcellular compartments such as dendrites (5, 8). Once inserted into the plasma membrane, channel receptor complexes can be internalized into both endosomes and lysosomes upon prolonged agonist exposure (9). This has also been shown to be true for GABAB receptors (10). Altogether, the formation of channel receptor complexes provides a means for: (1) optimizing signaling efficiency, (2) receptor-mediated channel targeting, and (3) receptor-mediated removal of channels from the membrane. The formation of channel receptor complexes is a novel means of regulating calcium channel function and density that can be examined with a variety of experimental approaches. Although there are numerous ways of addressing this issue, including FRET and BRET measurements, we shall focus here on techniques that have been successfully employed by our laboratory. It is worth noting, however, that their applicability goes well beyond calcium channels and G protein-coupled receptors and can be adapted to any type of ion channel. Moreover, the existence of channel–channel complexes is also beginning to emerge in the literature (11).

2. Materials

2.1. Cell Culture and Transient Transfection

1. HEK 293T cells.
2. 60 mm tissue culture dishes.
3. Dulbecco's Modified Eagle's Medium (DMEM; Invitrogen) supplemented with 10% fetal bovine serum (FBS) and 1% penicillin–streptomycin.
4. Trypsin/EDTA solution: 1 mM ethylenediamine tetraacetic acid (EDTA), 0.25% trypsin (Gibco).
5. Calcium phosphate solution: 250 mM $CaCl_2$. Dissolve 10.95 g $CaCl_2$ in 200 mL H_2O.
6. 2× HEPES-buffered saline (HBS): 8 g NaCl, 0.2 g $Na_2HPO_4 \cdot 7H_2O$, 0.5 g HEPES. Adjust pH to 7.0 and bring up to 500 ml with distilled water.
7. cDNA expression plasmids encoding the different VGCC subunits and G protein-coupled receptors of interest.

2.2. Immuno-luminometry on Cells Expressing Epitope-Tagged VGCC

1. Phosphate-Buffered Saline (PBS). 10× stock solution: 80.6 mM Na_2HPO_4, 19.4 mM KH_2PO_4, 27 mM KCl, and 1.37 M NaCl in high purity dH_2O. Adjust to pH 7.4 with HCl.
2. Poly-L-lysine solution: 0.1% (w/v) poly-L-lysine solution (Sigma) diluted to a working concentration of 0.002% in 1× PBS.

3. Glass coverslips (12 mm) treated with 0.002% poly-L-lysine solution.
4. Paraformaldhyde solution: 16% paraformaldehyde stock solution (Electron Microscopy Sciences) diluted to a 4% (w/v) working concentration in PBS 1×.
5. Blocking solution: PBS plus 2% goat serum.
6. Permeabilization solution: 0.1% (v/v) Triton X-100 in pre-blocking solution.
7. Rat monoclonal anti-HA IgG (3F10; Roche Molecular Biochemicals).
8. Horseradish Peroxidase (HRP)-conjugated goat anti-rat IgG (Jackson ImmunoResearch Laboratories).
9. 24-well white-bottomed microtiter plates (PerkinElmer).
10. Supersignal ELISA Femto Maximum Sensitivity Substrate (Pierce).
11. Aluminum foil.
12. Wallac 1420 Multilabel HTS counter or similar luminescence plate reader.

2.3. Co-immunoprecipitation from Brain Homogenates

1. Lysis Buffer: 20 mM Tris–HCl (pH 7.5), 0.5% (v/v) triton-X-100, 0.005% sodium deoxycholate, 150 mM NaCl, and protease inhibitor cocktail tablet (Roche).
2. DC Protein Assay (BioRad).
3. Protein A-sepharose 4 fast flow beads (GE Healthcare).
4. Anti-VGCC IgG for immunoprecipitation.
5. Protein A-sepharose beads or Protein G-sepharose beads.
6. Buffer B: 0.2% NP-40, 10 mM Tris–HCl (pH 7.5), 0.15 M NaCl, 2 mM EDTA.
7. Buffer C: 0.2% NP-40, 10 mM Tris–HCl (pH 7.5), 0.5 M NaCl, 3 mM EDTA.
8. Buffer D: 10 mM Tris–HCl (pH 7.5).
9. 2× sodium dodecyl sulfate (SDS) sample buffer: 130 mM Tris–HCl (pH8.0), 20% (v/v) glycerol, 4.6% (w/v) SDS, 0.02% bromophenol blue, 2% dithiothreitol (DTT).
10. Gels, buffers, and apparatus for SDS-Polyacrylamide Gel Electrophoresis (SDS-PAGE) and protein transfer.
11. Polyvinylidene fluoride (PVDF) immobilon membranes (Millipore).
12. Blocking buffer: 0.1% (v/v) Tween-20, 5% (w/v) nonfat dry milk in PBS.
13. Wash buffer (PBS-T): 0.1% (v/v) Tween-20 in PBS.
14. Primary antibodies against co-precipitated receptor(s) of interest.

15. HRP-conjugated goat IgG targeted against the primary antibody IgG species.
16. ECL detection reagent and X-ray film.

2.4. GST-Fusion Protein Pull-Down Assays

1. Glutathione S-transferase (GST)-fusion proteins containing the protein interaction domain.
2. Glutathione-sepharose 4B beads (Amersham).
3. Wash buffer (PBS-T): 0.1% (v/v) Tween-20 in PBS.
4. 2× SDS sample buffer containing 0.1 M DTT.
5. Primary anti-VGCC antibody.
6. HRP-conjugated goat IgG antibody targeted against the primary antibody IgG species.

2.5. Confocal Microscopy

1. 35 mm glass bottom culture dishes (MatTek Corporation).
2. cDNA expression plasmids encoding the different VGCC subunits and yellow fluorescent protein (YFP)-tagged G protein-coupled receptor of interest.
3. Rat monoclonal anti-HA IgG (3F10; Roche Molecular Biochemicals).
4. Hank's Balanced Salt Solution (HBSS): 0.185 g/L $CaCl_2 \cdot 2H_2O$, 0.097 g/L $MgSO_4$, 0.4 g/L KCl, 0.06 g/L KH_2PO_4, 0.35 g/L $NaHCO_3$, 8 g/L NaCl, 0.047 g/L Na_2HPO_4, 1 g/L D-glucose.
5. Antibody solution: HBSS plus 2% goat serum.
6. Paraformaldehyde solution: 4% (w/v) paraformaldehyde in PBS.
7. Permeabilization solution: 0.05% (v/v) Triton X-100 in antibody solution.
8. AlexaFluor 594-coupled goat anti-rat IgG (Molecular Probes).
9. Zeiss LSM 510 META confocal microscope with 63× 1.4NA oil immersion objective, or equivalent confocal fluorescence microscope.
10. Metamorph (Molecular Devices) and ImageJ (NIH) software for image analysis.

3. Methods

3.1. Cell Surface Immunoluminometry Assays to Measure Surface Expression of Ion Channels

1. HEK 293T cells are cultured in DMEM supplemented with 10% FBS and 1% penicillin–streptomycin. Cells are passaged using trypsin/EDTA solution.
2. Split cells at 60–70% confluence 8 h before transfection and plate onto 60 mm dishes.

3. Transfect cells overnight using the calcium phosphate transfection method. Use cDNA encoding the different VGCC subunits, e.g., HA-tagged Ca_v-α_1 subunit + β subunit + $\alpha_2-\delta_1$ subunit; ratio 2:1:1, alone and in the presence of the receptor.
4. Fourteen hours later, wash cells with PBS, detach from the plate using trypsin-EDTA, and re-seed in triplicate onto poly-L-lysine-coated coverslips.
5. Incubate cells for an additional 48 h at 37°C prior to agonist treatment.
6. For each condition, a single replicate (three coverslips) is treated for 30 min at 37°C with receptor agonist diluted in HBSS. After stimulation, wash cells with PBS and fix with 4% paraformaldehyde for 5 min at room temperature (RT). Wash cells with PBS.
7. Pre-block fixed cells at RT for 30 min in PBS containing 2% goat serum. At this stage, permeabilize a second replicate (three coverslips) from each condition by incubating in 0.1% Triton X-100 permeabilization solution for 7 min. Then wash the cells two times with PBS.
8. At this point mix solutions A and B of the Supersignal ELISA Femto Maximum Sensitivity Substrate (see Note 1). 200 μl/well of substrate is added to a 24-well, white-bottomed microtiter plate which is kept at RT in the dark until use.
9. Incubate coverslips in anti-HA antibody (0.5 μg/ml) in PBS/serum-blocking buffer for 1 h at RT.
10. Wash coverslips 3 × 10 min in PBS/serum-blocking buffer. The first wash consists of aspirating the solution from each coverslip. For subsequent washes, single coverslips are transferred to a clean 24-well dish containing blocking buffer.
11. Next incubate coverslips in HRP-conjugated secondary antibody (1:1,000) for 40 min at RT.
12. Wash coverslips 4 × 20 min in PBS/serum-blocking buffer. As before, the first wash consists of aspirating the solution from each coverslip. For subsequent washes, single coverslips are transferred to a clean 24-well dish containing blocking buffer (see Note 2).
13. After washing, immediately transfer individual coverslips into the 24-well white-bottomed microtiter plate containing substrate that has been sitting in the dark for 2 h.
14. Cover the plate with aluminum foil for 1 min, then measure luminescence at 492 nm using a Wallac 1420 Multilabel HTS counter or similar sensitive plate reader.
15. The ratio of cell surface channel expression, determined using unpermeabilized cells, to total cellular channel expression,

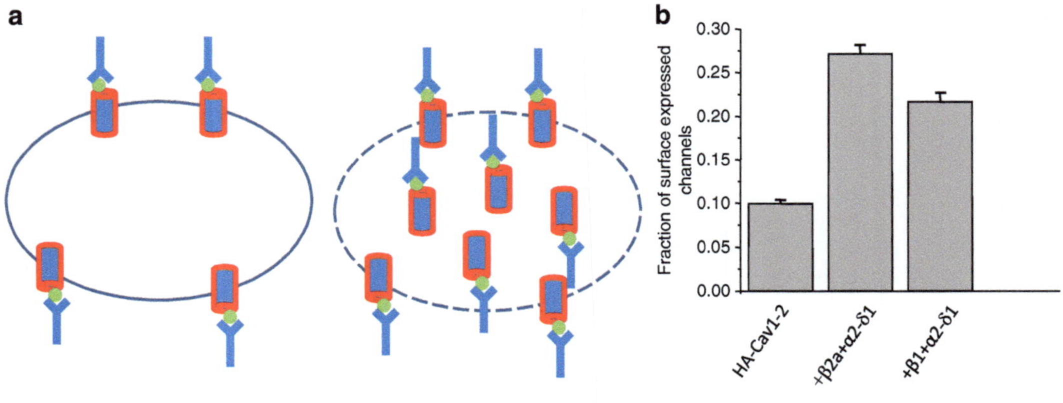

Fig. 1. Quantification of channel internalization by immunoluminometry. (**a**) Schematic representation of an immunoluminometry experiment. In the nonpermeabilized conditions only plasma membrane-inserted membrane HA-tagged channels are detected (*left*). In the permeabilized condition, all HA-tagged channels in the cell are detected (*right*). (**b**) The ratio of surface to total signal gives the fraction of HA-tagged channels expressed at the cell surface. The graph shows the surface expression of HA-Cav1.2 (L-type channel) alone or co-expressed with auxiliary subunits (i.e., $\beta2a+\alpha2-\delta1$ or $\beta1b+\alpha2-\delta1$).

determined using permeabilized cells, allows for comparisons between different conditions and between similar conditions assayed from different batches of cells on different days (see Note 3; Fig. 1).

3.2. Co-immuno-precipitation from Rat or Mouse Brain

1. Preparation of rat or mouse brain homogenates:
   (a) Immediately following euthanasia, remove the whole brain and homogenize in ice cold lysis buffer.
   (b) Centrifuge the crude homogenate at 10,000 × *g* for 10 min at 4°C. Collect the supernatant, re-homogenize, repeat the centrifugation step, and collect the supernatant.
   (c) Quantify total protein in the supernatant is quantified using the DC Protein Assay.
   (d) The supernatant is kept on ice and used immediately in co-immunoprecipitation experiments.
2. Perform co-immunoprecipitation using 200 μl of brain homogenate containing 100 μg protein/sample, diluted to final concentration in lysis buffer.
3. Pre-clear the homogenate with protein A-sepharose 4 fast flow beads for 1 h at 4°C.
4. Centrifuge the sample at 2,000 × *g* for 5 min at 4°C and collect the pre-cleared supernatant.
5. Incubate the pre-cleared supernatant with α-VGCC antibody (2 μg) overnight at 4°C.
6. Add 50 μl of Protein A-sepharose beads or Protein G-sepharose beads depending on the antibody manufacturer's recommendations and tumble for 1 h at 4°C.

7. Wash slurries twice with buffer B, once with buffer C, and once with buffer D, then solubilize immunoprecipitated proteins using SDS sample buffer.
8. Separate proteins by SDS-PAGE and transfer to PVDF immobilon membranes for immunoblot analysis.
9. Pre-block membranes overnight at 4°C in PBS/T with 5% milk prior to either overnight incubation at 4°C or 2 h incubation at RT with an appropriate primary antibody, e.g., against a co-precipitating receptor.
10. Wash membranes 3×10 min with PBS-T.
11. Incubate for 45 min at RT with an HRP-conjugated goat IgG antibody targeted against the primary antibody IgG species (1:5,000).
12. Wash membranes 3×20 min with PBS-T.
13. Visualize proteins after the addition of ECL reagents and exposure to X-ray film.

3.3. GST-Fusion Protein Pull-Down of Receptor-Associated Channel Proteins

An alternative to co-immunoprecipitation is to perform pull-down assays using Glutathione S-transferase (GST)-fusion proteins instead of antibodies for protein isolation. For pull-down experiments, GST-fusion proteins of the desired intracellular regions of the receptor must first be prepared (see Note 4).

1. Bind GST-fusion proteins to glutathione-sepharose 4B beads (Amersham) by incubation of one volume of 50% beads in PBS-T with 10 volumes of protein lysate for 1 h, 4°C.
2. Wash beads 3× with PBS-T.
3. Incubate beads for 3 h at 4°C with rat brain homogenate (see Subheading 3.2).
4. Wash beads washed 2×15 min, at 4°C with PBS-T. The resulting protein bound beads are prepared for western blot analysis by addition of 2× SDS sample buffer containing 0.1 M DTT.
5. Separate proteins by SDS-PAGE and transfer at 70 V for 3 h to PVDF immobilon membranes for immunoblot analysis.
6. Immunoblotting is carried out as described above for co-immunoprecipitation experiments (Subheading 3.2), using an α-VGCC antibody and HRP-conjugated goat IgG antibody targeted against the primary antibody IgG species (1:5,000).

3.4. Visualizing Receptor-Channel Co-trafficking Using Confocal Microscopy

1. Plate HEK293T cells at 60–70% confluence on MatTek dishes coated with poly-L-lysine solution 8 h before transfection.
2. Transfect HEK 293T cells using the calcium phosphate transfection method with cDNA encoding different subunits of voltage-gated calcium channels cDNA, e.g., HA-tagged Ca_v-α_1 subunit + β subunit + $\alpha_2-\delta_1$ subunit; ratio 2:1:1,

and a full-length G protein-coupled receptor that has been tagged at the C terminus with YFP.

3. Fourteen hours following transfection, wash cells with PBS and refeed with complete growth medium. Incubate cells at 37°C for 48–60 h.
4. To visualize HA-Cav1 internalization in response to receptor stimulation, remove the medium and immunostain cells for the HA-Ca_v1 channel using 1 μg/ml 3F10 anti-HA IgG in antibody solution for 30 min at 37°C.
5. Prepare the necessary receptor agonist by diluting to its final concentration in antibody solution.
6. Wash cells with antibody solution, then stimulate with receptor agonist at 37°C for 30 min.
7. Wash cells with antibody solution 2×5 min at RT, fix with 4% paraformaldehyde solution for 10 min at RT and then incubate with 0.05% Triton X-100 permeabilization solution for 10 min RT.
8. Wash 3×10 min with blocking solution, then incubate cells for 45 min at room temperature, in the dark, with AlexaFluor 594-coupled goat anti-rat IgG (1:1,000 dilution).
9. Wash 3×10 min with blocking solution prior to imaging.
10. Obtain images using a Zeiss LSM 510 META confocal microscope with a 63× 1.4NA oil immersion objective in the inverted configuration. The entire cell volume is imaged by a series of *z*-plane sections to produce a *z*-stack. Also obtain a regular phase transmission image for all confocal images (see Note 5).
11. Perform quantitative analysis using Metamorph and ImageJ software. Images are background subtracted and then median filtered (2×2 pixels, effective 1 pixel radius) to remove impulse-type noise and boost signal-to-noise.
12. For calculation of the relative fluorescence ratio, we select two optical sections above and below the largest membrane cross-section, as this will typically cover the effective cross-section of the cell. For each image plane, we manually trace the nucleus, the intracellular region, and cell outline, and compute the integrated fluorescence intensity. Therefore, the intracellular fluorescence intensity is corrected for the presence of the nucleus.
13. For each cell, using a series of regions of interest (ROIs), we compute the internalization ratio as defined by $Ri = I/M$, where I is the integrated fluorescence intensity within the cell (corrected for the nucleus), and M is the fluorescence intensity in the membrane region as obtained by subtracting the total intracellular fluorescence (including the nucleus) from the total integrated intensity (see Note 6; Fig. 2).

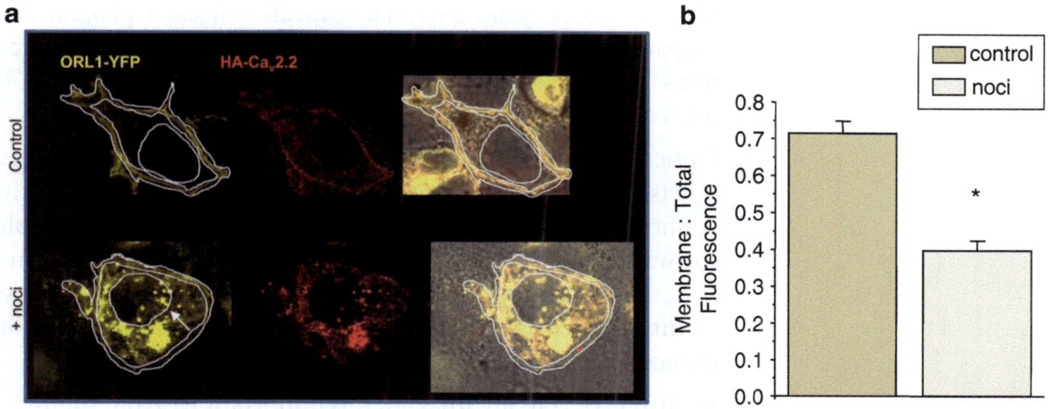

Fig. 2. Quantification of channel internalization by fluorescence confocal microscopy. (**a**) Representative confocal image of tsA-201 cells transfected with a YFP-tagged nociceptin receptor (ORL1-YFP) and an externally tagged N-type channel (HA-Cav2.2). To visualize HA-Cav2.2 internalization upon receptor agonist application, membrane-inserted calcium channels were labeled with anti-HA-antibody for 30 min at 37°C. Cells were then washed and stimulated with 100 nM nociceptin at 37°C for 30 min. Cells were washed with PBS and fixed with 4% PFA for 5 min at room temperature and then, permeabilized with 0.05% Triton X-100. Cells were then incubated for 45 min at room temperature with Alexa Fluor 594-coupled goat anti-rat antibody. Immunofluorescence microscopy was performed using a Zeiss LSM 510 META confocal microscope. White lines in the images depict the ROIs for the nucleus, cytoplasmic region, and plasma membrane. (**b**) Relative cell surface expression of HA-Cav2.2 co-expressed with the ORL1 receptor, without or with application of the agonist nociceptin, measured by confocal microscopy and image analysis as described in the method.

14. The cumulative intracellular fluorescence is corrected for the presence of the nuclei (*I*). *R*i is computed for each optical section and a mean value for all five planes are used for each cell for subsequent pooling of data and statistical analysis (see Note 7).

4. Notes

1. Mixing the chemiluminescent substrate 2 h before performing the final reading will decrease the background signal.
2. Alternatively, one can keep the coverslips in the same dish and use forceps to lift up the coverslip after each single wash. This way, antibody trapped under the coverslips will be washed away.
3. As the HA-tag on the $C_{av}1.2$ channel construct is extracellular, signal intensity from un-permeabilized cells are used as a measure of $C_{av}1.2$ channel surface expression, and signal intensity from permeabilized cells are used as a measure of total cellular channel expression.
4. To prepare GST-fusion proteins, *Escherichia coli* BL-21 bacterial cells transformed with a pGEX vector encoding the protein region of interest. Protein expression is induced by addition of 100 mM isopropyl-beta-thio galactopyranoside (IPTG)

for 2 h. Large-scale bacterial sonicates of each protein are prepared according to pGEX vector manufacturer's instructions (http://www6.gelifesciences.com/aptrix/upp01077.nsf/content/life_sciences_homepage).

5. Optical section thickness varies slightly between experiments but is typically near 0.3 μm. Internalization is qualitatively confirmed, before quantitative analysis (using metamorph software), by the use of orthographic image projections (i.e., *X*–*Z* plane projection), which can show the presence of internalized elements that are clearly localized to within intracellular regions.
6. If clusters of cells are imaged, it is important to avoid multiple inclusions of cell membranes. Hence, the total fluorescence of the cluster is determined, and all intracellular compartments are integrated and subtracted from total integrated fluorescence of the cluster to obtain total membrane fluorescence (*M*).
7. The ratio of these quantities is dimension-less and allows us to quantify the relative fluorescence intensity in the effective sections covering the membrane and intracellular regions without bias toward experimental variations including cell-to-cell protein expression level, detector gain, laser intensity, and cell shape.

Acknowledgments

We thank Dr. Bourinet for creating an HA-tagged version of the Cav2.2 channel. GWZ is Scientist of the Alberta Heritage Foundation for Medical Research, and a Canada Research Chair. Work described in this chapter was supported by grants from the Canadian Institutes of Health Research.

References

1. Catterall, W. A. (2000) Structure and regulation of voltage-gated Ca2+ channels. *Annu Rev Cell Dev Biol* **16**, 521–55.
2. Dolmetsch, R. E., Pajvani, U., Fife, K., Spotts, J. M. and Greenberg, M. E. (2001) Signaling to the nucleus by an L-type calcium channel-calmodulin complex through the MAP kinase pathway. *Science* **294**, 333–9.
3. Tedford, H. W. and Zamponi, G. W. (2006) Direct G protein modulation of Cav2 calcium channels. *Pharmacol Rev* **58**, 837–62.
4. Beedle, A. M., McRory, J. E., Poirot, O., Doering, C. J., Altier, C., Barrere, C., Hamid, J., Nargeot, J., Bourinet, E. and Zamponi, G. W. (2004) Agonist-independent modulation of N-type calcium channels by ORL1 receptors. *Nat Neurosci* **7**, 118–25.
5. Kisilevsky, A. E., Mulligan, S. J., Altier, C., Iftinca, M. C., Varela, D., Tai, C., Chen, L., Hameed, S., Hamid, J., Macvicar, B., A. and Zamponi, G. W. (2008) D1 receptors physically interact with N-type calcium channels to regulate channel distribution and dendritic calcium entry. *Neuron* **58**, 557–70.
6. Kisilevsky, A. E. and Zamponi, G. W. (2008) D2 dopamine receptors interact directly with

N-type calcium channels and regulate channel surface expression levels. *Channels* (Austin) **2**, 269–77.

7. Evans, R. M., You, H., Hameed, S., Altier, C., Mezghrani, A., Bourinet, E. and Zamponi, G. W. (2010) Heterodimerization of ORL1 and opioid receptors and its consequences for N-type calcium channel regulation. *J Biol Chem* **285**, 1032–40.
8. Altier, C. Khosravani, H., Evans, R. H., Hameedi, S., Peloquin, J. B., Vartian, B., Chen, L., Vartian, B., Beedle, A., Ferguson, S. S. G., Mezghrani, A., Dubel, S. J., Bourinet, E., McRory, J. E. and Zamponi, G. W. (2006) ORL1 receptor-mediated internalization of N-type calcium channels. *Nat Neurosci* **9**, 31–40.
9. Dale, L. B., Seachrist, J. L., Babwah, A. V. amd Ferguson, S. S. G. (2004) Regulation of angiotensin II type 1A receptor intracellular retention, degradation, and recycling by Rab5, Rab7, and Rab11 GTPases. *J Biol Chem* **279**, 13110–8.
10. Tombler, E., Cabanilla, N. J., Carman, P., Permaul, N., Hall, J. J., Richman, R. W., Lee, J., Rodriguez, J., Felsenfeld, D. P., Hennigan, R. F. and Diversé-Pierluissi, M. A. (2006) G protein-induced trafficking of voltage-dependent calcium channels. *J Biol Chem* **281**, 1827–39.
11. Berkefeld, H., Sailer, C. A., Bildl, W., Rohde, V., Thumfart, J. O., Eble, S., Klugbauer, N., Reisinger, E., Bischofberger, J., Oliver, D., Knau,s H. G., Schulte, U. and Fakler, B. (2006) BKCa-Cav channel complexes mediate rapid and localized Ca2+-activated K+signaling. *Science* **314**, 615–620.

Part IV

Receptor–Effector Coupling

Chapter 12

Multicolor BiFC Analysis of G Protein βγ Complex Formation and Localization*

Thomas R. Hynes, Evan A. Yost, Stacy M. Yost, and Catherine H. Berlot

Abstract

Cells co-express multiple G protein β and γ subunit isoforms, but the extent to which individual subunits associate to form particular βγ complexes is not known. This issue is important because in vivo knockout experiments suggest that specific βγ complexes may have unique functions despite the fact that most complexes exhibit similar properties when assayed in reconstituted systems. This chapter describes how multicolor bimolecular fluorescence complementation (BiFC) can be used in living cells to study the association preferences of β and γ subunits. Multicolor BiFC determines the association preferences of these subunits by quantifying the two fluorescent complexes formed when β or γ subunits fused to amino terminal fragments of yellow fluorescent protein (YFP-N) and cyan fluorescent protein (CFP-N) compete for interaction with limiting amounts of a common γ or β subunit, respectively, fused to a carboxyl terminal fragment of CFP (CFP-C). One means by which βγ complexes may differ from each other and thereby mediate unique functions in vivo is in the kinetics and patterns of their internalization responses to stimulation of G protein-coupled receptors (GPCRs). Methods are described for imaging and quantifying the internalization of pairs of βγ complexes in response to GPCR stimulation in living cells.

Key words: Multicolor bimolecular fluorescence complementation, Heterotrimeric G protein, G protein βγ complex, Spectrofluorometer, Yellow fluorescent protein, Cyan fluorescent protein, Fluorescence microscopy, Live cell imaging, Subcellular targeting, G protein-coupled receptor

1. Introduction

Although most combinations of the 5 G protein β subunits and 12 γ subunits known to be expressed in mammals form dimers with similar abilities to modulate the activities of effector proteins in vitro, specific αβγ combinations appear to be preferred for particular G protein-coupled receptor (GPCR)-G protein signaling

*This work was supported by National Institutes of Health Grant GM050369.

Louis M. Luttrell and Stephen S.G. Ferguson (eds.), *Signal Transduction Protocols*, Methods in Molecular Biology, vol. 756, DOI 10.1007/978-1-61779-160-4_12, © Springer Science+Business Media, LLC 2011

pathways in vivo (1). For instance, ribozyme-mediated depletion of γ_7 in HEK-293 cells leads to the selective loss of β_1 and results in decreased activation of adenylyl cyclase in response to stimulation of β-adrenergic receptors (2, 3). Mice lacking γ_7 exhibit increased startle responses and specific decreases in the levels of α_{olf} in the striatum (4). In addition, mice lacking γ_3, which are lean and display an increased susceptibility to seizures, display selective decreases in α_{i3} and β_2 (5).

In most cases, the αβγ heterotrimers that mediate GPCR signaling pathways and the βγ combinations that predominate in particular cell types are not known. The relative amounts of the βγ complexes formed in a cell will depend on the expression levels of the β and γ subunits and on their accessibilities to and relative affinities for each other. Multicolor BiFC enables quantification of the association preferences of β and γ subunits in intact cells.

Multicolor BiFC consists of the simultaneous visualization of the two fluorescent complexes formed when proteins fused to amino terminal fragments of YFP and CFP (YFP-N and CFP-N, respectively) interact with a common binding partner fused to a carboxyl terminal fragment of CFP (CFP-C). The amino terminal fragment of the fluorescent protein contains the chromophore and determines the spectral properties of the complex (6). Therefore, complexes of YFP-N and CFP-C fusion proteins are yellow, whereas those consisting of CFP-N and CFP-C fusion proteins are cyan (see Fig. 1). In the methods described here the fluorescent proteins are split at residue 158 such that the amino terminal fragment consists of residues 1–158 and the carboxyl terminal fragment consists of residues 159–238. For competition analysis, we use Cerulean, a modified version of ECFP that is 2.5-fold brighter than ECFP (7), to produce Cer-N fusion proteins, because Cer-N fusions compete more effectively with YFP-N fusions than do CFP-N fusions.

To compare the abilities of different γ subunits to compete for the same β subunit, one of the γ subunits (Red in Fig. 1a) is fused to the carboxyl terminus of YFP-N (yellow in Fig. 1a) and each of the γ subunits (green in Fig. 1b) is fused to the carboxyl terminus of Cer-N (cyan in Fig. 1b). The β subunit that is competed for (magenta in Fig. 1a, b) is fused to the carboxyl terminus of CFP-C (dark blue in Fig. 1a, b). Competition is quantified as the loss of yellow fluorescence of the CFP-C-β/YFP-N-γ complex upon co-expression of Cer-N-γ subunits (see Fig. 3). Conversely, to compare the abilities of different β subunits to compete for a common γ subunit, one of the β subunits (Red in Fig. 1c) is fused to the carboxyl terminus of YFP-N (yellow in Fig. 1c) and each of the β subunits (green in Fig. 1d) is fused to the carboxyl terminus of Cer-N (cyan in Fig. 1d). The γ subunit that is competed for (magenta in Fig. 1c, d) is fused

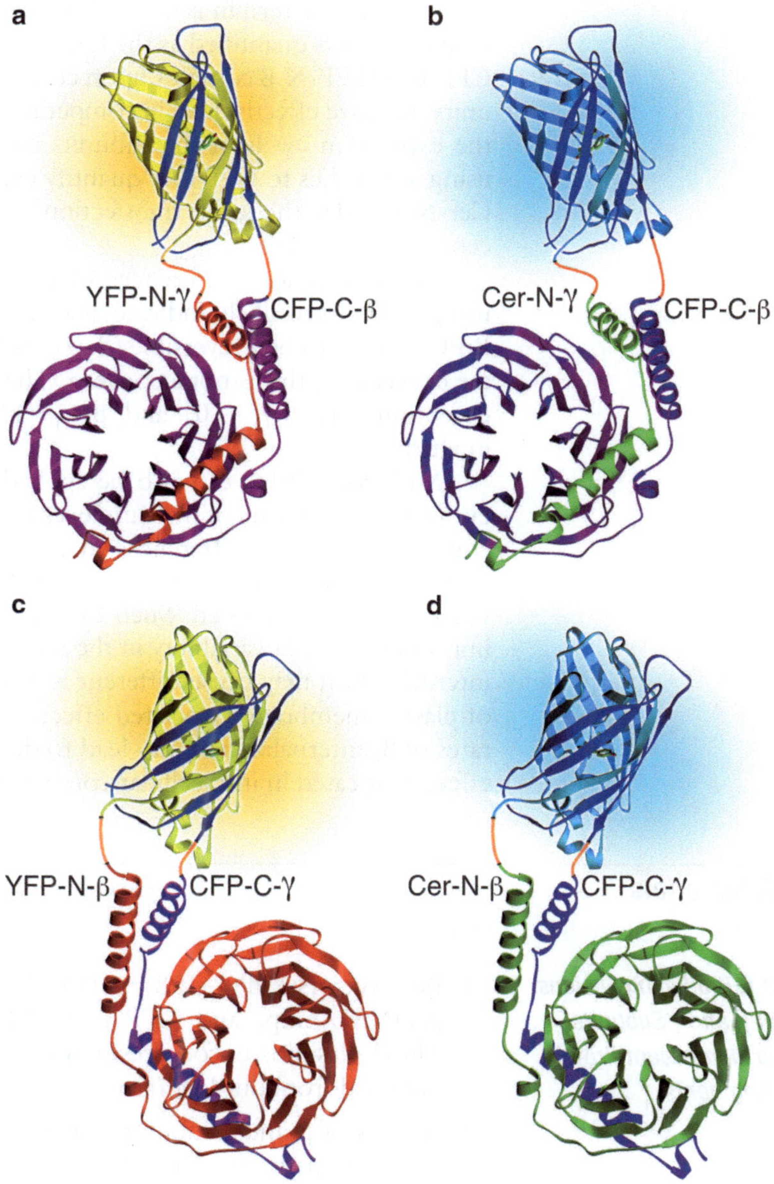

Fig. 1. Models of fluorescent βγ complexes produced with multicolor BiFC. The split fluorescent protein at the top of each model is joined by linkers (*orange*) to the βγ dimer at the bottom. The CFP-C fragment (*dark blue*) is combined with either the YFP-N fragment (*yellow*) to produce yellow fluorescence or the Cer-N fragment (*cyan*) to produce cyan fluorescence. (**a**, **b**) YFP-N-γ and Cer-N-γ compete for CFP-C-β. β is *magenta* and γ is Red (**a**) or *green* (**b**). (**c**, **d**) YFP-N-β and Cer-N-β compete for CFP-C-γ. γ is *magenta* and β is Red (**c**) or *green* (**d**). The structures of the fluorescent protein fragments are adapted from the structure of GFP (16). The structures of β and γ are from the structure of an α_t/α_{i1} chimera complexed with $\beta_t\gamma_t$ (17). Reprinted from (18) with permission from Elsevier.

to the carboxyl terminus of CFP-C (dark blue in Fig. 1c, d). Competition is quantified as the loss of yellow fluorescence of the CFP-C-γ/YFP-N-β complex upon co-expression of Cer-N-β subunits. Relative effectiveness in competition assays is normalized to the expression levels of the subunits by means of immunoblots using antibodies to GFP that quantify expression of Cer-N-β and Cer-N-γ under the same transfection conditions as the fluorescence measurements.

The interaction preferences of β and γ subunits identified using BiFC most likely indicate association preferences, because BiFC appears to be irreversible (8, 9). As β and γ generally associate irreversibly, this is not a concern. The only reported potential exceptions are $\beta_5\gamma_2$ (10) and $\beta_4\gamma_{11}$ (11), which are unstable in vitro.

Multicolor BiFC can also be applied to visualizing dynamic events involving pairs of βγ complexes using fluorescence microscopy. For example, differences in the kinetics and localization patterns of GPCR-stimulated βγ internalization responses can be visualized and quantified. Such differences may have functional importance in that variability in the rates of agonist-stimulated βγ internalization may cause differences in the deactivation kinetics of plasma membrane-associated effectors. Alternatively, different rates of βγ internalization may lead to different activation rates of effectors located in intracellular compartments.

2. Materials

2.1. Producing Fusions of β and γ Subunits to Fluorescent Protein Fragments

1. BiFC vectors: YFP(1–158)/pcDNAI/Amp, Cerulean(1–158)/pcDNAI/Amp, and CFP(159–238)/pcDNAI/Amp (12). These plasmids encode resistance to ampicillin and may be obtained from our laboratory (see Note 1).
2. cDNAs of β and γ subunits for which BiFC constructs have not been made (see Note 1).
3. TaqPlus Precision PCR System (Stratagene).
4. Qiaquick PCR purification and Qiagen MinElute Gel Extraction kits (Qiagen).
5. PCR thermal cycler.

2.2. Transient Transfections to Compare Association Preferences of β and γ Subunits

1. HEK-293 cells (American Type Culture Colleciton; CRL-1573).
2. Minimal Essential Medium with Earle's salts with L-glutamine (MEM) (Invitrogen/Life Technologies).
3. Fetal Bovine Serum (Hyclone).
4. Trypsin-EDTA solution: 0.05% trypsin, 0.53 mM EDTA (Invitrogen/Life Technologies).

5. Lipofectamine 2000 Reagent (Invitrogen/Life Technologies).
6. Opti-MEM I Reduced Serum Medium (Invitrogen/Life Technologies).
7. 60-mm tissue culture dishes.

2.3. Measurement of BiFC βγ Fluorescence Using a Spectrofluorometer

1. PC1 photon-counting spectrofluorometer with Vinci software (ISS, Inc.) or equivalent instrument. The spectrofluorometer is configured with motorized filter wheels on both the excitation path between the excitation monochrometer and the sample, and on the emission path between the sample and the emission monochrometer. The slits on the excitation and emission monochrometers are set to a 16 nm band-pass.
2. 430/25 and 492/18 band-pass filters, 1.3 OD neutral density filter, and 455, 515, and 590 long-pass filters (Chroma).
3. Glass fluorometer cuvettes with Teflon Covers (Cole-Palmer).
4. HBSS + $CaCl_2$ media: 20 mM Hepes, pH 7.2, 118 mM NaCl, 4.6 mM KCl, 10 mM D-glucose, 1 mM $CaCl_2$.
5. HBSS + EDTA media: 20 mM Hepes, pH 7.2, 118 mM NaCl, 4.6 mM KCl, 10 mM D-glucose, 0.5 mM EDTA.

2.4. Correcting for the Expression Levels of Cer-N-β and Cer-N-γ Subunits

1. Dulbecco's Phosphate-Buffered Saline (D-PBS) with calcium and magnesium (Invitrogen/Life Technologies).
2. Dulbecco's Phosphate-Buffered Saline (D-PBS) without calcium and magnesium (Invitrogen/Life Technologies).
3. Coomassie Plus Protein Assay Reagent Kit (Pierce Biotechnology).
4. 2 mg/ml Bovine Serum Albumin Standard Ampules (Pierce Biotechnology)
5. SDS-PAGE standards, low range (Bio-Rad Laboratories).
6. XCell SureLock Mini-Cell and XCell II Blot Module Kit CE Mark (Invitrogen/Life Technologies).
7. Nu-PAGE Bis-Tris gels, NuPAGE MES SDS Running Buffer, Nu-PAGE Antioxidant, NuPAGE LDS Sample Buffer, NuPAGE Transfer Buffer (Invitrogen/Life Technologies).
8. Dithiothreitol (DTT).
9. Nitrocellulose 0.45 μm pore size or Invitrolon PVDF (Invitrogen/Life Technologies).
10. Ponceau stain: 0.2% (w/v) Ponceau S in 3% (v/v) trichloroacetic acid.
11. SuperBlock T20 TBS Blocking Buffer (Pierce Biotechnology, Inc).
12. Rabbit polyclonal antibody to residues 3–17 of GFP (Anti-GFP, N-terminal; Sigma-Aldrich), which is used for Cer-N-γ subunits, and goat polyclonal antibody to full-length GFP

(Rockland Immunochemicals), which is used for Cer-N-β or Cer-N-γ subunits.

13. Goat anti-rabbit IgG-peroxidase (Sigma-Aldrich) to use with Anti-GFP, N-terminal antibody, and rabbit anti-goat IgG-peroxidase (Sigma-Aldrich) to use with full-length GFP antibody.
14. TBS-Tween: 50 mM Tris, pH 7.5, 150 mM NaCl, 0.05% Tween 20.
15. SuperSignal West Pico Chemiluminescent Substrate (Pierce Biotechnology).
16. FluorChem SP Imaging System (Alpha Innotech) or equivalent instrument.
17. IPLab software (BD Biosciences) or equivalent imaging program.

2.5. Imaging Dynamic Events Involving Pairs of βγ Complexes in Living Cells

1. Lab-Tek II, four-well chambered coverslips (Fisher Scientific).
2. A white light spinning-disc confocal microscope comprised of an Olympus IX81 inverted microscope, UIS2 60× 1.42 N.A. objective, IX2-DSU spinning-disc system, 100 watt mercury arc lamp, Hamamatsu C9100-02 electron multiplier camera, Ludl filter wheels, shutters, and *x–y–z* stage, under the control of IPLab software (BD Biosciences), or equivalent fluorescence microscope that can image live cells labeled with CFP, YFP, and mCherry (13).
3. Excitation and emission filters for CFP (438/24, 483/32), YFP (504/12, 542/27), Red (589/15, 632/22), and a triple dichroic (FF444/521/608) (Semrock).
4. CSMI stage incubator (Harvard Apparatus) for imaging at 37°C.
5. Minimal Essential Medium (MEM) powder with Earle's salts, with L-glutamine, without sodium bicarbonate (Invitrogen/Life Technologies). To prepare HEPES-buffered MEM, add HEPES to 20 mM and pH to 7.4, then sterilize by filtration.
6. mCherry-Mem, a membrane marker used for quantifying plasma membrane association of G protein subunits (14) that can be obtained from our lab.
7. Cintiq pen-based display screen (Wacom).

3. Methods

3.1. Producing Fusions of β and γ Subunits to Fluorescent Protein Fragments

1. Using PCR, add a linker sequence encoding Arg-Ser-Ile-Ala-Thr and a BamHI site to the 5′ end of the β or γ subunit cDNA and a Bgl II site to the 3′ end. See Fig. 2 for an example

```
Sequence of Primer 1 (5' Coding Primer)
New Bam HI site is indicated in bold type and β1 residues 1-8 are underlined.

            Ser Ile Ala Thr Met Ser Glu Leu Asp Gln Leu Arg
CGC GAC GGA TCC ATC GCA ACC ATG AGT GAG CTT GAC CAG TTA CG


Design of Primer 2 (3' Noncoding Primer)
This primer introduces a silent mutation (underlined) to eliminate a Bgl II site in the β1
sequence and introduces a Bgl II site (bold type) 3' to the stop codon.  Residues 340-346
of the β1 sequence are included in this primer.

Ser Trp Asp Ser Phe Leu Lys Ile Trp Asn
 CC TGG GAT AGC TTC CTC AAG ATT TGG AAC TAA AGA TCT TCA CGG
 GG ACC CTA TCG AAG GAG TTC TAA ACC TTG ATT TCT AGA AGT GCC

Sequence of Primer 2
CCG TGA AGA TCT TTA GTT CCA AAT CTT GAG GAA GCT ATC CCA GG


Sequence surrounding the fusion junction of YFP(1-158) and β1.
YFP residues 151-158 and β1 residues 1-8 are separated by a linker sequence,
(Arg-Ser-Ile-Ala-Thr).
Ligation of Bam HI/Bgl II sticky ends produces Arg-Ser (bold type) and does not
regenerate either restriction endonuclease site.

Val Tyr Ile Met Ala Asp Lys Gln Arg Ser Ile Ala Thr Met Ser Glu Leu Asp Gln Leu Arg
  C TAT ATC ATG GCC GAC AAG CAG AGA TCC ATC GCA ACC ATG AGT GAG CTT GAC CAG TTA CG
```

Fig. 2. PCR primers used to produce β_1 cDNA for subcloning into BiFC vectors. The human β_1 sequence (obtained from Janet Robishaw, Weis Center for Research) was used.

of the coding and noncoding primers used for β_1 BiFC constructs (see Note 2).

2. Digest the PCR product with Bam HI and Bgl II and subclone it into the Bgl II site of one of these BiFC vectors: YFP(1–158)/pcDNAI/Amp, Cerulean(1–158)/pcDNAI/Amp, or CFP(159–238)/pcDNAI/Amp (see Note 3).

3.2. Transient Transfections to Compare Association Preferences of β and γ Subunits

1. For each transfection, plate 1.6×10^6 HEK-293 cells per 60-mm dish in 4 mL of MEM containing 10% FBS. Incubate the cells at 37°C, 5% CO_2. Transfections are performed in duplicate and each experiment is repeated at least three times.
2. 24 h later, transfect the cells with BiFC β and γ plasmids. Transfect with a range of plasmid amounts (see Note 4). For each transfection, dispense plasmid into a sterile 1.5 mL microcentrifuge tube. In a sterile hood, add 400 μL of Opti-MEM I to each tube.
3. In a separate microcentrifuge tube, add 6 μL of Lipofectamine 2000 Reagent to 400 μL of Opti-MEM I. Mix well by inverting the tube several times.
4. After 5 min, add the Lipofectamine 2000 mixture to the plasmid mixture.
5. After 20 min, add the 800 μL plasmid-Lipofectamine 2000 mixture to the cells by dripping gently all over the plate. Incubate the cells at 37°C, 5% CO_2.

3.3. Measurement of BiFC βγ Fluorescence Using a Spectrofluorometer

1. Two days after the transient transfections, calibrate the spectrofluorometer as described in the instrument manual.
2. Make measurements of CFP and YFP fluorescence and of light scattering for each sample. For CFP measurements, the excitation monochrometer is set to 430 nm with a 430/25 band-pass filter, and the emission monochrometer is set to 480 nm with a 455 long-pass filter. For YFP measurements, the excitation is set to 492 nm with a 492/18 band-pass filter, and emission is set to 530 nm with a 515 long-pass filter. The cell density of each sample is determined from a light-scattering measurement at 650 nm. Excitation and emission monochrometers are set to 650 nm, and a 1.3 OD neutral density filter in combination with a 590 long-pass filter is used in the excitation filterwheel (see Note 5). Control of the monochrometers, motorized filterwheels, and data acquisition is done using the Vinci software program.
3. Run a buffer control using HBSS + EDTA media. Values from this control will be subtracted from all measurements of the cells.
4. To prepare cell suspensions, add 4 mL of HBSS + $CaCl_2$ media to the dishes, swirl slightly (to get rid of the phenol Red in the media), aspirate, and then add 2 mL of HBSS + EDTA media. Scrape the cells off with a rubber policeman, pass through a pipet several times to break up clumps, and suspend in a 1 cm square glass cuvette with a magnetic stir bar. Lightly flick the bottom of the cuvette to get bubbles out of the stir bar area.
5. Make dilutions of cells transfected with vector alone to produce an autofluorescence vs. light-scattering standard curve. Make three serial 1:2 dilutions of the cells in HBSS + EDTA media by adding 2 mL of cells to 2 mL of HBSS + EDTA. Measure YFP, CFP, and light scattering for the undiluted, 1:2, 1:4, and 1:8 dilutions. Fit a line to the data.
6. Measure the YFP, CFP, and light-scattering signals of the undiluted experimental samples. Subtract autofluorescence from the YFP and CFP signals of these samples using their light-scattering values and the autofluorescence standard curve.
7. Express the relative preferences of the limiting subunit for the competing subunits as the IC_{50} for inhibition by the Cer-N-subunits of the yellow fluorescence produced by the CFP-C-subunit/YFP-N-subunit complex. For example, for inhibition of association of YFP-N-γ_2 with CFP-C-β_5 by Cer-N-γ subunits, the IC_{50} is defined as μg of Cer-N-γ subunit plasmid that produces a 50% decrease in the intensity of CFP-C-β_5YFP-N-γ_2. To determine IC_{50} values, the data are fit to: $Y=(100)/(1+(X/a)^b)$, where X is μg of transfected Cer-N-γ plasmid, Y is the % of maximal fluorescence

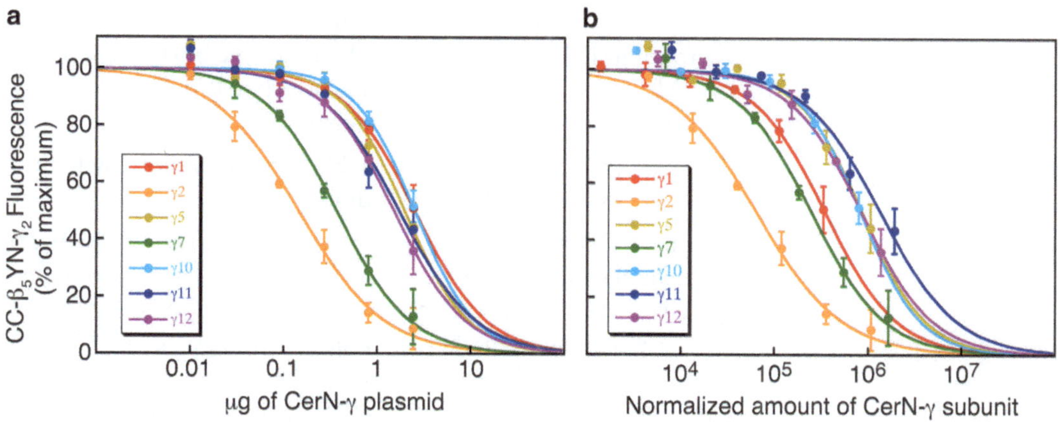

Fig. 3. β_5 interacts preferentially with γ_2 rather than γ_1, γ_5, γ_7, γ_{10}, γ_{11}, or γ_{12}. (**a**) Competition between Cer-N-γ subunits and YFP-N-γ_2 for limiting amounts of CFP-C-β_5. The intensity of CFP-C-β_5YFP-N-γ_2 was measured in the presence of each Cer-N-γ subunit or empty vector. HEK-293 cells were transfected with 0.6 μg each of plasmids expressing CFP-C-β_5 and YFP-N-γ_2, and the indicated μg of each Cer-N-γ plasmid. The total amount of plasmid in each transfection was maintained at 3.63 μg using pcDNAI/Amp. Values represent the means ± S.E. from three experiments performed in duplicate. (**b**) CFP-C-β_5YFP-N-γ_2 intensity is expressed as a function of the relative amounts of co-expressed Cer-N-γ. Expression levels were determined in HEK-293 cells transfected with 0.6 μg each of plasmids expressing CFP-C-β_5 and pcDNAI/Amp, and 0.03, 0.09, 0.27, or 2.43 μg of each Cer-N-γ plasmid. The total amount of plasmid in each transfection was maintained at 3.63 μg using pcDNAI/Amp. The expression levels of the Cer-N-γ subunits varied linearly and the data were fit by linear regressions. The plasmid amounts used in (**a**) were multiplied by Cer-N-γ expression/μg plasmid to yield the normalized amount of each Cer-N-γ subunit. CC indicates CFP-C and YN indicates YFP-N. Reprinted from (14) with permission from the American Society for Pharmacology and Experimental Therapeutics and from (18) with permission from Elsevier.

produced by CFP-C-β_5YFP-N-γ_2, *a* is the half-maximal inhibitory concentration (IC_{50}) of the Cer-N-γ subunit, and *b* is the slope factor. Figure 3a shows the results of competition between a set of Cer-N-γ subunits with YFP-N-γ_2 for association with CFP-C-β_5.

3.4. Correcting for the Expression Levels of Cer-N-β and Cer-N-γ Subunits (see Note 6)

1. Transfect HEK-293 cells with the same amounts of Cer-N-subunits and CFP-C-subunits as in Subheading 3.2, and substitute an equal amount of empty vector for the amount of transfected YFP-N-subunit. Keep the total amount of transfected plasmids constant using empty vector.
2. Two days later, aspirate media from the 60-mm dishes.
3. Gently add and remove 4 mL of ice-cold D-PBS with calcium and magnesium.
4. Add 4 mL of ice-cold D-PBS without calcium or magnesium and dislodge the cells from the dish using a rubber policeman.
5. Determine protein concentration in 50 μl of cells using the Coomassie Plus Protein Assay (Micro Test Tube Protocol) with a standard curve of 0, 2, 5, 10, and 20 μg of Bovine Serum Albumin.
6. Spin down 15 μg aliquots of cells in a refrigerated microcentrifuge and resuspend in 7 μl of D-PBS without calcium or

magnesium. Then add 3 μl of NuPAGE sample buffer containing DTT (2.5 μl of NuPAGE sample buffer plus 0.5 μl of 1 M DTT). Boil 5 min and run on a Nu-PAGE Bis-Tris gel with 0.5 μl of SDS-PAGE standards.

7. Transfer proteins from gels to Nitrocellulose or Invitrolon PVDF using XCell II Blot Module.
8. Incubate Nitrocellulose or Invitrolon PVDF with 0.2% Ponceau S in 3% trichloroacetic acid on shaker for a few minutes, rinse with H_2O, mark locations of molecular weight markers with a permanent marker, produce an image of the blot to have a record of the sample protein loadings, and then incubate the blot in SuperBlock T20 TBS Blocking Buffer for 30 min with shaking. Replace the Blocking Buffer and shake for another 30 min.
9. For Cer-N-γ subunits, incubate either with anti-GFP, N-terminal antibody at a dilution of 1:2,500, or with full-length GFP antibody at a dilution of 1:400 in TBS-Tween overnight on a shaker at 4°C. For Cer-N-β subunits, best results are obtained with the full-length GFP antibody at a dilution of 1:400.
10. For blots incubated in full-length GFP antibody, incubate for 1 h at room temperature in anti-goat IgG-peroxidase at a dilution of 1:40,000 in TBS-Tween. For blots incubated in anti-GFP, N-terminal antibody, incubate for 1 h at room temperature in anti-rabbit IgG-peroxidase at a dilution of 1:2,000.
11. Detect antigen–antibody complexes using SuperSignal West Pico Chemiluminescent Substrate and a FluorChem SP Imaging System or equivalent instrument.
12. Quantify expression levels using of IPLab software or equivalent imaging program.
13. Normalize the IC_{50} values calculated in Subheading 3.3.7 by multiplying these values (in μg of Cer-N-subunit plasmid) by Cer-N-subunit expression/μg plasmid to yield the normalized amount of each Cer-N-subunit (see Fig. 3b).

3.5. Imaging Dynamic Events Involving Pairs of βγ Complexes in Living Cells

1. Plate HEK-293 cells at a density of 1×10^5 cells per well in 500 μL of MEM on Lab-Tek II, four-well chambered coverslips.
2. 24 h later, transiently transfect the cells with plasmids encoding the CFP-C-subunit, YFP-N-subunit, and Cer-N-subunit of interest, along with mCherry-Mem, using 0.25 μL of Lipofectamine 2000 Reagent. Co-expressing plasmids encoding a GPCR and an associated α subunit enables investigations of agonist-stimulated internalization of βγ complexes.

The transfection procedure is the same as in Subheading 3.2 except that the plasmids and the Lipofectamine 2000 reagent are each suspended in 50 μL of Opti-MEM I (see Note 7).

3. Two days after the transfection, image the cells at 60× using an Olympus white light spinning-disc confocal microscope or equivalent instrument. At least 1 h before imaging, replace the bicarbonate-buffered medium with HEPES-buffered MEM (see Note 8).

4. Collect images of cells in the CFP, YFP, Red, and DIC channels at timed intervals before and after stimulation of a GPCR with an agonist. A motorized *x*–*y*–*z* stage makes it possible to collect images of cells located at 5–6 independent positions in the well during a single experiment. Cells selected for imaging should express all of the labeled proteins and have a clearly delineated plasma membrane. Individual exposure times should be optimized for each cell and color channel. Figure 4a shows images of $\beta_1\gamma_7$ and $\beta_1\gamma_{11}$ internalizing from the plasma membrane in response to stimulation of the β_2-adrenergic receptor.

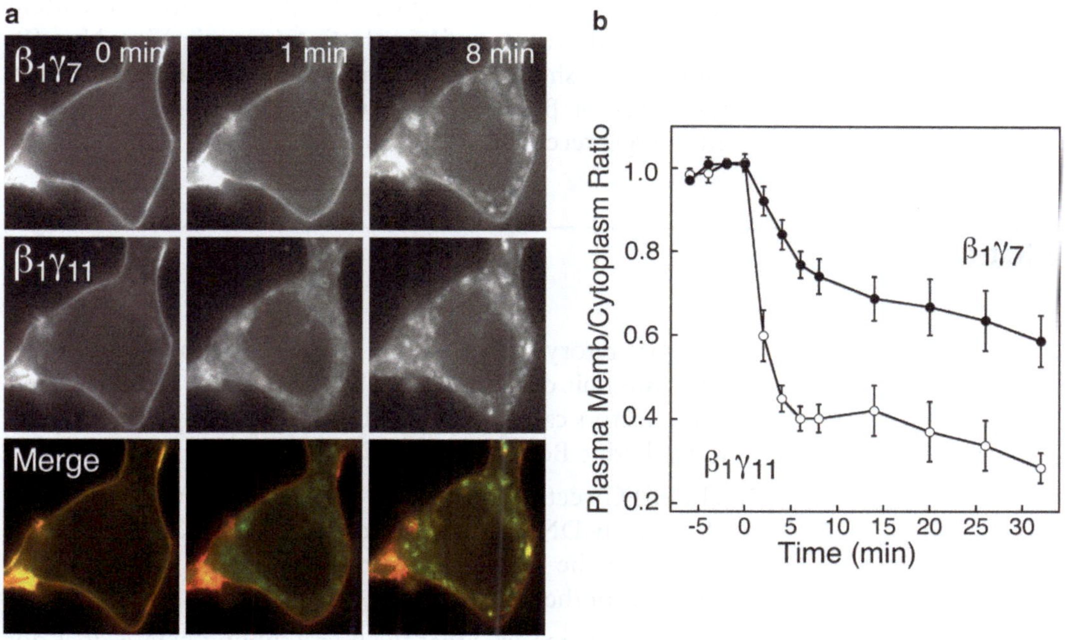

Fig. 4. The stimulus-induced localization pattern of $\beta_1\gamma_{11}$ differs from that of $\beta_1\gamma_7$. (**a**) YFP (*top*), CFP (*middle*), and merge (*bottom*) images from the same cell expressing CFP-C-β_1YFP-N-γ_7 and CFP-C-β_1Cer-N-γ_{11} acquired at the indicated times before and after stimulation with 10 μM isoproterenol. In the merge image, co-localization of CFP-C-β_1YFP-N-γ_7 (*Red*) and CFP-C-β_1Cer-N-γ_{11} (*green*) is indicated in *yellow*. Reprinted from (18) with permission from Elsevier. (**b**) Quantification of isoproterenol-stimulated decreases in plasma membrane/cytoplasm ratios of CFP-C-β_1YFP-N-γ_{11} and CFP-C-β_1Cer-N-γ_7 in HEK-293 cells. The cells were stimulated with 10 μM isoproterenol immediately after time = 0. Values represent the mean ± S.E of measurements in 7 cells.

5. Determine the plasma membrane to cytoplasm intensity ratios for the βγ complexes at each time point. Make a "high resolution" version of the mCherry-Mem image that has a flat background and peaks corresponding to in-focus features. First, blur the image with a 15 × 15 pixel filter to produce a "low resolution" image. Then, subtract the low resolution image from the original image, followed by a 5 × 5 pixel filter to smooth noise. Threshold the high resolution image with an intensity cut-off value that selects pixels that match the features seen in the unprocessed image. Use the same threshold value for all time points. Draw a border around the edge of the cell centered on the plasma membrane, 6–10 pixels wide (0.6–1.0 μm) using a Cintiq pen-based display screen. Pixels within this border that are above the threshold are counted as plasma membrane pixels. Also, circle the nucleus of each cell in the DIC image. The cytoplasm pixels are inside the border centered on the plasma membrane and outside of the nucleus. Determine the average intensities of the plasma membrane and cytoplasm pixels for the CFP-C-subunit/YFP-N-subunit complex, the CFP-C-subunit/Cer-N-subunit complex, and mCherry-Mem in the original YFP, CFP, and Red images, respectively. Divide the plasma membrane to cytoplasm intensity ratios of the βγ complexes by those of mCherry-Mem (normalized to a value of 1 for the first time point) (see Note 9). Figure 4b shows quantification of the internalization responses of $\beta_1\gamma_7$ and $\beta_1\gamma_{11}$ upon stimulation of the β_2-adrenergic receptor.

4. Notes

1. Our laboratory has produced a wide assortment of BiFC β and γ subunit constructs (12, 14, 15). These constructs and BiFC vectors can be obtained by sending a request by E-mail to Catherine Berlot (chberlot@geisinger.edu).
2. The BiFC vectors, YFP(1–158)/pcDNAI/Amp, Cerulean (1–158)/pcDNAI/Amp, and CFP(159–238)/pcDNAI/Amp fuse the fluorescent protein fragment to the amino terminus of the β or γ subunit.
3. This strategy requires that the β or γ subunit cDNA does not have internal Bam HI or Bgl II sites. If these sites are present, they will need to be removed using silent mutations that do not change the amino acid sequence (see Fig. 2). cDNAs digested with Bam HI and Bgl II have compatible sticky ends that when ligated do not regenerate either site (see Fig. 2). With this strategy, the fusion protein cDNAs can be moved to different vectors as Bam HI/Bgl II cassettes.

4. In competition experiments, the subunit that is competed for is expressed in a limiting amount compared to the competing subunits. Loss of YFP fluorescence from CFP-C-β/YFP-N-γ or YFP-N-β/CFP-C-γ complexes is measured in the presence of a range of amounts of Cer-N-γ or Cer-N-β subunits, respectively. The optimum amounts of plasmids to transfect need to be determined empirically by measuring competition between fusions of Cer-N and YFP-N to the same subunit. The initial fluorescence of the CFP-C-β/YFP-N-γ or YFP-N-β/CFP-C-γ complex needs to be high enough to provide a workable dynamic range of fluorescence intensities in the presence of the competing Cer-N-γ or Cer-N-β subunits. The β subunit fusions generally express at lower levels than do the γ subunit fusions, so optimal conditions for measuring competition of γ subunits for β subunits may differ from those for competition of β subunits for γ subunits. For example, to compare γ subunits competing for β_1 or β_5, HEK-293 cells were transfected with 0.6 μg each of plasmids expressing CFP-C-β_1 or CFP-C-β_5 and YFP-N-γ_2 and 0, 0.01, 0.03, 0.09, 0.27, 0.81, or 2.43 μg of each Cer-N-γ subunit (14, 15) (Fig. 3). The total amount of plasmid was maintained at 3.63 μg by making up the difference with empty vector (pcDNAI/Amp). The total amount of transfected plasmid needs to be constant, because promoter competition decreases fluorescence. In contrast, to compare β_1 and β_5 competing for γ_2, cells were transfected with 0.3 μg of CFP-C-γ_2 plasmid, 0.6 μg of YFP-N-β_1 plasmid, and 0.033, 0.1, 0.3, 0.9, 2.7, or 8.1 μg of plasmids encoding either Cer-N-β_1 or Cer-N-β_5. The total amount of plasmid was maintained at 9 μg using pcDNAI/Amp (14).
5. Two significant sources of background signal must be eliminated in order to measure fluorescent proteins in a suspension of cells accurately. One source of background signal is the strong light-scattering property of cells. The monochrometers found in most spectrofluorometers transmit a small amount of stray light outside the selected wavelength band. The band-pass filters commonly used for fluorescence microscopy block significantly more stray light. A band-pass filter between the excitation monochrometer and the sample will block any stray light in the emission wavelength range that would be scattered and detected. Additionally, a band-pass or long-pass filter between the sample and the emission monochrometer will prevent scattered excitation light from reaching the emission monochrometer. With filters in place, the signal from a nonfluorescent scattering sample, such as a dilute solution of glass beads (glass-milk) should be the same as that of a buffer control. A second background signal is autofluorescence from cellular proteins. Autofluorescence is

proportional to cell density, which is determined with the described light-scattering measurement.

6. In addition to determining the expression levels of BiFC β and γ constructs, it is important to assess the functionality of BiFC βγ complexes. We have demonstrated that YFP-N-β_1YFP-C-γ complexes potentiate α_s-mediated activation of adenylyl cyclase in COS-7 cells (12) and that CFP-C-β_5Cer-N-γ complexes activate phospholipase C-β2 expressed in HEK-293 cells (14).
7. The optimal amounts of transfected plasmids need to be determined empirically. For instance, when dually imaging CFP-C-β_1YFP-N-γ_7 and CFP-C-β_1Cer-N-γ_{11} for Fig. 4, more YFP-N-γ_7 than Cer-N-γ_{11} plasmid was used to normalize the fluorescence intensities of the two βγ complexes. For these images, HEK-293 cells were transfected with the following amounts (in μg) of plasmids: α_s, 0.3; CFP-C-β_1, 0.3; YFP-N-γ_7, 0.1125; Cer-N-γ_{11}, 0.0375; β_2AR, 0.1; and mCherry-Mem, 0.0025. Because CFP-C-subunit/Cer-N-subunit complexes are brighter than CFP-C-subunit/YFP-N-subunit complexes, it may be helpful to use a Cer-N fusion to the subunit with the lower expression level.
8. It is important to replace bicarbonate-buffered medium with HEPES-buffered medium to keep the pH constant when viewing cells in the room environment, because changes in pH can alter localization patterns. Alternatively, the dish can be kept in a 5% CO_2 atmosphere while imaging.
9. Changes in cell shape during time courses will alter the membrane to cytoplasm ratio. This is corrected for with the mCherry-Mem measurements, because the distribution of the membrane marker does not change in response to agonist stimulation. A bleach correction is not necessary because ratios of fluorescence are calculated.

References

1. Robishaw, J. D., and Berlot, C. H. (2004) Translating G protein subunit diversity into functional specificity. *Curr Opin Cell Biol* **16**, 206–209.
2. Wang, Q., Mullah, B., Hansen, C., Asundi, J., and Robishaw, J. D. (1997) Ribozyme-mediated suppression of the G protein gamma7 subunit suggests a role in hormone regulation of adenylylcyclase activity. *J Biol Chem* **272**, 26040–8.
3. Wang, Q., Mullah, B. K., and Robishaw, J. D. (1999) Ribozyme approach identifies a functional association between the G protein beta1gamma7 subunits in the beta-adrenergic receptor signaling pathway. *J Biol Chem* **274**, 17365–71.
4. Schwindinger, W. F., Betz, K. S., Giger, K. E., Sabol, A., Bronson, S. K., and Robishaw, J. D. (2003) Loss of G protein gamma 7 alters behavior and reduces striatal alpha(olf) level and cAMP production. *J Biol Chem* **278**, 6575–9.
5. Schwindinger, W. F., Giger, K. E., Betz, K. S., Stauffer, A. M., Sunderlin, E. M., Sim-Selley, L. J., Selley, D. E., Bronson, S. K., and Robishaw, J. D. (2004) Mice with deficiency of G protein gamma3 are lean and have seizures. *Mol Cell Biol* **24**, 7758–68.

6. Hu, C. D., and Kerppola, T. K. (2003) Simultaneous visualization of multiple protein interactions in living cells using multicolor fluorescence complementation analysis. *Nat Biotechnol* **21**, 539–45.
7. Rizzo, M. A., Springer, G. H., Granada, B., and Piston, D. W. (2004) An improved cyan fluorescent protein variant useful for FRET. *Nat Biotechnol* **22**, 445–9.
8. Kerppola, T. K. (2006) Visualization of molecular interactions by fluorescence complementation. *Nat Rev Mol Cell Biol* 7, 449–56.
9. Magliery, T. J., Wilson, C. G., Pan, W., Mishler, D., Ghosh, I., Hamilton, A. D., and Regan, L. (2005) Detecting protein-protein interactions with a green fluorescent protein fragment reassembly trap: scope and mechanism. *J Am Chem Soc* **127**, 146–57.
10. Jones, M. B., and Garrison, J. C. (1999) Instability of the G-protein beta5 subunit in detergent. *Anal. Biochem* **268**, 126–33.
11. McIntire, W. E., MacCleery, G., Murphree, L. J., Kerchner, K. R., Linden, J., and Garrison, J. C. (2006) Influence of differential stability of G protein betagamma dimers containing the gamma11 subunit on functional activity at the M1 muscarinic receptor, A1 adenosine receptor, and phospholipase C-beta. *Biochemistry* **45**, 11616–31.
12. Hynes, T. R., Tang, L., Mervine, S. M., Sabo, J. L., Yost, E. A., Devreotes, P. N., and Berlot, C. H. (2004) Visualization of G protein betagamma dimers using bimolecular fluorescence complementation demonstrates roles for both beta and gamma in subcellular targeting. *J Biol Chem* **279**, 30279–86.
13. Shaner, N. C., Campbell, R. E., Steinbach, P. A., Giepmans, B. N., Palmer, A. E., and Tsien, R. Y. (2004) Improved monomeric red, orange and yellow fluorescent proteins derived from Discosoma sp. red fluorescent protein. *Nat Biotechnol* **22**, 1567–72.
14. Yost, E. A., Mervine, S. M., Sabo, J. L., Hynes, T. R., and Berlot, C. H. (2007) Live cell analysis of G protein beta5 complex formation, function, and targeting. *Mol Pharmacol* **72**, 812–25.
15. Mervine, S. M., Yost, E. A., Sabo, J. L., Hynes, T. R., and Berlot, C. H. (2006) Analysis of G protein beta-gamma dimer formation in live cells using multicolor bimolecular fluorescence complementation demonstrates preferences of beta1 for particular gamma subunits. *Mol Pharmacol* **70**, 194–205.
16. Ormo, M., Cubbitt, A. B., Kallio, K., Gross, L. A., Tsien, R. Y., and Remington, S. J. (1996) Crystal structure of the Aequoria victoria green fluorescent protein. *Science* **273**, 1392–5.
17. Lambright, D. G., Sondek, J., Bohm, A., Skiba, N. P., Hamm, H. E., and Sigler, P. B. (1996) The 2.0 A crystal structure of a heterotrimeric G protein. *Nature* **379**, 311–9.
18. Hynes, T. R., Yost, E., Mervine, S., and Berlot. C. H. (2008) Multicolor BiFC analysis of competition among G protein beta and gamma subunit interactions. *Methods* **45**, 207–13.

Chapter 13

Real-Time BRET Assays to Measure G Protein/Effector Interactions

Darlaine Pétrin, Mélanie Robitaille, and Terence E. Hébert

Abstract

Advances in imaging assays based on resonance energy transfer (RET) have made it possible to study protein/protein interactions in living cells under physiological conditions. It is now possible to measure the kinetics of changes in these interactions in response to ligand stimulation in real time. Here we describe protocols for these assays focusing on the basal and ligand-stimulated interaction between tagged Gβγ subunits and adenylyl cyclase II. We describe relevant positive and negative controls and various experimental considerations for optimization of these experiments.

Key words: G protein-coupled receptor, Bioluminescence resonance energy transfer, G proteins, Protein–protein interaction assays

1. Introduction

Constitutive trafficking of some GPCR-regulated effectors, such as adenylyl cyclase isoforms or various ion channels, demonstrates that components of these signalling pathways can make their way to the membrane independently of the receptor or G protein. However, there is now significant evidence that, like GPCR dimers, effector/G protein complexes are also formed during or shortly after biosynthesis. A number of studies have demonstrated that effectors such as Kir3 channels and adenylyl cyclase (AC) interact with nascent Gβγ subunits in the endoplasmic reticulum (1, 2). The interactions between adenylyl cyclase or β_2AR and Gβγ, or between receptor equivalents in the β_2AR homodimer, were insensitive to dominant negative Rab 1 or Sar 1 constructs that regulate receptor trafficking (3). However, these studies also

Louis M. Luttrell and Stephen S.G. Ferguson (eds.), *Signal Transduction Protocols*, Methods in Molecular Biology, vol. 756, DOI 10.1007/978-1-61779-160-4_13, © Springer Science+Business Media, LLC 2011

indicated that Gα subunits were assembled into nascent receptor/Gβγ/effector complexes either at endoplasmic reticulum exit sites or in the Golgi as these interactions were blocked by dominant negative Sar1 and Rab 1 (2, 3). If these complexes are preformed during protein biosynthesis and maturation, they would need to be trafficked inside the cell as complexes and not necessarily as individual proteins.

Despite the classical model of G protein signalling, several authors emphasized both the lack of experimental evidence for heterotrimer subunit dissociation in vivo and evidence that intact heterotrimers can actually signal (4–6). In vitro experiments, which helped establish this dogma, were often carried out in the presence of high concentrations of Mg^{2+}, detergents, and other factors that could artificially favor dissociation. Several groups have since constructed Gα and Gβγ fusion proteins in order to test the subunit dissociation hypothesis using fluorescence or bioluminescence resonance energy transfer (FRET or BRET). Bünemann and colleagues reported that FRET between labeled mammalian $G\alpha_i$ and Gβ or Gγ subunits could increase after receptor activation, an observation that is difficult to reconcile with physical dissociation of heterotrimers (7). This report also showed that the direction of the FRET change was reversed if the position of the fluorescent reporter was changed from one end of the Gγ subunit to the other. The interpretation of these results was that activation produced a rearrangement of G protein subunits but not physical dissociation. Subsequent reports have shown similar activation-induced increases or decreases in BRET or FRET depending on the Gα isoform studied and the precise location of the reporter moieties (8–12). The latter study, in particular, demonstrated that agonist could induce either increases or decreases in resonance energy transfer depending on the position of the bioluminescent donor moiety in Gα when looking at interactions between the same two proteins, whether it was Gα/Gβγ or receptor/Gα interactions. These results strongly suggest that G protein subunits can maintain some degree of association both with receptor and as a heterotrimer after activation. Interestingly, a recent study has demonstrated that the magnitude of the physical dissociation between G protein heterotrimers depended on the Gα isoform studied, with $G\alpha_{i/o}$ isoforms dissociating more readily than $G\alpha_s$ isoforms – suggesting a spectrum of stability in this regard (13). Interactions between G proteins and effectors such as AC or Kir3 channels are also stable in the face of short-term agonist stimulation (1). Questions regarding the composition or stability of these complexes at different points during their ontogeny or in response to short- or long-term agonist stimulation can be addressed using imaging approaches such as BRET or FRET. Here, we describe some basic considerations to study the kinetics of G protein/effector interactions in living cells using BRET. When combined with modulation of protein trafficking and the use of membrane-permeable ligands, questions about where, when and with what stability receptors, G protein heterotrimers and effector molecules interact can be analyzed in real time and in different subcellular compartments.

2. Materials

2.1. Cell Culture and Transfection

1. Human embryonic kidney cells (HEK293-F; Invitrogen).
2. Dulbecco's Modified Eagle Medium (DMEM) High Glucose (Invitrogen) without supplements.
3. DMEM supplemented with 2.5% FBS.
4. Complete growth medium consisting of DMEM supplemented with 5% fetal bovine serum (FBS) and penicillin–streptomycin at final concentrations of 100 U/ml and 100 μg/ml, respectively.
5. Linear polyethylenimine, 25 kDa, (PEI; Polysciences, Inc.) is dissolved in Milli-Q water heated to 80°C to obtain a 1 mg/ml solution, neutralized with HCl to pH 7.0 and filter sterilized using a 0.20 μm filter. Store in aliquots at –80°C. Caution! Wear gloves, safety glasses, and appropriate protective clothing.
6. T-75 tissue culture flasks.
7. Six-well tissue culture plates.
8. 1.5 ml sterile Eppendorf tubes.
9. Constructs used were the following: As a basic BRET pair, we used EGFP-tagged Gβ1 (14) and adenylyl cyclase II (AC II) tagged with *Renilla* luciferase (ACII-RLuc; (1)). We have previously demonstrated that AC II interacts in a constitutive manner with Gβγ subunits and that conformational changes in the interaction occur in response to agonist stimulation (1, 2). In some experiments, we used EGFP-Gβ1 and Rluc-Gγ2 (12, 15) as a BRET pair. Other constructs were co-transfected along with the BRET1 pair in order to maintain protein stoichiometry for proper activation of adenylyl cyclase: 0.5 μg of Gαs, 0.5 μg of HA-tagged Gγ2 (HA-Gγ2), and 1 μg of HA-tagged β_2-adrenergic receptor (HA-β_2AR; see Note 1).

2.2. Receptor Ligands and Rluc Substrate

1. Phosphate-buffered saline (PBS) 10× stock solution: 1.37 M NaCl, 27 mM KCl, 189 mM Na_2HPO_4, 18 mM KH_2PO_4, pH adjusted to 7.4 with HCl. PBS stock is autoclaved and stored at room temperature. Working solutions are freshly prepared.
2. (–)-Isoproterenol hydrochloride freshly dissolved on the day of the experiment in a freshly prepared 100 μM l-ascorbic acid solution in 1× PBS.

3. (±)-Propanolol hydrochloride freshly dissolved in Milli-Q water.
4. Coelenterazine h (Invitrogen) reconstituted in ethanol at a concentration of 1 mg/ml. Protect from light and store dessicated at −20°C. It is important not to dissolve in DMSO as it is unstable in this solvent.

2.3. Data Collection

1. Optiplate-96, white opaque 96-well microplates.
2. In this article, we used the *Synergy 2* multimode microplate reader (BioTek) but where appropriate we comment on the use of other instruments.

2.4. Data Analysis

1. GraphPad Prism curve fitting software.

3. Methods

In recent years, a number of excellent reviews have appeared which provide detailed theoretical discussions and practical advice on how to design and perform BRET experiments (see (16–23)). Readers are directed to these reviews for more information on BRET and resonance energy transfer approaches in general. Here, we only briefly mention the most salient features required for the design of a real-time experiment. In order to generate BRET donors and acceptors, proteins of interest must be fused to either *Renilla* luciferase (Rluc) or GFP variants, respectively. In doing so, it is important to ensure that the insertion of the bioluminescent and fluorescent tags does not interfere with the proper folding of the protein and both correct functionality and localization must be verified. The position of the RLuc and GFP relative to the protein of interest (N-terminally, C-terminally or internally positioned) deserve a particular attention. Here, we performed real-time BRET using the BRET1 configuration (i.e., EGFP and RLuc) with the substrate coelenterazine h. BRET2 with RLuc2 or RLuc8 (which have much higher luciferase activities, see (24)), but not RLuc, could also be used with the substrate coelenterazine 400a due to its rapid decay.

3.1. Cell Culture and Transfection

1. HEK293F cells are maintained in T75 culture flasks at 37°C, 5% carbon dioxide. Cells are passaged 1:10 every 3–4 days when approaching confluence and only cells below passage 40 are used for BRET experiments. Generally, 48 h prior to transfection, HEK293F cells are appropriately plated in six-well plates in order to obtain 50% confluence on the day of transfection (see Note 2).
2. Pipette the desired plasmid DNA constructs in 1.5 ml Eppendorf tubes. The amount of EGFP-Gβ1 (1 μg) is kept constant but the amount of ACII-RLuc is varied from 0.25 to

1 μg depending on the experiment and stoichiometric amounts of Gγ2, β_2AR and Gαs were co-transfected.

3. Prepare two different sets of negative controls. In one case, ACII-RLuc is substituted by a truncated version of CD8-RLuc (100 ng) which localizes to same subcellular compartments. The second control is a donor-only control, where EGFP-Gβ1 is simply replaced by 1 μg of FLAG-Gβ1 construct (see Note 3).
4. Mix the combined plasmid DNA constructs in 100 μl DMEM (without FBS and penicillin–streptomycin).
5. In a second tube, a 2× dilution of PEI is prepared in DMEM. A 3:1 ratio of PEI:DNA is used (see Note 4).
6. Combine DNA and PEI mixtures and allow complex formation at room temperature for 15 min.
7. In the meantime, replace complete culture medium with 2 ml of DMEM supplemented with 2.5% FBS.
8. Add the 200 μl PEI-complexed plasmid DNA to culture dishes. Do not add the PEI/DNA solution directly to cells, as this will kill the cells. Place pipette on the side of the well and push out the solution slowly into the media surrounding the cells. Shake wells gently to be sure that the PEI/DNA solution is well mixed with the media.
9. Incubate for 24 h at 37°C with 5% CO_2 and replace the latter medium with complete culture medium. Incubate for an additional 24 h prior to BRET experiment (see Note 5).

3.2. Preparation of Cells, Receptor Ligands, and Rluc Substrate

1. On the day of the experiment, freshly prepare a 10× stock of isoproterenol (see Note 6).
2. Wash HEK293F cells twice in 1× PBS (see Note 7).
3. Add 500 μl of 1× PBS to each well and gently detach cells by trituration.
4. Transfer 90 μl of resuspended cells to a white opaque 96-well microplate. Plan series of wells for different treatments. For instance, one series of wells should be prepared for treatment with vehicle and another for stimulation with receptor ligands.
5. Dilute the substrate coelenterazine h 1:500 using 1× PBS.

3.3. Optimization of Synergy 2 Microplate Reader Settings

1. Using the Gen5 software (provided by the manufacturer of individual instruments), design a protocol to set excitation and emission filters needed for the measurement of fluorescence intensity.
2. EGFP-Gβ1 is excited using a 485/20 excitation filter and its emission detected with a 528/20 filter. The light source used is a Xenon Flash lamp, sensitivity is set to 35 and the optic position is defined as Top 50% (i.e., the readings are taken from the top of the plate and all wavelengths are collected but 50% of the emissions are lost in this collection using this mirror;

see Note 8). These settings need to be verified empirically and likely would need to be altered in a plate reader-specific fashion depending on the characteristics of the instrument used.

3. Total luminescence measurements are taken using an empty position on the emission filter wheel to collect all light generated by luciferase activation. Machine sensitivity is set to 135, and by default, the optic position is set to Top (no mirror is involved in collection of signals).
4. For BRET1 assays using our EGFP-tagged protein, the emission filter set is composed of 460/40 and 528/20 filters. Sensitivity is set to 150, and by machine default, the optic position is defined to Top.
5. Prime the dispenser module and its associated tubing with the diluted substrate using the first injector and the vehicle or agonist in the second injector (see Note 9). Sample injectors were used for both standard BRET assays (i.e., a single time point) or for real-time BRET assays as described below.

3.4. Running a Standard BRET Experiment

1. Measure fluorescence intensity of all samples to verify similar expression levels for BRET acceptor molecules. Include a nontransfected control to correct for background fluorescence (see Note 10). An example of typical acceptor expression values in a BRET experiment is shown in Fig. 1a. Altering the sensitivity of the instrument will alter the absolute values.
2. Measure total luminescence following substrate injection to ensure all samples express similar levels of BRET donor molecules and that this increases with increasing amounts of transfected cDNA. To do so, 10 μl of 10× concentrated substrate is injected, followed by a 55 s delay before the first reading is taken (see Note 10; Fig. 1b)
3. Immediately read samples using BRET1 emission filter set. The BRET ratios obtained represent basal BRET for the different samples (see Note 11).
4. Inject 11 μl of vehicle (100 μM ascorbic acid in 1× PBS). Readings are taken using BRET1 emission filter set after a 55 s delay.
5. Vehicle is now replaced by the β_2AR agonist isoproterenol. To do so, purge the dispenser tubing completely and prime it with the diluted agonist as previously described.
6. Repeat Subheading 3.4, steps 1–4 to obtain BRET1 ratios after stimulation with agonist (Fig. 2). It is important to insure that constructs that are used as negative controls need to be expressed at similar levels and in similar compartments as the experimental BRET pair. CD8-RLuc expresses at higher levels than ACII-RLuc (Fig. 1), which could impact on the interpretation of results. Expression levels could in fact

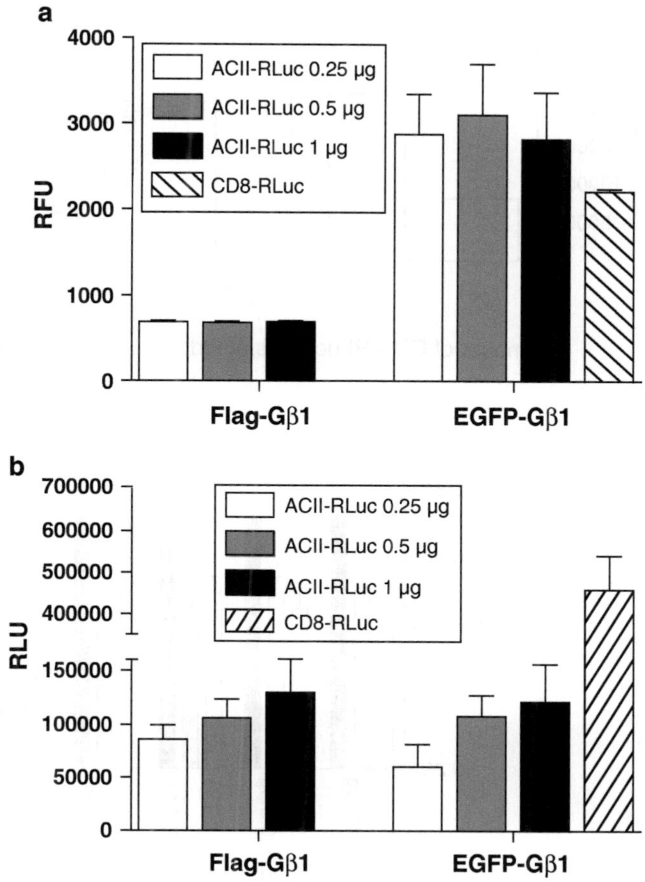

Fig. 1. Expression levels of BRET acceptors and donors. Indicated amounts of ACII-RLuc and 1 µg of EGFP-Gβ1 or Flag- Gβ1 were co-transfected with Gαs (0.5 µg), HA-Gγ2 (0.5 µg), and HA-β_2AR (1 µg) using PEI. For the negative control, ACII-RLuc was replaced by 100 ng of truncated CD8-RLuc. (**a**) Total fluorescence was measured using a 485/20 excitation filter and a 528/20 emission filter. Graph represents three independent experiments and data are expressed as relative fluorescence units (RFU). (**b**) Total luminescence was measured using an empty position on the emission filter wheel. Graph represents three independent experiments and data are expressed as relative luminescence units (RLU). Data are presented as mean ± SEM.

be titrated by varying the amount of the plasmids transfected. We show here that at different levels of CD8-RLuc expression, no BRET was detected with EGFP-Gβ1, providing an alternative means to showing the specificity of the interaction with ACII-RLuc (Fig. 2, *inset*). For the interaction between EGFP-Gβ1 and ACII-RLuc, the BRET signal was clearly above the background from the CD8-RLuc or the donor-only control and was sensitive to stimulation with isoproterenol (and not vehicle) in the presence of the β_2AR (Fig. 2). We have previously demonstrated that this interaction is saturable and competed by untagged partners

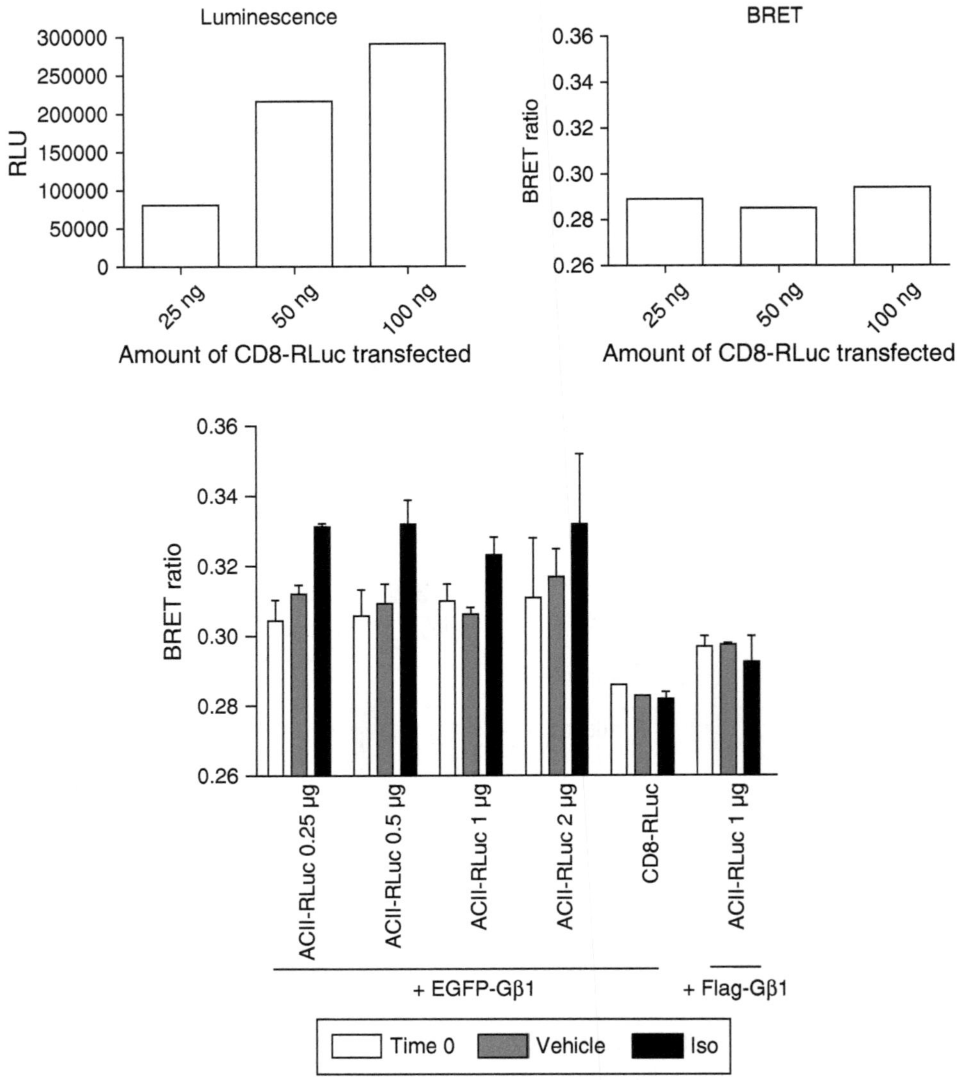

Fig. 2. Basic BRET assay measuring interactions between ACII-RLuc and EGFP-Gβ1. HEK293F cells were transfected with ACII-RLuc or CD8-RLuc, and EGFP-Gβ1 or Flag-Gβ1 in presence of Gαs, HA-Gγ2, and HA-β2AR. Luminescence and fluorescence were measured 55 s after addition of substrate (Time 0), and 55 s after addition of isoproterenol (10 μM) or vehicle, using 460/40 and 528/20 emission filters. BRET ratios were calculated by dividing values obtained with the 528/20 filter by those obtained with the 460/40 filter. Data shown were from at least two independent experiments. Data are presented as mean ± SEM. *Inset*: Representative experiment showing that absence of BRET signal (*right panel*) in the CD8-RLuc negative control is insensitive to the amount of donor present (*left panel*).

(1) using ACII-RLuc and GFP-Gγ2. Similar results were obtained in that study for the ACII-RLuc/EGFP-Gβ1 pair (data not shown). In the absence of an acceptor, the BRET value was higher than the CD8-RLuc control since they do not have the same luminescence in this particular experiment (i.e., they are two different proteins). Thus, they are not directly comparable but both provide a reliable indication that the experimental conditions are sound.

3.5. Running a Real-Time BRET Experiment

1. Verify how long it takes for the BRET ratios to stabilize after the initial addition of the substrate. Using the well mode, set up a loop (called KINETIC in the software platform for the Synergy 2) with a run time of 3 min, readings are taken at an interval of 00:02.00 (MM:SS:ss), with an integration time of 00:00.50 (MM:SS:ss). Inject 10 μl of substrate and run samples for 3 min and examine the traces obtained. In the experiment shown here, BRET ratios became stable by 30 s after substrate injection, thus 1 min after addition of substrate was the time point selected for initial injection of the agonist or vehicle (Fig. 3).
2. By decreasing the integration time, it is theoretically possible to decrease the interval between each reading to maximize the amount of kinetic information obtainable for a given BRET pair (see Note 12). However, the magnitude of variation between subsequent data appear to be inversely proportional to the length of the integration time and reading interval, as shown in Fig. 4. Other instruments, such as the Mithras LB 940 instrument (Berthold) are capable of collecting data more rapidly (see, e.g., (12)). This particular feature should be carefully considered when making the choice of which instrument to use for real-time BRET measurements.
3. It is clear that although the BRET between EGFP-Gβ1 and ACII-RLuc is clearly higher than that measured between EGFP-Gβ1 and CD8-Rluc, the real-time experiment reveals a significant degree of fluctuation in the signal. Does this reflect some sort of oscillatory behavior in the interaction itself or is it related to a peculiarity of the instrument's ability to measure the interaction in real time? To address this, it is recommended that the experimenter test known stable interactions. The interaction between EGFP-Gβ1 and RLuc-Gγ2 is an example of such an interaction and even in this case small oscillations are detected over the period when readings were taken (Fig. 5). This suggests that the oscillations seen with the EGFP-Gβ1/ACII-RLuc pair are also an artifact of the experimental paradigm. We advise that this be determined empirically no matter what instrument is used or interaction is measured. This could also be used to test interactions for their stability in the presence of ligand, as isoproterenol had no effect on BRET between EGFP-Gβ1 and RLuc-Gγ2 (data not shown).
4. Once basal conditions have been determined, a real-time assay with injection of a ligand can be performed. Inject 10 μl of substrate and read samples for a period of 1 min. Dispense 11 μl of 10× agonist or its vehicle and continue measuring for an additional minute (see Note 13; Fig. 6). The BRET values were reasonably stable in the absence of ligand, again high-

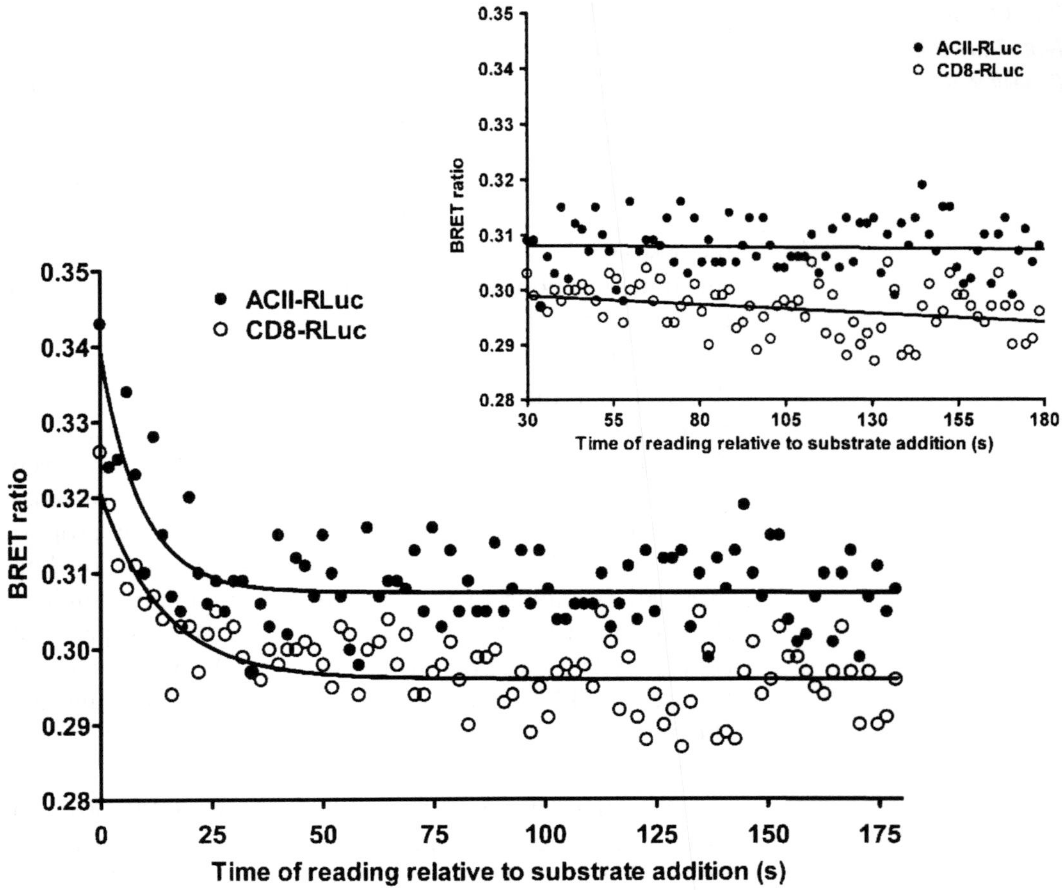

Fig. 3. Stabilization and stability of the BRET signal. ACII-RLuc or CD8-RLuc were transfected with EGFP-Gβ1, Gαs, HA-Gγ2, and HA-β_2AR. Diluted coelenterazine h was injected at time 0 and luminescence and fluorescence measurements were taken over a 3-min period as described for Fig. 2. Data are representative and were fit with an exponential function. After addition of coelenterazine h, emitted luminescence exhibits rapid decay during the first 30 s after which luminescence continues to decay, but at a much slower rate. *Inset*: The same data set was analyzed and fit by linear regression from 30 s after coelenterazine h injection and shows that BRET signals are stable after equilibration for up to 3 min.

lighting the notion that ACII and Gβγ are part of a preassembled complex with the receptor (1, 2). Addition of ligand, but not vehicle resulted in a rapid increase in BRET which again stabilized at the new value. This suggests that the complex rapidly re-equilibrates into a new conformation which remains stable at least for the length of the experiment. This has previously been shown for receptor/G protein complexes and for the G protein heterotrimer (12). Here, we are only dealing with relatively short-term treatments. In principle, given the stability of the BRET signal, the interaction between EGFP-Gβ1 and ACII-RLuc or other components of GPCR signalling systems could be measured on a much longer time scale. This may particularly be of interest for understanding events

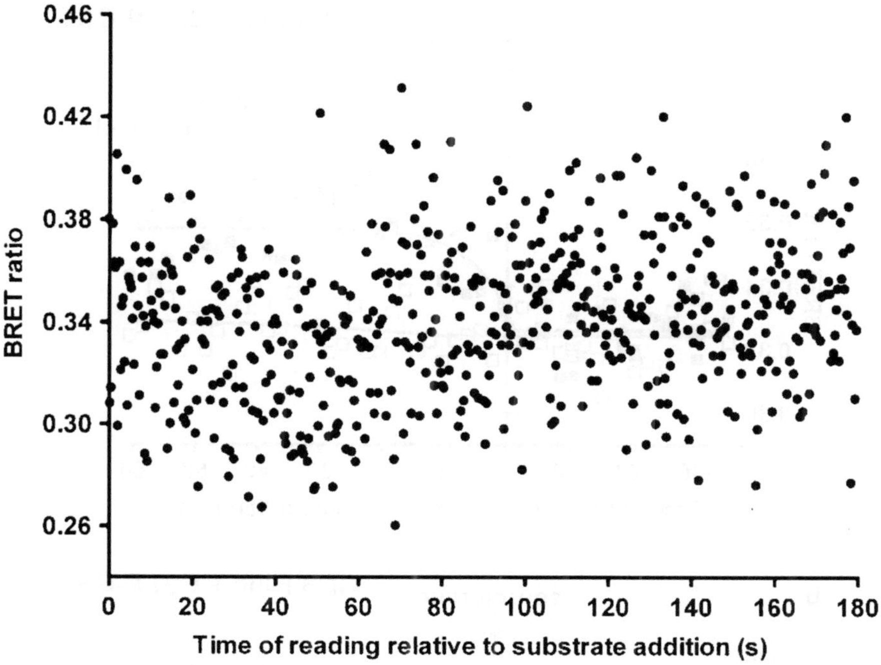

Fig. 4. Shortening integration time and reading intervals results in a noisier BRET signal. ACII-RLuc and EGFP-Gβ1 were transfected in presence of Gαs, HA-Gγ2, and HA-β2AR. Integration time was decreased to 00:00.06 MM:SS.ss which permitted recordings at every 00:00.32 MM:SS.ss. Diluted coelenterazine h was injected at time 0 and luminescence and fluorescence measurements were taken over a 3-min period.

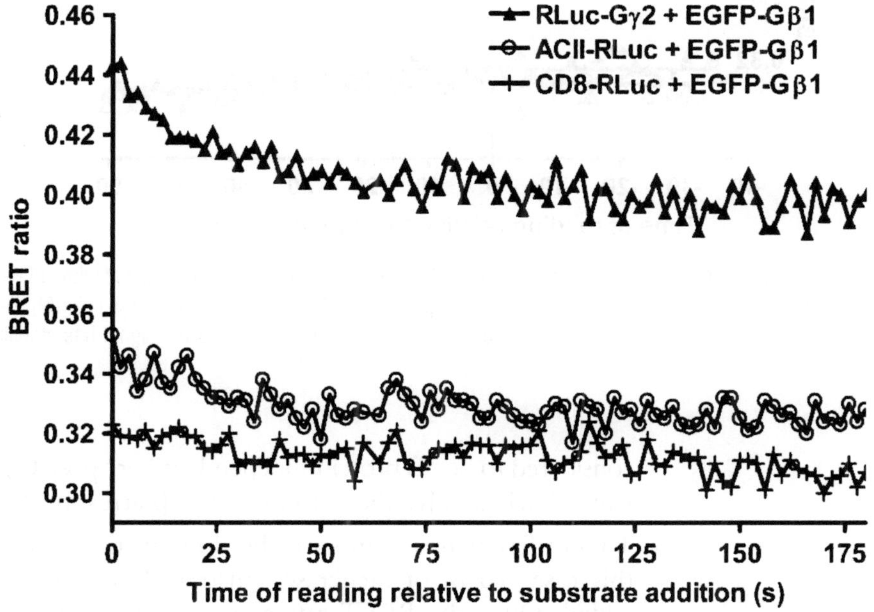

Fig. 5. Oscillation of BRET signals from point to point are due to instrumentation factors. ACII-RLuc, CD8-RLuc, or RLuc-Gγ2 were transfected with EGFP-Gβ1, Gαs, HA-Gγ2, and HA-β2AR. Diluted coelenterazine h was injected at time 0 and luminescence and fluorescence measurements were taken over a 3-min period. Data are from a representative experiment.

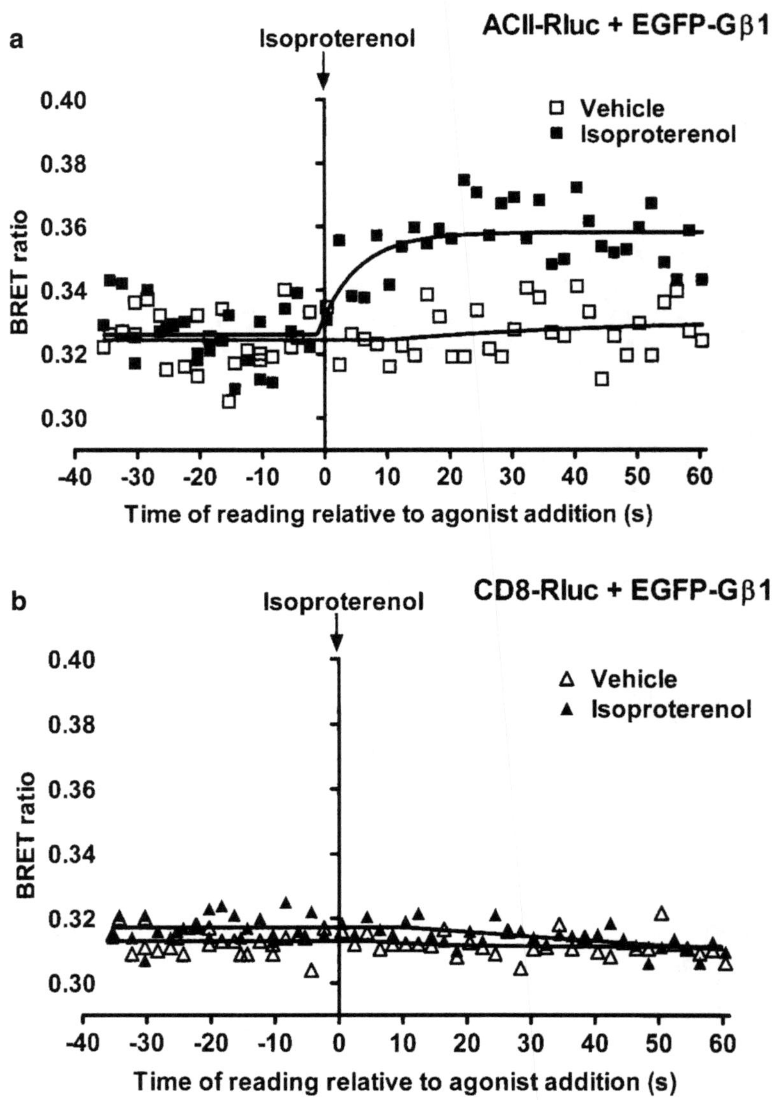

Fig. 6. Real-time BRET following β_2AR agonist stimulation. ACII-RLuc or CD8-RLuc were transfected with EGFP-Gβ1, Gαs, HA-Gγ2, and HA-β2AR. Coelenterazine h was injected 60 s prior to treatment with isoproterenol (10 μM). Data shown represent two independent experiments. Fitting of the curves was started after the BRET signal has stabilized in the absence of ligand (i.e., after 25 s).

considered to be "G protein-independent" or "post-G protein" which likely involve the recruitment of β-arrestin (25–30).

5. Two different treatments can be done in the same well. In this case, 10 μl of luciferase substrate has to be manually added. Once the BRET ratios are stabilized, drugs can be sequentially injected using both injectors. An example is shown in Fig. 7 where samples were preincubated for 1 min with 1 μM propanolol and then stimulated with 10 μM isoproterenol for an additional minute. This experiment

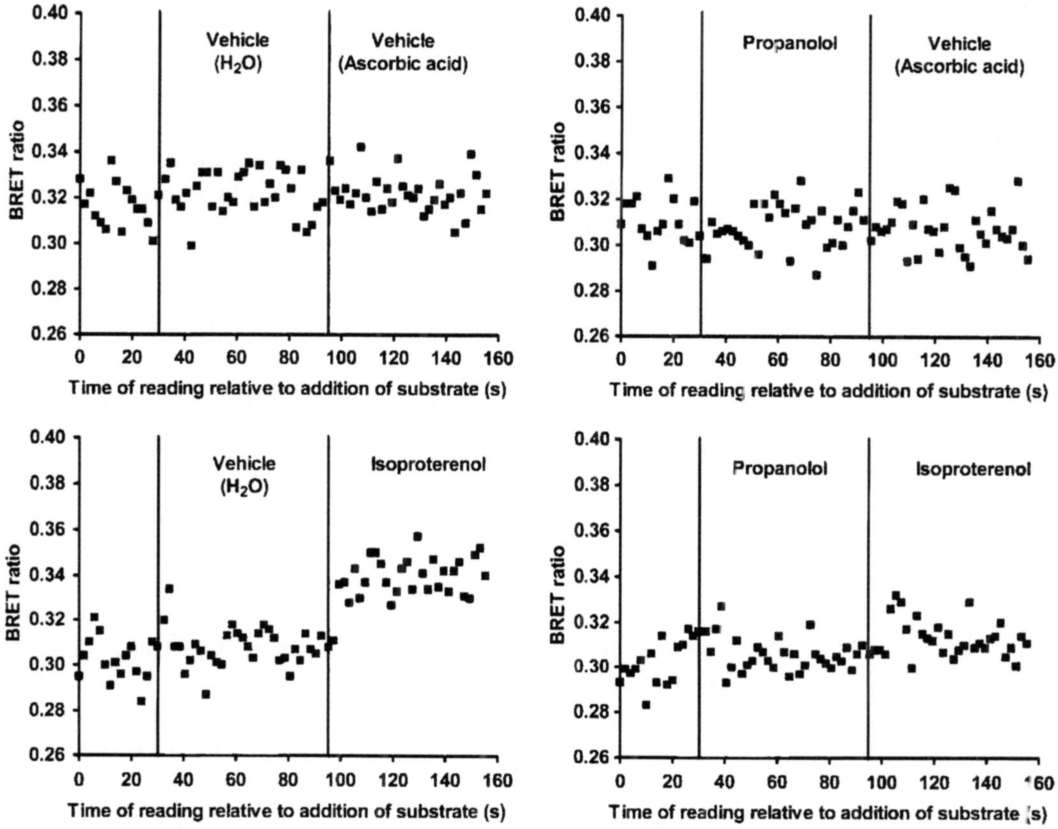

Fig. 7. Real-time BRET following β_2AR antagonist and agonist treatment. ACII-RLuc or CD8-RLuc were transfected with EGFP-Gβ1, Gαs, HA-Gγ2, and HA-β2AR. Coelenterazine h was manually injected 30 s prior to propanolol (1 μM) addition. One minute after addition of propanolol or its vehicle, isoproterenol (10 μM) or its vehicle was added. Data shown is from a representative experiment.

shows that the ligand-induced change in BRET is receptor specific, since pretreatment with a β-adrenergic receptor antagonist, propranolol, completely blocks the increase in BRET by subsequent treatment with isoproterenol (Fig. 7). Propanolol alone has no effect, suggesting that it acts as a neutral antagonist for this particular interaction. This system is amenable to addressing issues related to functional selectivity or biased signalling with different receptor/ligand pairs.

3.6. Data Analysis

1. To assess the stability of BRET signals as shown in Fig. 3, a one-phase exponential was used to fit curves while simple linear regression was used in the *inset* to demonstrate stability after the equilibration period. The data for real-time BRET experiments following isoproterenol treatment (Fig. 6) were fit using a one-phase exponential association equation selected with a plateau then an increase to top. Curve fitting was performed using GraphPad Prism.

4. Notes

1. Cloning and validation of the constructs can take several weeks. Consequently, advantages and disadvantages of the different BRET variations must be carefully considered prior to the selection of the donors and acceptors. There are several excellent reviews dedicated to the choice of BRET acceptors, the optimization of individual BRET pairs and potential problems to be encountered and how to overcome them (see (16–23)).
2. Stably transfected cell lines could also be used as they offer the advantage of constant expression levels of proteins of interest.
3. CD8-RLuc is selected as a negative control as its cellular localization is the same as ACII-RLuc and it was previously shown not to interact with Gβ1. The donor-only control is present in order to establish background signal and to ensure that this signal is not modified by treatments with β2AR agonist. In cases where the samples are treated with ligand, the vehicle control is absolutely critical as is a control with a BRET partner that does not interact with the receptor in question.
4. Depending on the cell line used and the cDNA transfected, transfection efficiency can by optimized by varying the DNA:PEI ratios or by trying a different transfection reagent. The aim is to be able to increase levels of expression and reduce cell toxicity.
5. For transiently expressed fusion proteins, the expression time can vary from 24 to 72 h.
6. Make sure the plate reader is ready to be used and the data recording protocols designed and tested before preparing the cells, receptor ligands and luciferase substrate as this step will be time consuming the first time. For the *Synergy 2*, and likely for other plate readers as well, it is also possible to validate software-driven protocols before running an actual experiment. The minimal volumes for injection should be noted and accommodated in planning experiments (i.e., a minimal volume of 5 μl is required for injection using the *Synergy 2* multimode microplate reader). Remember that isoproterenol and ascorbic acid solutions must be protected from light to prevent ascorbic acid from being degraded.
7. Cells are typically 90% confluent when ready to be used for BRET assay.

8. The sensitivity during fluorescence intensity measurements using a Xenon Flash light source can be increased. However, it is important to avoid saturation of the signal. This will vary according to the DNA construct, the transfection efficiency and the plate reader used.
9. The minimum recommended priming volume (in the case of the *Synergy 2*) is 1,000 μl. Tip priming is recommended before dispensing reagents into the 96-well microplate. Upon assay completion, it is important to carefully rinse the tubing using clean water, as some reagents may form crystals inside the dispenser apparatus.
10. Low fluorescence and luminescence counts may be explained by different factors. Check that the appropriate filter sets are defined on the instrument. Make sure the substrate was added to each well prior to luminescence measurement. Other possible problems may include insufficient cell number or low transfection efficiency.
11. Negative BRET results do not necessarily mean there is no interaction. If the validation of the fusion protein is correct, negative results could also be explained by an inappropriate distance or orientation between donor and acceptor moieties which may only be revealed by ligand stimulation. Another consideration might be the absence of key partners in stoichiometric amounts which foster resonance energy transfer between the tagged proteins (see (16–23)).
12. Setting the integration time to 00:00.06 (MM:SS.ss) and reading interval to 00:00.32 will allow collection of 565 values. The minimum integration time possible on Synergy 2 is 00:00.02 (MM:SS.ss) with reading intervals every 00:00.24. However, readings taken at an interval of 00:02.00 (MM:SS.ss), with an integration time of 00:00.50 (MM:SS.ss) is preferable.
13. It is important to work in well mode, meaning that all measurements and treatments must be complete in the first well before proceeding to the next sample.

Acknowledgments

This work was supported by grants from the Canadian Institutes of Health Research to T.E.H (MOP-36279) as well as the CIHR Team in GPCR Allosteric Regulation (CTiGAR). T.E.H. is a Chercheur National of the Fonds de la Recherche en Santé du Québec (FRSQ). We thank Vic Rebois (NIH) for helpful discussions.

References

1. Rebois, R. V., Robitaille, M., Gales, C., Dupre, D. J., Baragli, A., Trieu, P., Ethier, N., Bouvier, M., and Hebert, T. E. (2006) Heterotrimeric G proteins form stable complexes with adenylyl cyclase and Kir3.1 channels in living cells. *J Cell Sci* **119**, 2807–18.
2. Dupre, D. J., Baragli, A., Rebois, R. V., Ethier, N., and Hebert, T. E. (2007) Signalling complexes associated with adenylyl cyclase II are assembled during their biosynthesis. *Cell Signal* **19**, 481–9.
3. Dupre, D. J., Robitaille, M., Ethier, N., Villeneuve, L. R., Mamarbachi, A. M., and Hebert, T. E. (2006) Seven transmembrane receptor core signaling complexes are assembled prior to plasma membrane trafficking. *J Biol Chem* **281**, 34561–73.
4. Rebois, R. V., Warner, D. R., and Basi, N. S. (1997) Does subunit dissociation necessarily accompany the activation of all heterotrimeric G proteins? *Cell Signal* **9**, 141–51.
5. Levitzki, A., and Klein, S. (2002) G-protein subunit dissociation is not an integral part of G-protein action. *Chembiochem* **3**, 815–8.
6. Evanko, D. S., Thiyagarajan, M. M., Takida, S., and Wedegaertner, P. B. (2005) Loss of association between activated Gαq and Gβγ disrupts receptor-dependent and receptor-independent signaling. *Cell Signal* **17**, 1218–28.
7. Bunemann, M., Frank, M., and Lohse, M. J. (2003) Gi protein activation in intact cells involves subunit rearrangement rather than dissociation. *Proc Natl Acad Sci U S A* **100**, 16077–82.
8. Azpiazu, I., and Gautam, N. (2004) A fluorescence resonance energy transfer-based sensor indicates that receptor access to a G protein is unrestricted in a living mammalian cell. *J Biol Chem* **279**, 27709–18.
9. Hein, P., Frank, M., Hoffmann, C., Lohse, M. J., and Bunemann, M. (2005) Dynamics of receptor/G protein coupling in living cells. *EMBO J* **24**, 4106–14.
10. Gales, C., Rebois, R. V., Hogue, M., Trieu, P., Breit, A., Hebert, T. E., and Bouvier, M. (2005) Real-time monitoring of receptor and G-protein interactions in living cells. *Nat Methods* **2**, 177–84.
11. Gibson, S. K., and Gilman, A. G. (2006) Giα and Gβ subunits both define selectivity of G protein activation by α2-adrenergic receptors. *Proc Natl Acad Sci U S A* **103**, 212–7.
12. Gales, C., Van Durm, J. J., Schaak, S., Pontier, S., Percherancier, Y., Audet, M., Paris, H., and Bouvier, M. (2006) Probing the activation-promoted structural rearrangements in preassembled receptor-G protein complexes. *Nat Struct Mol Biol* **13**, 778–86.
13. Digby, G. J., Lober, R. M., Sethi, P. R., and Lambert, N. A. (2006) Some G protein heterotrimers physically dissociate in living cells. *Proc Natl Acad Sci U S A* **103**, 17789–94.
14. Robitaille, M., Ramakrishnan, N., Baragli, A., and Hebert, T. E. (2009) Intracellular trafficking and assembly of specific Kir3 channel/G protein complexes. *Cell Signal* **21**, 488–501.
15. Dupre, D. J., Robitaille, M., Richer, M., Ethier, N., Mamarbachi, A. M., and Hebert, T. E. (2007) Dopamine receptor-interacting protein 78 acts as a molecular chaperone for Gγ subunits before assembly with Gβ. *J Biol Chem* **282**, 13703–15.
16. Ayoub, M. A., and Pfleger, K. D. (2010) Recent advances in bioluminescence resonance energy transfer technologies to study GPCR heteromerization. *Curr Opin Pharmacol* **10**, 44–52.
17. Pfleger, K. D. (2009) Analysis of protein–protein interactions using bioluminescence resonance energy transfer. *Methods Mol Biol* **574**, 173–83.
18. Hebert, T. E., Gales, C., and Rebois, R. V. (2006) Detecting and imaging protein–protein interactions during G protein-mediated signal transduction in vivo and in situ by using fluorescence-based techniques. *Cell Biochem Biophys* **45**, 85–109.
19. Pfleger, K. D., Seeber, R. M., and Eidne, K. A. (2006) Bioluminescence resonance energy transfer (BRET) for the real-time detection of protein–protein interactions. *Nat Protoc* **1**, 337–45.
20. Pfleger, K. D., and Eidne, K. A. (2006) Illuminating insights into protein–protein interactions using bioluminescence resonance energy transfer (BRET). *Nat Methods* **3**, 165–74.
21. Kroeger, K. M., and Eidne, K. A. (2004) Study of G-protein-coupled receptor-protein interactions by bioluminescence resonance energy transfer. *Methods Mol Biol* **259**, 323–33.
22. Marullo, S., and Bouvier, M. (2007) Resonance energy transfer approaches in molecular pharmacology and beyond. *Trends Pharmacol Sci* **28**, 362–5.
23. Milligan, G., and Bouvier, M. (2005) Methods to monitor the quaternary structure of G protein-coupled receptors. *FEBS J* **272**, 2914–25.
24. Kocan, M., See, H. B., Seeber, R. M., Eidne, K. A., and Pfleger, K. D. (2008) Demonstration

of improvements to the bioluminescence resonance energy transfer (BRET) technology for the monitoring of G protein-coupled receptors in live cells. *J Biomol Screen* **13**, 888–98.

25. Wei, H., Ahn, S., Shenoy, S. K., Karnik, S. S., Hunyady, L., Luttrell, L. M., and Lefkowitz, R. J. (2003) Independent β-arrestin 2 and G protein-mediated pathways for angiotensin II activation of extracellular signal-regulated kinases 1 and 2. *Proc Natl Acad Sci U S A* **100**, 10782–7.
26. Tohgo, A., Choy, E. W., Gesty-Palmer, D., Pierce, K. L., Laporte, S., Oakley, R. H., Caron, M. G., Lefkowitz, R. J., and Luttrell, L. M. (2003) The stability of the G protein-coupled receptor-β-arrestin interaction determines the mechanism and functional consequence of ERK activation. *J Biol Chem* **278**, 6258–67.
27. Zidar, D. A., Violin, J. D., Whalen, E. J., and Lefkowitz, R. J. (2009) Selective engagement of G protein coupled receptor kinases (GRKs) encodes distinct functions of biased ligands. *Proc Natl Acad Sci U S A* **106**, 9649–54.
28. Lee, M. H., El-Shewy, H. M., Luttrell, D. K., and Luttrell, L. M. (2008) Role of β-arrestin-mediated desensitization and signaling in the control of angiotensin AT1a receptor-stimulated transcription. *J Biol Chem* **283**, 2088–97.
29. Smith, N. J., and Luttrell, L. M. (2006) Signal switching, crosstalk, and arrestin scaffolds: novel G protein-coupled receptor signaling in cardiovascular disease. *Hypertension* **48**, 173–9.
30. Gesty-Palmer, D., Chen, M., Reiter, E., Ahn, S., Nelson, C. D., Wang, S., Eckhardt, A. E., Cowan, C. L., Spurney, R. F., Luttrell, L. M., and Lefkowitz, R. J. (2006) Distinct β-arrestin- and G protein-dependent pathways for parathyroid hormone receptor-stimulated ERK1/2 activation. *J Biol Chem* **281**, 10856–64.

Chapter 14

Luminescent Biosensors for Real-Time Monitoring of Intracellular cAMP

Brock F. Binkowski, Frank Fan, and Keith V. Wood

Abstract

G-protein coupled, seven-transmembrane (7-TM) receptors (GPCRs) comprise a diverse class of signaling molecules involved in cellular physiology and pathology. In recent years, intracellular biosensors have been developed to monitor changes in intracellular cAMP in real time, facilitating studies on the mechanisms of GPCR activation and desensitization in living cells. However, methods based on fluorescence can show limitations in response dynamics together with being difficult to perform. Here we present the use of genetically encoded, luminescent biosensors that allow a facile, non-lytic assay format for monitoring cAMP dynamics in living cells.

Key words: cAMP, Live cell assay, GloSensor, Biosensor, G-protein-coupled receptor, Inverse agonist

1. Introduction

The GloSensor cAMP Assay provides an extremely sensitive and easy-to-use format for the interrogation of overexpressed or endogenous GPCRs that signal via changes in the intracellular concentration of cAMP. The assay utilizes genetically encoded biosensor variants with cAMP binding domains fused to mutant forms of *Photinus pyralis* luciferase (1–3). Upon binding to cAMP, conformational changes occur that promote large increases in light output. Following pre-equilibration with substrate, cells transiently or stably expressing a biosensor variant can be used to assay GPCR function using a live-cell, non-lytic assay format, enabling facile kinetic measurements of cAMP accumulation or turnover in living cells. Moreover, the assay offers a broad dynamic

Louis M. Luttrell and Stephen S.G. Ferguson (eds.), *Signal Transduction Protocols*, Methods in Molecular Biology, vol. 756, DOI 10.1007/978-1-61779-160-4_14, © Springer Science+Business Media, LLC 2011

range and is extremely sensitive, allowing detection of Gi-coupled receptor activation or inverse agonist activity in the absence of compounds such as forskolin.

2. Materials

2.1. Cell Culture Preparation

1. HEK293 or CHO cells (American Type Culture Collection).
2. Phosphate-buffered saline, Ca^{2+} and Mg^{2+}-free (PBS).
3. 0.05% trypsin-EDTA.
4. HEK293 cell growth medium: Dulbecco's modified Eagle medium (DMEM) supplemented with 10% fetal bovine serum (FBS).
5. CHO cell growth medium: F12 medium supplemented with 10% FBS.
6. T-75 tissue culture flasks.
7. Tissue culture-treated, solid white, 96-well assay plates.

2.2. Transient Transfection

1. pGloSensor-20F cAMP and/or pGloSensor-22F cAMP plasmids (Promega).
2. FuGENE HD transfection reagent (Promega).
3. Opti-MEM® I reduced-serum medium (Invitrogen).

2.3. Pre-equilibration

1. GloSensor cAMP Reagent (Promega).
2. GloSensor cAMP Reagent Stock Solution: resuspend GloSensor cAMP Reagent in 10 mM HEPES, pH 7.5 (25 mg/817 μl; 250 mg/8.17 ml). Store in limited-use aliquots at −80°C.
3. Equilibration medium: CO_2-independent medium (Invitrogen) supplemented with 10% FBS and containing 2% v/v GloSensor cAMP Reagent Stock Solution.

2.4. Compound Addition and Luminescent Measurements

1. To obtain a concentration–response curve, serially dilute the compound in storage solvent (aqueous solution or DMSO) to 100× stock solutions, followed by direct addition to the respective wells. Alternatively, serially dilute the compound in storage solvent to 1,000× stock solutions, followed by dilution to 10× aqueous stock solutions and delivery to the respective wells. We have found no deleterious effects associated with running assays at 1% v/v DMSO.
2. Forskolin: 100 mM solution in DMSO (Sigma).
3. Luminometer (see Note 1).

3. Methods

Two versions of the biosensor exist for use in the GloSensor cAMP Assay (Fig. 1). Following cell-free expression in vitro, the version encoded by the pGloSensor-22F cAMP construct (22F) shows an increased EC50 for activation together with increased S/B ratio at saturation relative to the version encoded by the pGloSensor-20F cAMP construct (20F). Moreover, the 22F construct is linearly proportional to cAMP over an increased range, encompassing the intracellular concentrations reported for living cells (4).

In general, we have observed similar relationships between the two constructs when their performance is compared in living cells. For Gs-coupled receptors, the 22F construct has shown markedly increased S/B and an enhanced ability to discriminate between the efficacy of full and partial agonists (Fig. 2) compared to the 20F construct, likely due to saturation effects associated with the 20F construct. For Gi-coupled receptors, the 22F

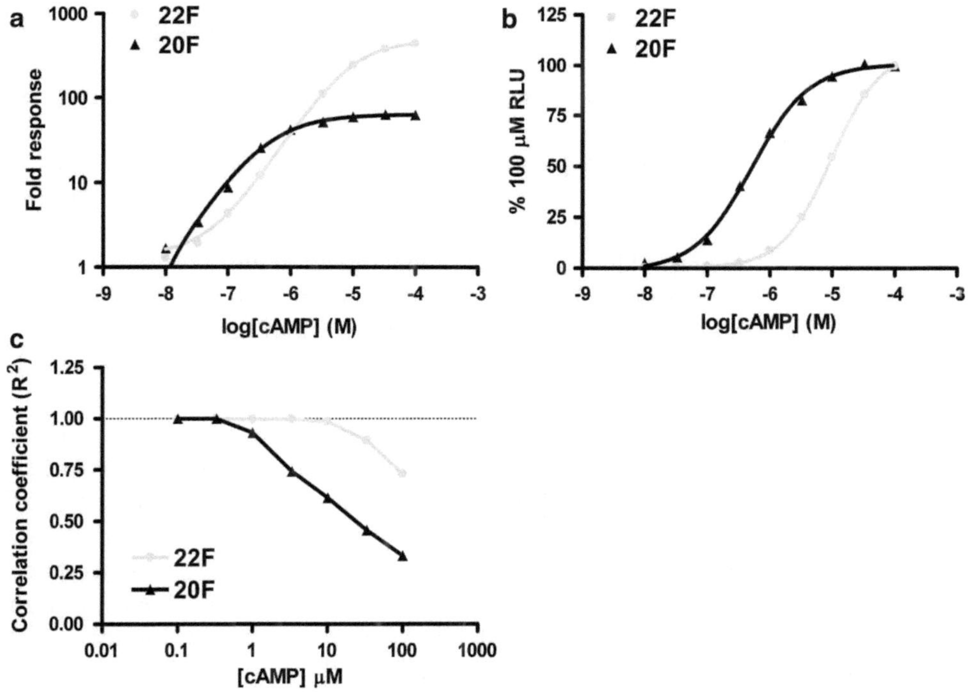

Fig. 1. Cell-free expression of biosensor variants and incubation with varying concentrations of cAMP in vitro. (**a**) The 22F construct shows an increased S/B at saturation and similar stepwise increases in fold response at lower concentrations of cAMP. (**b**) The data set of (**a**) normalized to the luminescence at 100 μM cAMP. The 22F construct has a higher EC50 value for activation relative to the 20F construct (9.9 μM vs. 0.53 μM, respectively). (**c**) Linear regression analysis performed on the data set of (**a**). The correlation coefficient was plotted as a function of the maximal concentration of cAMP used in the analysis (10 nM minimum for all). Fold response was calculated relative to a control sample containing vehicle alone. $n = 3$ per dose.

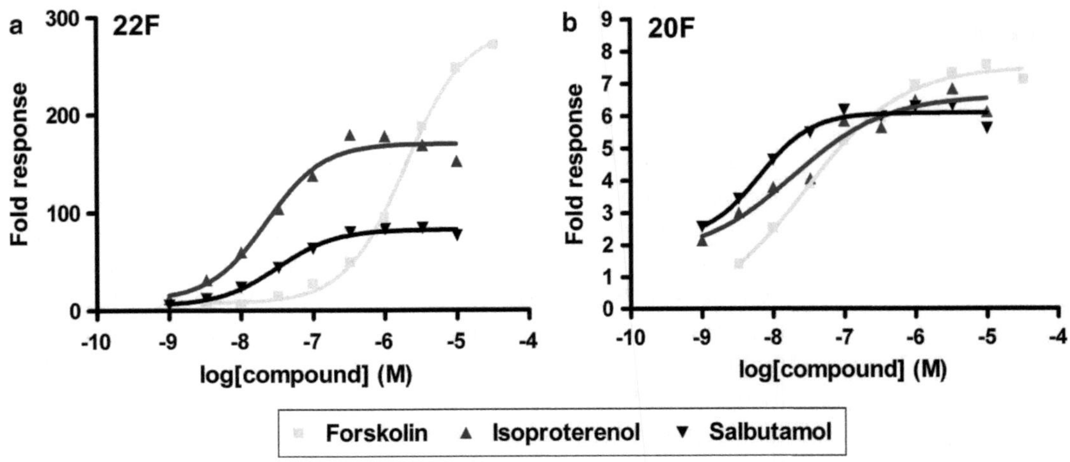

Fig. 2. Performance comparison of biosensor variants following activation of an endogenous Gs-coupled receptor in HEK293 cells. (**a**) HEK293 cells transiently expressing the 22F construct or (**b**) the 20F construct were assayed 10 min after addition of varying concentrations of the respective compounds, where this value of luminescence was divided by a preread measurement as a measure of fold response. Isoproterenol, full β2-adrenergic receptor agonist; salbutamol, partial β2-adrenergic receptor agonist; forskolin is an activator of endogenous adenylate cyclase. This experiment was performed in the absence of phosphodiesterase inhibitors. $n = 1$ per dose.

construct has shown increased S/B in the presence or absence of added forskolin, where saturation effects can hinder the 20F construct in the presence of forskolin in select experimental systems (Fig. 3).

The timing and order of addition of compound(s) will depend on the type of 7-TM receptor being assayed. For Gs-coupled receptors, simply add varying concentrations of agonist and acquire kinetic or end-point measurements. In general, we have seen maximal changes in light output within 2–10 min following addition of saturating concentrations of agonist (Fig. 4), depending on assay temperature (see Note 2). For targets that undergo rapid desensitization, phosphodiesterase inhibitors can be included to stabilize signals near maximal values. However, the use of phosphodiesterase inhibitors is not a standard requirement of the assay (see Note 3), in contrast to lytic assays for cAMP. If measuring antagonist activity, first determine the EC80 concentration of the agonist that will be used in the assay as described above. Once done, preincubate with varying concentrations of antagonist for 5–10 min, followed by addition of an EC80 concentration of agonist to all wells. Acquire kinetic or end-point measurements of luminescence for 10–20 min postagonist addition.

For Gi-coupled receptors, preincubate with varying concentrations of agonist for 5–10 min followed by addition of a fixed concentration of forskolin to all wells. The optimal concentration of forskolin for maximal signal-to-background ratio (S/B) of agonist is determined empirically, where doses between 0.1 and 10 μM are typical (depending on the cell line). Acquire kinetic or

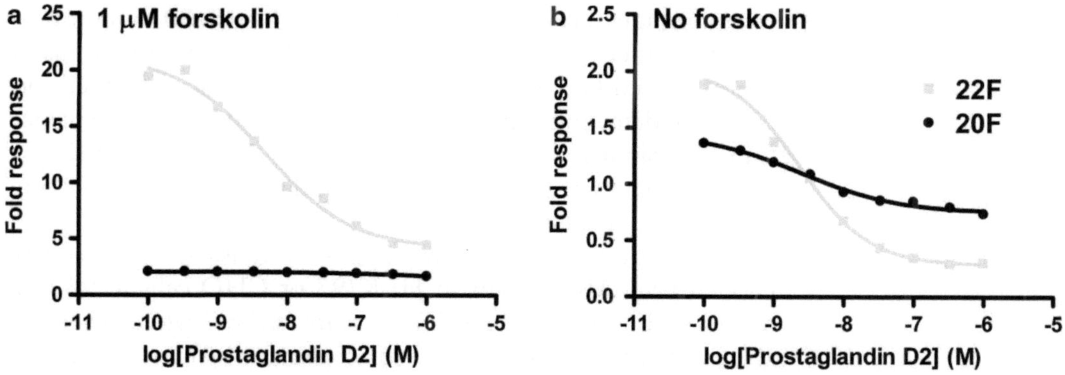

Fig. 3. Performance comparison of biosensor variants following activation of an overexpressed Gi-coupled receptor in HEK293T cells. HEK293T cells stably expressing the DP2/GPR44 receptor and transiently expressing the 20F or 22F constructs were pretreated with varying concentrations of prostaglandin D2 agonist for 5 min prior to the addition of either (**a**) 1 μM forskolin or (**b**) vehicle alone. Luminescence was measured 30 min after forskolin addition, and this value was divided by a preread measurement taken prior to compound delivery to determine fold response. This experiment was performed in the absence of phosphodiesterase inhibitors. $n=1$ per dose.

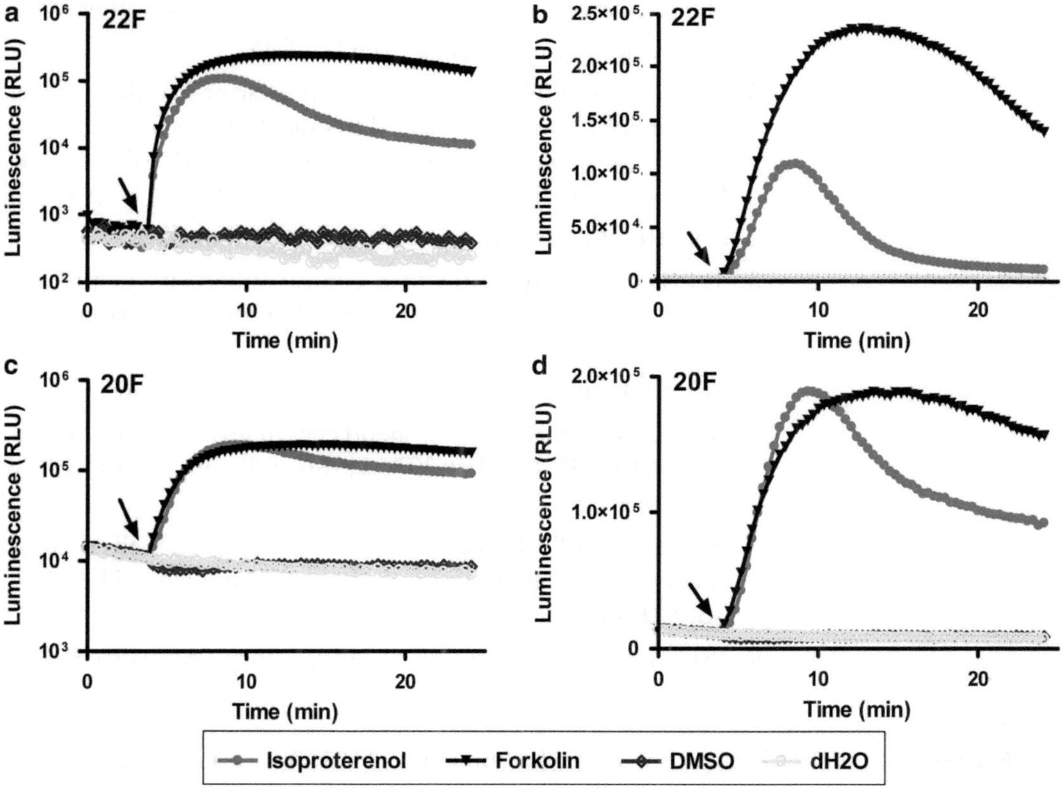

Fig. 4. Performance comparison of kinetic measurements taken following the activation of an endogenous Gs-coupled receptor in HEK293 cells. HEK293 cells transiently expressing the 20F or 22F constructs were assayed in real time at 28°C. Following pre-equilibration to the steady-state operating temperature of the luminometer, 10 μM isoproterenol (endogenous β2-adrenergic receptor agonist) or 10 μM forskolin (activator of adenylate cyclase) were added at the indicated time points. Kinetic traces from cells expressing the 22F construct were plotted on log (**a**) or linear (**b**) scales. Kinetic traces from cells expressing the 20F construct were plotted on log (**c**) or linear (**d**) scales. This experiment was performed in the absence of phosphodiesterase inhibitors. $n=1$ per trace.

end-point measurements of luminescence for 15–30 min postforskolin addition. In numerous cases, we have been able to assay overexpressed Gi-coupled receptors in the absence of added forskolin by detecting a decrease in the basal level of luminescence (Fig. 4). Similarly, we have been able to detect the activity of inverse agonists of overexpressed Gs- and Gi-coupled receptors in the absence of added forskolin.

The protocol listed below applies to transient expression of GloSensor cAMP variants in HEK293 or CHO cells in 96-well format (see Note 4 *for additional validated cell types*), where an abbreviated protocol can be used to assay cells stably expressing biosensor variants (see Note 5). Related protocols for bulk transient transfection and assay miniaturization and can be found at http://www.promega.com.

3.1. Cell Culture Preparation

The volumes listed in Steps 1–4 below are for propagation in a T75 flask. Scale volumes accordingly for flasks with different total surface area.

1. Harvest cells when the monolayer is at 50–90% confluence.
2. Wash cell monolayer using 10 ml of PBS.
3. Add 2 ml of 0.05% trypsin-EDTA prewarmed to 37°C. Coat the surface of the flask evenly. Dislodge the cells from the flask surface by rocking and gently tapping the side of the flask. Once cells are dislodged, proceed immediately to Step 4.
4. Add 10 ml of growth medium prewarmed to 37°C.
5. Transfer cell suspension to a conical tube. Mix gently and dislodge cell aggregates by slowly pipetting. Determine cell number using a hemacytometer.
6. Pellet cells at $250 \times g$ for 5 min at room temperature.
7. Aspirate supernatant. Resuspend cells in growth medium prewarmed to 37°C: HEK293, 1.5E6 cells/ml; CHO, 1.0E6 cells/ml.
8. Add 100 μl of cell suspension to the individual wells of a tissue culture-treated, 96-well flat bottom plate (HEK293, 15,000 cells; CHO, 10,000 cells).
9. Place plates in a 37°C tissue culture incubator with 5–10% CO_2, overnight.

3.2. Transient Transfection

This protocol can be applied to HEK293 or CHO cells and is sufficient for 20 wells (100 μl of medium per well prior to addition of FuGENE® HD transfection reagent/DNA complex).

1. Dilute the pGloSensor™-22F cAMP or pGloSensor™-20F cAMP plasmid to a final concentration of 12.5 ng/μl in Opti-MEM® I reduced-serum medium.

2. Add 6 µl of FuGENE® HD transfection reagent to 160 µl of diluted plasmid and mix carefully by gentle pipetting.
3. Incubate for 0–15 min at room temperature.
4. Add 8 µl of complex per well of a 96-well plate and gently mix without disturbing the cell monolayer.
5. Incubate 20–24 h in a 37°C tissue culture incubator with 5–10% CO_2.

3.3. Substrate Pre-equilibration

1. Carefully remove the medium from the individual wells. To accomplish this, place the pipet tip at the side of the well to minimize disruption of the cell monolayer. Move quickly to Step 2.
2. Add 100 µl of equilibration medium per well for a 96-well plate. Add medium to the side of each well; do not pipet directly onto the cell monolayer. The equilibration medium contains a 2% v/v dilution of the GloSensor cAMP Reagent stock solution in a buffered medium. Please note, a buffered medium is required to perform the assay under most conditions (see Note 6).
3. Incubate for 2 h at room temperature or until a steady-state basal signal is obtained. Incubation at higher temperatures can facilitate equilibration, but care must be taken to allow the entire plate to come to a uniform temperature prior to starting the assay. If the basal level of luminescence is not significantly above the luminometer background, increased concentrations of substrate can promote increased levels of light output (see Note 7).

3.4. Compound Addition and Luminescent Measurements

1. Add compounds following the guidelines discussed above, and measure luminescence continuously (kinetic measurements) or at a fixed time point (end-point measurements). We have found that normalization to a preread measurement (prior to the addition of compounds of interest) can reduce data variability associated with cell number or well-to-well variability in transient transfection.
2. For kinetic measurements, it is important to note that most luminometers operate above room temperature in kinetic modes of operation (even if the specified temperature is set lower). Therefore, it is important to allow the plate to pre-equilibrate to the steady-state operating temperature of the instrument prior to compound addition (see Note 8). Following pre-equilibration, remove the plate from the instrument and quickly add compounds from 10× to 100× stock solutions using a multichannel pipet. Quickly return the plate to the instrument and begin taking measurements. Alternatively, use a luminometer with injectors to deliver compound(s)

following the manufacturer's recommendations. Temperature artifacts can be readily identified by following the kinetic traces of control wells (see Note 9), and protocol modifications can be used to help buffer signal changes associated with temperature shifts (see Note 10).

3. For end-point measurements, add compounds from 10× to 100× stock solutions using a multichannel pipet, and measure luminescence at the desired time point(s). If desired, cells can be starved of serum prior to adding test/control compounds of interest. However, serum starvation will influence the basal signal, as expected.

4. Notes

1. We have routinely used GloMax 96 Microplate Luminometer (Promega), GloMax-Multi (Promega), GloMax-Multi + (Promega), Varioskan Flash (Thermo Electron), Mithras (Berthold Technologies), or BMG Pherastar (Imgen) instruments (integration times ranging between 0.1 and 1 s per well). We are also aware of the successful use of instruments commonly used in drug discovery and high-throughput screening, such as ViewLux, FLIPR tetra, FDSS7000, etc.
2. Changes in the assay temperature promote changes in the overall levels of light output, where changes in the basal level of cAMP in the cell may be a contributing factor. In general, increased temperatures will promote decreased basal and induced levels of light output, where these decreases are typically associated with increased S/B in most experimental systems.
3. Although they are not required, PDE inhibitors can be used in the assay. If PDE inhibitors are used, we recommend performing assay optimization in the presence of these compounds, as they can influence the magnitude and kinetics of the biosensor response. For example, we have found that pre-incubation with 500 μM isobutyl-methylxanthine promotes a decrease in S/B for the 20F version of the sensor owing to increased basal levels of luminescence (and likely saturation of the sensor).
4. Common laboratory cell lines validated by Promega: HEK293, HEK293T, CHO, HeLa, NIH3T3, and U2OS. A 2% v/v dilution of the GloSensor cAMP Reagent is viable for all except CHO, where a 6% v/v dose is recommended (following a 2 h pre-equilibration at room temperature). In general, when validating a new cell type, the optimal percent dilution of the GloSensor cAMP Reagent will need to be

determined empirically, together with viable equilibration times/temperatures.

5. For stable cell lines, plate cells on day 1 and follow Subheadings 3.3 and 3.4 on day 2. Stable cell lines can be generated using the pGloSensor-20F cAMP or pGloSensor-22F cAMP plasmids by selection using 200 μg/ml hygromycin.
6. Requirement for use of buffered medium. If the plates will be out of the CO_2 incubator for extended periods of time (such as during a kinetic read), the medium must be buffered to avoid the deleterious pH changes associated with equilibration to atmospheric conditions. This can be achieved using a commercially available buffered medium (e.g., CO_2-independent medium; Invitrogen).
7. We have found equilibration medium with 2% v/v GloSensor™ cAMP Reagent stock solution to be viable for a majority of cell type/luminometer combinations. However, if the basal level of luminescence is not significantly above the luminometer background, increased concentrations of substrate can promote increased levels of light output. For example, equilibration medium with 6% v/v GloSensor™ cAMP Reagent stock solution provides a significantly increased basal level of light output (up to 50-fold) from CHO cells transiently expressing the 22F construct following a 2 h pre-equilibration at room temperature.
8. This can typically be done by acquiring preread kinetic measurements for 15–20 min, where the basal level of luminescence can be monitored until a steady-state value is reached.
9. If present, cooling effects will be apparent as sharp increases in the kinetic traces of wells receiving vehicle alone (negative controls) or left untreated.
10. If performing experiments at elevated temperature (such as 37°C), it may be beneficial to increase the total volume to 200 μl per well (making the appropriate changes to compound stock solutions, etc.) and to include distilled water in the spaces between wells to buffer any temperature changes associated with removing the plate from the instrument.

References

1. Fan, F., Binkowski, B. F., Butler, B. L., Stecha, P. F., Lewis, M. K. and Wood, K. V. (2008) Novel genetically encoded biosensors using firefly luciferase. *ACS Chem Biol* **3**, 346–51.
2. Binkowski, B. F., Fan, F. and Wood, K. V. (2009) Live-cell luminescent assays for GPCR studies. *Gen Eng Biotech* **29**, 30–31.
3. Binkowski, B. F., Fan, F. and Wood, K.V. (2009) Engineered luciferases for molecular sensing in living cells. *Curr Opin Biotech* **20**, 14–18.
4. Willoughby, D. and Cooper, D. M. F. (2008) Live-cell imaging of cAMP dynamics. *Nat Methods* **5**, 29–36.

Chapter 15

Simultaneous Real-Time Imaging of Signal Oscillations Using Multiple Fluorescence-Based Reporters

Lianne B. Dale and Stephen S.G. Ferguson

Abstract

It is now well understood that G protein-coupled receptor (GPCR)-mediated cell signalling is subject to extensive spatial–temporal control, and that a meaningful understanding of this complexity requires techniques to study signalling at the molecular and sub-cellular level. This complexity in cell signal pattern begins with ligand binding to the receptor and its coupling to a variety of different effector systems. These signal transduction cascades within a cell involve a very complex series of molecular events requiring the generation of multiple second messenger responses and the activation a multiple effector proteins. In the present chapter, we will describe methodology for the simultaneous assessment of the spatial–temporal measurement of increases in intracellular Ca^{2+} concentrations and the activation of protein kinase C (PKC) in response to the agonist activation of a $G\alpha_{q/11}$-coupled GPCR. Specifically, we will describe a confocal imaging approach to simultaneously measure oscillilations in intracellular Ca^{2+} levels and PKC translocation to the plasma membrane in response to mGluR1 stimulation in transiently transfected human embryonic kidney (HEK293) cells. The changes in intracellular Ca^{2+} were imaged using the fluorescent indicator Oregon Green 488 BAPTA and a recombinant PKCβII-DsRed fusion protein was used to image the sub-cellular distribution of the PKCβII isoform.

Key words: Intracellular Ca^{2+}, Protein kinase C, Oscillation, Fluorescence, Laser scanning confocal microscopy

1. Introduction

Communication within a cell is a very complex and coordinated series of events that involves numerous signalling molecules. There are several approaches to imaging these signalling events and here we will describe a method to image the response of two signalling molecules following the stimulation of a G protein-coupled receptor (GPCR). More specifically, we follow changes

Louis M. Luttrell and Stephen S.G. Ferguson (eds.), *Signal Transduction Protocols*, Methods in Molecular Biology, vol. 756, DOI 10.1007/978-1-61779-160-4_15, © Springer Science+Business Media, LLC 2011

in intracellular Ca^{2+} levels simultaneously with changes in protein kinase C (PKC) distribution in response to group I metabotropic glutamate receptor 1 (mGluR1) stimulation.

The mGluR1 is a member of the GPCR superfamily and couples through the heterotrimeric G protein, $G\alpha_{q/11}$, to the hydrolysis of phosphatidylinositol 4,5-bisphosphate (PIP_2) to form inositol 1,4,5-trisphosphate (IP_3) and diacylglycerol (DAG) (1). IP_3 subsequently triggers the release of Ca^{2+} from intracellular stores and this increase in intracellular Ca^{2+} along with DAG recruits the conventional PKC isoforms (α, βI, βII, γ) from the cytosol to the plasma membrane (2, 3). Once activated, PKC phosphorylates various substrates involved in the signal transduction cascade (3, 4).

The pattern of activation varies between different receptors and PKC isoforms. For instance, the stimulation of the angiotensin AT_1R receptor leads to a single transient increase in intracellular Ca^{2+} and reversible redistribution of PKCβII to the membrane (5), whereas the activation of the Group I mGluRs, mGluR1, and mGluR5, produces an oscillatory pattern of changes (6). Interestingly, stimulation of either the AT_1R and mGluR1 receptors leads to variable patterns of translocation of the PKCβI isoform within individual cells, including an oscillatory pattern as well as persistent translocation to the plasma membrane (5, 7). Furthermore, isoform-selective translocation in response to thyrotropin-releasing hormone (TRH) receptor stimulation has been shown to involve a coordinated cascade of activation of the various PKC isoforms (8). It is thought that the amplitude and frequency of Ca^{2+} oscillations and the pattern of PKC activation ultimately translates into various biological responses (8, 9).

The level of Ca^{2+} within a cell is commonly studied using various fluorescent indicators that change with respect to their fluorescent properties when bound with Ca^{2+}. Generally, these indicators only bind free Ca^{2+}. In a non-signalling cell, the majority of the intracellular Ca^{2+} is not free but rather bound to various molecules or sequestered within intracellular stores (10). When the cell receives a signal to release Ca^{2+} from intracellular stores and/or allow the entry of extracellular Ca^{2+}, the level of free cytosolic Ca^{2+} increases, which subsequently binds to the indicator altering its fluorescent properties (10).

There are a number of fluorescent Ca^{2+} indicators commercially available and they can be classified as either single wavelength or ratiometric indicators and described as having high or low affinity for Ca^{2+}. The single wavelength indicators demonstrate a Ca^{2+}-dependent change in fluorescence intensity (without shifting excitation or emission wavelengths), whereas the ratiometric indicators change with respect to either their excitation or emission wavelengths when bound to Ca^{2+} (10). Each has its own advantages and disadvantages and a number of factors, such as the imaging equipment available and experimental approach, will determine the best indicator to choose. In general, the ratiometric indicators are typically used to quantify intracellular Ca^{2+} levels, since they can be precisely calibrated, whereas the single wavelength indicators are more difficult to quantify, but unlike the ratiometric ones they can be easily imaged with multiple fluorophores (10).

To image the activation or recruitment of PKC from the cytosol to the plasma membrane, we have engineered fluorescent PKC fusion proteins using PCR to remove the stop codon within the cDNA of PKC and subsequently cloning it in an expression vector in the same reading frame as a genetically encoded fluorescent protein. Fluorescent fusion proteins are routinely used to image the localization of specific proteins within a cell and one of their main advantages is that distribution and/or trafficking of specific proteins can be imaged within living cells and in "real-time" (11). The green fluorescent protein (GFP) from the jellyfish *Aequorea victoria* is the most commonly used, but there is a rainbow of fluorescent proteins expression vectors commercially available. Most of these fluorescent proteins are genetic variants of GFP, but there are also others that originate from other sources such as DsRed from a coral (*Discosoma*) (11).

When selecting the Ca^{2+} indicator and fluorescent protein, it is important to ensure that both are compatible with your imaging system with respect to the excitation wavelengths and separation of the emission. For this particular study, we used a LSM 510 META laser scanning confocal microscope that was equipped with 488 and 543 nm laser lines along with the single wavelength calcium indicator Oregon Green® 488 BAPTA and PKCβII fused to DsRed.

2. Materials

2.1. Cell Culture

1. Human embryonic kidney 293 (HEK293) cells (American Type Culture Collection).
2. Complete Minimum Essential Medium (MEM): MEM supplemented with 10% heat inactivated Fetal Bovine Serum (Invitrogen).
3. 0.25% (w/v) Trypsin and 0.05% (w/v) ethylenediamine tetraacetic acid (EDTA) solution (Invitrogen).
4. Phosphate-Buffered Saline (PBS): 137 mM NaCl, 2.7 mM KCl, 10 mM Na_2HPO_4, 1.76 mM KH_2PO_4, pH 7.4.
5. 100 mm cell culture dishes (BD Bioscience).
6. 35 mm glass bottom dishes (MatTek Corporation).

2.2. Calcium Phosphate Transfection

1. Sterilized water.
2. 2x HBS: 280 mM NaCl, 50 mM HEPES (free acid), 1.5 mM Ha_2HPO_4, pH 7.10, filter sterilized with a 0.45 μm filter (see Note 1).
3. 2.5 M $CaCl_2$, filter sterilized with 0.45 μm filter, may be stored at −20°C for several months.
4. cDNA for PKCβII-DsRed and mGluR1a (see Note 2).

2.3. Sample Preparation

1. Oregon Green® 488 BAPTA-AM cell permeant (Molecular Probes/Invitrogen), stored at ≤−20°C and protected from the light.
2. Anhydrous dimethyl sulfoxide (DMSO) or 20% (w/v) Pluronic F-127 in DMSO (Molecular Probes/Invitrogen).
3. Quisqualate (Tocris), aliquoted and stored at −80°C as a 10 mM solution in water.
4. HEPES-buffered saline solution (HBSS): 1.2 mM KH_2PO_4, 5 mM $NaHCO_3$, 20 mM HEPES, 11 mM glucose, 116 mM NaCl, 4.7 mM $MgSO_4$, 2.5 mM $CaCl_2$, pH 7.4.

2.4. Imaging

The images were acquired using a Zeiss LSM 510 META laser scanning confocal microscope with a 63x/1.4NA plan-apochromat oil immersion objective and equipped with an Argon laser for 488 nm excitation of the Oregon Green® 488 BAPTA and a HeNe laser for 543 nm excitation of the DsRed. During the image acquisition, the sample was maintained at 37°C with a heated stage insert.

3. Methods

3.1. Cell Culture

1. HEK 293 cells are maintained in complete MEM containing 10% FBS at 37°C, humidified 5% CO_2 atmosphere, with the medium replenished every 3–5 days.
2. For transient transfections, 3×10^6 cells are seeded in a 100 mm cell culture dish containing the appropriate volume of complete MEM for a final volume of 10 ml per dish. Incubate the cells for at least 24 h at 37°C, 5% CO_2 to allow sufficient time for the cells to adhere to the dish.

3.2. Transient Transfection Using a Modified Calcium Phosphate Precipitation Method

Cells should be transfected ~24 h after seeding into the 100 mm dishes and they should be 60–75% confluent at the time of transfection.

1. Dilute 2 μg of each of the pcDNA3.1 mGluR1a and DsRed1-PKCβII in 450 μl of sterile water.

2. Add 50 μl 2.5 M $CaCl_2$.
3. Drip 500 μl 2xHBS over DNA/$CaCl_2$ solution, mix immediately, and immediately drip onto the medium covering the cell monolayer ensuring even distribution.
4. Incubate the cells for ~18 h at 37°C in 5% CO_2.
5. Aspirate the transfection media, wash the cells once with PBS and allow the cells to recover from the transfection for ~8 h in 10 ml of complete MEM in the incubator at 37°C and 5% CO_2.

3.3. Sample Preparation

1. Aspirate the cell culture medium, wash 1× with PBS and add 1 ml Trypsin-EDTA.
2. Wait no more than 1–2 min and gently tap the sides of the dish to detach the cells.
3. Add 20 ml of complete MEM.
4. Rinse the cells off the bottom of the dish and break apart any clumps of cells gently with a pipette.
5. Seed 2 ml of the cell solution per 35 mm glass bottom dish and allow 18–24 h for the cells to adhere to the surface and for expression of the PKCβII-DsRed and mGluR1a cDNA.

3.4. Loading Cells with the Fluorescent Ca^{2+} Indicator

1. Aspirate the cell culture medium and wash the cells 3× in warm HBSS (~2 ml/dish).
2. Incubate the cells at 37°C for 30 min in HBSS (~2 ml/dish).
3. Following the manufacturer's instructions, prepare a stock solution and then a loading solution of Oregon Green® 488 BAPTA, AM. Briefly, prepare a 2 mM stock solution by diluting an aliquot of the fluorescent indicator in anhydrous DMSO or 20% (w/v) Pluronic F-127 in DMSO. Subsequently, prepare a 1 μM loading solution by diluting the stock solution in room temperature HBSS (see Note 3).
4. Load the cells with the fluorescent Ca^{2+} indicator by incubating with the Oregon Green® 488 BAPTA-AM loading solution for 20 min at room temperature (see Notes 4 and 5).
5. Wash cells 3× with HBSS warmed to 37°C to remove any unincorporated indicator and add 2 ml of warmed HBSS for imaging.

3.5. Microscope Configuration and Parameters for Image Acquisition

The tail of the emission spectrum of Oregon Green® 488 overlaps with DsRed, therefore the images for each fluorophore cannot be acquired simultaneously but must be acquired sequentially (multi-track) to prevent "bleed through" of the Oregon Green® 488 into the DsRed channel. Changes in intracellular Ca^{2+} levels and PKC activation are very fast events, therefore the image acquisition and switching between the tracks must be rapid. In order to accomplish this, set the configurations of the microscope, if possible, so that none of the components (beam splitters and emission filters) have to change between the two tracks and the only difference is the laser line and channel that is active. Turning on or off laser lines and detection channels is very quick, whereas changing beam splitters and/or emission filters is relatively slow.

1. For the Zeiss LSM 510 META, set the microscope configuration for a multi-track (sequential) acquisition with the tracks switching after each line (instead of frame) with the HFT 488/543 for the main dichroic beam splitter and NFT 545 for the secondary beam splitter.
2. For the first track, use the 488 nm laser line and BP 505–530 emission filter to excite and collect the emission of the Oregon Green® 488 with the detection channel receiving the emission wavelengths less than 545 nm from the secondary NFT 545 beam splitter. For the second track, use the other channel receiving the wavelengths longer than 545 nm with the 543 nm laser line and LP 560 emission filter to excite and collect the emission of the DsRed-PKCβII.

3.6. Image Acquisition

To follow the distribution of PKCβII-DsRed and intracellular Ca^{2+} levels, the cells will be imaged continuously for several minutes. As a result, photobleaching of one or both fluorophores is a potential problem. Although it is difficult to prevent bleaching completely, the scan parameters need to be optimized to minimize the extent of photobleaching while acquiring a quality image at a speed fast enough to capture the rapid changes in PKC redistribution and intracellular Ca^{2+} levels in response to receptor activation.

1. With the temperature of the heated stage set to 37°C, place the sample in position and wait a few minutes for the temperature of the dish to equilibrate with the stage to minimize any focal drift.
2. Select a cell or group of healthy cells that does not have any compartmentalization of the indicator and that expresses PKCβII-DsRed diffusely in the cytosol. The changes in fluorescence intensity for both the Ca^{2+} indicator and PKCβII-DsRed will be measured in the cytosol, therefore try to select cells with an adequate cytosolic area to draw a region of interest for the measurement.
3. Focus through the middle of the cell(s) and crop in on the cell(s) of interest if necessary. Optimize the intensity and background levels by adjusting the laser intensity as well as the detector gain and offset. Adjust the settings so that the full dynamic range of the detector will be used with minimal

laser intensity to minimize photobleaching. The signal for the fluorescent Ca^{2+} indicator will increase as the intracellular Ca^{2+} concentration increases, therefore make sure any increase in fluorescence of the Ca^{2+} indicator will not be out of range of the detector.

4. Acquire as a series of images (512×512, 12 bit data depth, with an averaging function of 4) with no delay between images. Capture 1–2 min of images to establish "baseline" distribution of DsRed-PKCβII and Oregon Green® 488 fluorescence (see Note 6), then stimulate mGluR1a by adding 50 μl of a 1.2 mM stock of quisqualate (30 μM final concentration) and continue imaging once every 6–12 s for 20 min.

3.7. Image Analysis

To analyze PKCβII activation and changes in intracellular Ca^{2+} levels, select a region of interest (ROI) in the cytosol of the cell using the Zeiss image analysis software, avoiding the plasma membrane. The intensity of Oregon Green 488 BAPTA increases as the level of intracellular Ca^{2+} increases in response to mGluR1a activation, whereas the intensity of PKCβII-DsRed in the cytosol decreases as it is activated and translocates to the plasma membrane. As the level of intracellular Ca^{2+} decreases and PKCβII redistributes back to the cytosol, the fluorescence intensity of Oregon Green® 488 BAPTA decreases and PKCβII-DsRed intensity in the cytosol increases (Fig. 1).

4. Notes

1. 2× HBS may be stored at −20°C for several months or at 4°C for up to a month.
2. More recent versions of DsRed are now commercially available that can be substituted and offer reduced oligomerization properties and potentially more desirable fluorescence emission spectra. The mGluR1 can be exchanged for any $G\alpha_{q/11}$-coupled GPCR.
3. Either DMSO or 20% (w/v) Pluronic F-127 solution in DMSO can be used to prepare the stock solution. Pluronic F-127 is a non-ionic, surfactant polyol that functions as a dispersing agent and may help solubilize the hydrophillic dye.
4. According to the manufacturer, the Oregon Green® 488 BAPTA-AM stock solution in DMSO may be stored desiccated and protected from light at −20°C for several months, but it is best to prepare the stock solutions just prior to use since the AM esters are susceptible to hydrolysis in solution. Long-term storage of AM esters prepared in 20% (w/v) Pluronic F-127 is not recommended. The loading solution of

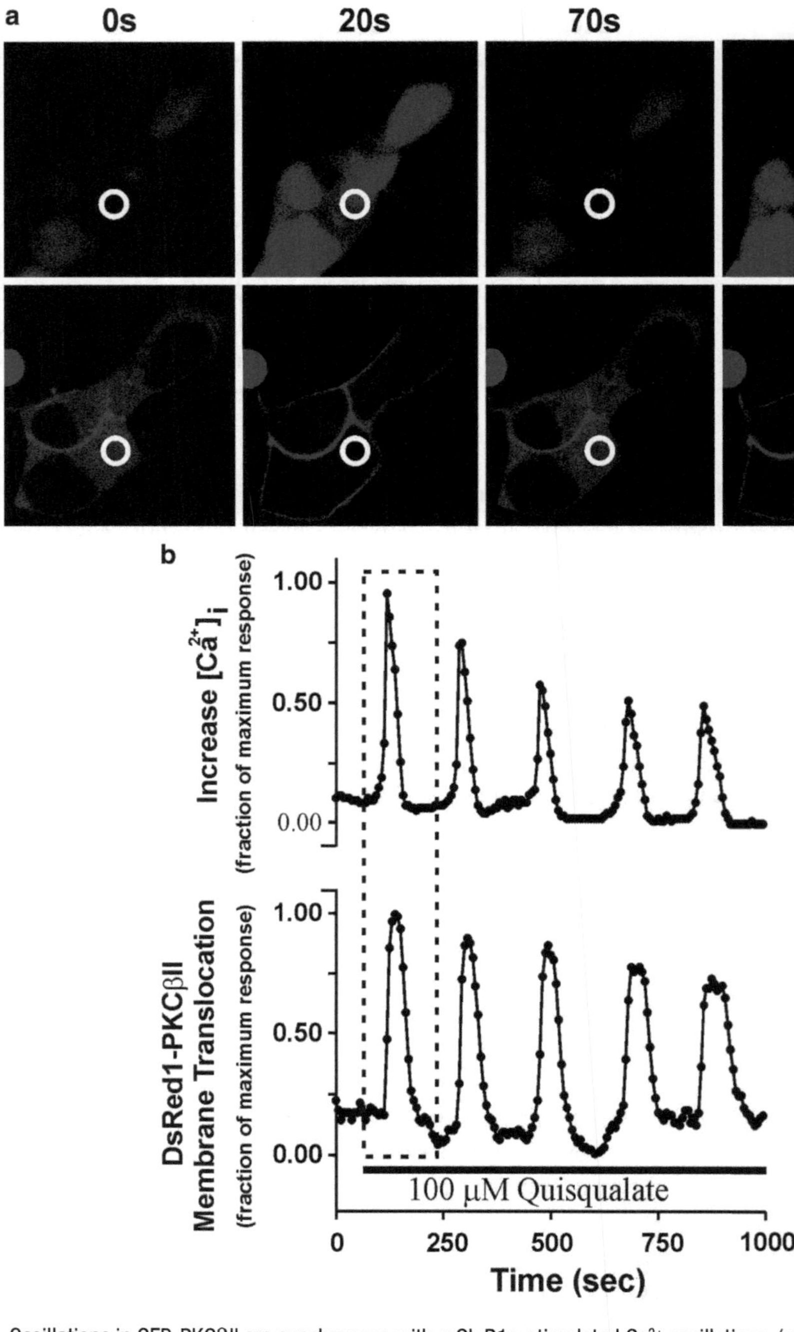

Fig. 1. Oscillations in GFP-PKCβII are synchronous with mGluR1a-stimulated Ca^{2+} oscillations. (**a**) Representative images selected from a time series of laser scanning confocal microscopic images showing synchronous oscillations in intracellular free Ca^{2+} ($[Ca^{2+}]_i$) (Oregon Green BAPTA-1AM) and DsRed1-PKCβII oscillations in response to mGluR1a activation with 100 μM quisqualate. (**b**) Example of the time course and frequencies of synchronous oscillations in $[Ca^{2+}]_i$ (Oregon Green BABTA-1AM) and DsRed1-PKCβII in response to activation of mGluR1a. The bar represents the time of exposure to 100 μM quisqualate. This research was originally published in *The Journal of Biological Chemistry*. Babwah, A.V., Dale L.B., and Ferguson S.S. Protein kinase C isoform-specific differences in the spatial-temporal regulation and decoding of Metabotropic glutamate receptor1a-stimulated second messenger responses. *J Biol Chem* 2003; **278**: 5419–5426.

the indicator is not stable and therefore should be prepared fresh each time.

5. The concentration of the stock solution and loading solution should be optimized for your cell system. The manufacturer suggests a range of 2–5 mM for the stock solution and 1–10 μM for the loading solution. As well, the loading time and temperature should also be optimized. The manufacturer suggests 20 min to 1 h to load the cells and indicates that the cells may be loaded at 37°C; however ,fluorescent Ca^{2+} indicators may compartmentalize within the cell and lowering the loading temperature to room temperature or below may minimize this.
6. This will allow you to assess the extent of fluorescence bleaching of your sample.

References

1. Niswender, C.M. and Conn, P.J. (2010) Metabotropic Glutamate Receptors: Physiology, Pharmacology, and Disease. *Annu Rev Pharmacol Toxicol* **50**, 295–322.
2. Woehler, A. and Ponimaskin, E.G. (2009) G Protein-mediated Signaling: Same Receptor, Multiple Effectors. *Curr Mol Pharmacol* **2,** 237–48.
3. Steinberg, S.F. (2008) Structural Basis of Protein Kinase C Isoform Function. *Physiol Rev* **88**, 1341–1378.
4. Newton, A.C. (2010) Protein Kinase C: poised to signal. *Am J Physiol Endocrinol Metab* **298**, E395–E402.
5. Policha, A., Daneshtalab, N., Chen, L., Dale, L.B., Altier, C., Khosravani, H., Thomas, W.G., Zamponi, G.W. and Ferguson, S.S. (2006) Role of angiotensin II type 1A receptor phosphorylation, phospholypase D, and extracellular calcium. *J Biol Chem* **281,** 26340–26349.
6. Dale L.B., Babwah A.V., Bhattacharya M., Kelvin D.J. and Ferguson S.S. (2001) Spatial-temporal patterning of metabotropic glutamate receptor-mediated inositol 1,4,5-triphosphate, calcium, and protein kinase C oscillations: protein kinase C-dependent receptor phosphorylation is not required. *J Biol Chem* **276**, 35900–8.
7. Babwah, A.V., Dale L.B. and Ferguson S.S. (2003) Protein kinase C isoform-specific differences in the spatial-temporal regulation and decoding of Metabotropic glutamate receptor1a-stimulated second messenger responses. *J Biol Chem* **278**, 5419–26.
8. Collazos, A., Diouf, B., Guérineau, N.C., Quittau-Prévostel, C., Peter, M., Coudane, F., Hollande, F. and Joubert, D. (2006) A spatiotemporally coordinated cascade of protein kinase C activation controls isoform-selective translocation. *Mol Cell Biol* **26,** 2247–61.
9. Uhlén, P. and Fritz, N. (2010) Biochemistry of calcium oscillations. *Biochem Biophys Res Commun* **396,** 28–32.
10. Paredes, R.M., Etzler, J.C., Watts, L.T., Zheng, W. and Lechleiter, J.D. (2008) Chemical calcium indicators *Methods* **46,** 143–51
11. Wiedenmann, J., Oswald, F. and Nienhaus, G.U. (2009) Fluorescent proteins for live cell imaging: opportunities, limitations and challenges *IUBMB Life* **61,** 1029–42.

Part V

Spatial Control of Signal Transduction

Chapter 16

Using FRET-Based Reporters to Visualize Subcellular Dynamics of Protein Kinase A Activity

Charlene Depry and Jin Zhang

Abstract

The ubiquitous Protein Kinase A (PKA) signaling pathway is responsible for the regulation of numerous processes including gene expression, metabolism, cell growth, and cell proliferation. This method details how to monitor real-time PKA activity dynamics in mammalian cells using fluorescence resonance energy transfer (FRET)-based reporters.

Key words: Protein kinase A, Kinase activity reporter, Fluorescence resonance energy transfer, Live-cell imaging

1. Introduction

1.1. Protein Kinase A

Protein kinase A (PKA), also known as cAMP-dependent protein kinase, is ubiquitously expressed and regulates key cellular functions including gene expression, metabolism, growth, and proliferation (1). The PKA holoenzyme is tetrameric and consists of a regulatory subunit dimer and two catalytic subunits. Upon cAMP binding to the former, the catalytic subunits are released and able to phosphorylate numerous substrate proteins throughout the cell (2). Given that PKA regulates a myriad of different signaling events, it is vital that proper phosphorylation occurs in a specific temporal and spatial pattern. Four regulatory subunit isoforms (RIα, RIβ, RIIα, and RIIβ) and three catalytic subunit isoforms (Cα, Cβ, Cγ), which are differentially expressed in cells and have distinct biological and physical properties, play a role in achieving signaling specificity (1, 3). Additionally, A-kinase anchoring proteins (AKAPs) assemble

Louis M. Luttrell and Stephen S.G. Ferguson (eds.), *Signal Transduction Protocols*, Methods in Molecular Biology, vol. 756, DOI 10.1007/978-1-61779-160-4_16, © Springer Science+Business Media, LLC 2011

signaling complexes containing PKA and its specific substrates and regulators at distinct subcellular locations, thereby facilitating specific phosphorylation and regulation of PKA substrates (4–6). All AKAPs anchor PKA via its regulatory domain, bind other signaling molecules to form multiprotein complexes, and target these signaling complexes to distinct subcellular locations (6).

1.2. FRET-Based Protein Kinase A Activity Reporter

Using fluorescence microscopy, genetically encodable fluorescence resonance energy transfer (FRET)-based A-Kinase Activity Reporters (AKARs) allow for live-cell visualization of endogenous PKA activity dynamics with high spatiotemporal resolution. AKARs consist of a molecular switch sandwiched between a FRET pair, which undergoes a conformational change when phosphorylated by PKA, leading to a change in FRET (Fig. 1). The molecular switch is comprised of a surrogate substrate for PKA and a phospho-amino acid-binding domain (PAABD) (e.g., forkhead-associated domain, FHA). Upon phosphorylation of the surrogate substrate by PKA, the PAABD binds the phosphorylated substrate.

The FRET response generated by kinase activity reporters depends on the fluorescent proteins used. FRET takes place when an excited donor fluorophore (e.g., Cyan Fluorescent Protein, CFP) transfers energy to an acceptor fluorophore (e.g., Yellow Fluorescent Protein, YFP) in close molecular proximity (i.e., <10 nm). For FRET to occur, the donor emission spectrum must overlap with the acceptor excitation spectrum (7, 8). CFP and YFP are commonly used as a FRET pair because the emission spectrum of

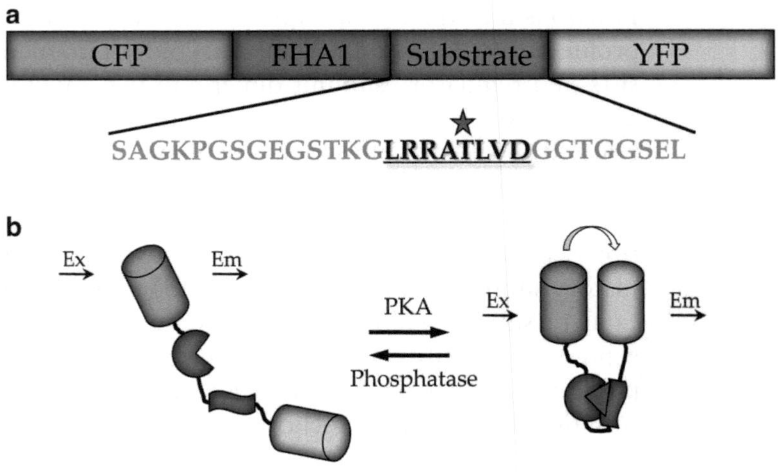

Fig. 1. Design and Mechanism of AKAR. (**a**) AKAR uses CFP as the donor FP and YFP as the acceptor FP with the phospho-amino acid-binding domain, FHA1, and a PKA substrate sandwiched in between. The star indicates the phosphorylation site. (**b**) Once PKA phosphorylates AKAR, FHA1 binds the phosphorylated substrate, inducing a conformational change that brings the FPs into closer proximity. This action is reversed by phosphatases. The *triangle* in the closed conformation represents the phosphorylated threonine.

CFP significantly overlaps with the excitation spectrum of YFP, while their excitation spectra have minimal overlap.

The energy transfer process causes donor emission intensity to be quenched and acceptor emission intensity to increase. As the stoichiometry between the donor and acceptor is fixed in AKAR (i.e., since it is a unimolecular reporter), the changes in emission ratio directly correlate to changes in FRET (7). On the other hand, FRET efficiency can be directly determined by acceptor photobleaching. This technique destroys the acceptor, thus abolishing FRET, which results in dequenched donor fluorescence intensity (7, 8).

For all experiments using AKAR it is important to use a negative control to ensure that the observed changes in FRET are due to phosphorylation of the reporter. The negative control AKAR is mutated at the phosphorylation site and can no longer be phosphorylated.

1.3. Studying Various PKA Activities

1.3.1. Basics of Studying PKA Activity with AKARs

AKAR is most commonly used to visualize changes in PKA activity dynamics over time. Such changes are typically induced by drugs or other perturbations to stimulate or inhibit PKA under various conditions (see Note 1). For example, AKAR was used to demonstrate that chronic insulin treatment induces a delay in β-adrenergic receptor (β-AR) stimulated PKA activity in adipocytes. The study used isoproterenol, a β-AR stimulant, and forskolin, an adenylyl cyclase activator, in the presence and absence of insulin to show that the time delay is specific to β-AR-stimulated PKA (9).

1.3.2. Studying Discrete Domains of PKA Activity with Subcellularly Targeted AKARs

PKA activity dynamics at specific subcellular locations can be monitored using AKAR. In order to study PKA activity at a discrete location, AKAR may be targeted to that location. Subcellularly targeted AKARs are created by adding N- or C-terminal localization motifs, and then verified with colocalization studies using specific subcellular markers. For instance, proper targeting of AKAR to the nucleus can be validated by checking the colocalization of the reporter with a DNA stain. A study demonstrating the ability of AKAR to monitor PKA activity in discrete subcellular locales targeted AKAR: (1) to the plasma membrane via the addition of a C-terminal lipid modification, (2) to the nucleus via a C-terminal nuclear localization signal, (3) to the cytoplasm via a C-terminal nuclear export signal, and (4) to the outer membrane of mitochondria via an N-terminal localization sequence derived from a mitochondria-targeted protein. Subsequently, the mitochondria-targeted AKAR was used to show that PKA activity at mitochondria and global PKA activity are differentially regulated (10).

1.3.3. Studying Spatially Localized PKA Activity

AKAR can be used to study the spatial organization of PKA activity during different cellular processes. For example, using a plasma membrane targeted AKAR it was found that PKA activity is spatially organized in migrating Chinese Hamster Ovary (CHO) cells. In this study, cell migration was initiated by scratching a wound into a monolayer of CHO cells. Increased PKA activity was observed at the leading edge of migrating cells, but not at the trailing edge (11).

The following method details how to maintain Human Embryonic Kidney (HEK) 293T and Chinese Hamster Ovary (CHO) cells, transfect these cells with AKAR plasmid DNA, prepare the cells for imaging, prepare the imaging setup, image live-cell kinase activity, and analyze acquired data to quantify observed changes in FRET.

2. Materials

2.1. Cell Culture and Transfection

1. Cell lines: Human Embryonic Kidney – SV40 T Antigen (HEK293T) and Chinese Hamster Ovary (CHO) (American Type Culture Collection).
2. Dulbecco's phosphate-buffered saline without Mg^{2+} and Ca^{2+} (DPBS)
3. T-25 cm^2 tissue culture flasks.
4. 35 mm glass-bottom imaging dishes (MatTEK).
5. Dulbecco's Modified Eagle's Medium (DMEM) supplemented with 10% fetal bovine serum (FBS) and 1% penicillin–streptomycin (DMEM-HEK 293T) to use with HEK 293T cells. Supplement this medium with 1% nonessential amino acids (DMEM-CHO) to use with CHO cells (see Note 2).
6. Solution of trypsin (0.05%) and ethylenediamine tetraacetic acid (EDTA, 0.53 mM).
7. Fibronectin from human plasma lyophilized powder is dissolved in DPBS to 5 μg/mL and stored in 200 μL aliquots at −20°C.
8. Bovine Serum Albumin lyophilized powder (BSA) is dissolved in DPBS to make a 1% BSA solution and then heat-denatured by boiling for 5 min. Be sure to let this solution cool to room temperature before using.
9. Lipofectamine 2000 (Invitrogen)
10. OPTI-MEM I Reduced Serum Medium (Opti-MEM; Gibco).
11. AKAR and pm-AKAR plasmid DNA.

2.2. Epifluorescence Microscopy

1. All the described experiments are performed on an Axiovert 200 M microscope using a 40×/1.3NA oil-immersion objective lens equipped with an Aqua Stop to prevent liquid from running down the objective (Zeiss). Images are captured using a MicroMAX BFT512 cooled charge-coupled device camera (Roper Scientific).

2. Xenon lamp: XBO 75W (Zeiss).
3. Neutral density filters 0.6 and 0.3 (Chroma Technology).
4. Filter sets for individual channels (All from Chroma Technology):

   FRET – 420DF20 excitation filter, 450DRLP dichroic mirror, 535DF25 emission filter.

   CFP – 420DF20 excitation filter, 450DRLP dichroic mirror, 475DF40 emission filter.

   YFP – 495DF10 excitation filter, 515DRLP dichroic mirror, 535DF25 emission filter.

   YFP photobleaching – 525DF40 excitation filter, 560DRLP dichroic mirror.

   A Lambda 10-2 filter changer (Sutter Instruments) alternates the filters being used.
5. Immersol® 518F fluorescence free immersion oil (Zeiss).
6. METAFLUOR 6.2 software (Molecular Devices).

2.3. Preparing Cells for Imaging

1. Hanks' Balanced Salt Solution for Imaging (HBSS*): 10× Hanks' Balanced Salt Solution (Gibco), 20 mM HEPES, 2.0 g/L d-glucose; adjust pH to 7.4, then filter sterilize using a 0.22 μm filter. Keep a 50 mL aliquot at room temperature in the microscope room and store the rest at 4°C (see Note 3).

2.4. Cell Stimulation and Image Acquisition

1. Forskolin (Fsk) dissolved at 50 mM in dimethyl sulfoxide (DMSO) and stored at −20°C.
2. 200 μL pipet tip to scratch and wound CHO cell monolayer.

2.5. Image and Data Analysis

1. Spreadsheet application (e.g., Microsoft Office Excel).

3. Methods

3.1. Cell Culture and Transfection

1. The cells are maintained in T-25 cm² flasks at 37°C with 5% CO_2 and passaged when they are 85–95% confluent (every 2–3 days) into flasks or 35 mm imaging dishes.
2. HEK293T cells can be plated on uncoated imaging dishes, but CHO cells must be plated on fibronectin-coated dishes. To coat the imaging dishes, add 200 μL of 5 μg/mL fibronectin solution to the glass cover-slip in the imaging dish and incubate at room temperature for 30–45 min. Then aspirate off the fibronectin solution and add 200 μL of 1% BSA solution to the glass cover-slip of the imaging dish for 1 h at room temperature (see Note 4).

3. To passage cells, aspirate the culture medium from the flask and wash cells with 2 mL of DPBS. Add 300 μL of trypsin/EDTA solution and gently rock the dish from side to side to disperse the solution, then let it sit for 2–5 min (see Note 5). Add 4.7 mL of fresh medium (make certain to use cell line appropriate medium) into the flask and mix well. Perform a 1:10 split of CHO cells and a 1:20 split of HEK 293T cells into 35 mm glass-bottom imaging dishes. Both cell lines should reach 60–70% confluence in approximately 24 h (see Note 6). Transfect the cells at this confluence.
4. For each 35 mm dish to be transfected, prepare two separate microcentrifuge tubes. Tube 1 contains 1 μg AKAR or pm-AKAR plasmid DNA and 50 μL Opti-MEM. Tube 2 contains 2 μL Lipofectamine 2000 and 50 μL Opti-MEM. Let these tubes incubate at room temperature for 5 min. Then add Tube 1 drop-wise to Tube 2 and mix well with a pipet (see Note 7). Incubate the transfection solution at room temperature for 20 min.
5. Gently add the AKAR transfection solution to HEK293T cells and the pm-AKAR transfection solution to CHO cells drop-wise and lightly rock the dish from side to side to get even distribution. Incubate at 37°C with 5% CO_2 for 18–24 h.

3.2. Preparing the Epifluorescence Microscope

1. Turn on the lamp, microscope, filter changer, camera, and computer. Load the METAFLUOR 6.2 application and a protocol to acquire a time series of sets of images for the FRET, CFP, and YFP channels (see Note 8). Check that all of the appropriate filters are in place.
2. Set the excitation exposure times for the FRET, CFP, and YFP channels to 500, 500, and 50 ms, respectively. The time lapse between each set of acquisitions is set between 10 and 120 s, typically 30 s.
3. Apply a small drop of immersion oil directly onto the objective. Make sure not to use an excess amount (i.e., 1 drop from the attached applicator should suffice).

3.3. Preparing Cells for Imaging

3.3.1. HEK 293T Cells

1. Aspirate the medium from transfected cells in the imaging dish and wash twice with 1 mL HBSS*.
2. Gently add 1–2 mL HBSS* to the imaging dish, while holding the dish on a slight angle. Slowly return the dish to a level position and place securely on microscope stage (see Note 9).
3. Raise the objective until the drop of the oil comes into full contact with the glass cover-slip and then examine the cells using the eyepiece and focus.

4. In the dark, use the FRET or CFP channel to select cells with good morphology and good AKAR expression, meaning intermediate to high emission intensity and a uniformly distributed fluorescence (see Note 10).

3.3.2. CHO Cells

1. Using a 200 μL pipet tip, scratch the glass cover-slip in the imaging dish with the transfected monolayer of CHO cells (see Note 11). Carefully wash the cells twice with 1 mL of HBSS* to remove cell debris. Gently add 1–2 mL HBSS* to the imaging dish, securely fasten dish on microscope stage, raise the objective, focus on cells near the scratch, and let cells migrate for 15 min before imaging.
2. In the dark, use the FRET or CFP channel to select cells with good morphology and good AKAR expression, meaning intermediate to high emission intensity and plasma membrane-restricted fluorescence distribution.

3.4. Cell Stimulation and Image Acquisition

3.4.1. Chemically Induced PKA Activity in HEK 293T Cells

1. Select several regions of interest to follow during the course of the experiment (see Note 12). A background region consisting of an untransfected cell must also be selected to correct for cell autofluorescence and other background fluorescence.
2. Acquire 3–5 min of data (all three channels) from the unstimulated cells to establish a baseline for the experiment. Pipet ~300 μL of HBSS* out of the imaging dish, mix with a 1–2 μL aliquot of 50 mM Fsk in a 1.5 mL tube, then gently pipet this solution back into the imaging dish. The final concentration of Fsk should be 50 μM. Be sure to note the time of the drug addition (see Note 13). The yellow to cyan emission ratio (FRET channel emission/CFP channel emission) should rapidly increase, indicating a change in PKA activity.
3. At the end of the experiment remove all neutral density filters, use the YFP photobleaching excitation filter, and then excite for 5 min. This should sufficiently photobleach YFP, but it is important to verify this by acquiring the YFP channel. The acquired data can be used to calculate absolute FRET efficiency using the following formula:

$$\text{FRET Efficiency} = 1 - \frac{\text{CFP Emission (before YFP photobleaching)}}{\text{CFP Emission (after acceptor photobleaching)}}$$

3.4.2. Localized PKA Activity in Migrating CHO

1. Select several regions of interest that are specific to the leading edge and trailing edge of a migrating cell to follow during the course of the experiment. Regions within a nonmigrating

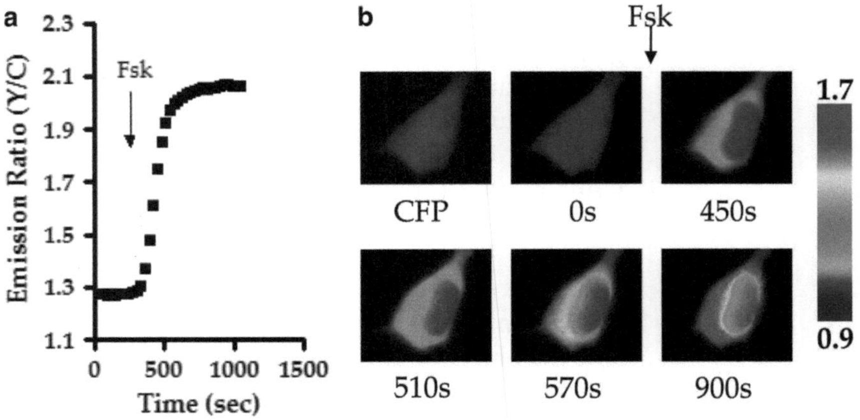

Fig. 2. AKAR response in live cells. HEK 293 cells were transfected with AKAR and imaged 24 h later. The cells were stimulated with 50 μM Fsk, an adenylyl cyclase activator. (**a**) *Yellow to cyan* emission ratios plotted against time represent changes in PKA activity over time. (**b**) Pseudocolor images showing AKAR response.

cell should also be selected to compare PKA activity between these two types. A background region must also be selected to correct for cell autofluorescence and other background fluorescence.

3.5. Image and Data Analysis

1. Use METAFLUOR 6.2 to generate pseudo-colored images for each acquisition where a pseudocolor is used to indicate the yellow to cyan emission ratio (FRET channel emission/CFP channel emission) (see Note 14). These images can be strung together in a movie clip or a selection of them can be used to visually represent the observed real-time changes. An example is shown in Fig. 2.
2. Using a spreadsheet application, calculate emission ratios from the logged data using the following formula for each time point:

$$\text{Yellow to Cyan Emission Ratio} = \frac{\text{FRET channel Emission Intensity} - \text{FRET channel Emission Intensity of Background}}{\text{CFP channel Emission Intensity} - \text{CFP channel Emission Intensity of Background}}$$

3. Plot the ratio time course (ratios vs. time).

4. Notes

1. It is important to verify effective drug concentrations and conditions that could affect their function. For example, when stimulating the PKA pathway using a G-Protein-Coupled Receptor agonist, first make certain that the specific

receptor is expressed in the cell line being used. Western blots using antibodies against the receptor of choice can determine its presence. Additionally, western blots employing an anti-phospho-PKA substrate antibody (Cell Signaling) can be used to determine effective drug concentrations.

2. All solutions should be made under sterile conditions in a tissue culture hood and cell culture media should be warmed to 37°C before using with cells.
3. All solutions should be prepared with water that has an 18.2 MΩ cm resistivity unless otherwise noted.
4. It is unlikely that fibronectin coating of the imaging dish will be 100% effective, meaning uncoated spots will be present for other secreted adhesion proteins to bind to. In order to control for these situations 1% BSA is used to block the potential non-fibronectin adhesion sites.
5. Be sure the cells are fully detached before continuing. Gently swaying the flask side to side should help.
6. This protocol can be adapted for other cell lines by following recommended cell culture and transfection guidelines for the cell line of choice. Additionally, cell growth rates may vary, so it is important to verify doubling times for each cell line used.
7. Be sure to mix gently. Do not vortex the solution.
8. The FRET channel logs YFP emission intensity when CFP is excited, the CFP channel logs CFP emission intensity upon direct CFP excitation, and the YFP channel logs YFP emission intensity upon direct YFP excitation. The YFP channel serves to control for YFP photobleaching and is not used to determine emission ratios.
9. Securing the dish to the stage is important as it minimizes slight movement of the dish that may occur while imaging.
10. Key criteria for proper cell selection: first, cell morphology is important to verify before starting an experiment as healthy cells are required for successful imaging experiments. For instance, when imaging HEK 293 cells, select cells that are spread out and lying flat rather than balled-up and rounded, as the latter could indicate unhealthy cells. Second, the fluorescence intensity level of AKAR should be closely monitored, though a recommended range cannot be given as the intensity values will vary with microscope setups. However, cells with a moderate- to high-intensity level are typically used. Cells with very dim fluorescence intensities will have a low signal-to-noise ratio, thus changes in FRET will be difficult to visualize, whereas cells with very high fluorescence intensities may have perturbed endogenous signaling pathways because of excessive expression of AKAR. Third, if using a targeted

AKAR to look at specific effects at a subcellular locale, verification of proper AKAR distribution within the cell is critical. For example, diffusible AKAR should fluoresce uniformly throughout cells, whereas fluorescence of pm-AKAR should be limited to the plasma membrane of cells.

11. It is important to verify that the CHO cells have formed a confluent monolayer before inflicting the wound.
12. The selected regions of interest will need to remain in the same cellular region throughout the time series, and thus may be adjusted should the cells move. Alternatively, cell tracking software (e.g., Imaris Track) can be used to overcome this problem.
13. All different types of PKA agonists and antagonists can be used, either alone or together in a single imaging dish.
14. If the software being used does not have this feature, an image processing application (e.g., ImageJ) can be used to create pseudo-colored ratiometric images using the raw emission intensity images from the individual channels.

References

1. Tasken, K., and Aandahl, E. M. (2004) Localized Effects of cAMP Mediated by Distinct Routes of Protein Kinase A. *Physiol Rev* **84**, 137–167.
2. Taylor, S. S., Yang, J., Wu, J., Haste, N. M., Radzio-Andzelm, E., and Anand, G. (2004) PKA: A Portrait of Protein Kinase Dynamics. *Biochim Biophys Acta* **1697**, 259–269.
3. Skalhegg, B. S., and Tasken, K. (2000) Specificity in the cAMP/PKA Signaling Pathway. Differential Expression,Regulation, and Subcellular Localization of Subunits of PKA. *Front Biosci* **5**, D678–93.
4. Wong, W., and Scott, J. D. (2004) AKAP Signalling Complexes: Focal Points in Space and Time. *Nat Rev Mol Cell Biol* **5**, 959–970.
5. Smith, F. D., and Scott, J. D. (2002) Signaling Complexes: Junctions on the Intracellular Information Super Highway. *Curr Biol* **12**, R32–40.
6. Beene, D. L., and Scott, J. D. (2007) A-Kinase Anchoring Proteins Take Shape. *Curr Opin Cell Biol* **19**, 192–198.
7. Ananthanarayanan, B., Ni, Q., and Zhang, J. (2008) Chapter 2: Molecular Sensors Based on Fluorescence Resonance Energy Transfer to Visualize Cellular Dynamics. *Methods Cell Biol* **89**, 37–57.
8. Miyawaki, A., and Tsien, R. Y. (2000) Monitoring Protein Conformations and Interactions by Fluorescence Resonance Energy Transfer between Mutants of Green Fluorescent Protein. *Methods Enzymol* **327**, 472–500.
9. Zhang, J., Hupfeld, C. J., Taylor, S. S., Olefsky, J. M., and Tsien, R. Y. (2005) Insulin Disrupts Beta-Adrenergic Signalling to Protein Kinase A in Adipocytes. *Nature* **437**, 569–573.
10. Allen, M. D., and Zhang, J. (2006) Subcellular Dynamics of Protein Kinase A Activity Visualized by FRET-Based Reporters. *Biochem Biophys Res Commun* **348**, 716–721.
11. Lim, C. J., Kain, K. H., Tkachenko, E., Goldfinger, L. E., Gutierrez, E., Allen, M. D., Groisman, A., Zhang, J., and Ginsberg, M. H. (2008) Integrin-Mediated Protein Kinase A Activation at the Leading Edge of Migrating Cells. *Mol Biol Cell* **19**, 4930–4941.

Chapter 17

Genetically Encoded Fluorescent Reporters to Visualize Protein Kinase C Activation in Live Cells

Lisa L. Gallegos and Alexandra C. Newton

Abstract

Protein kinase C (PKC) signaling drives many important cellular processes and its dysregulation results in pathophysiologies such as cancer (Gokmen-Polar et al., Cancer Res 61:1375–1381, 2001). Because PKC is activated acutely and allosterically, it is difficult to monitor the cellular activity of endogenous PKC by conventional methodologies (Newton, Methods Enzymol 345:499–506, 2002). Rather, PKC signaling is best studied in situ using biosensors such as FRET-based reporters. We have generated several FRET-based reporters for studying PKC signaling in real time in live cells (Violin and Newton, IUBMB Life 55:653–660, 2003). Using these reporters, we have demonstrated phase-locked oscillations in Ca^{2+} release and membrane-localized endogenous PKC activity in response to histamine (Violin et al., J Cell Biol 161:899–909, 2003), as well as distinct signatures of endogenous PKC signaling at specific organelles in response to uridine-5 -triphosphate (UTP; Gallegos et al., J Biol Chem 281:30947–30956, 2006). Here we describe methods to image cells expressing the reporters and elaborate on data analyses, control experiments, and variations for imaging the activity of expressed PKC.

Key words: Protein kinase C, Diacylglycerol, Förster resonance energy transfer, Targeted reporter, Live-cell imaging

1. Introduction

Signal transduction relies greatly upon the regulated enzymatic activity of protein kinases (1). The protein kinase C (PKC) family is a group of ten mammalian isozymes sharing a highly conserved kinase core that is "matured" by a series of priming phosphorylation events promoted by the upstream kinases PDK-1 and mTORC2 (reviewed in (2)). PKC isozymes contain membrane-targeting domains, C1 and C2, which sense levels of second messengers, diacylglycerol (DAG), and Ca^{2+}, respectively, that are

Louis M. Luttrell and Stephen S.G. Ferguson (eds.), *Signal Transduction Protocols*, Methods in Molecular Biology, vol. 756, DOI 10.1007/978-1-61779-160-4_17,

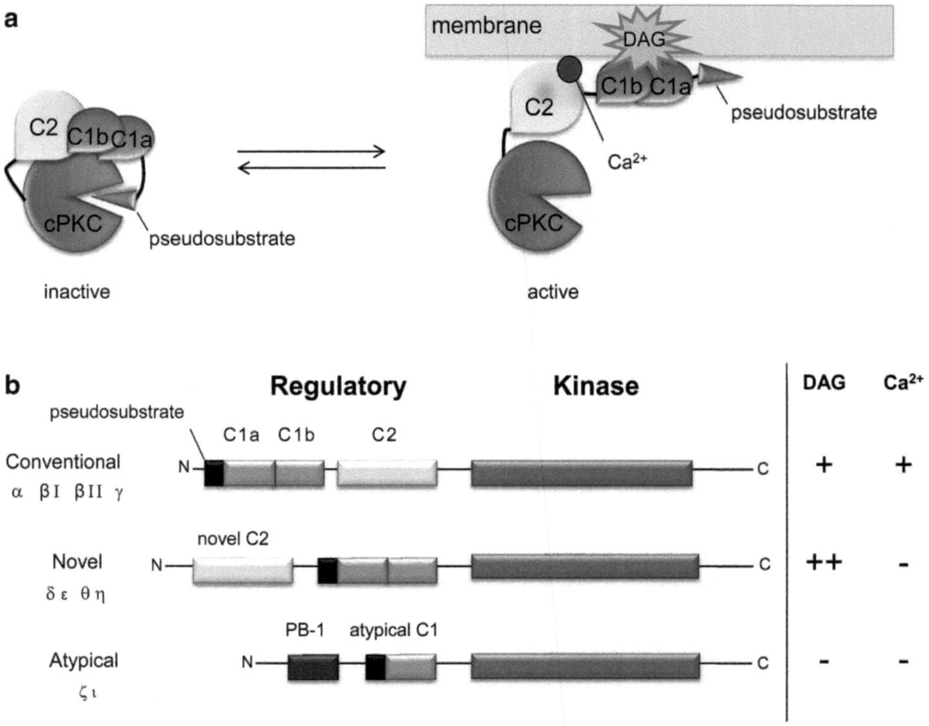

Fig. 1. Regulation of PKC isoforms. (**a**) Activation of PKC. In unstimulated cells, PKC (cPKC shown here) exists in a fully phosphorylated, closed conformation (*left species*) with an N-terminal autoinhibitory pseudosubstrate peptide resting in the substrate-binding cavity of the kinase core; this prevents phosphorylation of cellular substrates. Upon receptor stimulation and subsequent DAG and Ca^{2+} elevation, PKC translocates to cellular membranes (*right species*) using a Ca^{2+}-sensitive C2 domain and DAG-sensitive C1 domains. Membrane binding relieves autoinhibition and permits phosphorylation of cellular substrates. (**b**) Domain structure of the PKC family. Conventional PKC isoforms contain a Ca^{2+}-sensitive C2 domain and DAG-sensitive C1 domains. Novel PKC isoforms contain a Ca^{2+}-insensitive C2 domain and DAG-sensitive C1 domains; however, these C1 domains bind membranes with two orders-of-magnitude higher affinity in the presence of DAG compared to those present in cPKC isozymes. Atypical PKC isoforms contain a DAG-insensitive C1 domain and a PB-1 domain; both of these serve as protein–protein interaction modules.

produced upon receptor activation (Fig. 1). In response to changing levels of second messengers, PKC isozymes are allosterically and locally activated to phosphorylate downstream substrates (Fig. 1a). Conventional PKC isoforms (cPKC: α, βI/II, γ) are activated by coincident elevation in intracellular Ca^{2+} and membrane-bound DAG; Novel PKC isoforms (nPKC: δ, ε, θ, η) respond robustly to DAG alone; Atypical PKC isoforms (aPKC: ζ, ι) respond to neither second messenger, although they are spatially localized by protein:protein interactions (3) and appear to be regulated at the level of priming phosphorylation events (for extensive review of PKC signaling, see (4, 5)). Phosphorylation of PKC substrates is tightly regulated by both spatio-temporally localized PKC activation and the opposing actions of cellular phosphatases (6, 7).

Traditionally, the method of choice for demonstrating protein kinase activation has been monitoring the phosphorylation of residues that prime or activate kinases using Western blotting with phospho-specific antibodies that recognize these sites (as in (8)). However, conventional and novel PKC isoforms are most often constitutively phosphorylated as a maturation step, rendering analysis of phosphorylation sites an ineffective measure of prior cell activation. Rather, PKC is acutely activated by allosteric mechanisms resulting from binding membrane-embedded diacylglycerol (reviewed in (9)). As a result, the classic method for examining PKC activity has been to probe for the presence of PKC in membrane fractions of lysed cells, or by staining fixed cells with PKC antibodies as a measure of membrane translocation (as in (10–12)). But, these methods can fall short of achieving precise spatial and temporal resolution of signaling events because they require fractionation or staining of cells that have been lysed or fixed at defined time points. The advent of green fluorescent protein (GFP; recently reviewed in (13)) presented PKC researchers with the ability to monitor membrane translocation of fluorescently tagged PKC in live cells as a readout of activation, generating numerous elegant examples of PKC translocation over time in response to the addition of natural receptor agonists (14, 15) or the addition of ligands which activate PKC by directly engaging the C1 domains of cPKC isozymes and nPKC isozymes, such as tumor-promoting phorbol esters (16). But, these experiments rely on overexpressed, tagged PKC. Thus, these studies are blind to the activity of endogenous PKC and, importantly, the balance between PKC activity and the activity of cellular phosphatases. Therefore, live-cell imaging using biosensors or reporters to read out endogenous activity is ideal for examining PKC signaling.

Our lab has developed and characterized förster resonance energy transfer (FRET)-based reporters for studying PKC signaling (Fig. 2) which rely on changes in FRET from donor cyan fluorescent protein (CFP; ECFP variant) to acceptor yellow fluorescent protein (YFP; Citrine variant) to reflect changes in signaling activity. C Kinase Activity Reporter, CKAR, is a tool to measure directly the activity of PKC (6), and is similar in structure to the prototypical kinase reporter, A Kinase Activity Reporter, AKAR (17). CKAR consists of CFP at the N terminus, followed by an FHA2 phosphopeptide-binding domain linked to a substrate sequence that is specific for PKC, followed by YFP at the C terminus (Fig. 2a). CKAR is phosphorylated by all conventional and novel protein kinase C isozymes tested, although with varying efficiency (23). Importantly, CKAR is not phosphorylated by other kinases predicted to have similar substrate specificity, such as PKA and CAMKII, in vitro (6).

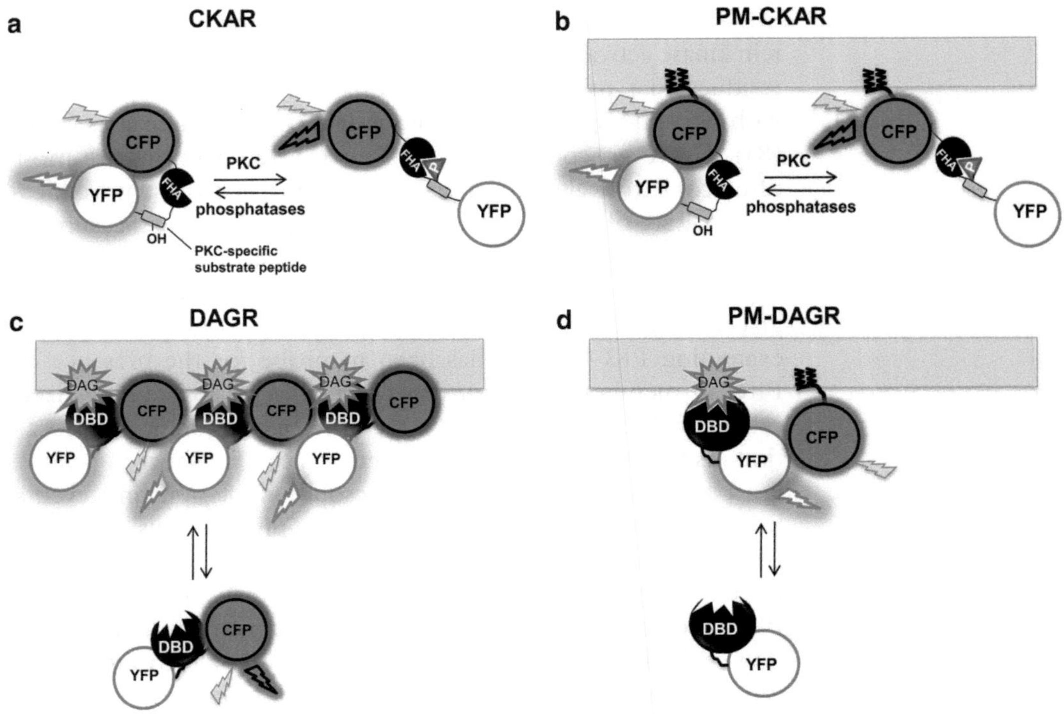

Fig. 2. FRET-based reporters for PKC signaling pathways. (**a**) CKAR, C kinase activity reporter. CFP is linked to YFP by a substrate peptide specific for PKC and an FHA2 phosphopeptide-binding domain. Basal intramolecular FRET from CFP to YFP is reduced upon phosphorylation of the reporter by PKC, but can be restored upon dephosphorylation of the reporter by cellular phosphatases. (**b**) PM-CKAR, plasma membrane-targeted CKAR. The N-terminal 7 residues of Lyn kinase encode sequences for myristoylation and palmitoylation; when these residues are fused to the N terminus of CKAR, the reporter expressed in cells is enriched at the cytoplasmic side of the plasma membrane. (**c**) DAGR, diacylglycerol reporter. CFP and YFP flank a diacylglycerol-binding domain (DBD). Intermolecular FRET between reporters increases as they come in close proximity upon translocation to cellular membranes in response to DAG production. FRET decreases as DAG is metabolized and the reporters re-localize to the cytosol. (**d**) PM-DAGR, plasma membrane-targeted DAGR. PM-DAGR consists of two separate constructs transfected together encoding YFP-tagged DBD and CFP targeted to the plasma membrane using the N-terminal 7 residues of Lyn kinase as described above. Intermolecular FRET from CFP to YFP increases as the reporter translocates to membranes in response to stimulated DAG production, and decreases upon DAG turnover and re-localization of YFP-DBD to the cytosol. Note that the DBD in (**b**) and in (**c**) are different (see text).

Because CKAR is genetically encoded, we have been able to fuse short sequences to the N or C terminus of CKAR to target or enrich the reporter at specific regions of cells to monitor localized signaling (6, 7) (plasma membrane-targeted reporter shown in Fig. 2b). We also generated FRET-based tools to measure the production of the upstream second messenger, DAG. Diacylglycerol reporter, DAGR, consists of CFP and YFP flanking a diacylglycerol-binding domains (DBD), in this case, the C1 domain of PKCβ ((6); Fig. 2c). In addition to this original DAGR, we also generated targeted versions of DAGR. These consist of separate constructs encoding CFP targeted either to the plasma membrane or to the Golgi co-transfected with a

YFP-tagged DBD, in this case, a mutated form of the C1b domain of PKCβ ((7); Fig. 2d). These tools have allowed us to monitor with great precision when and where PKC activity is elevated in live cells in response to agonist-mediated signaling, and to correlate this activity to regional differences in second messenger production and phosphatase activity.

2. Materials

2.1. Cell Culture

1. Dulbecco's Modified Eagle's Medium (DMEM) supplemented with 10% fetal bovine serum (FBS).
2. 1% (w/v) trypsin and 1 mM ethylenediamine tetraacetic acid (EDTA) solution (GIBCO/BRL).
3. COS7 or other adherent cell line (American Type Culture Collection).
4. 10 cm tissue culture dishes.
5. Uncoated 35 mm glass-bottom imaging dishes (MatTek).
6. FuGENE 6 transfection reagent (Roche Pharmaceuticals).
7. Plasmid DNA encoding CKAR, CKAR T/A (Ala at phosphoacceptor site), DAGR or targeted versions of these reporters (Addgene).

2.2. Cell Stimulation and Data Acquisition

1. Hanks' balanced salt solution (HBSS) supplemented with 1 mM $CaCl_2$ just prior to imaging.
2. Phorbol myristate acetate (PMA) or phorbol-12,13-dibutyrate (PdBu) dissolved in dimethyl sulfoxide (DMSO) to a stock concentration of 200 μM (dilute 1:1,000 for working concentration) (see Note 1).
3. Calyculin A at a stock concentration of 50 μM in DMSO (dilute 1:1,000 for working concentration) in DMSO.
4. Gö6976 in solution at a stock concentration of 500 μg/ml (add 0.77 μL to 2 ml saline for working concentration of 500 nM).
5. Gö6983 at a stock concentration of 250 μM in DMSO (dilute 1:1,000 for working concentration).
6. Test agonist, such as UTP at a stock concentration of 100 mM in distilled H_2O to be diluted 1:1,000 for working concentration, or other receptor agonist.
7. Zeiss Axiovert microscope (Carl Zeiss Microimaging, Inc.) or similar wide-field inverted epifluorescence microscope equipped with appropriate optical filters and cold CCD camera (see Notes 2 and 3).

3. Methods

It is important to note that the maximal FRET ratio change possible with CKAR is 20%, and typical agonist-stimulated responses reach 5%. Because of this relatively low signal-to-noise ratio, it is critical to average data from multiple cells across three or more dishes and calculate error before attempting to compare PKC responses. For initial characterization, the full range of the CKAR response should be determined (as in (7); Fig. 3). The response of test agonists will fall within this range, and the detection limit will be apparent. Other experiments that must be done to ensure confidence in positive CKAR responses are to reverse or prevent the PKC response using PKC inhibitors, such as Gö6983 and Gö6976. One should also image the response of a mutant form of the reporter containing an Ala residue at the phosphoacceptor site (CKAR T/A), which should not change upon addition of agonist.

Expression of these reporters in cells has the inherent possibility of altering signaling by competing with endogenous signaling molecules for access to PKC (CKAR) or DAG (DAGR). Therefore, when carrying out these experiments, it is important to select cells that express low levels of CKAR and DAGR. In fact, we have calibrated the YFP fluorescence intensity with reporter concentration using purified protein (6), and consistently image cells expressing no more than approximately 1 μM CKAR. This is well below concentrations of an abundant cellular substrate, MARCKS, which can reach concentrations of 20 μM (18). In addition, CKAR expression levels ranging from approximately 0.5–5 μM report consistent FRET ratio changes in response to agonist, supporting that this level of expression does not appreciably interfere with signaling (6). Expression of DAGR also has the potential to alter signaling by acting as a "sponge" to soak up agonist-stimulated DAG, preventing endogenous downstream effectors from responding. Indeed, we have experimentally verified this effect by expressing high levels of C1 domain and monitoring effects on PKC activity (7). Thus, for both CKAR and DAGR, it is critical to evaluate signaling only in cells expressing low reporter levels.

While the strength of this imaging approach is the ability to measure signaling carried out by endogenous signaling molecules in real time, it is also possible to measure the activity of expressed PKC isoforms. For example, one might want to compare the activity of a mutant PKC to wild-type PKC. The variation described in Subheading 3.5 describes how to explore these differences in activity using CKAR. The challenges with this method are (1) dealing with unequal expression of PKC amongst individual cells and (2) separating the activity contributed by overexpressed PKC from that contributed by the endogenous PKC.

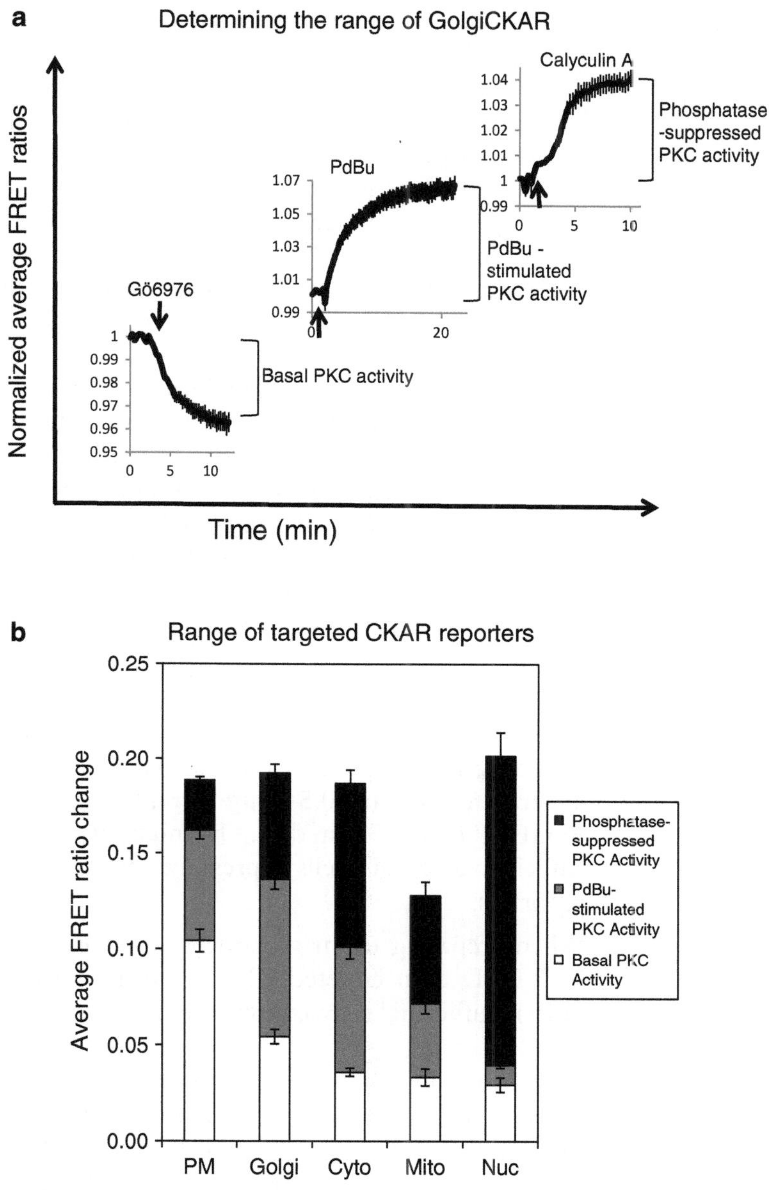

Fig. 3. Experimentally verifying the range of targeted CKARs. (**a**) The range of a kinase activity reporter is determined in three experiments. "Basal PKC activity" is the magnitude of the decrease in FRET ratio upon adding PKC inhibitor after acquiring baseline FRET ratios. "PdBu-stimulated activity" is the FRET ratio increase from baseline upon addition of PdBu. "Phosphatase-suppressed PKC activity" is the FRET ratio increase from the plateau of the maximal response to PdBu upon addition of Calyculin A. FRET ratio changes from Golgi-CKAR depicted here are normalized to 1, and represent the average responses of multiple COS7 cells across three or more dishes; maximal FRET ratio changes are determined by fitting the data to monoexponential curves. (**b**) Experimentally determined ranges for targeted CKAR responses in COS7 cells. Note that Mito-CKAR has a decreased range compared to all other reporters, which all exhibit an approximately 20% maximal FRET ratio change; this likely results from slight proteolysis of Mito-CKAR in COS7 cells. Data shown in (**b**) were originally published in *The Journal of Biological Chemistry*. Gallegos, L. L., Kunkel, M. T., and Newton, A. C. Targeting protein kinase C activity reporter to discrete intracellular regions reveals spatiotemporal differences in agonist-dependent signaling. *J Biol Chem* 2006; **281**, 30947–56. © the American Society for Biochemistry and Molecular Biology.

The first challenge is easily overcome by tagging PKC with a spectrally compatible fluorophore (i.e., mCherry (19)) whose fluorescence intensity can be monitored as a measure of PKC expression level without interfering with the FRET reporter readout. The second challenge can be overcome by generating a "dose–response" curve for the endogenous PKC vs. the expressed PKC using the agonist of choice. This will allow the determination of an agonist concentration that is sufficient to activate the overexpressed and more abundant PKC independently from the endogenous PKC.

3.1. Cell Culture

1. COS7 and other adherent cell lines are passaged just prior to confluence using trypsin/EDTA to dissociate cells. The cells are diluted in fresh medium in 10 cm dishes for maintenance of the culture and split into 35 mm glass-bottom dishes for experimental setup. Cells are plated sparsely for imaging; for example, a 1:40 split of a 70–80% confluent 10 cm dish of COS7 cells provides conditions that are optimal for efficient transfection and imaging of individual cells (see Note 4).
2. Once cells have adhered to the glass (either later on the day of plating or on the following day), cells are transfected according to manufacturers' protocols with plasmid DNA encoding the reporter of interest. For example, COS7 cells are transfected with 0.5–1 μg of reporter DNA using 3 μL FuGENE 6 per 35 mm dish. Overnight expression is typically sufficient to obtain cells expressing appropriate levels of the reporters.
3. When preparing to image targeted DAGR constructs, more YFP-DBD than targeted CFP is transfected (typically a 3:1 ratio is sufficient) to maximize the range of responses.

3.2. Imaging CKAR/DAGR

1. Cells expressing the reporter of interest are rinsed once with HBSS/$CaCl_2$ and imaged in 2 ml of this solution (see Note 5). Once cells expressing an optimal level of the reporter are selected, a series of CFP, FRET, and YFP images are acquired.
2. Background signal is subtracted from areas of the image lacking cells or from areas with untransfected cells.
3. If using Metafluor software, one region per cell is selected for monitoring FRET ratios in real time (see Notes 6 and 7).
4. Acquisition of CFP, FRET, and YFP images over fixed time intervals, typically 10–15 s, is carried out through the entire experiment. Integration times are 200 ms for CFP and FRET, and 50 ms for YFP. If using Metafluor software, the average FRET ratios (CFP/FRET for CKAR or FRET/CFP for DAGR) for the selected cellular regions are plotted as a readout of the signaling response. The YFP intensity is also graphed to monitor for photobleaching.

5. Baseline FRET ratio readings are acquired for 5–15 min. Agonist or inhibitor is not added until the baseline is either flat or has a consistent linear slope (see Note 8).
6. To add agonist or inhibitor, approximately 0.5–1 ml of HBSS is withdrawn from the dish and used to pre-dilute the drug taken from the stock. This pre-diluted drug is added carefully in between image acquisitions, and the time of addition is noted (see Notes 9 and 10).

3.3. CKAR Data Analysis

An example of the stepwise procedure for CKAR data analysis is shown in Fig. 4a.

1. The average FRET ratio for each region selected is plotted over time using a graphing program such as Microsoft Excel.
2. For each average FRET ratio, the slope of the baseline approximately 5 min before the addition time point is determined. Ideally, this slope will be zero, but occasionally the baseline FRET ratio exhibits drift. The drift often has a steeper slope initially and then levels off to a fairly linear slope that is consistent throughout the course of the experiment. One can mathematically determine the slope by graphing the linear

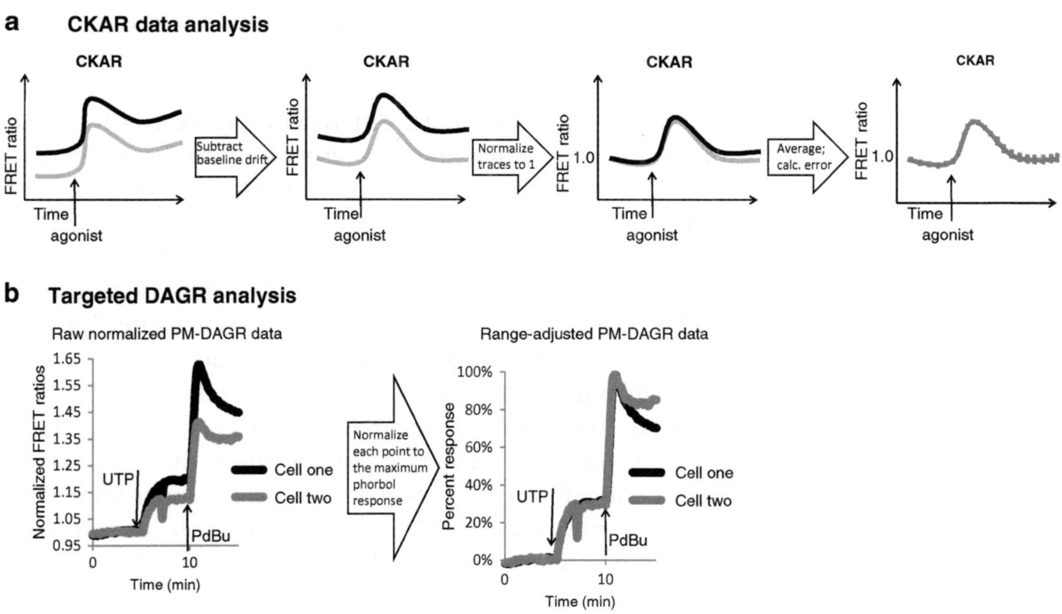

Fig. 4. Data analysis. (**a**) Flow chart depicting typical analysis steps that yield average FRET ratios suitable for comparison across different scenarios. The *far left panel* is an illustration of raw FRET ratio values that drift over time. The *middle left panel* depicts drift-corrected FRET ratios. The *middle right panel* shows normalized FRET ratios, and the *far right panel* shows the normalized average FRET ratios with calculated error. This analysis step is described in detail in Subheading 3.3. (**b**) Normalization step for targeted DAGR analysis. The *left plot* contains raw FRET ratios from two cells transfected with PM-DAGR and stimulated with UTP (100 μM) followed by PdBu (200 nM). The *right plot* contains these same data normalized for the ranges of the individual cells. This analysis step is described in detail in Subheading 3.4.

portion of the baseline FRET ratio drift. Then, the linear drift (i.e., slope x time) is simply subtracted systematically from each data point in the response curve noted (see Note 11).

3. Responses (corrected for drift, if necessary) are all normalized to 1 by dividing by the initial FRET ratio.
4. The normalized FRET ratio responses are averaged across all cells imaged, referencing data from different dishes to the agonist or inhibitor addition time point. If agonist or inhibitor was added at different time points for different dishes, baseline FRET ratios can be deleted from the beginning of the experiment. The normalized average FRET ratio from all cells imaged is plotted against time in a new graph.
5. The standard error of the mean is calculated and plotted for each data point in the average FRET ratio.

3.4. DAGR Data Analysis

1. Original DAGR is imaged and analyzed in a similar manner as described above for CKAR.
2. For targeted versions of DAGR, variability will exist in the range of responses for each cell (Fig. 4b; *left panel*). This results from the variations in relative expression levels of CFP and YFP. It is possible to work around these variations by selecting cells with similar expression levels (as was done for the data in Fig. 5a). However, because phorbol esters such as PMA and PdBu cause an overriding maximal translocation of DAG-binding C1 domains to cellular membranes, it is also possible to correct for these differences. Figure 4b shows two different cells on the same dish responding to the addition of a natural agonist (UTP) and the subsequent stimulus of PdBu. Note that in the left panel, the cell with the better response to UTP also responds better to PdBu (*Cell one*). Because both cells are responding to the same concentration of PdBu, this indicates that "Cell one" has a greater capacity to respond because of a more favorable ratio of CFP to YFP. Normalizing for the range of each cell by taking the baseline to be "0% response" and the maximal PdBu-stimulated FRET ratio change to be "100% response" causes the responses of individual cells to UTP to become nearly super-imposable (Fig. 4b; *right panel*). This correction controls for variations in the range of responses resulting from cell-to-cell variability in CFP relative to YFP expression levels (20), providing a meaningful way to compare differences in the magnitudes of the responses to different ligands using membrane-targeted DAGR.

3.5. Variation: Using CKAR to Measure the Activity of Expressed PKC

A variation of the procedures described above can be employed to measure the activity of exogenously expressed PKC isozymes or mutants (Fig. 5b).

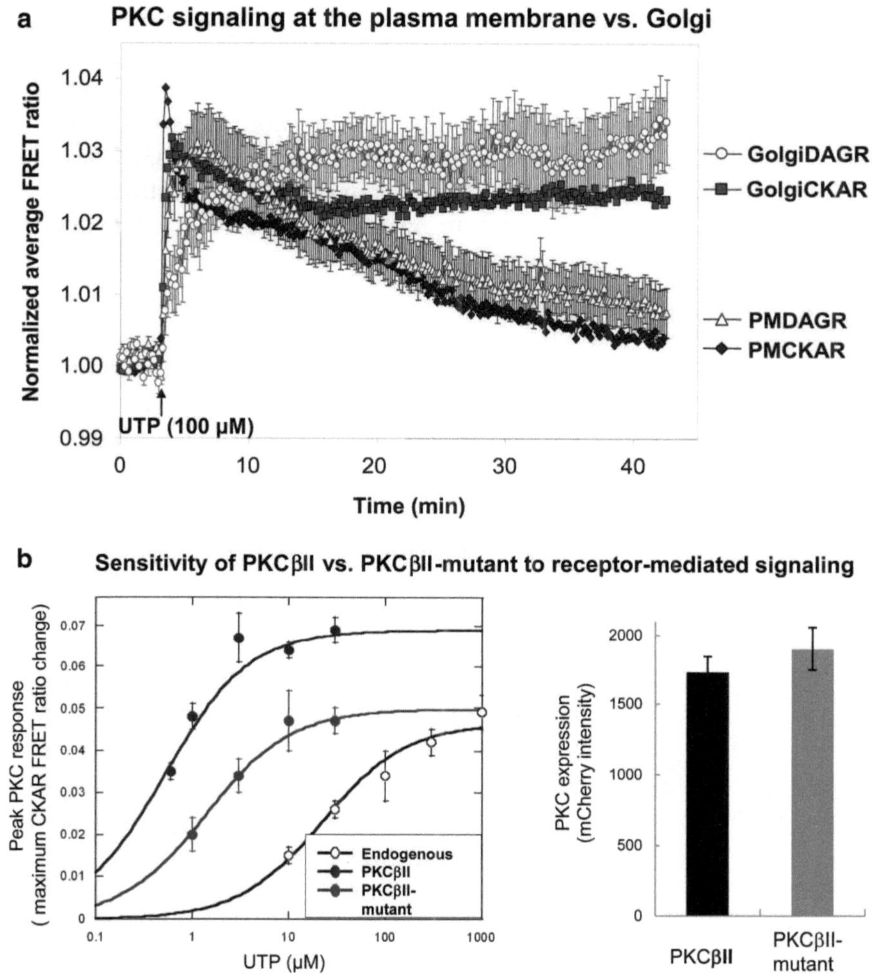

Fig. 5. Examples of data generated using FRET-based reporters for PKC signaling. (**a**) Localized PKC signaling in response to UTP in COS7 cells. COS7 cells were transfected with PMCKAR (*closed black diamonds*), PM-DAGR (*open black triangles*), Golgi-CKAR (*gray squares*) or Golgi-DAGR (*open gray circles*), and stimulated with UTP (100 μM); the average FRET ratio change was plotted over time. Here, targeted DAGR responses were not normalized for range, but cells expressing similar levels of CFP and YFP were imaged. Error bars represent SEM of data obtained from over 10 cells across three dishes per response. Note that, in this experiment, PKC activity and DAG production persist longer at the Golgi compared to the plasma membrane. Data shown in (**a**) were originally published in *The Journal of Biological Chemistry.* Gallegos, L. L., Kunkel, M. T., and Newton, A. C. Targeting protein kinase C activity reporter to discrete intracellular regions reveals spatiotemporal differences in agonist-dependent signaling. *J Biol Chem* 2006; **281**, 30947–56. © the American Society for Biochemistry and Molecular Biology. (**b**) Using CKAR to test the effects of mutation on exogenous PKC. COS7 cells were transfected with CKAR alone (to monitor endogenous PKC activity, *open circles*), CKAR and mCherry-WT PKCβII (*closed black circles*), or CKAR and mCherry PKCβII-mutant (*closed gray circles*). The *left panel* shows the maximal FRET ratio change in response to increasing concentrations of UTP. Each data point represents the maximum response determined by curve-fitting the average FRET ratio change within the first 2 min as described in Subheading 3.5. Note that, in this case, the mutant PKC is desensitized to receptor-mediated signaling compared to the wild-type PKC. The *right panel* is a graph of mCherry intensity values, demonstrating equivalent expression of mCherry-tagged PKC constructs.

1. The maximum response to a low concentration of agonist in cells transfected with CKAR alone is determined. In practice, the maximum response is calculated by curve-fitting the average response from nine or more cells over three or more dishes. The maximum PKC response to GPCR agonists typically occurs within 2 min of agonist addition. Determination of the maximum response is carried out for increasing concentrations of agonist.
2. The maximum PKC responses are plotted on the *y*-axis and concentration of agonist on the *x*-axis using a semi-log scale. This generates a "dose–response" curve for endogenous PKC.
3. The "dose–response" curve for expressed PKC is similarly determined, selecting cells expressing similar levels of tagged PKC. Because the exogenous PKC is expressed more highly than the endogenous, or, in some cases, perhaps because particular PKC isoforms are better at phosphorylating CKAR than the endogenous isoforms present, the expressed PKC will respond to lower concentrations of agonist, and the response curve will shift to the left. The difference between the response curves for the expressed wild-type PKC and the endogenous isoforms provides a "window" for monitoring effects of mutations on PKC activity.
4. The "dose–response" curve for the tagged mutant PKC is determined next, selecting cells that express equivalent levels of mutant PKC compared to the wild type. Mutations that decrease the ability of PKC to be activated will shift the curve to the right (as in Fig. 5b), while mutations that sensitize PKC to agonist will shift the curve leftward. These dose–response curves may be fitted with Michaelis–Menten kinetics, and can provide a quantitative measure of differences in PKC activity amongst mutants by revealing the concentration of agonist yielding half-maximal response.

4. Notes

1. Phorbol esters, Gö6976, Gö6983, and Calyculin A are toxic substances that should be handled with care and disposed of according to institutional guidelines. Stock solutions of these compounds are stored at −20°C in small aliquots; repeated thawing and freezing is not recommended. UTP is stored at −20°C in one large aliquot, and thawed on ice for use.
2. The minimal setup for carrying out these experiments consists of a wide-field inverted epifluorescence microscope equipped with an automated filter wheel, appropriate filters, and a CCD camera controlled by image acquisition software.

Table 1
Filters for imaging CKAR and DAGR

Fluorochrome	Excitation (nm)	Dichroic mirror (nm)	Emission (nm)
CFP	420/20	450	475/40
YFP (FRET-based)	420/20	450	535/25
YFP (Direct excitation)	495/10	505	535/25

The filters must enable sequential collection of CFP and YFP fluorescent emission upon excitation of CFP; collecting YFP emission upon direct excitation of YFP is also beneficial as a way to monitor photobleaching. These experiments are carried out in our lab on a Zeiss Axiovert microscope (Carl Zeiss Microimaging, Inc.). Excitation and emission filters are switched in filter wheels (Lambda 10–2; Sutter), and images are acquired through a 10% neutral density filter. Optical filters (Chroma Technologies) are listed in Table 1. Images are captured using a MicroMax digital camera (Roper-Princeton Instruments) controlled by MetaFluor software (Universal Imaging, Corp.). Using this setup, cells having approximate YFP intensity of 1,000 contain a cellular concentration of approximately 1 μM reporter; these levels of reporter provide consistent responses that are subject to regulation by endogenous receptors, PKC isoforms, and phosphatases.

3. If using a different setup than that described above, one needs to determine conditions that will provide sufficient fluorescence intensity for consistent FRET ratio changes over time without photobleaching. This is determined by varying neutral density filters (typically less than 30%) and exposure time (typically less than 1 s) until a good signal-to-noise ratio with minimal photobleaching is achieved over the appropriate period of time.

4. It is difficult to image cells that adhere loosely to glass or that move during the duration of the experiment (usually less than 30 min). In particular, treatment of loosely adherent cells with phosphatase inhibitors such as Calyculin A often results in release of the cells from the dish. One can attempt to improve conditions for imaging by plating cells on glass that has been coated with a matrix such as Matrigel or poly-lysine. In our experience, coating glass coverslips with matrix has not interfered with the FRET ratio readout of CKAR.

5. The experiments described here were carried out at room temperature and in HBSS supplemented with Ca^{2+}, but the use of other media and temperature controls may improve the quality of the data for certain experiments. In addition, other accessories such as controlled flow perfusion systems may enable additional types of experiments to be carried out, such as pulse-addition of agonist.
6. For untargeted reporters, selection of a dim cytoplasmic region (rather than the entire cell) provides the most robust and consistent responses, since CKAR in the nucleus does not often respond to stimulation (see Fig. 3b). The ratiometric nature of the reporter controls for effects of minor movement of the population of reporters in and out of the selected region. However, it is important that the cell itself does not move in and out of the region selected.
7. For imaging targeted reporters in new cell lines, visual verification and/or extensive validation of targeting should be performed. Dyes that highlight the mitochondria and Golgi are available which are spectrally compatible with CKAR (Invitrogen), or one may co-localize the reporter with immunofluorescence staining for organelle marker proteins. Additionally, one should always keep in mind that for CKAR, CFP, and YFP are linked in the same molecule; therefore, CFP intensity should co-localize with YFP. Failure of these two images to co-localize indicates cleavage of the reporter resulting from targeting it near a cellular protease (as can happen on the surface of the mitochondria, for example; (7)). This can also be tested by western blot.
8. With these described acquisition settings and analyses, baseline FRET ratios are not quantitatively meaningful; only stimulated changes in FRET give a quantitative readout of changes in activity.
9. For each new application of CKAR (i.e., testing a new agonist), it is very important to test that the control CKAR T/A reporter does not yield a FRET ratio change and that the PKC inhibitor blocks and/or reverses the response.
10. If the agonist or inhibitor has intrinsic fluorescence, it may interfere with the CKAR readout. The ability of a fluorescent drug to interfere with CKAR readout can be tested by either selecting a region of the image containing no cells and monitoring CFP and FRET, or by monitoring the FRET ratio of cells transfected with CKAR T/A. If these values consistently change upon addition of the agonist or inhibitor, then the agonist or inhibitor cannot be used for FRET imaging experiments. The exception is that if the fluorescence of an inhibitor is fairly low and very minimally affects the FRET ratio of CKAR T/A, it is appropriate to pretreat samples with this

inhibitor to verify that the inhibitor blocks the response to an agonist.

11. During data analysis, it is important to not oversubtract drift. Adding inhibitor to reverse the agonist-induced response back to baseline not only controls for specificity but also provides a good readout of whether the baseline has been oversubtracted. The FRET ratio over time after PKC inhibition should level off to a line with a flat slope if the subtraction has been carried out properly.

    For targeted DAGR constructs, it is difficult to compare the magnitude of the FRET ratio change between the plasma membrane and Golgi. This is because the C1 domain may already be prelocalized to a particular subset of membranes and therefore the stimulated response will appear to be small in magnitude. We have only used these reporters to test the kinetics of the response at a particular region of the cell, correlating the duration of the DAG response with duration of the PKC response to the same agonist ((7); Fig. 5a).

Acknowledgments

We thank Maya Kunkel for critical reading of this manuscript. This work was supported by National Institutes of Health grants GM43154 and PO1 DK54441.

References

1. Manning, B. D. (2009) Challenges and opportunities in defining the essential cancer kinome. *Sci Signal* **2**, pe15.
2. Newton, A. C. (2003) Regulation of the ABC kinases by phosphorylation: protein kinase C as a paradigm. *Biochem J* **370**, 361–71.
3. Suzuki, A., and Ohno, S. (2006) The PAR-aPKC system: lessons in polarity. *J Cell Sci* **119**, 979–87.
4. Gallegos, L. L., and Newton, A. C. (2008) Spatiotemporal dynamics of lipid signaling: protein kinase C as a paradigm. *IUBMB Life* **60**, 782–9.
5. Newton, A. C. (2008) Protein kinase C. *IUBMB Life* **60**, 765–8.
6. Violin, J. D., Zhang, J., Tsien, R. Y., and Newton, A. C. (2003) A genetically encoded fluorescent reporter reveals oscillatory phosphorylation by protein kinase C. *J Cell Biol* **161**, 899–909.
7. Gallegos, L. L., Kunkel, M. T., and Newton, A. C. (2006) Targeting protein kinase C activity reporter to discrete intracellular regions reveals spatiotemporal differences in agonist-dependent signaling. *J Biol Chem* **281**, 30947–56.
8. Mattingly, R. R. (2003) Mitogen-activated protein kinase signaling in drug-resistant neuroblastoma cells. *Methods Mol Biol* **218**, 71–83.
9. Newton, A. C. (2002) Analyzing protein kinase C activation. *Methods Enzymol* **345**, 499–506.
10. Pan, T. T., Neo, K. L., Hu, L. F., Yong, Q. C., and Bian, J. S. (2008) H2S preconditioning-induced PKC activation regulates intracellular calcium handling in rat cardiomyocytes. *Am J Physiol Cell Physiol* **294**, C169-77.
11. Hosoda, K., Saito, N., Kose, A., Ito, A., Tsujino, T., Ogitat, K., Kikkawat, U., Onot, Y., Igarashif, K., Nishizukat, Y., and Tanaka, C. (1989) Immunocytochemical localization of the beta I subspecies of protein kinase C in rat brain *Proc Natl Acad Sci U S A* **86**, 1393–7.
12. Saito, N., Kose A, Ito A, Hosoda, K., Mori, M., Hirata, M., Ogitat, K., Kikkawat, U., Onot, Y.,

Igarashif, K., Nishizukat, Y., and Tanaka, C. (1989) Immunocytochemical localization of beta II subspecies of protein kinase C in rat brain. *Proc Natl Acad Sci U S A* **86**, 3409–13.

13. Miyawaki, A. (2008) Green fluorescent protein glows gold. *Cell* **135**, 987–90.
14. Sakai, N., Sasaki, K., Ikegaki, N., Shirai, Y., Ono, Y., and Saito, N. (1997) Direct visualization of the translocation of the gamma-subspecies of protein kinase C in living cells using fusion proteins with green fluorescent protein. *J Cell Biol* **139**, 1465–76.
15. Oancea, E., and Meyer, T. (1998) Protein kinase C as a molecular machine for decoding calcium and diacylglycerol signals. *Cell* **95**, 307–18.
16. Wang, Q. J., Bhattacharyya, D., Garfield, S., Nacro, K., Marquez, V. E., and Blumberg, P. M. (1999) Differential localization of protein kinase C delta by phorbol esters and related compounds using a fusion protein with green fluorescent protein. *J Biol Chem* **274**, 37233–9.
17. Zhang, J., Ma, Y., Taylor, S. S., and Tsien. R. Y. (2001) Genetically encoded reporters of protein kinase A activity reveal impact of substrate tethering. *Proc Natl Acad Sci U S A* **98**, 14997–5002.
18. Wang, J., Gambhir, A., Hangyas-Mihalyne, G., Murray, D., Golebiewska, U., and McLaughlin, S. (2002) Lateral sequestration of phosphatidylinositol 4,5-bisphosphate by the basic effector domain of myristoylated alanine-rich C kinase substrate is due to nonspecific electrostatic interactions. *J Biol Chem* **277**, 34401–12.
19. Shaner, N. C., Campbell, R. E., Steinbach, P. A., Giepmans, B. N., Palmer, A. E., and Tsien, R. Y. (2004) Improved monomeric red, orange and yellow fluorescent proteins derived from Discosoma sp. red fluorescent protein. *Nat Biotechnol* **22**, 1567–72.
20. Dries, D. R., Gallegos, L. L., and Newton, A. C. (2007) A single residue in the C1 domain sensitizes novel protein kinase C isoforms to cellular diacylglycerol production. *J Biol Chem* **282**, 826–30.
21. Gokmen-Polar, Y., Murray, N. R., Velasco, M. A., Gatalica, Z., and Fields, A. P. (2001) Elevated protein kinase C betaII is an early promotive event in colon carcinogenesis. *Cancer Res* **61**, 1375–81.
22. Violin, J. D., and Newton, A. C. (2003) Pathway illuminated: visualizing protein kinase C signaling. *IUBMB Life* **55**, 653–60.
23. Kajimoto, T., Sawamura, S., Tohyama, Y., Mori, U., and Newton AC. (2010) Protein kinase C δ-specific activity reporter reveals agonist-evoked nuclear activity controlled by Src family of kinases. *J Biol Chem* **285**, 41896–910.

Chapter 18

Visualizing Receptor Endocytosis and Trafficking

Ali Salahpour and Larry S. Barak

Abstract

G-protein-coupled 7 transmembrane domain receptors (GPCR-7TMR) represent the largest class of membrane protein drug targets. They respond to a plethora of ligands ranging from small molecules to polypeptide hormones. Upon activation, almost all GPCR-7TMRs undergo desensitization followed by receptor internalization and resensitization. This cycle is crucially important for regulating the signal emanating from the receptor and is tightly linked to the receptor and/or the ligands studied. In this chapter, we describe some of the technical approaches that can be used to study GPCR internalization and trafficking.

Key words: G-protein-coupled receptor, Green fluorescent protein, Immunofluorescence, Internalization, Trafficking

1. Introduction

The internalization of plasma membrane receptors in response to drugs, hormones, mutations, and posttranslational modifications is now a well recognized biological phenomenon (1). However, as recently as the early 1970s, the architecture of the plasma membrane was still the subject of speculation and the distributions of receptors in them even less well understood (2). Our understanding of the biology of cell surface receptors has advanced considerably, in no small measure due to techniques which have permitted their visualization in both space and time and in quantities as small as single molecular complexes (3). The older and more successful techniques for receptor visualization predominantly utilized antibodies against the receptors themselves or labeled them using high affinity ligands and toxins tagged with fluorophores (4, 5). While successful in many instances, these approaches lacked the generality to apply them on a routine basis to uncharacterized receptors.

Louis M. Luttrell and Stephen S.G. Ferguson (eds.), *Signal Transduction Protocols*, Methods in Molecular Biology, vol. 756, DOI 10.1007/978-1-61779-160-4_18, © Springer Science+Business Media, LLC 2011

In the past 15 years, recombinant DNA techniques have enabled almost any protein to remain active after tagging by epitopes for which commercial antibodies are available (6, 7). The availability of commercial plasmids containing fluorescent proteins has facilitated not only compartmentalization studies but also the routine assessment of dynamic receptor properties (8, 9). We will present guidelines for applying these latter techniques to the visualization of membrane receptors during the study of clathrin-mediated endocytosis.

2. Materials

2.1. Cell Culture

1. HEK 293 cells or U2OS cells (American Type Culture Collection) (see Notes 1 and 2).
2. Complete Minimum Essential Medium (MEM): MEM supplemented with 10% fetal bovine serum (FBS).
3. Complete Dulbecco's Modified Eagle Medium (DMEM): DMEM supplemented with 10% FBS.
4. Serum starving medium: MEM or DMEM supplemented with 0.1% bovine serum albumin (BSA) and 10 mM HEPES (pH 7.4).
5. Trypsin-EDTA: 0.05% Trypsin with EDTA for HEK 293 cells (Invitrogen).
6. Trypsin-EDTA: 0.25% Trypsin with EDTA for U2OS cells (Invitrogen).
7. 10 cm tissue culture dishes.
8. 35 mm glass-bottomed culture dishes (MatTek).

2.2. Culture of Stably Transfected Cell Lines

There are many different compounds available to maintain selection pressure on cell lines possessing the corresponding plasmid. We commonly utilize geneticin (G418), zeocin, and puromycin. These drugs can be used either singly or in combination, but the employment of two selection markers over a single one can dramatically accelerate the development of clones. Running a dose–response killing curve with the intended drugs to determine cell loss rate on the untransfected cell lines can provide valuable information on what drug concentrations will provide optimal selection results.

1. Same culture media as above with the addition of antibiotics for selection.
2. Antibiotic stocks: Puromycin, G418 or zeocin, depending on the cell line.

2.3. Calcium/Phosphate Transfection

Transient or stable expression of recombinant proteins is accomplished by the use of transfection reagents or electroporation equipment.

The least expensive yet reliable technique for protein expression in many cell types, including the efficient expression in HEK-293 cells, is the calcium phosphate method. Three stock solutions are required to carry out the procedure. Store the HBS at 4°C and the other solutions at room temperature.

3. 2× HBS solution: Combine 2.8 ml of 2 M NaCl solution, with 2 ml of 0.5 M HEPES solution (pH 7.0), 0.06 ml of 0.5 M Na_2HPO_4 (pH 7.1), and 15.2 ml of water to a final volume of 20 ml. Adjust pH to 7.1 ± 0.05. Preparing the 2× HBS solution at the proper pH is the most critical aspect of the procedure.
4. 2.5 M $CaCl_2$ solution.
5. Deionized water.
6. Sterile 15 ml polypropylene conical screw cap culture tubes.

2.4. Labeling Epitope-Tagged Receptors with Antibodies

For double labeling of surface proteins, two compatible short epitopes are the Flag and HA sequences (see Note 3).

1. cDNA plasmid encoding the GPCR of interest bearing either an HA-epitope (YPYDVPDYA) or Flag epitope (DYKDDDDK) at the N terminus (see Note 4).
2. Rat monoclonal anti-HA 3F10 antibody (Roche): working dilution 1:400 (see Note 5).
3. Monoclonal anti-flag M2 antibody (Sigma): working dilution 1:500.
4. Alexa Fluor 488 (or other wavelength) conjugated goat anti-rat (Invitrogen): working dilution 1:1,000.
5. Alexa Fluor 488 (or other wavelength) conjugated goat anti-mouse (Invitrogen): working dilution 1:1,000–2:000.
6. Plain MEM or DMEM: no additives.
7. MEM containing 1–2% BSA.
8. Fixing solution: freshly prepared 1–4% paraformaldehyde in phosphate-buffered saline (PBS).
9. Permeabilization solution: 0.5% Triton X-100, 2% BSA in PBS.
10. Blocking solution: PBS supplemented with 1–2% BSA.
11. Vectashield® mounting medium (Vector Laboratories).
12. MEM or DMEM supplemented with 25 mM HEPES (pH 7.4).

2.5. Confocal Microscopy

This is a superior technique for visualizing the time-dependent behavior of labeled receptors. Live cells grown in 35 mm glass-bottomed dishes and expressing the tagged proteins of interest can be observed over extended periods using HEPES-buffered media (see Note 6).

1. Zeiss or equivalent confocal fluorescence scanning microscope platform with 40× or greater magnification high numerical aperture oil objectives.
2. Heated microscope stage.

3. Methods

Many endocytosis studies are carried out using a clonal cell line stably expressing the receptor of interest. There are several advantages and considerations in using stable cell lines (see Note 7). Nevertheless, some internalization/endocytosis studies may require a transient transfection approach (see Note 8). It is therefore important to know from the onset what major questions need to be addressed and then choose whether to carry out the studies on stable or transient cell lines expressing the receptor of choice. However, assessing the behavior of receptors in transient experiments during the weeks/months it takes to establish permanent clonal lines usually enables the permanent line studies to be more focused.

3.1. Transfection of Cells Using Calcium/Phosphate DNA Precipitation

1. On day 1, seed 3×10^6 HEK-293 cells in a 10-cm tissue culture dish (see Note 9).
2. On day 2, transfect cells. To a 15-mL sterile polypropylene culture tube add 450 μl of sterile water containing 5 μg of plasmid DNA. Then add 50 μl of 2.5 M $CaCl_2$ solution and mix. The total volume is 500 μl. Add 500 μl of 2× HBS solution dropwise to the DNA/CaCl mix. Bubble the solution for 30 s after the addition of the HBS, then immediately add the 1 ml of DNA/CaCl/HBS solution to the 10-cm culture dish containing the cells.
3. On day 3, trypsinize the cells and seed 1×10^5 cells per 35 mm MatTek glass-bottomed dishes.
4. On day 4, change medium on cells and serum starve overnight.
5. On day 5, carry out the immunofluorescence experiment (see Subheadings 3.3–3.6).

3.2. Generation of Stable Cell Lines

1. The same transfection conditions are used to generate stable HEK 293 and U2OS cells.
2. On day 1, seed 3×10^6 HEK293 or 1.5×10^6 U20S cells in a 10 cm Petri dish.
3. On day 2, transfect cells using the Calcium/Phosphate method (see Subheading 3.1). The expression plasmid must bear a selection marker for eukaryotic cells. We recommend

genticin (G418), zeocin, and puromycin, but other mammalian selectable markers can be used.

4. It is important to determine the concentration of the drug needed to get efficient selection. This can be achieved by performing killing curves for each drug using untransfected HEK293 or U2OS cells. In our experience, a final concentration of 2 μg/mL of puromycin is sufficient to allow selection of stable clones for HEK 293 cells. For U2OS cells a combination of G418/zeocin at 400/200 μg/ml is used for the selection and 100/50 μg/ml of G418/zeocin is used for maintenance of the line.
5. On day 3, change medium, replacing it with medium not containing the selection antibiotic.
6. On day 4, change medium, replacing with medium containing the selection antibiotic. From this point, all cultures will be carried with medium containing the antibiotic to induce selection pressure.
7. On day 6, trypsinize the cells and transfer cells to six 10-cm dishes. Continue to change medium twice weekly. In about 7–14 days most cells may be dying or dead and distinct colonies will be visible in the culture dishes. To isolate colonies, place the dish under a microscope, use a P200 micropipette to aspirate individual colonies, and seed into a 24-well plate for expansion.
8. Select 24 clones per line and receptor of interest. Once the cells have grown sufficiently, determine receptor expression levels using immunofluorescence (see Note 10).

3.3. Antibody Labeling of Surface Receptors and Stimulation with Ligands for Visualization of Receptors on Fixed Cells

Prior to stimulation of cells, receptors on the cell surface may be labeled with antibody. This ensures visualization of only endocytosed receptors. If labeling is carried out after stimulation and cell permeabilization, all species of receptors, including those newly synthesized and transiting the ER/Golgi apparatus will be labeled. By labeling surface receptors prior to stimulation one avoids such complications and only visualizes the internalization of surface receptors. The protocol below is for a receptor bearing an HA-epitope at the N terminus.

1. Place cells in 35-mm glass-bottomed culture dishes onto ice to arrest vesicle trafficking.
2. Wash cells twice gently with plain MEM.
3. Block nonspecific sites with serum-free MEM containing 1–2% BSA for 30 min. This step can be omitted when using high quality monoclonal antibodies.
4. Rinse with serum-free MEM.

5. Incubate for 1 h on ice with rat monoclonal anti-HA 3F10 antibody (dilution: 1:400) in serum-free MEM containing 1% BSA.
6. Wash three times 5 min with serum-free MEM.
7. Return cells in 1 ml serum-free MEM to the 37°C incubator for stimulation.
8. While labeling with primary antibody, prepare the ligand of choice in serum-free MEM. The receptors should be stimulated with a concentration of 10–100-fold above the Kd of the drug in order to ensure maximum occupancy.
9. Once the ligand is prepared, add to the 35-mm dishes at the desired concentration. This can be done either by replacing the medium with 1 ml of the ligand solution at the final concentration or by adding 100 μl of a 10× concentrated solution of the ligand.
10. Cells should be stimulated for at least 15 min. In some cases, longer stimulations (30–60 min) may be required to observe complete endocytosis of the receptor (see Note 11).
11. After stimulation, transfer the dish to ice.
12. Rinse cells twice with serum-free MEM.
13. Fix cells on ice with ice cold 4% paraformaldehyde in PBS for 15 min.
14. Wash cells twice for 5 min with PBS.
15. Permeabilize cells using 0.5% Triton X-100 in PBS at 4°C for 20–30 min.
16. Wash three times with PBS for 5 min. The subsequent steps can be carried out at room temperature.
17. Block nonspecific sites with 1% BSA blocking solution for 30 min.
18. Incubate with secondary antibody: Alexa Fluor 488 conjugated goat anti-rat (dilution 1:1,000) in 1% BSA blocking solution for 30 min.
19. Wash three times 5 min with 1% BSA blocking solution.
20. Rinse with twice with PBS.
21. Apply Vectashield® mounting medium and visualize on microscope.

3.4. Antibody Labeling of Surface Receptors and Stimulation with Ligands for Visualization of Receptors on Live Cells

In this section, the surface receptors will be labeled with both primary and secondary antibody prior to stimulation. This allows for real-time observation of endocytosis of surface receptors in live cells. Obviously, this procedure precludes fixation and permeabilization. Note, however, that when following this procedure the receptors can become crosslinked by the secondary antibody, which may interfere with endocytosis.

Alternatively, using a fluorescent-labeled primary antibody to label receptors with a single epitope tag will avoid excessive cross-linking.

1. Transfer cells onto ice to stop any vesicle trafficking.
2. Wash cells twice with serum-free MEM.
3. Block nonspecific sites with serum-free MEM containing 1–2% BSA solution for 30 min. This step may be optional when using high quality monoclonal antibodies.
4. Rinse with serum-free MEM.
5. Incubate for 1 h on ice with rat monoclonal anti-HA 3F10 antibody (dilution: 1:400) in serum-free MEM containing 1% BSA.
6. Wash three times 5 min with serum-free MEM containing 1% BSA.
7. Incubate with secondary antibody: Alexa Fluor 488 conjugated goat anti-rat (dilution 1:1,000) in serum-free MEM containing 1% BSA for 30 min. If a fluorescent primary antibody is utilized there is no need for secondary antibody incubation.
8. While labeling with antibodies, prepare the drug of choice in serum-free MEM as described above (Subheading 3.3).
9. Maintain the cells on ice until the time of the experiment, then warm them quickly by adding 30–37°C medium containing the drug of choice and using a heated microscope stage.
10. Place the dish containing cells under the microscope. Image cells and find a cell or a group of cells on which to image the endocytosis.
11. Aspirate the medium and replace with medium-containing ligand to start stimulation.
12. Continuously image the cells (see Subheading 3.6).
13. Stimulation times can be as long as 60–90 min in order to observe full endocytosis.
14. Figure 1 shows an example of results that can be obtained using this approach.

3.5. Visualization of GFP-Tagged Receptors on Live Cells

In this section, the receptors are genetically engineered to harbor a GFP protein at their C terminus (8). The modification with GFP precludes the need for antibody labeling and therefore reduces experimental steps and manipulation of the cells. However, all receptor species, including those in intracellular compartments (endoplasmic reticulum and Golgi), will be visualized since they will all be labeled with GFP. This is the major drawback of this technique compared to the selective surface labeling described

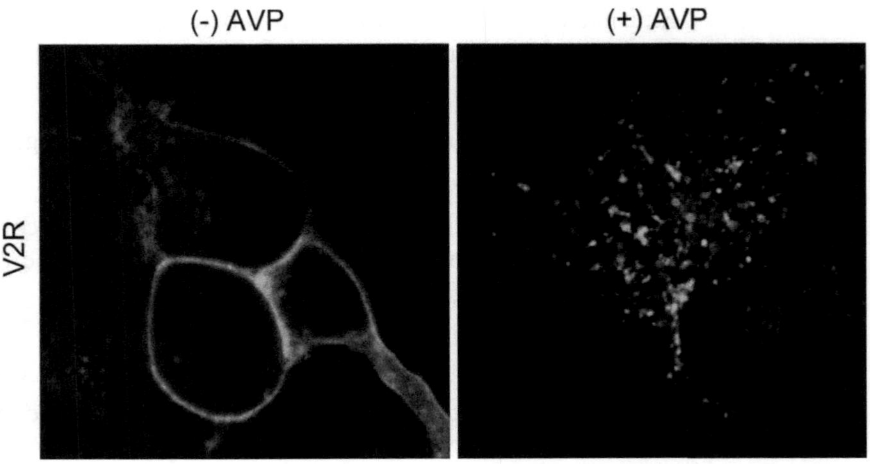

Fig. 1. Expression of HA-tagged vasopressin V2 receptors (V2R) in HEK-293 cells. Plasma membrane receptors were labeled with rhodamine-tagged mouse-monoclonal anti-HA antibody. The *left panel* shows receptor distribution in the absence of agonist. The *right panel* shows cells that were labeled with antibody before treatment with 100 nM arginine-vasopressin (AVP) for 30 min at 37°C.

above. Nevertheless, in cases where the plasma membrane contains the majority of receptors the loss of membrane definition with drug treatment is easily recognized as is the concentration of redistributed receptor in the endosomal compartment

1. Stable cell lines expressing GFP-tagged receptors are preferable for these experiments; however, experiments on cells transiently expressing the receptors are also feasible.
2. Prepare the drug of choice in serum-free MEM as described above (Subheading 3.3).
3. Place the dish containing cells under the microscope, image the cells and find a cell or a group of cells on which to image the endocytosis.
4. Aspirate medium and replace with medium-containing drug to start stimulation.
5. Continuously image cells (see Subheading 3.6).
6. Stimulation times can be as long as 60–90 min in order to observe full endocytosis.
7. Figure 2 shows an example of results that can be obtained using this approach.

3.6. Live Cell Imaging Using a Confocal Fluorescence Microscope

These experiments are most easily done on an inverted microscope using cells plated in 35-mm culture dishes containing a central well whose bottom is formed by a 0.17 mm or thinner glass coverslip, enabling the use of 40× or greater magnification high numerical aperture oil objectives (see Note 12). The following steps are appropriate for imaging receptor endocytosis over extended periods to produce a high quality movie. The procedure is written for a Zeiss or equivalent confocal scanning microscope platform.

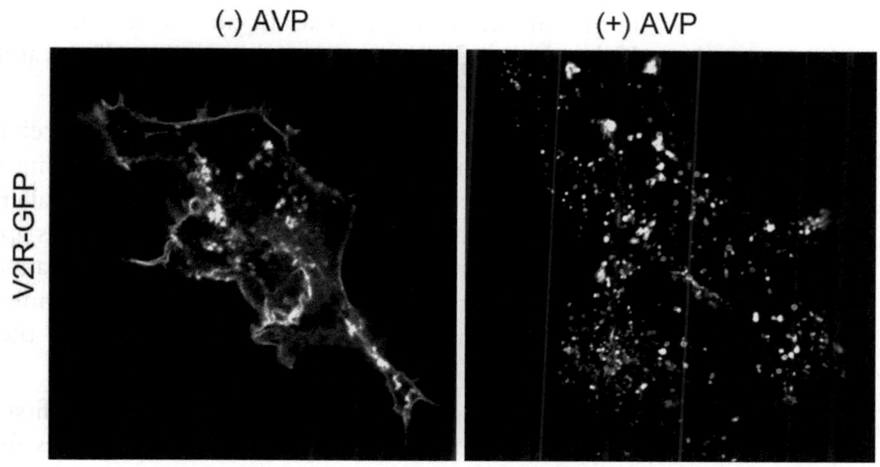

Fig. 2. Fluorescence images of Vasopressin V2R-GFP receptor in HEK-293 cells. Cells expressing V2R-GFP were treated with vehicle or AVP for 30 min at 37°C. The agonist induce redistribution of the receptor from the plasma membrane (*left panel*) to endocytic vesicles (*right panel*).

1. Imaging should be performed in medium containing an appropriate buffering system such as MEM/HEPES. Avoid serum or BSA at standard concentrations, as they will fluoresce at visible wavelengths and overwhelm the specific signal of the probe.
2. Set the temperature control stage to 30–37°. At the higher end of this temperature range the spontaneous cell movement can be appreciable enough to remodel or defocus the cell. The observed receptor kinetics at 30°C remain much more robust than at room temperature and provide a more stable environment for live cell imaging.
3. Adjust the image resolution to a full field of 1,024 × 1,024 pixels. This higher resolution provides more detailed images than 512 × 512 pixels, yet permits faster acquisition than using 2,048 × 2,048 pixels, and also takes considerably less memory. For fast processes a 512 × 512 pixel frame may provide better results by reducing acquisition time fourfold.
4. Prepare to acquire 8–10 frames for each 1 s of movie. Thus, a 15 s movie would be composed of 120–150 images. For a slow process that occurs with a time constant of 30 min this corresponds to one frame every 30 s for a 60 min study.
5. A major limiting factor in imaging live cells over extended periods is photobleaching of the probe molecule. Therefore, attempt to minimize the cell exposure to the excitation wavelength. This can be accomplished by first maximizing the gain of the primary amplifier, reducing the intensity of the exciting light to a minimum, and increasing the pinhole aperture to as large a size as possible without sacrificing the confocality of the image. Following these adjustments, reduce the dwell time per pixel or the number of averages over the image to

the point the noise per pixel begins to become noticeable. If signal averaging is used average over scan lines rather than frames to minimize motion artifacts.

6. Adjust the baseline offset of the signal so that the cell fluorescence falls within the dynamic range of the imaging system. Internalized receptor tends to cluster into a smaller endosomal compartment so that internalized receptors appear as bright cytoplasmic spots that may exceed the dynamic range of the imaging system. In this case, reduce the pinhole size and/or the illuminating laser intensity to bring the signal back into the linear range of the system.
7. Before adding drug to the system verify that the first two to three scans of the time series produce cell images from the same plane. Then add a 10× concentration of the drug dropwise surrounding the center well in a volume of 1/10 the dish volume (100 μl in 1 ml) before the next image is acquired.
8. File sizes of 50–200 megabytes can quickly fill file storage capacity. Stacks of .tif files are convenient to store images since they are nonproprietary and can be used to create movies on many different software platforms. Looping the movie to repeat once or twice can also make it much easier for the observer to appreciate your results.

4. Notes

1. HEK293 and U2OS cells have desirable properties for transient and permanent cell line development, respectively. They can be grown in similar media and do not require specialized culture techniques. To shorten development time, establish permanent clones by re-plating the unused cells that would otherwise be disposed of after transient transfection experiments.
2. HEK293 cells tend not to adhere well, especially to glass, but they also come off plastic surfaces with repetitive washes. U2OS cells are quite the opposite in their adhesive properties. Take advantage of this by detaching HEK293 cells from substrates with only cold PBS using 5 ml pipettes to avoid situations where membrane proteins may be degraded by trypsin-containing solutions. Additionally, HEK cells transfected directly in 35 mm dishes with glass-coverslip bottoms a day after splitting are easier to prepare and provide superior viewing for fluorescence microscopy compared to those cells that are transfected and then split into dishes with glass-bottomed wells.
3. Monoclonal mouse, rat, and polyclonal rabbit antibodies are commercially available for each epitope. Superior imaging

sensitivities can be achieved with secondary antibodies conjugated with one of the many bright Alexa fluor dyes that span the visible wavelengths.

4. We have had good results placing the tag after the native start codon. However, adding a new start codon before the first epitope tag works as well, especially for receptors that are engineered with multiple copies of the epitope.
5. For those in need of copious amounts of antibody, a monoclonal mouse hybridoma clone to produce the 12CA5 species can be found in most academic settings.
6. The use of solutions containing serum should be avoided due to autofluorescence, but media with and without phenol red work equally well at visible wavelengths. Additionally, paraformaldehyde may be maintained in buffered media at levels below 1% to preserve specimens without significantly quenching the fluorescence signal.
7. The time needed to generate stable cell lines can be anywhere from 4 to 6 weeks depending on the selection pressure used (Puromycin, Gentamycin, Zeocin). It is important to verify the observations in two distinct clonal lines ensuring that the observed effects are not unique to a particular clone. However, this is relatively straightforward if the clones are picked under the microscope while undergoing treatment by a receptor agonist or agonist/antagonist combination. The major advantage of stable clonal cell lines is that they afford a uniform receptor expression and internalization pattern among all the cells, allowing for qualitative and quantitative analysis. Furthermore, a stable clonal cell line made for internalization studies could also be useful for other studies.
8. Reasons for using transiently transfected cells may include the need to express receptors that are toxic to the cell and therefore not amenable to generate a stable line, and studies where a varied level of receptor expression is desired. In many cases, receptor-induced cell toxicity can be circumvented by using a tetracycline or doxycycline-inducible expression system to suppress the basal expression of the receptor. The main characteristic of transient transfection is that expression of the receptor may vary widely between individual transfected cells. While the internalization pattern and kinetics may be qualitatively similar between the cells, quantification may be more difficult due to the differences in receptor levels between cells. In fluorescent protein studies, some of this variability may be eliminated by utilizing cells where the receptor fluorescence falls within a defined intensity window. With transient transfection studies it is also possible to cotransfect other proteins that may modulate the internalization of the receptor and therefore this technique allows the rapid assessment of the

effects of ancillary proteins that could be missed in studies carried out on stable cell lines.

9. Direct transfection into 35-mm MatTek glass bottomed dishes can also be done. For this, seed 3×10^5 cells into 35-mm MatTek dishes on day 1. Make the same mix of DNA as described for Day 2 (Subheading 3.1) but use only 125–200 ml of the DNA/CaCl/HBS solution per dish. On Day 3, change medium on the cells and serum starve overnight. Carry out the immunofluorescence labeling and experimentation on Day 4.

10. We routinely perform colony selection using an inverted fluorescence microscope with 10–20× air objective by either prelabeling N-terminal epitope-tagged receptors on live cells in HEPES-buffered MEM with primary followed by secondary Alexa conjugated antibodies, or by directly visualizing the fluorescent protein-tagged receptors. This technique considerably enhances the yield of positive clones and reduces the time necessary to develop homogeneous colonies of receptor expressing cells.

11. It is recommended that both the concentration and duration of stimulation for any given receptors be determined empirically, as these conditions tend to be variable for cell lines, receptor levels, and individual receptors. The parameters described above are a good starting point but optimization should be carried out for each experimental condition.

12. The ability to image cells for extended periods and avoid image defocusing depends critically upon avoiding thermal gradients that may be produced by the addition of media to the dish, air currents in the room, or differences in temperature between the lens system and the dish bottom. In particular, adding drugs in solutions that differ by only a few degrees from the media surrounding the cell will cause a rapid change in the focus and is to be avoided at all costs. For this reason, we equilibrate our microscope components that include a heated stage for at least 30 min prior to performing an experiment. Moreover, when possible we store reagents and cell culture dishes on the same temperature control stage on which they are used.

References

1. Marchese, A., Chen, C., Kim, Y. M., and Benovic, J. L. (2003) The ins and outs of G protein-coupled receptor trafficking. *Trends Biochem Sci* **28**, 369–76.
2. Singer, S. J., and Nicolson, G. L. (1972) The fluid mosaic model of the structure of cell membranes. *Science* **175**, 720–31.
3. Saxton, M. J., and Jacobson, K. (1997) Single-particle tracking: applications to membrane dynamics. *Annu Rev Biophys Biomol Struct* **26**, 373–99.
4. Anderson, R. G., Goldstein, J. L., and Brown, M. S. (1980) Fluorescence visualization of receptor-bound low density lipoprotein in human fibroblasts. *J Recept Res* **1**, 17–39.
5. Ravdin, P., and Axelrod, D. (1977) Fluorescent tetramethyl rhodamine derivatives of alpha-bungarotoxin: preparation, separation,

and characterization. *Anal Biochem* **80**, 585–92.

6. von Zastrow, M., and Kobilka, B. K. (1992) Ligand-regulated internalization and recycling of human beta 2-adrenergic receptors between the plasma membrane and endosomes containing transferrin receptors. *J Biol Chem* **267**, 3530–8.
7. Barak, L. S., Tiberi, M., Freedman, N. J., Kwatra, M. M., Lefkowitz, R. J., and Caron, M. G. (1994) A highly conserved tyrosine residue in G protein-coupled receptors is required for agonist-mediated beta 2-adrenergic receptor sequestration. *J Biol Chem* **269**, 2790–5.
8. Barak, L. S., Ferguson, S. S., Zhang, J., Martenson, C., Meyer, T., and Caron, M. G. (1997) Internal trafficking and surface mobility of a functionally intact beta2-adrenergic receptor-green fluorescent protein conjugate. *Mol Pharmacol* **51**, 177–84.
9. Ferguson, S. S. (1998) Using green fluorescent protein to understand the mechanisms of G-protein-coupled receptor regulation. *Braz J Med Biol Res* **31**, 1471–7.

Chapter 19

Investigating G Protein-Coupled Receptor Endocytosis and Trafficking by TIR-FM

Guillermo A. Yudowski and Mark von Zastrow

Abstract

G protein-coupled receptors (GPCRs) represent the largest and most versatile family of signaling receptors. Their actions are highly regulated, both under physiological conditions and in response to clinically relevant drugs. A key element in this regulation is control of the number of functional receptors at the cell surface. Major processes that mediate this regulation are vesicular endocytosis and exocytosis of receptors. These trafficking events involve a concerted series of steps, some of which occur on a rapid timescale similar to that of functional signaling itself. Here, we describe and discuss an optical imaging approach, based on evanescent field or total internal reflection-fluorescence microscopy (TIR-FM), to investigate receptor endocytosis and recycling at the level of discrete membrane fission and fusion events. TIR-FM facilitates the study of receptor trafficking events near the plasma membrane with much greater spatial and temporal resolution than afforded by traditional methods. TIR-FM has already provided new insight to GPCR regulation, and we believe that this method has great potential for addressing a variety of questions in GPCR biology.

Key words: Fluorescence microscopy, Live-cell imaging, Total internal reflection microscopy, Trafficking, Endocytosis, Recycling, Receptor

1. Introduction

Membrane trafficking of signaling receptors is critical to many aspects of animal physiology. Rapid internalization of surface receptors is often stimulated by agonist-induced activation and is thought to control signaling both from the plasma membrane and intracellular compartments (1). The functional importance of endocytic trafficking has been particularly well established for various members of the large GPCR family. Many physiological responses mediated by GPCRs, particularly in the nervous system,

Louis M. Luttrell and Stephen S.G. Ferguson (eds.), *Signal Transduction Protocols*, Methods in Molecular Biology, vol. 756,
DOI 10.1007/978-1-61779-160-4_19, © Springer Science+Business Media, LLC 2011

occur on a relatively rapid time scale (seconds to minutes). This time scale is significantly shorter than the kinetics of most receptor trafficking events estimated using traditional methods. It is increasingly clear that certain GPCR trafficking events, particularly those occurring in the endocytic pathway, can occur with kinetics similar to those of acute signaling. It is also known that rapid membrane trafficking contributes to physiologically important regulation of receptor number in specific surface domains, such as spatially separated chemical synapses (2). These realizations have motivated increased interest in methods for examining GPCR trafficking with higher temporal and spatial resolution than afforded by traditional methods. Developments in equipment and reagents for live fluorescence imaging have greatly facilitated progress in achieving direct visualization of receptor trafficking.

We will focus on the application of total internal reflection-fluorescence microscopy (TIR-FM) to study GPCR trafficking in the endocytic pathway. Reflection of light at a refractive interface generates an evanescent field that diminishes exponentially with distance from the interface. The evanescent field creates a shallow field of illumination, extending in practice 100 nm from the reflective surface. If this surface is a cover slip supporting dissociated cells in a culture preparation, the evanescent field of illumination is useful for selectively exciting fluorescent probes located in the basal plasma membrane and extending a short distance into the cytoplasm (Fig. 1). TIR-FM thus facilitates observation of events occurring in the plasma membrane, and in a shallow region of cytoplasm immediately adjacent to the plasma membrane, with high signal-to-background ratio because fluorescent molecules located deeper within cells or in the culture medium are not

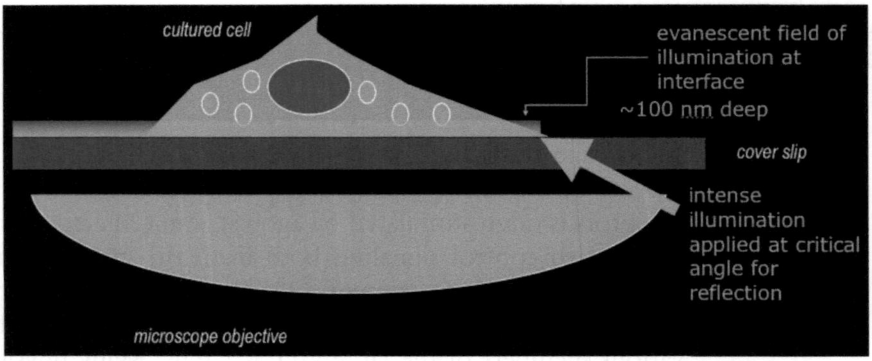

Fig. 1. Schematic view showing the main features of a TIR-FM imaging system. A standard wide-field microscope is used. The evanescent illumination field is generated by total internal reflection at the cover slip/sample interface. This requires illuminating the cover slip with a collimated light source at the critical angle, and is achieved in a typical "through-the-objective" system by focusing a laser beam near the edge of the back focal plane of a high numerical aperture objective. The evanescent field generated at the reflective interface falls off rapidly with distance, selectively exciting fluorophores located at or near the plasma membrane. This results in a signal-to-background ratio that is substantially higher than can be achieved in wide-field imaging using standard epifluorescence illumination, and generally higher than that obtainable using confocal fluorescence microscopy.

excited (3–5). Combined with continued improvements in rapid electronic sensor technology, such as intensified and electron-multiplying CCD cameras, TIR-FM is capable of investigating membrane events involving small numbers of receptors and with practical time resolution on the order of tens of milliseconds. Additional improvements have been made in automated focusing systems and thermoelectric control. This combination of technological advances, once in the domain of only highly specialized laboratories, has become widely available and provides a highly useful platform with which to study surface receptor trafficking at the single cell level and at physiological temperature.

The availability of a wide range of biologically compatible fluorescent probes, including genetically encoded fluorescent proteins, enable molecular specificity combined with spatio-temporal resolution that is useful for analyzing surface receptor dynamics. These probes and their application have been extensively reviewed elsewhere (6–8). Here we focus on imaging a pH-sensitive variant of the green fluorescent protein called superecliptic phluorin (SpH or SEP) (9, 10) fused to the amino-terminal extracellular domain of the human beta-2 adrenergic receptor (SEP-β2AR). SEP-β2AR is highly fluorescent at the neutral pH of the extracellular media, but its fluorescence is rapidly and reversibly quenched in the mildly acidic environment of the endocytic and recycling pathways. This property of SEP-β2AR facilitates the detection of discrete endocytic and exocytic events mediating surface receptor removal and insertion.

2. Materials

2.1. Cell Culture

1. HEK-293 cells passage 20–50 (American Type Culture Collection: CRL-1573).
2. 35 mm disposable MatTek glass bottom dishes.
3. Complete DMEM: Dulbecco's Modified Eagle's Medium-high glucose (DMEM) supplemented with 10% fetal bovine serum.
4. Lipofectamine 2000 (Invitrogen).
5. Opti-MEM imaging buffer supplemented with 20 mM HEPES (Invitrogen).
6. Sterile deionized water.
7. Poly-D-Lysine (Sigma).
8. Isoproterenol (Sigma).

2.2. Imaging Equipment

1. Inverted fluorescence microscope (Nikon TE2000E) with Perfect Focus and TIRF. objectives: 60×/1.45 Oil – Plan

Apo TIRF; 100×/1.49 Oil – Plan Apo TIRF. Nikon TIRF system with 440, 488, 514, and 561 nm lasers.

2. EM-CCD cameras Photometrics Quant EMCCD (http://www.photomet.com) or iXon^{EM}+EMCCD 897 Camera (Andor; http://www.andor.com).
3. Objective and Petri dish heaters with temperature controller to maintain media temperature at 37 °C (Bioscience Tools).
4. TYPE DF immersion oil (Cargille).

3. Methods

3.1. Cell Preparation

1. Dissolve poly-D-lysine in sterile water (50 μg/ml) and place 2 ml in each culture dish overnight at room temperature. Wash away residual poly-D-lysine PDL with sterile water (3 washes) and dry the culture dishes.
2. Seed HEK-293 cells onto the coated dishes.
3. Transfect with SEP-β2AR (11) construct (1 μg per dish) using lipofectamine 2000 following manufacturer's protocol 72 h prior to imaging.
4. The day of the imaging, replace incubation media 15–30 min before experiments with Opti-MEM or a low fluorescence media and return cells to the incubator (see Note 1).

3.2. Live-Cell Imaging

1. Start by initializing microscope, lasers, camera, and temperature control devices 30–45 min before any data acquisition.
2. Excitation and emission settings for TIRF: GFP = 488 nm laser excitation (2 mW); mCherry = 561 nm laser excitation (2–4 mW); 525/50 band pass, 527/21 and 645/24 nm dual bandpass emission filter.
3. Exposure time: continuous 100 ms exposure for receptor recycling, camera EM gain is set constant to obtain comparable results: X299, binning: 1 × 1, image: 512 × 512, pre-amp-gain = 4.90, horizontal readout = 10 vertical readout time = 3.3, temperature = –75. BitDepth = 14 bits for Andor iXon^{EM}+.
4. Select the proper TIRF objective and add a small amount of immersion oil on the objective and fit the glass bottom dish on the stage of the microscope and to the heating ring element (see Note 2).
5. First, find cells using transmission light to minimize photobleaching, and get them into focus. Second, illuminate cells in epifluorescent mode to find cells expressing tagged receptors and then switch to TIRF illumination. Move the laser

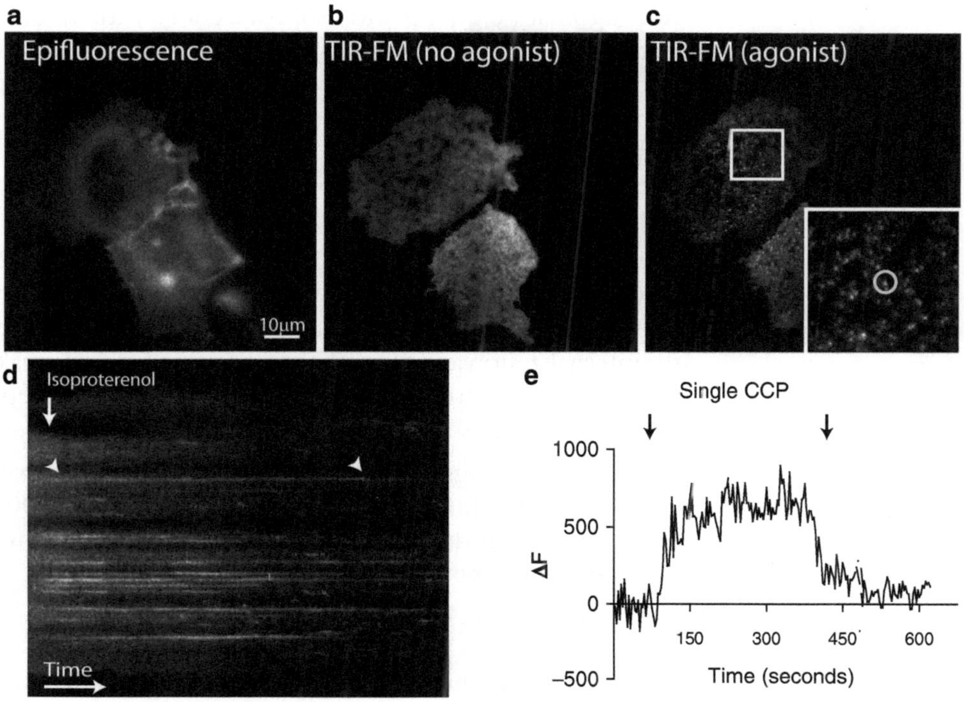

Fig. 2. Examples of GPCR localization observed by TIR-FM. (**a**) Example of SEP-β2AR-expressing HEK293 cells imaged using epifluorescence illumination. Two adjacent cells are shown. (**b**) TIR-FM view of the same field, showing the distinct "footprints" of each cell on the cover slip. (**c**) TIR-FM view of the same field acquired 1 min after adding agonist (1 μM isoproterenol) to the imaging bath. The region outlined by the white square is shown at higher magnification in the inset. The fluorescent spot surrounded by the circle represents a clathrin-coated pit containing SEP-β2ARs. (**d**) Kymograph showing SEP-β2AR dynamics in these representative cells, with increasing time going from left to right in the image. The *vertical arrow* indicates the addition of isoproterenol to the culture medium. The SEP-2AR fluorescence intensity pattern shifts from a diffuse appearance to defined horizontal lines, representing receptor clustering into clathrin-coated pits. An example is indicated by the *arrowhead at left*. The lines disappear shortly after endocytic scission of coated pits, as the SEP-β2AR-containing endocytic vesicles produced by this scission event move rapidly out of the evanescent illumination field. An example is indicated by the *arrowhead at right*. (**e**) Plot of the time course of maximum fluorescence intensity measured in the circled region indicated in (**c**), called ΔF because the value measured in an adjacent (nonclustering) region of the plasma membrane is subtracted. *Left arrow* indicates the time at which isoproterenol was added, showing the time course of SEP-β2AR concentration in the coated pit. *Right arrow* indicates the time at which the spot of SEP-β2AR fluorescence disappears from the evanescent illumination field following endocytic scission.

away from the center of the optical path and continue to achieve total internal reflection. Find the plasma membrane by adjusting the focal plane. Optimize the TIRF image as shown in Fig. 2a–b (see Note 3).

6. Find cells in the ideal fluorescence range for your experiments and begin data acquisition (see Note 4).
7. Acquisition settings for imaging agonist-induced clustering and endocytosis of receptors: Intermittent illumination and acquisition of 100 ms exposures every 3 s. Total time: 10 min.
8. Initiate data acquisition and acquire 10–30 frames before agonist addition.

9. Add agonist (e.g., 10 μM isoproterenol) with minimum disturbance to the cell either using an automated perfusion system or by careful addition of the agonist diluted in pre-warmed imaging media. Manual agonist addition should not be performed directly on top of the imaging area/cells but outside of the imaging area. See Fig. 2c for a representative effect of agonist addition on receptor clustering.
10. For resolving discrete fusion events mediating SEP-β2AR recycling, cells are exposed to the presence of 10 μM isoproterenol for 10 min in the incubator. This step induces receptor internalization and "loads" the endocytic pathway, making the fluorescence produced by vesicular insertion of receptor-containing vesicles more readily observed.
11. Acquisition settings for observing discrete recycling events: Continuous illumination and acquisition of serial 100 ms exposures, using the CCD in frame-transfer readout mode. Total imaging time: 60 s.
12. Save acquired data (see Note 5).

3.3. Analysis

Data management and analysis are critical steps in live-cell microscopy. Detailed discussion of image analysis methods is beyond the present scope and is addressed elsewhere (5, 12, 13). Examples include orthogonal views of image series as kymographs, useful for visually representing the time dependence of trafficking events (Fig. 2d), and intensity-vs.-time measurements to follow the dynamics of individual events (Fig. 2e). Additional examples can be found in the recent literature; e.g., (11, 14–16). Practical image analysis has been greatly aided by the development of computer software specifically intended for this application (see Note 6).

4. Notes

1. Remove phenol red, serum, folic acid, and riboflavin and other possible interference from the imaging media. EGFP photostability should also be taken into account during media selection (17).
2. Temperature of the imaging media must be monitored and kept constant when dishes are imaged, changes in temperature will affect trafficking kinetics.
3. Finding the exact angle for TIRF is the most critical step in this protocol. Cells illuminated in TIRF will present sharp edges and a increased signal-to-noise ratio when compared with out of TIRF or oblique illumination (*compare* Fig. 2a, b).

4. TIR-FM is intrinsically very sensitive, due to the low level of background fluorescence. We generally strive for the lowest expression level and illumination intensity that is sufficient for later analysis. In our experiments, using an Andor iXon^{EM} + electron-multiplying CCD camera, we have found that an intensity signal of ~2,000 (out of a maximum of 16,384 in the 14-bit readout mode) to be more than adequate. We recommend using a polyclonal cell line that stably expresses your receptor of interest, to allow rapid identification of cells in a suitable range of fluorescence intensity.
5. Because of the large amount of data generated, careful consideration must be given to file management and storage. Tags, metadata, and thorough indexing will help future data retrieval and analysis. See http://www.openmicroscopy.org for open source tools to support data management.
6. We typically use ImageJ, an excellent open source program developed by the NIH, which is supported by additional code written by an extensive user base and is available to the scientific community free of charge (http://www.rsbweb.nih.gov/ij/). We also recommend Micromanager (http://www.micro-manager.org) for controlling the microscope and peripheral devices during image acquisition. Micromanager is an open source program that is remarkably powerful and flexible, so is readily adapted to a variety of microscope systems, and it runs as an integrated plug-in linked to ImageJ.

Acknowledgments

The authors thank members of the von Zastrow laboratory and Dr. Kurt Thorn, Director of the UCSF/Nikon Imaging Center, for valuable discussion. The work discussed was supported by research grants from the NIH (DA023444 to G.A.Y. and DA010711 to M.v.Z.).

References

1. Sorkin, A., and von Zastrow, M. (2009) Endocytosis and signalling: intertwining molecular networks. *Nat Rev Mol Cell Biol* **10**, 609–22.
2. Shcherbakova, O. G., Hurt, C. M., Xiang, Y., Dell'Acqua, M. L., Zhang, Q., Tsien, R. W., and Kobilka, B. K. (2007) Organization of beta-adrenoceptor signaling compartments by sympathetic innervation of cardiac myocytes. *J Cell Biol* **176**, 521–33.
3. Steyer, J. A., and Almers, W. (2001) A real-time view of life within 100 nm of the plasma membrane. *Nat Rev Mol Cell Biol* **2**, 268–75.
4. Schmoranzer, J., Goulian, M., Axelrod, D., and Simon, S. M. (2000) Imaging Constitutive Exocytosis with Total Internal Reflection Fluorescence Microscopy. *J Cell Biol* **149**, 23–32.
5. Goldman, R. D., and Spector. D. L. (2005) Live cell imaging : a laboratory manual. Cold

Spring Harbor, N.Y. *Cold Spring Harbor Laboratory Press.*

6. Lippincott-Schwartz, J., Altan-Bonnet, N., and Patterson, G. H. (2003) Photobleaching and photoactivation: following protein dynamics in living cells. *Nat Cell Biol* **Suppl:**S7–14.
7. Miyawaki, A. (2005) Innovations in the imaging of brain functions using fluorescent proteins. *Neuron* **48,** 189–99.
8. Giepmans, B. N., Adams, S. R., Ellisman, M. H., and Tsien, R. Y. (2006) The fluorescent toolbox for assessing protein location and function. *Science* **312,** 217–24.
9. Miesenbock, G., De Angelis, D. A., and Rothman, J. E. (1998) Visualizing secretion and synaptic transmission with pH-sensitive green fluorescent proteins. *Nature* **394,** 192–5.
10. Sankaranarayanan, S., De Angelis, D., Rothman, J. E., and Ryan, T. A. (2000) The use of pHluorins for optical measurements of presynaptic activity. *Biophys J* **79,** 2199–208.
11. Yudowski, G. A., Puthenveedu, M. A., and von Zastrow, M. (2006) Distinct modes of regulated receptor insertion to the somatodendritic plasma membrane. *Nat Neurosci* **9,** 622–7.
12. Waters, J. C. (2009) Accuracy and precision in quantitative fluorescence microscopy. *J Cell Biol* **185,** 1135–48.
13. Bolte, S., and Cordelières, F. P. (2006) A guided tour into subcellular colocalization analysis in light microscopy. *J Microsc* **224,** 213–32.
14. Yudowski, G. A., Puthenveedu, M. A., Leonoudakis, D., Panicker, S., Thorn, K. S., Beattie, E. C., and von Zastrow, M. (2009) Real-time imaging of discrete exocytic events mediating surface delivery of AMPA receptors. *J Neurosci* **27,** 11112–21.
15. Puthenveedu, M. A., and von Zastrow ,M. (2006) Cargo regulates clathrin-coated pit dynamics. *Cell* **127,** 113–24.
16. Pucadyil, T. J., and Schmid, S. L. (2008) Real-time visualization of dynamin-catalyzed membrane fission and vesicle release. *Cell* **135,** 1263–75.
17. Bogdanov, A. M., Bogdanova, E. A., Chudakov, D. M., Gorodnicheva, T. V., Lukyanov, S., and Lukyanov, K. A. (2009) Cell culture medium affects GFP photostability: a solution. *Nat Methods* **6,** 859–60.

Chapter 20

Visualizing G Protein-Coupled Receptor Signalsomes Using Confocal Immunofluorescence Microscopy

Sudha K. Shenoy

Abstract

The heptahelical G protein-coupled receptors (GPCRs) receive and transmit a wide range of extracellular stimuli and induce a wide array of cellular responses by activating signaling kinases. It has become increasingly evident that the agonist-stimulated GPCR complexed with the adaptor protein, β-arrestin, serves as a focal point to recruit, activate, and target kinases to discrete subcellular compartments. This chapter describes a protocol to visualize the changes in the subcellular distribution of activated extracellular signal-regulated kinases 1 and 2 (ERK1/2) when induced by the angiotensin II type 1a receptor.

Key words: G protein-coupled receptor, Confocal microscopy, Extracellular signal-regulated kinase, Arrestin, Endocytosis, Signaling, Angiotensin, G protein, Scaffold

1. Introduction

Cellular signals transduced by the members of the G protein-coupled receptor (GPCR; a.k.a. seven transmembrane-spanning domain receptor) superfamily regulate most physiological functions in humans and therefore receive central attention in drug discovery research (1). GPCR signaling is propagated via the heterotrimeric G proteins as well as by the G protein-coupled receptor kinases and β-arrestins by either independent or coordinated mechanisms (2, 3). GPCR signal transduction involves activation of different signaling kinases including c-Src, AKT, and mitogen-activated protein kinases (MAPKs), such as the extracellular signal-regulated kinases 1 and 2 (ERK1/2) and c-Jun N-terminal kinases (2). In a MAPK cascade, phosphorylation of downstream

Louis M. Luttrell and Stephen S.G. Ferguson (eds.), *Signal Transduction Protocols*, Methods in Molecular Biology, vol. 756, DOI 10.1007/978-1-61779-160-4_20, © Springer Science+Business Media, LLC 2011

MAPK (e.g., ERK1/2) is mediated by a MAPK-kinase (e.g., MEK1), which is in turn phosphorylated by a MAPK-kinase-kinase (e.g., c-Raf). Upon agonist-stimulation of GPCRs, β-arrestins are not only recruited to the receptor to mediate endocytosis of the receptor (4), but also act as scaffolds for various components of the MAPK cascades (5).

Among the MAPKs, ERK1/2 are widely utilized by both G protein- and β-arrestin-dependent pathways (2). However, in each pathway the activated phosphorylated ERK1/2 (pERK) has distinct temporal and cellular localization signatures: in the G protein pathway, it is transiently activated upon agonist stimulation and mostly localized in the nucleus, while β-arrestin-mediated signals are sustained and mostly localized in the cytoplasm and endosomal membranes (2). Emerging evidence suggests that β-arrestin-mediated signals regulate diverse processes including receptor trafficking, cellular chemotaxis, apoptosis, dopaminergic behavior, cardiac contractility, and bone formation (2, 6–8).

The β-arrestin–GPCR interaction follows two patterns in general: the "Class A" pattern or loose complex formation where receptor and β-arrestins bind only at the plasma membrane and upon receptor internalization into endosomes these complexes fall apart, and the "Class B" pattern or tight complex formation resulting in cotrafficking and colocalization in endosomes (9). Activated GPCRs induce conformational changes and ubiquitination of β-arrestins that facilitates recruitment of endocytic and signaling proteins to the receptor complex, thus leading to the formation of multiprotein complexes or "signalsomes" (10, 11). Class A receptors form transient signalsomes, whereas Class B receptors form stable signalsomes.

This chapter outlines detailed methods for visualizing pERK as discrete units as well as in multiprotein complexes when bound to activated GPCRs and β-arrestins. The Class B angiotensin type II 1a receptor ($AT_{1a}R$) that robustly activates both G protein and β-arrestin-dependent pERK hosted by HEK-293 cells serves as an ideal system for conducting the analyses as described here; however, these methods should be applicable to other GPCRs, cell types, and kinases. Activated ERK1/2 proteins become phosphorylated at two residues; threonine-183 and tyrosine-185, and highly specific anti-pERK antibodies that recognize this phosphorylated domain are commercially marketed. This antibody has poor reactivity toward singly phosphorylated and unphosphorylated species. Therefore, immunostaining of pERK combined with high resolution confocal microscopy has become a useful approach to define signals in space and time as induced by GPCR activation (12–14).

2. Materials

2.1. Cell Culture

1. HEK-293 cells from American Type Culture Collection (ATCC, #CRL-1573).
2. Eagle's Minimum Essential Medium (MEM) supplemented with 10% fetal bovine serum (FBS) and 1% penicillin streptomycin solution (see Note 1).
3. Trypsin/EDTA 0.05%.
4. Phosphate-buffered saline (PBS) calcium and magnesium-free.
5. 100 mm tissue culture dishes.

2.2. Transfection

1. FuGENE 6 (Roche Applied Science).
2. Plasmids purified using Qiagen plasmid DNA purification kit, at ~1 μg/μL DNA concentration; plasmid DNA used here are pCDNA3-$AT_{1a}R$, pEGFP-β-arrestin2.
3. MEM medium without supplements.
4. Sterile 1.5 mL microfuge tubes.

2.3. Plating Cells for Confocal Microscopy

1. PBS containing calcium and magnesium (see Note 2).
2. Rat tail collagen solution (Roche; see Note 3): 10 mL sterile PBS with calcium and magnesium along with 50 μL 10 N acetic acid is added to the lyophilized collagen and allowed to dissolve overnight.
3. 35 mm glass-bottom dishes (MatTek).

2.4. Stimulation of Cells and pERK Activation

1. Serum-free starvation medium: Eagle's MEM supplemented with 10 mM HEPES pH 7.5 and 0.1% BSA (0.5 g/500 mL media), filter sterilized with a 0.2 μm vacuum filtration unit.
2. Angiotensin II (AngII) at 100 μM concentration is dissolved in 0.1% w/v Bovine Serum Albumin (BSA) solution prepared in sterile water and stored in single use aliquots at –80°C. Also, 0.1% BSA aliquots are stored and used for vehicle treatments.
3. Phorbol 12-myristate 13-acetate (PMA) dissolved in dimethylsulfoxide (DMSO, Sigma) as a 2 mM stock solution and stored as single use aliquots at –80°C.

2.5. Fixation and Permeabilization of Cells

1. Fixing solution: 4% paraformaldehyde dissolved in PBS with calcium and magnesium (see Notes 4 and 5).
2. Blocking Solution: 2% w/v BSA solution dissolved in PBS containing calcium and magnesium and filter sterilized with a 0.2 μm vacuum filtration unit. This solution is stable for several weeks at 4°C.

3. Permeabilization solution: 0.05% v/v Triton X-100 dissolved in blocking solution (see Note 6).

2.6. Immunostaining and Washing Steps

1. Antibody dilution buffer: 2% BSA solution from above.
2. Primary antibodies: Rabbit polyclonal anti-pERK antibody (Cell Signaling Technology) is recommended due to its sensitivity and specificity. Dilution has to be determined by trial and error (see Note 7). Generally, 1:300 dilution results in sufficient binding and immunodetection. Mouse monoclonal anti-HA antibody 12CA5 (Roche Applied Science) at a dilution of 1:500 is used for detecting the HA-AT_{1a}R.
3. Secondary antibodies (see Note 8): AlexaFluor® 633 donkey anti-rabbit IgG (H+L) 2 mg/mL (Molecular Probes), Alexa Fluor® 594 goat anti-mouse IgG (H+L) 2 mg/mL (Molecular Probes).
4. PBS with calcium and magnesium is used in all wash steps.

2.7. Image Acquisition

1. Zeiss LSM510 Meta confocal scanning microscope equipped with LSM510 imaging software; 488, 568, and 633 nm excitation; 515–540, 585–615, and 650 nm emission filter sets; and 100× oil objective lens.

3. Methods

Experimental steps are detailed below for the simultaneous detection of three proteins, namely HA-AT1aR, GFP-β-arrestin2, and pERK in HEK-293 cells. Three different flurophores are chosen to represent each protein component such that there is little spectral overlap. Bleed through or crosstalk between different channels is prevented by single channel acquisition as described below.

3.1. Cell Culture

1. Early passage HEK-293 cells are ideal to carry out these assays. During the propagation of cell cultures, care should be taken not to let cells grow to 100% confluence. Such cells will have altered morphology, will appear smaller and densely packed, and should not be used.
2. HEK-293 cells are cultured in MEM supplemented with 1% penicillin streptomycin and 10% FBS (MEM-complete). Inclusion of the antibiotics is not mandatory but will not affect the mammalian cells and will prevent bacterial contamination.
3. Medium from a 100 mm dish containing 60–70% confluent monolayer of cells is carefully aspirated and PBS without

calcium and magnesium is added to wash the cells. One quick but gentle wash is required.

4. After removal of PBS, 1 mL Trypsin-EDTA solution is added and the solution is evenly spread over the monolayer of cells, by gently tilting the dish side to side.
5. After 1 min, 5.5 mL of MEM-complete medium is added and cells are gently collected and transferred to a 15 mL sterile polypropylene tube.
6. The cell suspension is gently subjected to repeated up–down pipetting to ensure minimal clumping of cells.
7. 1.5 mL of cell suspension is transferred to a fresh 100 mm tissue culture dish containing 10 mL of MEM-complete. The covered dish is gently and briefly swirled to ensure even spreading of cells and returned to a 37°C, 5% CO_2 incubator for growth and recovery. 48 h later cells will be ~40% confluent and ready to be transfected with plasmid DNA.

3.2. Transient Transfection

1. Although expensive, the transfection reagent FuGENE 6 (Roche Applied Science) is preferred due to low toxicity and minimal effects on cell morphology. The downside is that this gentle method of transfection may not produce high efficiency of DNA uptake, but in general, transfection efficiency >60% is expected.
2. The 100 mm culture plate is examined under light microscopy to ensure even spreading of cells on the dish. If the cells are grouped or mounded they are not suitable for transfection.
3. Medium from the dish is gently aspirated and replaced with 5 mL of fresh warm MEM-complete. This change of media is performed at least 2–3 h before the transfection procedure. The reduction of medium volume improves transfection efficiency.
4. 3 μg of HA-AT1aR and 1.5 μg of β-arrestin2-GFP plasmid DNA are aliquoted into a sterile 1.5 mL microfuge tube.
5. 1 mL of unsupplemented MEM is added to the DNA and the tube is gently tapped to ensure even suspension of DNA.
6. After this, FuGENE 6 is added at a ratio of 4 μL per μg of DNA used. For the above setup 18 μL of FuGENE6 is added to the microfuge tube and the transfection mixture is incubated at room temperature for 10 min.
7. Next the transfection mixture is gently added drop-wise to the cells, which are then returned to the incubator for 24 h. Subsequently, the transfected cells are subcultured and plated on 35 mm glass-bottom culture dishes.

3.3. Seeding of Cells in Glass-Bottom Dishes

1. Each 35 mm glass-bottom dish is coated with rat tail collagen solution to ensure tight adherence of the cells and to prevent cells from lifting off the plate during the wash steps. One milliliter of the collagen solution is added to the plate, spread around to coat evenly.
2. The collagen solution is not discarded but retrieved into a sterile container for coating additional dishes and can be reused multiple times.
3. After removing collagen, dishes are left to air dry in the tissue culture hood for 15 min and then washed twice with PBS before plating cells.
4. The transfected cells from the 100-mm dish are detached using trypsin as described in Subheadings 3.1, step 3–6 and cells suspended in a total volume of 12 mL of MEM-complete.
5. 2 ml of cell suspension are transferred to each 35 mm dish and six dishes are prepared for different treatments.

3.4. Stimulation of Cells and pERK Activation

1. 24 h after plating, medium in the confocal dishes is replaced with 2 ml serum-free starvation medium. The removal of serum helps to reduce basal MAPK activity that might be induced by growth factors contained in the serum.
2. Two hours after the medium change, treatment of cells is performed as follows: Dish #1: no stimulation or mock treatment with 20 μL of 0.1% BSA for 30 min; Dish #2: 1 μM angiotensin (20 μL of 100 μM stock) 2 min; Dish #3: 1 μM angiotensin 5 min; Dish #4: 1 μM angiotensin 30 min; Dish #5: 1 μM PMA 30 min; and Dish #6: 1 μM angiotensin 30 min. Dish #6 will serve as a control to check the specificity of immunostaining and will be processed only for secondary antibody staining.
3. After addition of ligands, the cells are placed in the 37°C tissue culture incubator for the indicated times.

3.5. Fixation and Permeabilization of Cells

1. Each treatment is performed with careful timing and at the end of the time point, the medium is removed and fixing solution is added. Gentle suction with a vacuum trap is used to remove the culture medium and care is taken not to disturb cells in the cover-slip area.
2. 1 ml of fixing solution is added to each dish and cells are fixed for 30 min at room temperature.
3. At the end of the fixation period the paraformaldehyde solution is removed and placed in a biohazard waste container for disposal and three washes are performed using PBS with calcium and magnesium.
4. Each wash step involves adding of 1 mL PBS, gentle swirling for 30 s and aspiration. Alternatively, the dishes can be placed

for 1–2 min on a shaker platform and rotated at a low speed (<50 rpm) before aspirating wash solution.

5. After the third wash, 1 ml of permeabilization solution is added and cells are incubated for 60 min at room temperature.

3.6. Immunostaining and Wash Steps

1. The permeabilization solution is aspirated and cells are washed with PBS (with Calcium and Magnesium) two times as in Subheading 3.5, step 4.
2. Primary antibody solution is prepared by diluting (1:300) anti-pERK antibody in 2% BSA. A 500 μL minimum volume of antibody solution is needed in order to cover the cells evenly (see Note 7). Primary antibody is not added to dish #6. Instead 500 μL of 2% BSA is added.
3. The dishes are placed at 4°C overnight.
4. Next day the pERK antibody solution is aspirated and three PBS wash steps are performed as in Subheading 3.5, step 4. Alexa Fluor® 633 donkey anti-rabbit IgG diluted 1:300 in 2% BSA is added and the dishes are incubated at room temperature for 2 h. Since the secondary antibodies are light sensitive, from this step onward, dishes should be covered with aluminum foil and washes should be carried out under dim light.
5. At the end of incubation, the secondary antibody is aspirated and three washes with PBS solution is performed as in Subheading 3.5, step 4.
6. Next the primary antibody anti-HA 12CA5 diluted 1:500 in 2% BSA is added and incubation is carried out overnight at 4°C. No primary antibody is added to the control dish #6.
7. Next day the anti-HA antibody solution is aspirated and three PBS wash steps are performed as in Subheading 3.5, step 4. Alexa Fluor® 594 donkey anti-mouse IgG diluted 1:500 in 2% BSA is added and the dishes are incubated at room temperature for 2 h.
8. At the end of incubation, the secondary antibody is aspirated and three washes with PBS solution is carried out as in Subheading 3.5, step 4. 1 ml PBS is added to each dish and it is kept covered with aluminum foil.
9. It is ideal to scan confocal images immediately following the staining procedure. Alternately, the samples can be kept in an air tight container up to 2 weeks at 4°C. However, some deterioration of staining will occur and hence the samples should be imaged as soon as possible.

3.7. Image Acquisition and Processing

1. After the above staining procedure, images (Fig. 1) are obtained with a Zeiss LSM510 Meta confocal scanning microscope using LSM510 imaging software (see Note 9).

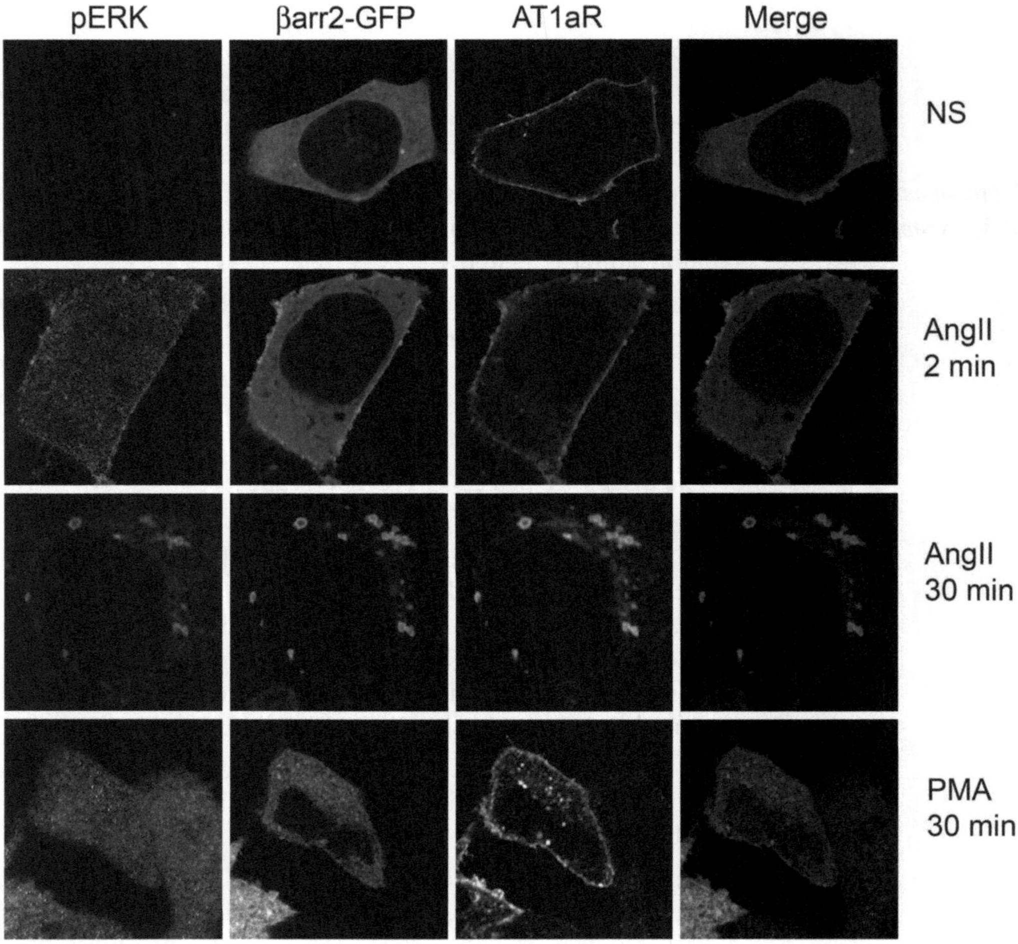

Fig. 1. Angiotensin II-stimulated colocalization of phosphorylated ERK, $AT_{1a}R$ and β-arrestin2 in signalsomes. HEK-293 cells transiently expressing the HA-AT1aR along with β-arrestin2-GFP were starved in serum-free media for 2 h and stimulated with 1 μM AngII at 37°C for the indicated periods of time. Fixed and permeabilized cells were then incubated with pERK polyclonal antibody followed by secondary Alexa-633-conjugated anti-rabbit IgG. This was followed by treatment with 12CA5 monoclonal antibody, which recognizes the HA epitope on the receptor and Alexa-594-conjugated anti-mouse IgG. Fluorescent confocal images were obtained on a Zeiss LSM-510 microscope using multitrack sequential excitation (488, 568, 633 nm) and emission filter sets at 515–540 nm for detecting GFP (β-arrestin, *green*), at 585–615 for Alexa-594 (receptor, *blue*), and at 650 nm for the Alexa-633 (pERK, *red*). The signals for pERK and HA were not detected in the absence of primary antibody incubation. The lowest row of images shows the pERK activation and distribution upon PMA stimulation, where nuclear pERK persists after a 30 min treatment. *NS* nonstimulated.

2. To minimize spectral bleed-through (channel cross-talk), multitracking and line-switching are employed. In this configuration, the scan for each flurophore is acquired separately and in succession using multitrack sequential excitation (488, 568, and 633 nm) and emission (515–540 nm for GFP; 585–615 nm for Alexa 594; 650 nm for Alexa-633) filter sets.
3. Each scan is set for 1 μm optical slice and a 100× oil objective is used.

4. The detector gain and amplifier offset are adjusted for individual wavelength channels to minimize saturation.
5. A 1,024 × 1,024 frame size and scan speed 4 are used to collect the final image.
6. Dishes from one experiment should be examined side by side and images for individual treatments collected using the same acquisition settings.
7. The LSM image is exported as a tagged image file (TIF file without any compression) which can be processed (separation of channels, placing multiple panels side by side, adding text, etc.) using Adobe Photoshop software. Image processing should not involve any changes to the original image. If any increase or decrease of brightness and contrast is required to visualize details, this change should be applied to the entire image and steps taken should be included in the methods section of the resulting publication.

4. Notes

1. Early passage HEK-293 cells are employed. If any changes in morphology are noted, for example, if cells spread out in a triangular shape with pointed edges, they should not be used. Use of penicillin streptomycin helps to minimize bacterial contamination.
2. PBS supplemented with calcium and magnesium is used for all washes except when trypsinization of cells is required.
3. Collagen coating is performed just before plating cells. Fibronectin 5 mg/mL solution can be used for coating dishes instead of collagen. 1 mL of diluted Fibronectin (1:500 in PBS) is added to the dish and aspirated after 30 min incubation at room temperature. Dishes are washed once with PBS before plating cells.
4. Dissolving paraformaldehyde requires careful stirring on a heated plate and should be performed by placing the stirring device inside a fume hood. The vapors are toxic and should not be inhaled.
5. Our laboratory has also routinely used 5% formaldehyde solution instead of paraformaldehyde for fixing cells. The 37% formaldehyde stock solution is diluted with PBS to obtain 5% formaldehyde.
6. Permeabilization is dependent on the nature of cell membrane, as well as the purity of detergent used. Recently, we have optimized this to be between 0.05 and 0.1% Triton X-100 for HEK-293 cells. The triton concentration and/or length of

permeabilization can be increased (e.g., 0.5% and 90 min) if no pERK is detected in the nucleus with PMA treatment. If the concentration of Triton X-100 is increased, the cells have to be washed three times with PBS, to thoroughly remove traces of Triton X-100 before antibody incubation.

7. The diluted anti-pERK antibody is used only once. However, the anti-HA 12CA5 antibody can be stored at 4°C and reused up to three times.
8. The secondary antibodies are stored at 4°C. Care should be taken not to expose the antibody solution to bright light.
9. Since each instrument setup is slightly different, the reader is advised to seek the guidance of the imaging facility manager, if available, for assistance in optimizing instrument settings.

Acknowledgments

This work was supported by National Institutes of Health grant HL 080525.

References

1. Lefkowitz, R. J. (2004) Historical review: a brief history and personal retrospective of seven-transmembrane receptors. *Trends Pharmacol Sci* **25**, 413–22.
2. DeWire, S. M., Ahn, S., Lefkowitz, R. J., and Shenoy, S. K. (2007) Beta-arrestins and cell signaling. *Annu Rev Physiol* **69**, 483–510.
3. Gesty-Palmer, D., and Luttrell, L. M. (2008) Heptahelical terpsichory. Who calls the tune? *J Recept Signal Transduct Res* **28**, 39–58.
4. Ferguson, S. S. (2001) Evolving concepts in G protein-coupled receptor endocytosis: the role in receptor desensitization and signaling. *Pharmacol Rev* **53**, 1–24.
5. Kendall, R. T., and Luttrell, L. M. (2009) Diversity in arrestin function. *Cell Mol Life Sci* *66*, 2953–73.
6. Patel, P. A., Tilley, D. G., and Rockman, H. A. (2009) Physiologic and cardiac roles of beta-arrestins. *J Mol Cell Cardiol* **46**, 300–8.
7. Beaulieu, J. M., and Caron, M. G. (2008) Looking at lithium: molecular moods and complex behaviour. *Mol Interv* **8**, 230–41.
8. Gesty-Palmer, D., Flannery, P., Yuan, L., Corsino, L., Spurney, R., Lefkowitz, R. J., and Luttrell, L. M. (2009) A β-Arrestin–Biased Agonist of the Parathyroid Hormone Receptor (PTH1R) Promotes Bone Formation Independent of G Protein Activation. *Science Translational Medicine* **1**, ra1.
9. Oakley, R. H., Laporte, S. A., Holt, J. A., Caron, M. G., and Barak, L. S. (2000) Differential affinities of visual arrestin, beta arrestin1, and beta arrestin2 for G protein-coupled receptors delineate two major classes of receptors. *J Biol Chem* **275**, 17201–10.
10. Lefkowitz, R. J., and Shenoy, S. K. (2005) Transduction of receptor signals by beta-arrestins. *Science* **308**, 512–7.
11. Shenoy, S. K. (2007) Seven-transmembrane receptors and ubiquitination. *Circ Res* **100**, 1142–54.
12. Luttrell, L. M. (2002) Activation and targeting of mitogen-activated protein kinases by G-protein-coupled receptors. *Can J Physiol Pharmacol* **80**, 375–82.
13. Shenoy, S. K., and Lefkowitz, R. J. (2005) Receptor-specific ubiquitination of beta-arrestin directs assembly and targeting of seven-transmembrane receptor signalosomes. *J Biol Chem* **280**, 15315–24.
14. Shenoy, S. K., Barak, L. S., Xiao, K., Ahn, S., Berthouze, M., Shukla, A. K., Luttrell, L. M., and Lefkowitz, R. J. (2007) Ubiquitination of beta-arrestin links seven-transmembrane receptor endocytosis and ERK activation. *J Biol Chem* **282**, 29549–62.

Part VI

Protein–Protein Interactions

Chapter 21

Detection and Characterization of Receptor Interactions with PDZ Domains

Stefanie L. Ritter and Randy A. Hall

Abstract

Many transmembrane receptors are regulated by associations with scaffold proteins containing PDZ domains, which interact with receptor carboxyl-termini to control receptor signaling, trafficking, and localization. We describe here approaches for detecting and characterizing interactions between receptors and PDZ scaffolds. These approaches include the construction and screening of proteomic arrays, blot overlays using fusion proteins, and co-immunoprecipitation studies to assess interactions in cells.

Key words: G protein-coupled receptor, PDZ domain, Scaffold, Affinity, Growth factor, Tyrosine kinase

1. Introduction

Cell-surface receptors mediate intercellular signaling and are common therapeutic targets for drug development (1). Receptor function is often regulated by receptor-interacting proteins, which can profoundly influence receptor signaling, trafficking, and/or pharmacology (2). One of the most well-studied classes of receptor-interacting partners is the family of PDZ domain-containing scaffold proteins. PDZ domains are specialized protein–protein interaction modules, which derive their name from the first three proteins in which they were identified: the postsynaptic density protein of 95 kDa (PSD95), the *Drosophila* protein disc large tumor suppressor A (DlgA), and the tight junction protein zona-occludens 1(zo-1). PDZ domains are approximately 90 amino acids in length and typically recognize target motifs at the extreme C terminus of interacting proteins (3), although some PDZ

Louis M. Luttrell and Stephen S.G. Ferguson (eds.), *Signal Transduction Protocols*, Methods in Molecular Biology, vol. 756, DOI 10.1007/978-1-61779-160-4_21, © Springer Science+Business Media, LLC 2011

proteins can bind to an internal PDZ ligand not found at the extreme C terminus (4). Most PDZ-interacting proteins possess a C-terminal motif consisting of a hydrophobic amino acid at the terminal position and either a hydrophobic amino acid, a hydroxyl-bearing amino acid (S or T), or an acidic amino acid (D or E) at the –2 position (5).

PDZ domain-mediated interactions are often of very high affinity and therefore amenable to detection by a number of different screening approaches. For example, many receptor/PDZ interactions have been first detected in yeast two-hybrid screens (6). Other screening approaches include phage display (7), fluorescence polarization (8), and pull-down studies from tissue samples followed by mass spectrometry (9). When novel PDZ interactions are detected via these types of unbiased screening approaches, it is natural to wonder about the specificity of the association: Was a particular PDZ scaffold detected as an interacting partner for a receptor of interest simply because the PDZ protein was very abundant in the yeast two-hybrid library or tissue sample that was chosen for the initial screen? Might there be other PDZ proteins that actually have much higher affinities for interacting with the receptor of interest? To address such questions, over the past few years a number of groups have developed screening approaches involving the creation of proteomic arrays of PDZ domains, which can be screened in a rapid and comprehensive manner for their binding to any target C terminus of interest. Commercially available PDZ domain arrays have been developed by Panomics, and several academic laboratories have also developed their own PDZ domain arrays (5, 10, 11). Here we describe a standard approach for screening a PDZ domain array with a receptor C terminus of interest to detect novel interactions, then confirming these interactions via reverse overlay and co-immunoprecipitation.

2. Materials

2.1. Screening of a PDZ Proteomic Array

1. Purified S- and hexahistidine-tagged fusion proteins of candidate PDZ domains (see Note 1).
2. Purified receptor C-terminal (CT) GST fusion proteins (see Note 2).
3. 96-well plates (plastic).
4. Parafilm.
5. Multiblot replicator – “spotter” (V&P Scientific, Inc.).
6. Absolute (200 proof) Ethanol.
7. Aluminum foil.

8. Nytran SuperCharge 96-grid nylon membranes, 0.45 μM (Whatman).
9. Blot Buffer: 2% (w/v) nonfat powdered milk, 0.1% (v/v) Tween-20, 50 mM NaCl, 10 mM Hepes, in dH_2O, pH 7.4.
10. Blot trays.
11. Rocking platform.
12. Anti-GST HRP Conjugate (GE Healthcare Life Sciences).
13. Enhanced chemiluminescence (ECL) kit (ThermoScientific).
14. Autoradiography cassette.
15. X-ray film for ECL detection.

2.2. Reverse Blot Overlays

1. SDS-PAGE mini-gel apparatus (Invitrogen).
2. Western blot transfer apparatus (Invitrogen).
3. Power supply (BioRad).
4. SDS-PAGE 4–20% tris-glycine mini gels (Invitrogen).
5. SDS-PAGE buffer: 30.5 g Tris, 144.8 g glycine, 10 g SDS, diluted up to 10 L of dH_2O.
6. Transfer buffer: 5.8 g Tris, 28.8 g glycine, 800 ml methanol, diluted up to 4 L of dH_2O.
7. 6× Sample Buffer: 12% (v/v) β-mercaptoethanol, 12% (w/v) SDS, 30% (v/v) glycerol, 100 mM TRIS, 5 mg of bromophenol blue, dH_2O up to 30 ml. Six times stock may be diluted with dH_2O to generate 2× and 1× stock as needed. Store out of light.
8. Anti-S-protein HRP Conjugate (Novagen).
9. Anti-Hexahistidine HRP Conjugate (Miltenyi Biotec Inc).
10. ImageJ or similar image analysis software.

2.3. Confirmation of Receptor/PDZ Interaction by Co-precipitation

1. HEK-293T cells (American Type Culture Collection).
2. Complete DMEM: 10% qualified fetal bovine serum (Invitrogen), 1% penicillin and streptomycin (Invitrogen), DMEM (1×) high glucose (Invitrogen).
3. 100 mm tissue culture dishes.
4. Lipofectamine™ 2000 (Invitrogen).
5. cDNAs expression plasmids encoding the full-length receptor and candidate PDZ interactor (FLAG-tagged receptor and HA-tagged PDZ scaffold).
6. PBS/Ca^{2+}: phosphate-buffered saline (Invitrogen) supplemented with 0.9 mM calcium chloride.
7. Harvest Buffer: 50 mM NaCl, 20 mM Hepes, 5 mM EDTA, 1 protease inhibitor cocktail tablet (Roche Applied Science), 1% (v/v) Triton-X-100, diluted with dH_2O up to 50 ml, pH 7.4.

8. 1.5 mL capped microcentrifuge tubes.
9. High-speed microcentrifuge in a cold room (4°C).
10. M2 anti-FLAG agarose beads (Sigma).
11. Table-top microcentrifuge.
12. Heat block.
13. Anti-HA (12CA5) mouse monoclonal antibody (Roche Diagnostics).
14. Anti-mouse IgG HRP Conjugate (GE Healthcare Life Sciences).

3. Methods

3.1. Preparation and Screening of the PDZ Proteomic Array

To construct the proteomic array, recombinant PDZ domains are spotted onto a nitrocellulose grid and, similar to a far Western blot, purified receptor CT fusion proteins are subsequently overlaid onto the membranes. For the generation of the PDZ domains mentioned in the below protocol, the bacterial expression vector pET30A was used to yield purified recombinant PDZ domains containing an S-tag and a hexahistidine tag. Screening multiple PDZ domain candidates simultaneously can increase the odds of detecting a candidate interaction. If using a commercially available PDZ domain array, skip to step 10.

1. Pipette 50 μl of each purified S-tagged PDZ protein (1 μg/μl) into its respective well of a 96-well plate. Keep the 96-well plate on ice while distributing PDZ proteins, in order to minimize protein degradation.
2. Cover the 96-well plate with parafilm and store at −80°C until required for experimental use. Because multiple freeze/thaw cycles can enhance the degradation or precipitation of the purified PDZ domains, limit the number of freeze–thaws.
3. To construct the proteomic array, remove the 96-well plate from the −80°C freezer and thaw on ice. It is important that all proteins are completely thawed before spotting.
4. Soak the spotter in fresh absolute ethanol.
5. Spread aluminum foil onto the bench-top and place the unused 96-grid nylon membranes on top of the foil, keeping the blue backing as a separation between the membranes and the foil. Remove the top blue covers to expose the membranes. Set the blue covers aside. It is helpful to construct multiple arrays at one time, as they may be stored for up to 1 year at 4°C.

6. Use a multichannel pipetter to thoroughly mix the thawed PDZ proteins. If any wells are lacking in volume, replenish the well(s) with purified PDZ protein (1 μg/μl).
7. Remove the spotter from the ethanol, shake off excess, and allow it to dry completely before use.
8. Dip the spotter into the 96-well plate and swish gently. The fusion proteins will be drawn up into the spotter. Then, lay the spotter onto one of the nitrocellulose membranes, beginning on one side and rolling to the other side to equally distribute the PDZ proteins. If some of the proteins do not transfer, manually pipette 1 μl of the PDZ domain stock (1 μg/μl) onto the appropriate grid block (Fig. 1).
9. Allow membranes to dry for at least 30 min at room temperature (RT), or overnight, until the spots are no longer visible.

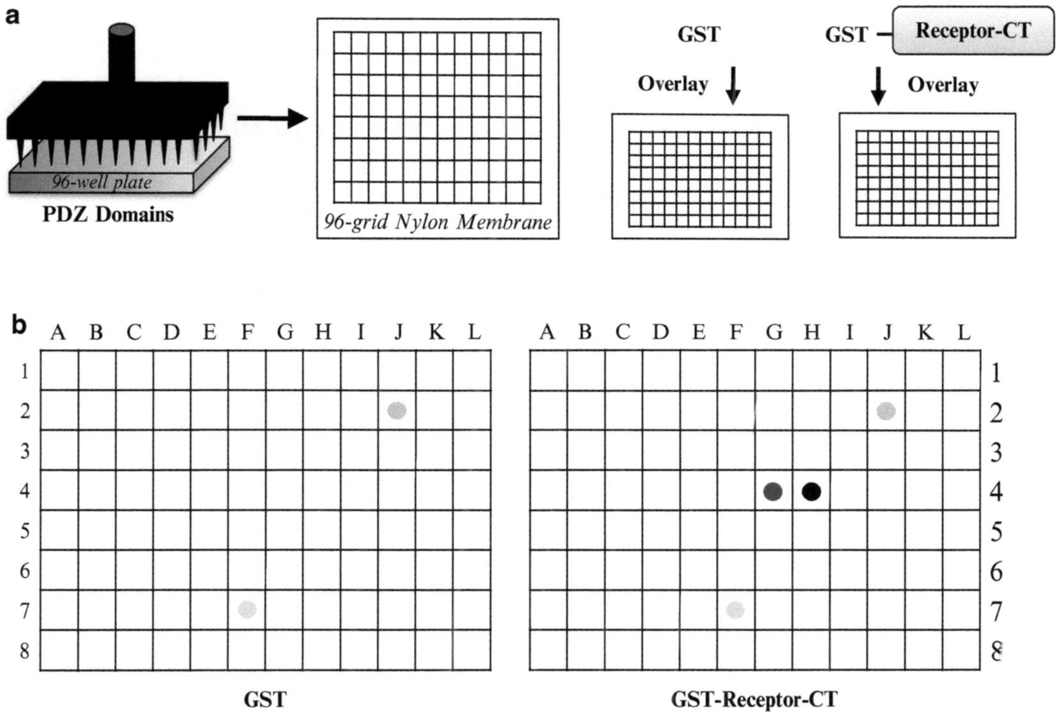

Fig. 1. Screening of the PDZ proteomic array. (**a**) A schematic is shown depicting the construction and screening of the PDZ proteomic array. A multireplicator "spotter" is used to equally distribute the PDZ domain fusion proteins onto gridded nitrocellulose. After drying overnight, purified recombinant GST alone or GST–receptor–CT fusion proteins are overlaid onto the membranes and incubated to detect PDZ interactions. (**b**) Representative data for screening GST alone (*left*) and GST–receptor–CT fusion proteins (*right*) are shown. On the GST alone blot, bins J2 and F7 depict the nonspecific binding of the GST fusion protein to the array. However, when 100 nM of GST–receptor–CT is overlaid onto the membrane, two additional positive hits are seen in bins G4 and H4, indicating that the corresponding PDZ domains most likely interact with the receptor–CT. Conversely, the intensity of the signal in bin J2 remains the same, which is representative of a "false positive" on the array. Therefore, only the PDZ proteins identified in bins G4 and H4 should be pursued further and validated using a reverse overlay and co-immunoprecipitation approaches.

Replace the blue membrane top cover and stack arrays together. Return arrays to the packaging envelope and store at 4°C until needed for use. Return the PDZ proteins in the 96-well plate for storage at −80°C.

10. To screen the array, remove membranes from the refrigerator and block them in blot buffer for 1 h at RT on a gently shaking platform.
11. Overlay 100 nM of the purified GST–receptor–CT fusion protein and 100 nM of the purified GST protein alone, diluted in 10 ml of blot buffer, onto duplicate blocked PDZ domain membranes (see Note 3).
12. Incubate membranes and GST fusion proteins for 1 h, at RT, or overnight at 4°C.
13. Wash membranes five times using 10 ml of blot buffer per 5-min wash.
14. Incubate membranes with an HRP-conjugated anti-GST monoclonal antibody (1:4,000) for 1 h at RT, with gentle shaking (see Note 4).
15. Wash membranes five times using 10 ml of blot buffer per 5-min wash.
16. Use an ECL kit to visualize the HRP.
    (a) Incubate membranes with freshly prepared ECL solution for 1–5 min.
    (b) Drain excess ECL solution and gently pat membranes dry.
    (c) Transfer membranes to a clear plastic sheet protector and tape into an autoradiography cassette.
    (d) Expose film for various time points and develop using a film developer.
    (e) Compare GST alone and GST–receptor–CT fusion proteins to determine specific binding (Fig. 1).

3.2. Reverse Blot Overlays and Receptor/PDZ Affinity Calculations

After screening for candidate receptor/PDZ interactions using a PDZ domain proteomic array or other screening technique, a common next step is to confirm any putative interactions and estimate their affinity. Candidate interactions should be confirmed using a different technique than was used for screening. For example, an overlay assay should be done in reverse, or a pull-down assay (9) should be performed instead. The following protocol will describe a reverse overlay approach for confirming a novel receptor/PDZ interaction.

1. Load 2 μg of purified receptor GST fusion proteins into individual wells of a SDS-PAGE gel and subject purified proteins to gel electrophoresis at 120 V for approximately 100 min.

2. Transfer separated GST fusion proteins to nitrocellulose paper using 25 V for 120 min.
3. Cut blot into vertical strips to separate groups of GST alone (negative control) and GST–receptor–CT.
4. Overlay increasing concentrations of candidate PDZ domain (S- and hexahistidine-tagged) in order to assess specificity of binding (i.e., 1, 3, 10, 30, 100, 300, 1,000, 3,000 nM).
5. Wash blots three times using 10 ml of blot buffer per 5-min wash.
6. Incubate blots with HRP-conjugated anti-S-protein antibody diluted in 10 ml of blot buffer (1:4,000) or HRP-conjugated anti-hexahistidine protein antibody diluted in 10 ml of blot buffer (1:4,000) for 1 h at RT, with gentle shaking.
7. Wash membranes three times using 10 ml of blot buffer per 5-min wash.
8. Use ECL kit to visualize HRP (see Subheading 3.1; step 16).
9. Scan blots into ImageJ (freeware from NIH.gov) or another suitable program, and calculate the optical densitometry (OD value) of each immunoreactive band.

In order to generate a binding curve, OD values must be converted into percentages of maximal binding and plotted on a graph. "Maximal binding" is defined when the amount of PDZ domain binding (OD value) does not change between two increasing concentrations (OD values) of the amount of PDZ domain overlaid. As shown in Fig. 2, the curve may then be used to estimate the affinity constant (K_D) of the interaction (see Note 5).

3.3. Confirmation of Receptor/PDZ Interaction in a Cellular Context

The first two sections describe approaches for screening a receptor C terminus for potential interactions with a large number of candidate PDZ domains and then confirming and estimating the affinities of any detected interactions. An important next step is to assess whether a given interaction actually occurs when the full-length PDZ scaffold and full-length receptor are expressed in a cellular context. This can be done using BRET or FRET approaches (12), or, as described below, by co-immunoprecipitation.

1. Maintain HEK-293T cells in Complete DMEM in a humidified incubator at 37°C, 5% CO_2/95% air mixture. For transfection and immunoprecipitation experiments, culture HEK-293T cells on 100 mm tissue culture-treated sterile plates.
2. Use Lipofectamine™ 2000 to transfect HEK-293T cells with 1 µg each of a FLAG-tagged receptor cDNA and an HA-tagged PDZ scaffold cDNA. It is also necessary to have a condition in which just the HA-tagged PDZ scaffold is expressed, as well as 1 µg of pcDNA3.1 (mock cDNA) to control

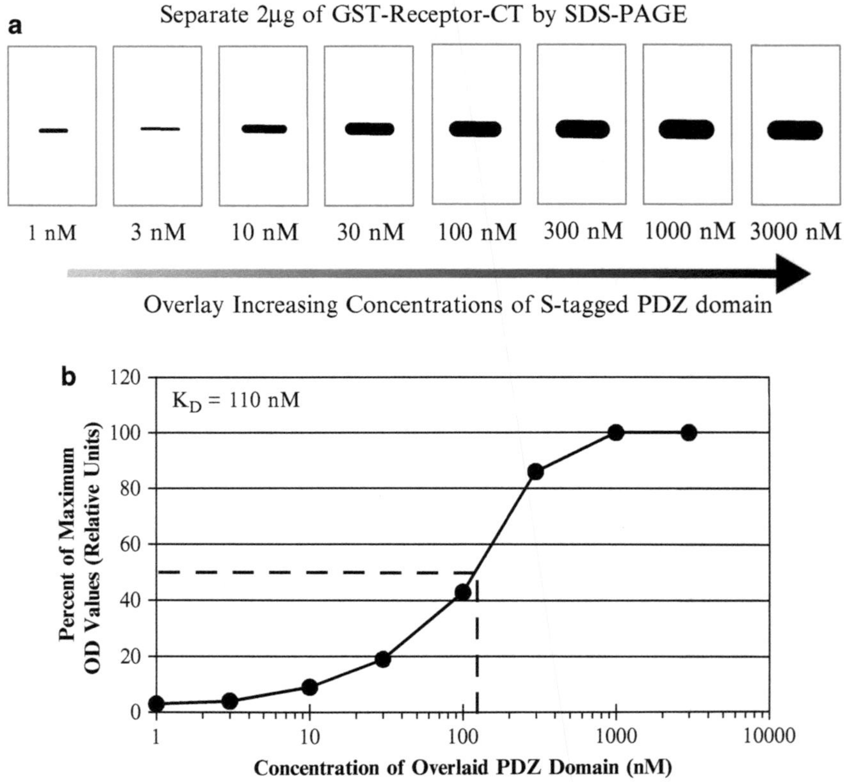

Fig. 2. Reverse blot overlays and receptor–PDZ affinity estimations. (**a**) Representative data shown for reverse blot overlay experiments. Briefly, 2 µg of GST–receptor–CT are separated by SDS-PAGE, transferred to nitrocellulose and then cut into strips. Membranes are then overlaid with increasing concentrations of an S-tagged PDZ domain that was identified as a positive hit in the original screen of the PDZ array. Importantly, no binding is seen for overlay of GST alone (*data not shown*). (**b**) After converting the immunoreactive bands in the overlay experiments into OD values, the maximum OD value can be identified as that value does not change between two increasing concentrations of S-tagged PDZ domain overlay. The remaining OD values are converted into a percentage of the maximal OD value and plotted onto a dose–response graph, in which the concentration of the PDZ domain is on a logarithmic scale. The curve can then be used to estimate the K_D of the receptor/PDZ domain interaction, or the concentration of PDZ domain required for 50% of maximum binding (*dashed line*).

for any HA-tagged PDZ scaffold that may be nonspecifically pulled down by the FLAG beads used in the immunoprecipitation (see Note 6).

3. After 24–48 h, transfer cells to ice in order to slow protein degradation, aspirate old media, and wash cells two times with 5 ml of ice-cold PBS/Ca^{2+} per 5-min wash.
4. Add 500 µl of ice-cold Harvest Buffer to cells and scrape cells into an Eppendorf tube.
5. Solubilize proteins for 30 min at 4°C, with end-over-end rotation.
6. Microcentrifuge the samples for 20 min at 18,000 *g* to separate insoluble membrane fraction (pellet) from the soluble lysate (supernatant).

7. Remove 50 µl of the soluble lysate to check the efficiency of the transfection and receptor solubilization. Add 10 µl 6× sample buffer to achieve a 1× sample buffer final concentration, denature proteins, and facilitate sample storage.
8. Incubate the remaining soluble lysate with 30 µl of M2 FLAG agarose beads for 2 h at 4°C, with end-over-end rotation.
9. Spin down beads using a table-top centrifuge (20 s) and carefully aspirate the supernatant. Wash beads three times with 1 ml of ice-cold lysis buffer, spinning down beads in-between washes.
10. Resuspend beads in 60 µl of 2× Sample Buffer.
11. Boil samples at 100°C for 10 min to elute proteins from beads.
12. Load 20 µl of each IP supernatant into an SDS-PAGE gel and separate proteins by gel electrophoresis at 120 V for 100 min. Load two separate gels in order to probe with an antibody for the receptor, as well as an antibody for the PDZ domain scaffold.
13. Transfer proteins to nitrocellulose at 25 V for 120 min.
14. Perform a Western blot.
    (a) Block membranes with 10 ml blot buffer for 30 min at RT.
    (b) Probe one membrane with an antibody directed against the receptor to confirm that the FLAG-tagged receptor was immunoprecipitated by the FLAG beads. Use the anti-HA 12CA5 antibody (1:4,000) to probe the other blot for the HA-tagged PDZ scaffold. Dilute primary antibodies in 10 ml of blot buffer and incubate membranes for 1 h at RT, with gentle shaking.
    (c) Wash membranes three times using 10 ml of blot buffer per 5-min wash.
    (d) Incubate membranes with the appropriate secondary HRP-conjugated antibody, directed against the host species of the primary antibody. Use the anti-mouse HRP-conjugated secondary antibody (1:4,000) to detect if the HA-tagged PDZ scaffold was co-immunoprecipitated.
    (e) Wash membranes three times using 10 ml of blot buffer per 5-min wash.
    (f) Use an ECL kit to visualize the HRP (see Subheading 3.1; step 16).

An example of a successful immunoprecipitation between a receptor and a PDZ domain-containing scaffold using a heterolgous expression system is shown in Fig. 3 (see Note 7).

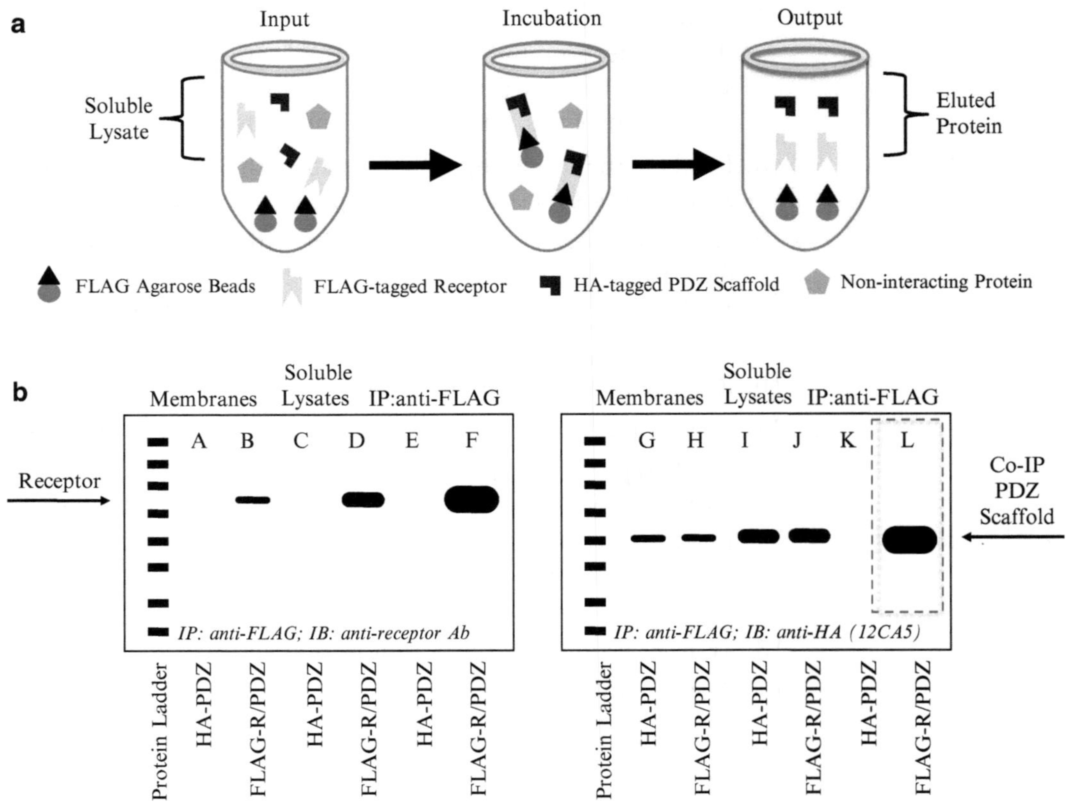

Fig. 3. Immunoprecipitation experiments to validate receptor/PDZ interactions in a cellular context. (**a**) Schematic diagram illustrating the immunoprecipitation (IP) of a FLAG-tagged receptor using FLAG agarose beads. Solublized cell lysates containing FLAG-tagged receptor and HA-tagged PDZ scaffold are incubated with FLAG-conjugated agarose beads. The FLAG antibody (*triangle*) that is covalently attached to the beads (*circle*) immunoprecipitates the FLAG-tagged receptor (*gray shape*), while the noninteracting protein (*pentagon*) does not associate. The HA-tagged PDZ scaffold (*L-shape*) also binds to the FLAG-tagged receptor and is co-immunoprecipitated by the beads. After the incubation, the beads are incubated with 2× Sample Buffer and the receptor and the PDZ scaffolds are eluted. (**b**) Representative data showing a successful immunoprecipitation (IP) of a FLAG-tagged receptor and the specific co-immunoprecipitation (co-IP) of a HA-tagged PDZ scaffold. Samples from the membrane, soluble lysate, anti-FLAG IP fractions are run on an SDS-PAGE gel, transferred to nitrocellulose, and blotted with an antibody corresponding to the receptor (*left blot*) and the HA-tag on the PDZ scaffold (*right blot*). An efficient solubilization of the receptor from the membrane is shown (*lane D* vs. *lane B*), and incubation of this soluble lysate with FLAG beads results in a robust immunoprecipitation of the FLAG-tagged receptor (*lane F*). Likewise, the HA-PDZ scaffold is solubilized efficiently (*right blot, lanes I and J* vs. *G and H, respectively*) and a band corresponding to the predicted molecular weight of the HA-PDZ scaffold is only seen in the lane in which the receptor was immunoprecipitated (*right blot, lane L*). As a negative control, the FLAG beads do not pull-down the HA-PDZ scaffold when the receptor is not co-transfected (*right blot, lane K*).

4. Notes

1. We typically prepare PDZ domain fusion proteins using the pET30A bacterial expression vector, which creates proteins containing a hexahistidine tag and an S-tag, separated by a thrombin cleavage site. The modular nature of the PDZ

domain makes it quite amenable to recombinant protein expression. The domain itself contains six anti-parallel β sheets sandwiched between two α helices, that together form a hydrophobic pocket in which the last amino acid of the PDZ consensus sequence can interact (3). PDZ domain fusion proteins should be generated using this entire domain.

2. Target protein C-termini are typically prepared as GST fusion proteins. The length of the CT fragment is important. It should be no less than 25 amino acids in length, in order to allow proper spacing between the PDZ-interacting motif and the GST moiety. Also, there must not be any sort of tag on the C-terminal end of the CT fragment. The GST moiety must be on the N-terminal side and the CT fragment must have a free C terminus. Please see Vikis and Guan for a detailed protocol describing the generation of GST fusion proteins (13).

3. When purifying GST proteins used in screening the proteomic array, it is important to prepare a GST alone control, alongside of the GST–receptor–CT, in order to assess non-specific background in the overlay. This will help to determine which hits, if any, are "false positives" in the screening of the PDZ proteomic array.

4. For initial experiments involving antibodies, it can be helpful to perform a pilot experiment to titrate the amount of the antibody required for optimal detection of the bands of interest. Concentrations of commercially available antibodies can vary across lots; therefore, a range of different dilutions can be employed in pilot studies to determine an optimal dilution.

5. The estimated affinities of PDZ domain-mediated interactions can vary greatly depending on the method used to assess the affinity of the interaction (3). In addition to the saturation-binding approach described here, other approaches that can be used include fluorescence polarization, ELISA, and surface plasmon resonance.

6. For the described immunoprecipitation experiment, the receptor has an N-terminal FLAG-tag, while the PDZ protein has an N-terminal HA-tag. The addition of two distinct tags will allow for the specific pull-down of the receptor with M2 FLAG agarose and subsequent detection of the PDZ domain for the co-immunoprecipitation experiment. Because most receptor/PDZ interactions occur at the extreme C terminus of the receptor, it is important to avoid having any tag on the C terminus of the receptor of interest. A tag on the C terminus of the receptor will almost certainly interfere with the binding of PDZ domain-containing partners.

7. The receptor can also be immunoprecipitated with an antibody raised against the receptor, as opposed to using tagged receptor constructs. This type of experimental approach is especially

useful when determining if the receptor/PDZ interaction occurs in native cells and/or tissues. However, it is important to avoid using an antibody that recognizes an epitope at the extreme C terminus of the receptor, as such an antibody may compete for C-terminal binding with the PDZ protein that is being assessed for potential co-immunoprecipitation.

Acknowledgments

The authors' research is funded by the National Institutes of Health, USA.

References

1. Overington, J. P., Al-Lazikani, B., and Hopkins, A. L. (2006) How many drug targets are there? *Nat Rev Drug Discov* **5**, 993–6.
2. Ritter, S. L., and Hall, R. A. (2009) Fine-tuning of GPCR activity by receptor-interacting proteins. *Nat Rev Mol Cell Biol* **10**, 819–30.
3. Sheng, M., and Sala, C. (2001) PDZ domains and the organization of supramolecular complexes. *Annu Rev Neurosci* **24**, 1–29.
4. Hillier, B. J., Christopherson, K. S., Prehoda, K. E., Bredt, D. S., and Lim, W. A. (1999) Unexpected modes of PDZ domain scaffolding revealed by structure of nNOS-syntrophin complex. *Science* **284**, 812–5.
5. Stiffler, M. A., Chen, J. R., Grantcharova, V. P., Lei, Y., Fuchs, D., Allen, J. E., Zaslavskaia, L. A., and MacBeath, G. (2007) PDZ domain binding selectivity is optimized across the mouse proteome. *Science* **317**, 364–9.
6. Fields, S., and Song, O. (1989) A novel genetic system to detect protein-protein interactions. *Nature* **340**, 245–6.
7. Bair, C. L., Oppenheim, A., Trostel, A., Prag, G., and Adhya, S. (2008) A phage display system designed to detect and study protein-protein interactions. *Mol Microbiol* **67**, 719–28.
8. Park, S. H., and Raines, R. T. (2004) Fluorescence polarization assay to quantify protein-protein interactions. *Methods Mol Biol* **261**, 161–6.
9. Brymora, A., Valova, V. A., and Robinson, P. J. (2004) Protein-protein interactions identified by pull-down experiments and mass spectrometry. *Curr Protoc Cell Biol* **Chapter 17**, Unit 17 15.
10. Fam, S. R., Paquet, M., Castleberry, A. M., Oller, H., Lee, C. J., Traynelis, S. F., Smith, Y., Yun, C. C., and Hall, R. A. (2005) P2Y1 receptor signaling is controlled by interaction with the PDZ scaffold NHERF-2. *Proc Natl Acad Sci U S A* **102**, 8042–7.
11. He, J., Bellini, M., Inuzuka, H., Xu, J., Xiong, Y., Yang, X., Castleberry, A. M., and Hall, R. A. (2006) Proteomic analysis of beta1-adrenergic receptor interactions with PDZ scaffold proteins. *J Biol Chem* **281**, 2820–7.
12. Ciruela, F. (2008) Fluorescence-based methods in the study of protein-protein interactions in living cells. *Curr Opin Biotechnol* **19**, 338–43.
13. Vikis, H. G., and Guan, K. L. (2004) Glutathione-S-transferase-fusion based assays for studying protein-protein interactions, *Methods Mol Biol* **261**, 175–186.

Chapter 22

Tandem Affinity Purification and Identification of Heterotrimeric G Protein-Associated Proteins

Syed M. Ahmed, Avais M. Daulat, and Stéphane Angers

Abstract

Heterotrimeric G proteins are the main signal-transducing molecules activated by G protein-coupled receptors. Their GTP-dependent activation leads to the regulation of different effectors such as adenylyl cyclases, phospholipases, and RhoGEFs. To understand the full biological consequences of GPCR signalling and to further understand the cross-talk with other signalling pathways, the complement of proteins associating with heterotrimeric G proteins needs to be identified. Here we describe our mass spectrometry-based proteomic approaches for the study of Gβγ and Gα protein complexes. This approach is predicated on the establishment of mammalian cell lines constitutively or inducibly expressing affinity-tagged versions of Gβγ or wild-type and constitutively active Gα subunits, respectively.

Key words: Heterotrimeric G protein, Tandem affinity purification, Proteomics, Protein complex, Gα subunit, Gβγ subunit

1. Introduction

G protein-coupled receptors (GPCRs) form the largest family of integral membrane receptors, and regulate tissue homeostasis in response to a myriad of extracellular stimuli. Given their diverse functions, their deregulation is predictably associated with several human diseases ranging from psychological disorders, to cancer, pain, and heart disease. GPCRs are thus preferred drug targets, with 30–40% of prescribed pharmaceuticals acting through the modulation of their activity. Agonist binding to GPCRs induces a conformational change within the receptor that triggers the exchange of GDP for GTP on the α subunit of the heterotrimeric G protein. This exchange is followed by a conformational change within the α and βγ subunits that allows the activation of their

Louis M. Luttrell and Stephen S.G. Ferguson (eds.), *Signal Transduction Protocols*, Methods in Molecular Biology, vol. 756, DOI 10.1007/978-1-61779-160-4_22,

respective effectors including phospholipases, adenylate cyclases, kinases, ion channels, and small GTPases that ultimately impinge on cellular functions.

As mentioned above, heterotrimeric G proteins are composed of α (39–49 kDa), β (35–39 kDa) and γ (6–8 kDa) subunits and play pivotal roles following G-protein-coupled receptor (GPCR) activation by agonists. To date 28 α subunits are known to be translated from 16 different genes and their splice variants. In addition, 5 β subunits and 12 γ subunits have been identified to date (1). Based on sequence homology between the different subunits, Gα proteins can be further subdivided into four families: G_s (G_s and G_{olf}), G_i (G_{tr}, G_{tc}, G_g, $G_{i1\text{-}3}$, G_o and G_z), G_q (G_q, G_{11}, G_{14} and $G_{15/16}$) and G_{12} (G_{12} and G_{13}).

Tandem Affinity Purification (TAP) schemes optimized to isolate protein complexes were pioneered using yeast (2) but have since been extended to mammalian cells (3). These methods consist of fusing two short affinity tags to a bait protein of interest to enable its purification along with its interacting proteins and to minimize the carryover of abundant or intrinsically sticky proteins that may have low affinity either to the tags or chromatography matrices. The original TAP protocol established in the laboratory of Dr. Séraphin uses the protein A IgG binding motif and the Calmodulin-Binding Peptide (CBP) as affinity tags (2). Recently, we have screened multiple short affinity tags to adapt the TAP method to the isolation of mammalian protein complexes and have settled on a combination of Streptavidin Binding Peptide (SBP) and CBP for our affinity cassette (4–6). The SBP tag has high affinity for streptavidin beads and bound proteins can be efficiently eluted in the presence of biotin (7). Eluted proteins are then further purified using calmodulin beads through binding to the CBP tag (8). Pure protein complexes can then be eluted in the presence of EGTA since the binding of CBP to calmodulin requires Ca^{2+} ions (Fig. 1). Eluted proteins are then directly digested in solution with trypsin and the resulting mixture of peptides analyzed by mass spectrometry.

We have recently utilized this novel proteomic approach to analyze the proteins associating with the Gβ and Gγ subunits of heterotrimeric G proteins (9) and are now extending our analysis to the four Gα families. This chapter describes the methods that we have optimized for the study of heterotrimeric G proteins complexes.

2. Materials

2.1. Reagents and Buffers

1. HEK293T cells (American Type Culture Collection).
2. HEK293 Flp-in™ T-Rex™ system (Invitrogen).

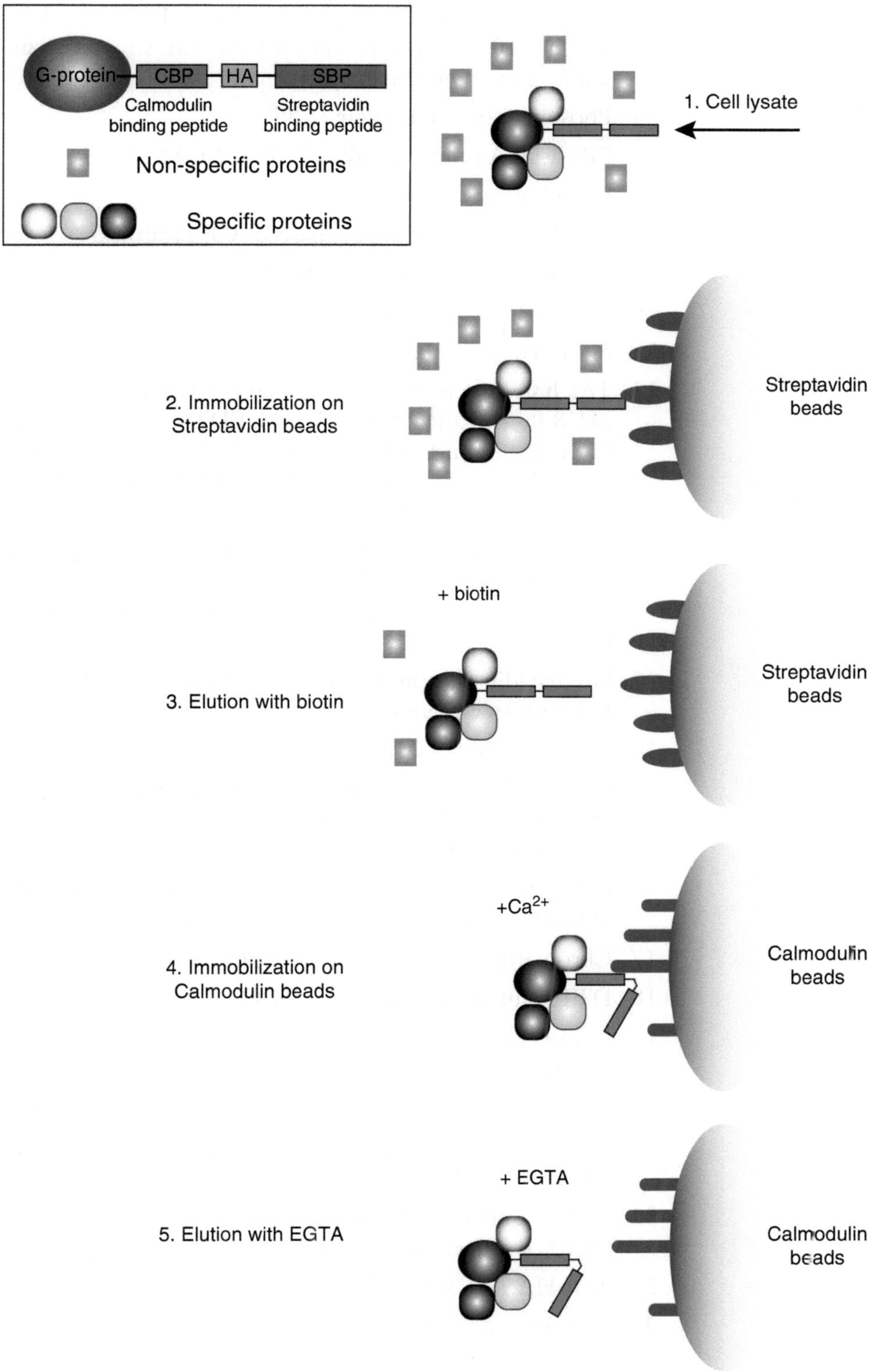

Fig. 1. Tandem affinity purification of G proteins with their associated proteins. The baits of interest are expressed as fusion proteins with streptavidin binding peptide (SBP), Human influenza hemagglutinin tag (HA), and calmodulin-binding peptide (CBP) sequences fused to the N-terminus of their coding sequences. Stable cells expressing the bait are lyzed in TAP lysis buffer and purified by two-steps affinity chromatography as shown.

3. Dulbecco's modified Eagle's medium (DMEM): 4.5 g/L glucose, 10 U/mL penicillin, 0.1 mg/mL streptomycin, and 1 mM glutamine (Invitrogen).
4. Phosphate-buffered saline (PBS).
5. PBS-EDTA solution: PBS, pH 7.4, 2 mM EDTA.
6. Liquid nitrogen.
7. Fast-Flow Streptavidin Sepharose (GE Healthcare).
8. Calmodulin-Sepharose 4B (GE Healthcare).
9. Protease inhibitor cocktail (Sigma).
10. Phosphatase inhibitors (Calbiochem).
11. TAP lysis buffer: 10% (v/v) glycerol, 50 mM HEPES-NaOH, pH 8.0, 150 mM NaCl, 2 mM EDTA, 0.1% (v/v) Igepal CA-630, 2 mM dithiothreitol (DTT), protease inhibitor cocktail, 10 mM NaF, 0.25 mM $NaOVO_3$, 100 mM β glycerophosphate (see Notes 1–3).
12. Calmodulin-binding buffer (CBB): 10 mM β-mercaptoethanol, 50 mM HEPES-NaOH, pH 8.0, 150 mM NaCl, 1 mM MgOAc, 1 mM imidazole, 0.1% Igepal CA-630 and 2 mM $CaCl_2$.
13. Streptavidin-elution buffer: CBB supplemented with 10 mM D-biotin (see Note 4).
14. $CaCl_2$ solution: 1 M $CaCl_2$ in dH_2O.
15. Calmodulin-rinsing buffer: 50 mM ammonium bicarbonate, pH 8.0, 75 mM NaCl, 1 mM MgOAc, 1 mM imidazole, and 2 mM $CaCl_2$.
16. Calmodulin-elution buffer: 50 mM ammonium bicarbonate, pH 8.0, and 25 mM EGTA (see Note 5).
17. Chromatography mini spin columns (Bio-Rad).
18. DTT solution: 1 M DTT in dH_2O.
19. Iodoacetamide solution: 500 mM iodoacetamide in dH_2O (prepare just prior to use).
20. Sequencing grade modified porcine trypsin, frozen (Promega).
21. HPLC buffer A: 95% (v/v) water, 5% (v/v) acetonitrile (Burdick & Jackson), and 0.1% formic acid (JT Baker).
22. HPLC buffer B: 5% (v/v) water, 95% (v/v) acetonitrile (Burdick & Jackson), and 0.1% (v/v) formic acid (JT Baker).
23. Anti-HA.11 Clone 16B12 antibody (Covance).
24. All chemicals are molecular biology grade from Sigma or other vendors.
25. Equipment: Refrigerated table-top centrifuge, shaker/roller, speed-vac, protein quantification kit (Bradford Reagent), spectrophotometer, mass spectrometer.

2.2. Mammalian Expression Vectors

2.2.1. Construction of Constitutive and Inducible Expression Plasmids

For the analysis of Gβ and Gγ protein complexes, we cloned Gβ2 and Gγ2 cDNAs in frame with the N-terminal SBP-HA-CBP affinity cassette present in the pGLUE plasmid that we described elsewhere (4). Briefly, the pGLUE plasmid contains the SBP and CBP affinity tags to allow for the TAP purification, a HA epitope to monitor the level of expression and the efficiency of purification by western blot (Fig. 2), as well as a multiple cloning site (MCS) downstream of the affinity tags to insert the cDNA coding for the bait protein of interest. The gene of interest is amplified by polymerase chain reaction using primers containing restriction sites compatible for its insertion into the MCS using standard molecular biology procedures. The vector also contains an internal ribosome entry site (IRES) driving the expression of the puromycin-resistance gene needed for the establishment of cell lines stably expressing the engineered fusion proteins.

For the four families of Gα proteins, we are interested in comparing the proteins associating with the wild-type and constitutively active versions of the proteins. Indeed, well-described mutations of glutamine (Q) to leucine (L) within the conserved GTP binding region of Gα proteins have been demonstrated to inhibit their intrinsic GTPase activity and to result in constitutively active Gα proteins (10–12). The constitutive activity of

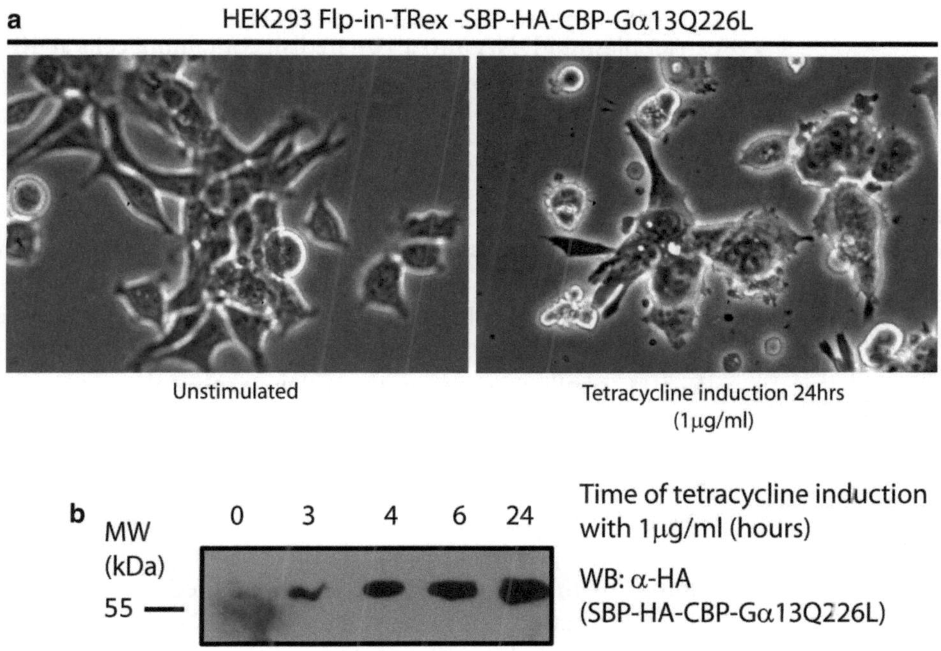

Fig. 2. Expression of wild-type and constitutively active Gα13 using the inducible cell system. (**a**) HEK293 Flp-in T-rex cells stably transfected with the constitutively active mutant SBP-HA-CBP-Gα13-Q226L were left un-induced (*left*) or treated with tetracycline (1 μg/ml) (*right*). A marked change in morphology can be observed when the constitutively active Gα13 is expressed. HEK293 Flp-in T-rex cells stably expressing SBP-HA-CBP-Gα13 were induced with tetracycline (1 μg/ml) for different times as indicated. (**b**) The induction of protein expression was monitored using western blot using anti-HA antibodies.

these proteins, however, precludes the establishment of stable cell lines, possibly due to the sustained activation of their effectors leading to cell death (Fig. 2). To circumvent this problem, we took advantage of the Flp-in™ T-Rex™ system developed by Invitrogen that allows the inducible control of protein expression. To do so, we modified the pcDNA5/FRT/TO plasmid to incorporate the SBP-HA-CBP cassette and the MCS from the pGLUE plasmid within the Frt recombination sites. For the proof of principle described here we tagged wild-type and constitutively active (Q226L) versions of the Gα13 protein at their N-termini, as this does not appear to perturb Gα13 activity (13, 14). However, for the other Gα proteins there is some evidence (albeit limited) that suggests that tagging Gα proteins at the extreme N terminus may affect their functions and that tagging the protein internally may represent a better strategy (15–17).

2.2.2. Verification of Protein Expression by Western Blot

Once the sequence integrity of the gene of interest inserted in the appropriate (constitutive or inducible) plasmid is confirmed, we generally test for protein expression by western blot analysis following transient transfections of mammalian cells. Although any mammalian cells can be used, we routinely use HEK293T cells for this purpose since they can be efficiently transfected using the calcium phosphate precipitation method (see Note 6). We typically transfect 2 μg of the bait plasmid together with 8 μg of carrier plasmid DNA into a 10-cm dish containing cells at 40–50% confluence. Using standard Western blotting techniques, cell extracts are then probed for the expression of the fusion protein using anti-HA antibodies.

2.3. Mammalian Stable Cell Lines

2.3.1. Establishing Cell Lines Constitutively Expressing the Affinity-Tagged Baits

To establish a stable cell line expressing the desired SBP-HA-CBP-tagged protein as in the case of Gβ2 and Gγ2 (9), we normally transfect 5 μg of the appropriate pGLUE plasmid DNA together with 5 μg of carrier plasmid (one that does not contain puromycin resistance marker) in a 10-cm dish containing HEK293T cells at 40–50% confluence using the calcium phosphate transfection procedure. Forty-eight hours posttransfection, the cells are rinsed once in PBS and dissociated using 1 ml of trypsin-EDTA. Cells are resuspended in 10 ml of Dulbecco's modified Eagle's medium (DMEM) supplemented with 10% (v/v) fetal bovine serum (FBS) and the resuspended cells are transferred into a 15-cm dish containing 20 ml of DMEM-FBS supplemented with 2 μg/ml of puromycin (effective concentration for HEK293T cells; independent kill curves should be performed for other cell lines). Stable integrants are then isolated by replacing the selective media every 2–3 days (during the first few days the media may need to be replaced more often if a large amount of cell death is observed). Since the gene driving the puromycin resistance is driven by an IRES present on the same messenger RNA (mRNA) as the fusion protein, all the cells selected also express the protein of interest.

2.3.2. Inducible Expression System

To generate the inducible stable cell line we used the Flp-in^{TM} T-Rex^{TM}-293 cells and the pcDNA5/FRT/TO plasmids containing SBP-HA-CBP tagged versions of wild-type and constitutively active (Q226L) human Gα13. The cells are initially maintained in selection media containing zeocin (100 μg/ml) and blasticidin (15 μg/ml) (selection for the Tet repressor). 1 μg of the individual pcDNA5/FRT/TO vector is then co-transfected with 9 μg of the pOG44 plasmid, expressing the Flp recombinase, in a 10-cm dish of Flp-in^{TM} T-Rex^{TM}-293 cells at 60–70% confluence. Forty-eight hours post-transfection cells are trypsinized and passed in media containing hygromycin B (200 μg/ml) to select for cells having recombined the insert flanked by the Frt sites (also containing the hygromycin resistance) on the pcDNA5 expression vector. Integrants become sensitive to zeocin (which is present between the Frt sites in the parental cell line and thereby excised by the recombinase) and resistant to hygromycin (the *hyg* gene is recombined into the Frt sites along with the cDNA encoding the desired protein) (see Note 7).

2.3.3. Validation of Stable Cell Lines

A polyclonal stable cell line is normally obtained after 2 weeks of selection when using puromycin with cells transfected with the pGlue plasmid (constitutive system). For the Flp-in T-Rex system it usually takes a little longer (3–5 weeks) since the recombination event induced by the Flp recombinase is less efficient. Soon after the establishment of the stable cell line, we generally freeze a subsequent passage of cells for future use. We recommend reassessing the level of expression of the fusion proteins in the newly developed stable cell lines before performing a large-scale purification. For the inducible system it is also recommended to optimize the concentration of tetracyclin and the time required to induce the expression of the bait protein (Fig. 2). As a starting point, protein expression is induced for 16 h with 1 μg/mL of tetracycline (see Note 8).

3. Methods

3.1. Amplification of Cells for Large-Scale Purification

To obtain adequate quantity of protein complexes for successful analysis using mass spectrometry, it is recommended to start with a relatively large amount of cells. Although the protocol has been optimized, every purification step results in loss of material. We start the expansion of cells from one confluent 10-cm cell-culture dish to two 15-cm cell-culture dishes. From there we usually obtain five to ten 15-cm dishes (see Note 9).

3.2. Preparation of Cell Extract

Although different procedures can be applied to harvest the cells, for HEK293T cells we use the following procedure. Note that if the sample is to be processed for mass spectrometry, steps should be taken to minimize keratin contamination from the surrounding environment, and all samples, buffers and tubes should be handled with gloves.

1. For the inducible system, induce expression of the bait proteins with the optimized concentration of tetracycline and for the correct time.
2. Remove the media from the cells, wash the cells with 10 ml of PBS and add 10 ml of PBS-2 mM EDTA to each 15-cm dish.
3. Within 5–10 min the cells will detach and can be collected into 50 ml conical tubes. The cells are then pelleted by centrifugation at 800 × *g* for 5 min, combined and washed in 50 ml of PBS.
4. After pelleting the cells for 5 min at 800 × *g*, the cells are lysed in 10 mL of TAP lysis buffer supplemented with protease and phosphatase inhibitors cocktails. We normally lyse the cells in a 15-ml conical tube at 4°C in a rotator for 30 min to 1 h.
5. To ensure complete cell lysis, proceed to one or two freeze–thaw cycles by immersing the tube in liquid nitrogen (see Note 10).
6. Thaw the lysate, aliquot into ten 1.5 ml-microfuge tubes, and clear by centrifugation at 16,100 × *g* for 10–15 min in a refrigerated microcentrifuge. We recommend retaining an aliquot of the pelleted (nonsoluble) material and a fraction (40 μl) of the cleared lysate (input). Western blot analysis with anti-HA antibody can then be performed to monitor the efficiency of the lysate preparation. If the majority of the protein of interest is in the pellet, the solubilisation conditions or alternative method of lysate preparation needs to be optimized.

3.3. Tandem Affinity Chromatography

All procedures are performed at 4°C and all buffers are prechilled on ice.

3.3.1. Streptavidin Affinity Chromatography

1. Packed streptavidin-sepharose beads (50 μl) are first equilibrated by three 800 μl washes with TAP lysis buffer (protease and phosphatase inhibitors are not necessary at this point). The beads are sedimented by centrifugation at 800 × *g* for 1 min using a microcentrifuge. We use a clean 27-gauge needle attached to a vacuum pump to remove the buffer during the washes (see Note 11).
2. Transfer the beads to a 15-ml conical tube to which the cleared lysates from the 10 microcentrifuge tubes is added. Although an incubation of 2 h at 4°C with rocking on a rotator is sufficient to isolate the majority of the proteins, we normally prepare the lysate in the afternoon of day 1 and leave the lysate in the streptavidin beads overnight.

3. After the incubation, the beads are sedimented by centrifugation at 800×*g* and the supernatant is discarded (keep an aliquot of the discarded supernatant to evaluate proteins that did not bind). At that point, the beads are transferred to a microcentrifuge tube to perform the washes.
4. Perform two washes with 800 μl of TAP lysis buffer and three washes with 800 μl of calmodulin-binding buffer (CBB). After the last wash when the beads are resuspended in 800 μl of CBB, keep a 20 μl aliquot to evaluate material bound to streptavidin beads.
5. Elute the protein complexes from the streptavidin beads using three consecutive elutions with 200 μl of biotin elution buffer (CBB supplemented with 10 mM D-biotin). Because of the very high affinity of biotin for streptavidin, the elution of the proteins from beads is instantaneous.
6. The 600 μl of eluted material is supplemented with an additional 400 μl of CBB making a total volume of 1,000 μl. Add 5 μl of 1 M $CaCl_2$ and apply to calmodulin-sepharose beads (100 μl packed material). An aliquot of eluted material (20 μl of the 1,000 μl) and of streptavidin beads post-elution (resuspend beads in 100 μl CBB and save 20 μl) can be saved for troubleshooting.

3.3.2. Calmodulin Affinity Chromatography

1. Equilibrate 100 μL of packed calmodulin-sepharose beads with three washes of 800 μL of CBB. Spin down beads by centrifugation at 800×*g* for 1 min at each step and discard the buffer by aspiration using a 27-gauge needle.
2. Incubate the biotin eluate with calmodulin-sepharose beads for 2 h with gentle agitation at 4°C.
3. After the incubation, the calmodulin beads are centrifuged and the supernatant is removed. The beads are washed twice with 800 μl of CBB and three times with calmodulin-rinsing buffer. In the last rinse save a 20 μl aliquot to analyze the material bound to beads before elution. Spin down the beads and discard the entire buffer by aspiration.
4. Resuspend the beads with 100 μl of calmodulin-elution buffer and incubate at 37°C for 5–10 min to ensure efficient elution. Spin down the beads at 800×*g* for 30 s and carefully collect the supernatant into a new microcentrifuge tube without disturbing the beads. Repeat these steps again to obtain a final 200 μl elution volume. A 10 μl aliquot of the final elution and a 20 μl aliquot of beads resuspended in 200 μl of calmodulin-rinsing buffer after the finale elution can then be saved to assess the efficiency of elution. The microcentrifuge tube containing the eluted protein complexes can be spin once more to prevent the carryover of calmodulin beads (see Note 12).

3.4. Tryptic Digestion of Protein Complexes

Once eluted, the sample is directly processed for trypin digestion.

1. The sample is reduced with the addition of 5 μl of 1 M DTT (25 mM final) and heated to 50°C for 20 min.
2. The free sulphhydrl groups are then alkylated by adding 40 μl of freshly prepared 500 mM iodoacetamide (100 mM final concentration) in the dark for 40 min.
3. The sample is then digested overnight at 37°C by adding 1 μg of sequence-grade trypsin.

3.5. Liquid Chromatography and Tandem Mass Spectrometry Analysis

3.5.1. Sample Analysis by Mass Spectroscopy

The tryptic mixture is injected on the analytical column using an autosampler but can also be manually loaded using a pressure valve. The analytical columns are made of 75 μm inner diameter (ID) fused silica and the tip is pulled either manually or using a laser puller (Sutter Instruments). We normally use 15–20 cm long columns that are packed with 14–19 cm of reverse phase material (Jupiter 4 μm Proteo 90A; Phenomenex, Inc.) using a pressure valve. The volume of the sample is reduced to 40 μl using a Speed-Vac. Half of the sample (20 μl) is then loaded on the column. In our setup, the analytical column is placed online with a LTQ linear ion-trap mass spectrometer and samples are loaded into the column through backpressure from an HPLC machine. The peptides are then eluted using a 2 h gradient method where the aqueous buffer A is progressively mixed with higher proportion of the organic buffer B by the HPLC (see Note 13). To reach nanoflow capabilities, a flow split system is used to reduce the flow of 150 μl/min coming out of the HPLC to 20–50 nl/min on the analytical column. Peptide ions are dynamically selected for fragmentation using data-dependent acquisition by the operating software (the five more intense precursor ions of each mass spectrometry (MS) scan are selected for subsequent MS/MS).

3.5.2. Representative Results

Using the described method, the protein complexes of wild-type and constitutively active (Q226L mutation) Gα13 were purified, processed, and analyzed by LC-MS/MS (Fig. 3). Several hundred peptides corresponding to each Gα13 baits could be detected (Table 1). Several peptides for the Ric8A protein, previously described to be a guanine nucleotide exchange factor for most Gα proteins (18), were also detected in both complexes. Interestingly, peptides corresponding to different Gβ and Gγ subunits were only detected in the wild-type protein complex (Table 1). Although this result is consistent with the dissociation of Gα13 protein from Gβγ dimers following activation, it may be that the Q226L mutant has lower affinity for Gβγ dimers, preventing their co-affinity purification. Strikingly, numerous peptides attributed to the three RhoGEFs effectors of Gα13, p115-RhoGEF (19), PDZ-RhoGEF (14), and LARG (20), were identified only in the Q226L protein complex (Table 1).

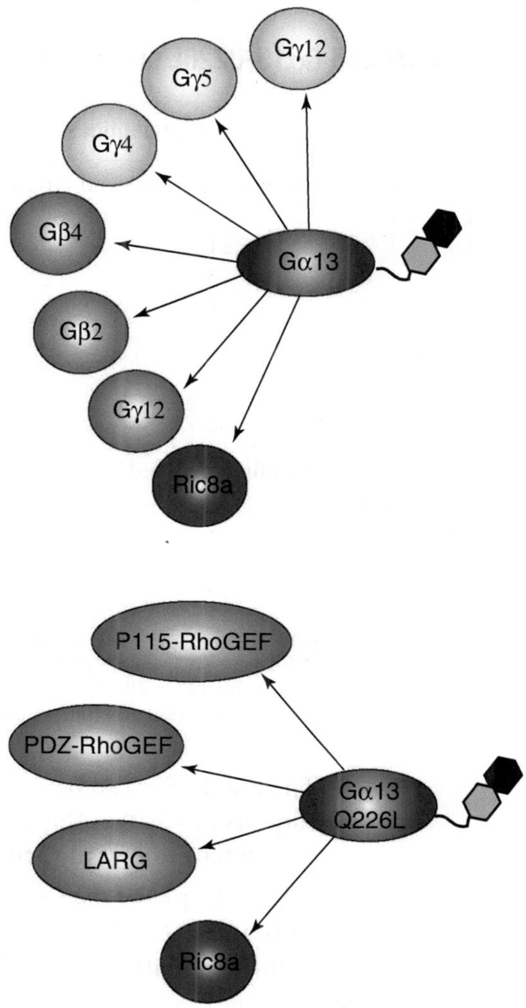

Fig. 3. Schematic representation of wild-type (*top*) and constitutively active (*bottom*) Gα13 protein–protein interaction network.

Although no novel effectors of Gα13 were identified, extending such approach to the other families or cell system may unravel novel hetrotrimeric G proteins regulated pathways.

4. Notes

1. Filter all buffers with a 0.44 μm filter to reduce contamination from dust and keratins.
2. During the wash steps the protease and phosphatase inhibitors can be left out of the buffers.

Table 1
Analysis of Gα13-WT and Gα13-Q226L protein complexes by LC-MS/MS

Gene ID	Protein	Unique peptides	Total peptides	% Coverage
A. LC-MS/MS analysis of tandem affinity purified wild-type Gα13				
10672	Gα13	59	1,187	71.9
2782	Gβ1	11	45	39.1
2783	Gβ2	8	33	33.8
59345	Gβ4	4	4	28.8
2786	Gγ4	1	2	22.7
2787	Gγ5	5	48	76.5
55970	Gγ12	2	3	31.9
60626	Ric8a	38	464	46.4
B. LC-MS/MS analysis of tandem affinity purified constitutively activated mutant Gα13Q226L				
10672	Gα13Q226L	39	558	61.8
9138	P115-RhoGEF	7	13	9.1
9826	PDZ-RhoGEF	14	75	13.4
23365	LARG	25	65	19.1
60626	Ric8a	17	232	36.9

Summary of the proteins identified in the respective affinity-purified protein complexes by LC-MS/MS. The total number of peptides, the number of unique peptides identified and the percent sequence coverage are listed. The table is representative of two independent affinity purification experiments for each bait protein

3. The lysis buffer can be stored with all the inhibitors in 10-mL aliquots at −20°C. Aliquots can be thawed at the time of the experiment to lyse the cells.
4. Adjust the pH of the streptavidin-elution buffer to pH 8.0 to let biotin into solution. We typically use 30 μL of NaOH 1 N to make 500 μL volume of solution.
5. The pH of the calmodulin-elution buffer rises upon storage. We recommend to always measure the pH before use. Ensure that the pH is at 8.0 to retain maximum activity of trypsin.
6. We recommend using the calcium phosphate precipitation procedure, which is inexpensive and efficient for the transfection of HEK293 cells. However, any other transfection reagent may be used. To avoid false-positive interacting proteins and undesired heat-shock proteins associating with the bait protein, clones expressing lower levels of the fusion proteins may help.
7. For more details about the Flp-in™ T-Rex™-293 cells, we recommend consulting the manufacturer's manual available at http://tools.invitrogen.com/content/sfs/manuals/flpintrex_man.pdf.
8. The tetracycline concentration (0.1–1 μg/mL) and the time of induction (12–48 h) may be varied to optimize or modulate the expression of your desired proteins.
9. If expression level of your bait protein is low, more starting material may be necessary.

10. This could represent a good stopping point as the cells can be stored in liquid nitrogen or at –80°C for several weeks without any decrease in protein complex isolation.
11. To prevent proteins coming out of solution avoid drying off the beads.
12. Generally 70–90% of the bait protein is eluted, however, for some bait proteins this step can be very inefficient. The addition of 0.1% (w/v) of the acid-cleavable detergent RapiGest to the calmodulin-elution buffer can improve the elution in these cases. After elution, the RapiGest needs to be cleaved using 1 M HCl before LC-MS/MS analysis. Alternatively, when elution is especially inefficient, the protein complex can be digested directly on the beads instead of eluting with calmodulin-elution buffer. To do this, resuspend the beads in 50–100 μl of ammonium bicarbonate buffer and add trypsin overnight. The next day, sediment the beads by centrifugation, collect the supernatant and proceed to the alkylation and reduction steps as described.
13. We always use the same glass cylinder to make up the HPLC buffer and only rinse it with milliQ H_2O water. Never use detergents because this will be a source of contamination in the mass spectrometer.

References

1. Cabrera-Vera, T. M., Vanhauwe, J., Thomas, T. O., Medkova, M., Preininger, A., Mazzoni, M. R., and Hamm, H. E. (2003) Insights into G protein structure, function, and regulation, *Endocr Rev* **24**, 765–81.
2. Rigaut, G., Shevchenko, A., Rutz, B., Wilm, M., Mann, M., and Seraphin, B. (1999) A generic protein purification method for protein complex characterization and proteome exploration. *Nat Biotechnol* **17**, 1030–2.
3. Gingras, A. C., Aebersold, R., and Raught, B. (2005) Advances in protein complex analysis using mass spectrometry. *J Physiol* **563**, 11–21.
4. Angers, S. (2008) Proteomic analyses of protein complexes in the Wnt pathway. *Methods Mol Biol* **468**, 223–30.
5. Angers, S., Thorpe, C. J., Biechele, T. L., Goldenberg, S. J., Zheng, N., MacCoss, M. J., and Moon, R. T. (2006) The KLHL12-Cullin-3 ubiquitin ligase negatively regulates the Wnt-beta-catenin pathway by targeting Dishevelled for degradation. *Nat Cell Biol* **8**, 348–57.
6. Angers, S., Li, T., Yi, X., MacCoss, M. J., Moon, R. T., and Zheng, N. (2006) Molecular architecture and assembly of the DDB1-CUL4A ubiquitin ligase machinery. *Nature* **443**, 590–3.
7. Keefe, A. D., Wilson, D. S., Seelig, B., and Szostak, J. W. (2001) One-step purification of recombinant proteins using a nanomolar-affinity streptavidin-binding peptide, the SBP-Tag. *Protein Expr Purif* **23**, 440–6.
8. Klevit, R. E., Blumenthal, D. K., Wemmer, D. E., and Krebs, E. G. (1985) Interaction of calmodulin and a calmodulin-binding peptide from myosin light chain kinase: major spectral changes in both occur as the result of complex formation. *Biochemistry* **24**, 8152–7.
9. Ahmed, S. M., Daulat, A. M., and Angers, S. (2010) G protein betagamma subunits regulate cell adhesion through Rap1a and its effector Radil. *J Biol Chem* **285**, 6538–51.
10. Kalinec, G., Nazarali, A. J., Hermouet, S., Xu, N., and Gutkind, J. S. (1992) Mutated alpha subunit of the Gq protein induces malignant transformation in NIH 3T3 cells. *Mol Cell Biol* **12**, 4687–93.
11. Landis, C. A., Masters, S. B., Spada, A., Pace, A. M., Bourne, H. R., and Vallar, L. (1989) GTPase inhibiting mutations activate the alpha chain of

Gs and stimulate adenylyl cyclase in human pituitary tumours. *Nature* **340**, 692–6.

12. Wong, Y. H., Federman, A., Pace, A. M., Zachary, I., Evans, T., Pouyssegur, J., and Bourne, H. R. (1991) Mutant alpha subunits of Gi2 inhibit cyclic AMP accumulation. *Nature* **351**, 63–5.
13. Xu, N., Bradley, L., Ambdukar, I., and Gutkind, J. S. (1993) A mutant alpha subunit of G12 potentiates the eicosanoid pathway and is hig hly oncogenic in NIH 3T3 cells. *Proc Natl Acad Sci U S A* **90**, 6741–5.
14. Fukuhara, S., Murga, C., Zohar, M., Igishi, T., and Gutkind, J. S. (1999) A novel PDZ domain containing guanine nucleotide exchange factor links heterotrimeric G proteins to Rho. *J Biol Chem* **274**, 5868–79.
15. Hynes, T. R., Hughes, T. E., and Berlot, C. H. (2004) Cellular localization of GFP-tagged alpha subunits, *Methods Mol Biol* **237**, 233–46.
16. Hughes, T. E., Zhang, H., Logothetis, D. E., and Berlot, C. H. (2001) Visualization of a functional Galpha q-green fluorescent protein fusion in living cells. Association with the plasma membrane is disrupted by mutational activation and by elimination of palmitoylation sites, but not be activation mediated by receptors or AlF4. *J Biol Chem* **276**, 4227–35.
17. Yu, J. Z., and Rasenick, M. M. (2002) Real-time visualization of a fluorescent G(alpha)(s): dissociation of the activated G protein from plasma membrane. *Mol Pharmacol* **61**, 352–9.
18. Tall, G. G., Krumins, A. M., and Gilman, A. G. (2003) Mammalian Ric-8A (synembryn) is a heterotrimeric Galpha protein guanine nucleotide exchange factor. *J Biol Chem 278*, 8356–62.
19. Kozasa, T., Jiang, X., Hart, M. J., Sternweis, P. M., Singer, W. D., Gilman, A. G., Bollag, G., and Sternweis, P. C. (1998) p115 RhoGEF, a GTPase activating protein for Galpha12 and Galpha13. *Science* **280**, 2109–11.
20. Suzuki, N., Nakamura, S., Mano, H., and Kozasa, T. (2003) Galpha 12 activates Rho GTPase through tyrosine-phosphorylated leukemia-associated RhoGEF. *Proc Natl Acad Sci U S A* **100**, 733–8.

Chapter 23

Study of G Protein-Coupled Receptor/β-arrestin Interactions Within Endosomes Using FRAP

Benjamin Aguila, May Simaan, and Stéphane A. Laporte

Abstract

β-arrestins, through their scaffolding functions, are key regulators of G protein-coupled receptor (GPCR) signaling and intracellular trafficking. However, little is known about the dynamics of β-arrestin/receptor interactions and how these complexes, and complexes with other regulatory proteins, are controlled in cells. Here, we use yellow fluorescent protein (YFP)-tagged β-arrestin 2 and a fluorescence recovery after photobleaching (FRAP) imaging approach to probe the real-time interaction of β-arrestin with a GPCR, the bradykinin type 2 receptor (B2R). We provide a detailed protocol to assess the avidity of β-arrestin2-YFP for B2R within endosomes in HEK293 cells. β-arrestin2-YFP associated with internalized receptors is photobleached with intense light, and fluorescence recovery due to the entry of nonbleached β-arrestin2-YFP is monitored over time as a measure of the rate exchange of β-arrestin2-YFP within the endosome. This approach can be extended to other GPCR/β-arrestin complexes and their putative regulators to provide information about the kinetics of similar protein–protein interactions in cells. Moreover, these techniques should provide insight into the role of β-arrestins in the intracellular trafficking and signaling of GPCRs.

Key words: Beta-arrestin, G protein-coupled receptor, Fluorescence recovery after photobleaching, Confocal microscopy, Yellow fluorescent protein

1. Introduction

G protein-coupled receptors (GPCRs) represent one of the most important targets in drug discovery research. Approximately one quarter of clinically available drugs act on these receptors (1). At the cellular level, GPCRs are under tight regulation to control their responsiveness. Following agonist binding and the onset of signaling, GPCRs are rapidly desensitized through phosphorylation by GPCR kinases (GRKs) (2). Central to receptor desensitization

Louis M. Luttrell and Stephen S.G. Ferguson (eds.), *Signal Transduction Protocols*, Methods in Molecular Biology, vol. 756, DOI 10.1007/978-1-61779-160-4_23,

are β-arrestins that bind GRK-phosphorylated GPCRs, uncouple them from their cognate G proteins, and promote receptor internalization (3). Not only does internalization play an important role in desensitization and subsequent resensitization of GPCRs, but also in the propagation of intracellular signaling through endosomal scaffolding of different signaling effectors by β-arrestins. A major step toward understanding the role of β-arrestins in GPCR trafficking and signaling was taken through the use of β-arrestin fused to green fluorescent protein (4). This tool provided the means to monitor, in live cells and in real-time, the role of β-arrestin in GPCR trafficking, and to classify GPCRs based on their interaction with β-arrestins. For instance, Class B GPCRs, like the angiotensin II (AngII) type I receptor (AT1R), traffic with β-arrestins into endosomes (5, 6). The prevailing model suggests that the high avidity interaction of β-arrestins with Class B GPCRs restricts them to endosomes, preventing receptors from fast recycling to the plasma membrane. We have previously shown that the interaction of β-arrestin2 with the bradykinin B2 receptor (B2R) is more labile than with the AT1R, enabling B2R to escape endosomes and recycle back to the plasma membrane (7, 8). The exact molecular mechanisms regulating endosomal receptor/β-arrestin interactions amongst different GPCRs are not fully understood. We recently reported the use of a fluorescence recovery after photobleaching (FRAP) approach to assess the lifetime of GPCR/β-arrestin complex within endosomes. In contrast to conventional biochemical techniques (e.g., immunoprecipitation and immunofluorescence), FRAP is noninvasive and allows examining multiprotein complexes in their native environment. Thus, FRAP can be used to probe the underlying mechanisms regulating β-arrestin/GPCR interactions, which should facilitate understanding receptor trafficking and intracellular signaling. To illustrate this method, we have used here the B2R and β-arrestin2-YFP.

2. Materials

2.1. Cell Culture

1. Human embryonic kidney (HEK293) cells.
2. Complete MEM growth medium: Minimum Essential Medium (MEM) supplemented with 10% heat-inactivated fetal bovine serum (FBS), 2 mM L-glutamine, and 20 μg/mL Gentamicin.
3. MEM/HEPES medium: MEM supplemented with 20 mM 4-(2-hydroxyethyl)-1-piperazineethanesulfonic acid (HEPES), pH 7.4.
4. 35-mm glass bottom dishes No. 1.0 (MatTek Corp).

2.2. cDNA Expression Constructs

1. cDNAs for rat β-arrestin2-YFP, human HA-tagged B2R, B2R-YFP and B2R-β-arrestin2-YFP chimera (sequences are available upon request).

2.3. Transient Transfection

1. Lipofectamine™ 2000 Transfection Reagent (Invitrogen).
2. Opti-MEM Reduced-Serum Medium (Gibco).
3. Sterile 14 ml disposable polypropylene culture tubes.

2.4. Confocal Microscopy

1. LSM-510-META Laser Scanning Confocal Microscope (Zeiss), equipped with a 40× oil-immersion objective, and an Argon 2 laser with single line excitation at 514 nm, and emission BP 530–600 nm filter sets.
2. LSM 5 software; version 4.2 (Carl Zeiss Microscope System).
3. Climate chamber (e.g., Zeiss XL-3) with heat source and controller.
4. GPCR agonist ligand at 10× final concentration in MEM/HEPES medium.

2.5. Software for Data Analysis

1. Adobe Photoshop CS3; version 10.0.1.
2. Metamorph; version 7.0 (Molecular Devices).
3. GraphPad Prism 4 (GraphPad Software).
4. Microsoft Excel.

3. Methods

Here we provide a simplified step-by-step method to quantify the lifetime of the interaction of β-arrestin with GPCRs within endosomes. This approach uses the fluorescence recovery of β-arrestin-YFP onto a photobleached endosome containing internalized receptors to infer the avidity of GPCR/β-arrestin complexes. The method is adapted from our recent work on the AT1R and B2R (8) and can be applied to other GPCRs that internalize with β-arrestins into endosomes (e.g., Class B GPCRs; (5, 6)). It is based on the premise that following endosome photobleaching, the fluorescence recovers as bleached β-arrestin-YFP dissociates from the receptors and is replaced by new unbleached-fluorescent β-arrestin-YFP from the cytosolic pool, proximal to the endosome (Fig. 1). This simple approach can also be used to study the effect of different regulators on receptor/β-arrestin complexes, which could impact the trafficking and signaling of GPCRs.

3.1. Cell Culture

1. HEK293 cells are grown in complete MEM growth medium at 37°C and 5% CO_2 environment. Cells are propagated on average every 3–4 days.

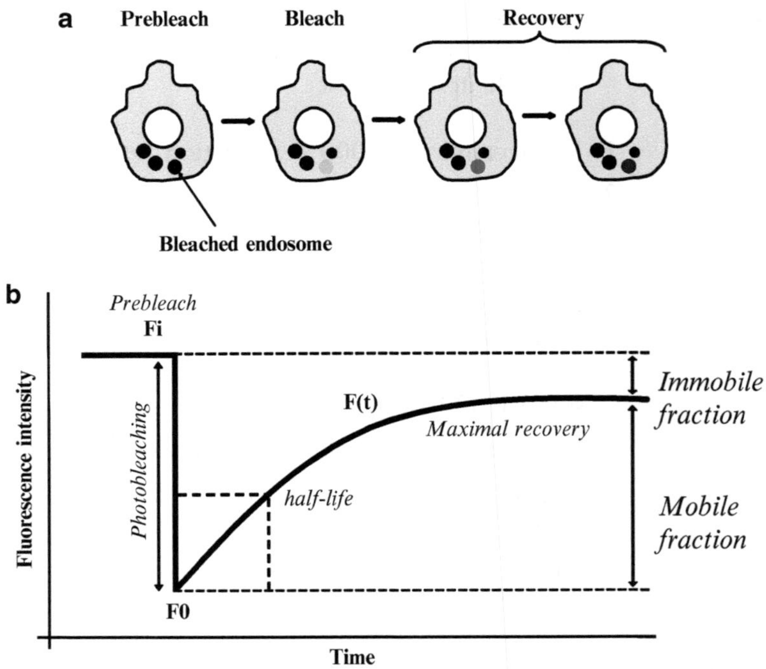

Fig. 1. Principles of fluorescent recovery after photobleaching on endosomes. (**a**) Depiction of the FRAP experiment. A single endosome containing β-arrestin2-YFP and internalized receptor (*circle*) is photobleached by repetitive scanning (bleached endosome). Over time, the fluorescence of the bleached endosome recovers by exchange with cytosolic non-bleached β-arrestin2-YFP (recovery). (**b**) Characteristics of FRAP and the recovery curve. The fluorescence recovery of the bleached endosome provides information on the recovery time (half-life), the mobile fraction [F_0 to maximal $F(t)$], and the immobile fraction [$Fi - \max F(t)$] (9) (see Subheading 3.4 for calculations).

2. On day 1 of the transfection protocol, HEK293 cells are trypsinized and split into 35-mm glass bottom dishes at a density of 1.5×10^5 cells per dish in 2 mL of complete medium. Cells are then incubated overnight at 37°C in a 5%-controlled CO_2 incubator.
3. On day 2, cells are transfected using Lipofectamine™ 2000 as described below and maintained in the 37°C, 5%-controlled CO_2 incubator until day 4 (see Note 1).
4. On day 4, the medium is replaced with 1.8 ml of preheated MEM/HEPES medium and transfected cells in 35-mm glass bottom dishes are placed on the preheated microscope stage to perform the FRAP experiment.

3.2. Transient Transfection

1. Dilute 0.5 μg/dish of the β-arrestin2-YFP plasmid and 2 μg/dish of HA-B2R (or other receptor cDNA) in 250 μL/dish of Opti-MEM in a 14-ml sterile polypropylene culture tube and mix gently.
2. Dilute 5 μL/dish of Lipofectamine™ in 250 μL/dish of Opti-MEM in a second sterile polypropylene culture tube and mix gently. Incubate the mixture at room temperature for 5 min.

3. Add the diluted Lipofectamine™ mix into the diluted DNA tubes and mix gently. Incubate the Lipofectamin/DNA mix at room temperature for 20 min.
4. Add the 500 μL/dish of the Lipofectamin/DNA mix drop-by-drop onto cells contained in the 35-mm glass bottom dish.
5. Place cells at 37°C overnight in a 5%-controlled CO_2 incubator. Cells are used for imaging 36–48 h posttransfection.

3.3. Confocal Microscopy

1. The 35-mm glass bottom dish containing transfected cells in MEM/HEPES is placed on the preheated microscope stage set at 37°C (operated through the AxioVision software). Allow 10–15 min for the chamber and cells to equilibrate. In the example presented here (Fig. 2a), HEK293 cells were co-transfected with HA-B2R and β-arrestin2-YFP and imaged using a 40× oil-immersion objective. YFP is detected using the Argon laser set at 5% and 514 nm excitation and BP 530–600 nm emission filter sets.
2. Add the 10× concentrated agonist in 200 μl of MEM/HEPES buffer (e.g., bradykinin at 1 μM final concentration in the example shown) to cells (see Note 2).
3. Cells expressing distinct YFP-endosomes are isolated after 15 min of agonist treatment using the *Crop* function. Selected cells should contain several distinguishable endosomes, some of which will be used as nonbleached reference endosomes (see Note 3). Select one β-arrestin2-YFP endosome as the "bleached endosome." Using the *Scan* menu, rapidly adjust the levels of the detector *Gain* and *Offset* to ensure the optimal dynamic range for image acquisition (see Note 4).
4. In the *Edit bleach* menu, set parameters as follows: 514 nm, 100% laser power and 100 iterations (see Note 5).
5. In the *Edit Bleach/Define Region* menu, choose a circle region of interest (ROI), and selected a size that encompasses exactly the endosome in order to minimize depletion of adjacent cytosolic β-arrestin2-YFP.
6. A first image is taken as the "prebleach condition" using a scanning size of 1,024 × 1,024 pixels at *Scan Speed 9* which yields a scan time of 1.97 s/image (Prebleach; Fig. 2a inserts) (see Note 6).
7. The selected endosome is then repetitively scanned 100 times (iterations) at 100% laser power to photobleach the β-arrestin2-YFP within the endosome (~80% bleaching; Fig. 2a, b). As a control for bleaching, an image is taken immediately after bleaching, to ensure that the fluorescence intensity of the selected endosome has strongly decreased and that a nearby nonbleached endosome displays similar fluorescence intensity than before bleaching (Bleach; Fig. 2a insets).

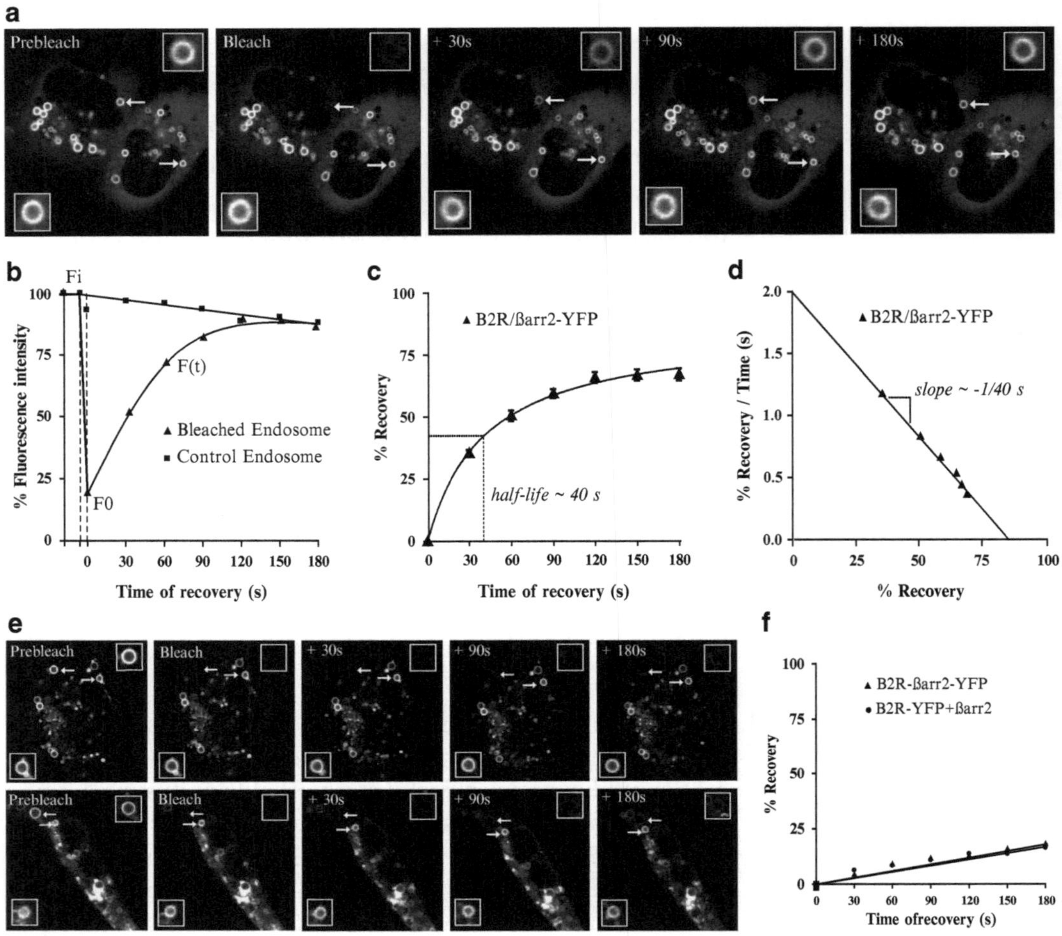

Fig. 2. FRAP analysis on the β-arrestin2/B2R endosomal complexes. (**a**) HEK293 cells were transiently transfected with β-arrestin2-YFP and B2R, and treated with 1 μM bradykinin. After 15 min stimulation, B2R is co-internalized with β-arrestin2-YFP into endosomes (*left panel*). A specific endosome is then selected for photobleaching (second panel; *top arrow* and *top inset*), and fluorescence recovery is followed over time every 30 s (+30 s, +90 s, and +180 s; panels 3–5). As a control, fluorescence recovery is recorded for a nonbleached endosome (panels 1–5*; bottom arrows and insets*). (**b**) Graph of FRAP quantification corrected for background. Represented are the data for the bleached endosome (*triangle)* and the control endosome (*square*). Baseline intensity fluorescence is collected (F_i) before bleaching and represents 100%, while F_0 represents maximal bleach fluorescence (~80%). Rapid fluorescence recovery of a bleached endosome is observed in time, which reaches a plateau (max $F(t)$). Also shown is the decay of fluorescence intensity over the same time period for a control nonbleached endosome (reaching a maximum of 10% after 180 s; *closed squares*). (**c**) Fluorescence recovery as calculated from Eq. (1) (*see* Subheading 3.4). This curve gives a maximal fluorescence recovery of β-arrestin2-YFP of ~85% and half-life recovery time of ~40 s. (**d**) Transformation of fluorescence recovery data from (**c**) to linear regression. (**e**) HEK293 cells were transiently transfected with either nontagged β-arrestin2 and B2R-YFP (*top panels*), or with the B2R-β-arrestin2-YFP chimera construct (*bottom panels*). Yellow fluorescent endosomes were observed in B2R-YFP/β-arrestin2 transfected cells after 15 min of agonist treatment, while for the B2R-β-arrestin2-YFP the fluorescence was observed in endosomes in the absence of stimulation, as the chimeric receptors were constitutively internalized. In these two conditions, a specific endosome was selectively photobleached (*top right squares*), and the fluorescence recovery monitored every 30 s, and compared to a nonbleached control endosome (*bottom left squares*). (**f**) Quantification of the FRAP data showed in (**e**). A slow and linear fluorescence recovery over time was observed with both experiments (reaching a maximum of ~20% after 3 min recovery) (see Note 8).

8. Subsequently, images are acquired every 30 s (image size of 1,024 × 1,024 pixels at Scan Speed 9) for a total of 180 s (see Note 7). These images will be used to calculate the fluorescence recovery (Fig. 2a–d).
9. A rapid fluorescence recovery of β-arrestin2-YFP is observed in a time-dependent manner with the B2R (Fig. 2a–d), due to the exchange of bleached β-arrestin2-YFP on the receptor by nonbleached cytosolic β-arrestin2-YFP. This contrasts with conditions where either the B2R-YFP or the B2R-β-arrestin2-YFP chimera are expressed in cells, and subjected to the same photobleaching protocol (Fig. 2e, f). Under these conditions, only weak and slow linear fluorescence recovery is observed (see Note 8).

3.4. FRAP Data Analysis

1. Once the FRAP experiment is completed, images are saved as TIFF files and exported for analysis using Metamorph software.
2. In Metamorph, open images (i.e., (1) Prebleach, (2) Bleach, (3) +30 s, (4) +60 s, (5) +90 s, (6) +120 s, (7) +150 s, (8) +180 s] from the same experiment, and build a stack of images using the *File menu/Open Special/Build Stack/User Defined* commands.
3. Designate the following three elliptical ROIs: (1) the bleached endosome, (2) a nonbleached control endosome, and (3) a blank region corresponding to the background.
4. Quantify the integrated intensity from these 3 ROIs using the *Measure menu/Region Measurement* functions, for the eight different images of the stack. Export the integrated intensity data in Excel and calculate the percentage of fluorescence recovery using Eq. (1):

$$\%\ \text{Recovery}\ (t) = \frac{\left[\dfrac{F(t)_{\text{ROI}} - F_{\text{BG}}}{F(t)_{\text{cont.}} - F_{\text{BG}}}\right] - \left[\dfrac{F_{0\_\text{ROI}} - F_{\text{BG}}}{F_{0\_\text{cont.}} - F_{\text{BG}}}\right]}{\left[\dfrac{F_{\text{i\_ROI}} - F_{\text{BG}}}{F_{\text{i\_cont.}} - F_{\text{BG}}}\right]} \times 100 \qquad (1)$$

where $F_{\text{i\_ROI}}$ vs. $F_{\text{i\_cont.}}$ is the initial fluorescence of the bleached vs. the control endosome before bleaching.

F_{0_ROI} vs. $F_{0\_\text{cont.}}$ is the fluorescence intensity of the bleached vs. the control endosome immediately after bleaching.

$F(t)_{\text{ROI}}$ vs. $F(t)_{\text{cont.}}$ is the fluorescence intensity of the bleached vs. the control endosome x seconds after bleaching (see Note 9).

F_{BG} is the background fluorescence, before, immediately after and x seconds after bleaching.

5. Fluorescence Intensity data are expressed as % recovery as a function of time (s) using GraphPad Prism 4 (Fig. 2c) and can be converted as nonlinear regression curve fit to obtain maximal recovery and half-life interaction of β-arrestin2-receptor complexes (Fig. 2c) (i.e., around 85% and half-life ~40 s, respectively). Data can also be converted into linear regression by expressing results from Fig. 2c as % recovery/time (s) as a function of % recovery (Fig. 2d).
6. For statistical data analysis, 10–20 bleached endosomes (from at least three different experiments) should be collected (9).

4. Notes

1. Forty-eight hours posttransfection, cells should display a flat, cobble stone morphology, and β-arrestin2-YFP should be expressed homogenously in the cytoplasm, excluded from the nucleus. If using β-arrestin1-YFP, the fluorescence should be distributed both in the cytosol and in the nucleus.
2. Cells should display a rapid plasma membrane translocation of β-arrestin2-YFP within 30–60 s of agonist exposure. This response confirms optimal expression of the receptor in cells.
3. Both the selected β-arrestin2-YFP endosome to be bleached and the control nonbleached β-arrestin2-YFP endosome should have the same size and approximately the same fluorescence intensity. These two parameters can be estimated by using the overlay toolbar (i.e., *Profile* for the fluorescence intensity and *Measure* for size).
4. Here, selected endosomes have a β-arrestin2-YFP intensity of ~200 (at a setup *Pinhole* < 1.0 μm (set at 74.6); *Gain* 600 (usually the gain range is between 550 and 600); and *Offset* of −0.035)]. It is important to select cells that still show cytosolic β-arrestin2-YFP. Also, during prebleaching and postbleaching image acquisition, illumination intensity and scanning time should be at lowest possible to minimize bleaching of the sample.
5. Bleaching of the fluorescence should be approximately 80%. Inadequate photobleaching may occur when either the number of iterations or the strength of the laser is set too low. Thus, it is recommended that photobleaching be performed with the laser set to full power (100%) and varying the number of iterations to avoid general photobleaching and to minimize the amount of time taken to perform the operation.
6. We found that an acquisition time of 1.97 s per image generates sufficient usable information. Scanning time may be increased to obtain better images. However, doing so will also increases both the minimal time of acquisition between

images (potentially reducing the recording of fast recovery events) and the decay of the overall fluorescence signal due to photobleaching.

7. A limitation in continuous image acquisition is the movement of cells and endosomes, which affects the Z-section over time. Focal "drift" should be manually corrected using the "*fast scan*" mode to recover the initial Z position (as estimated by the diameter of the selected endosome).
8. Fluorophores such as YFP can undergo reversible photobleaching, which could account in part for the fluorescence recovery observed with the B2R-YFP or the B2R-β-arrestin2-YFP chimera. Alternatively, fusion of proximal nonbleached endosomes with the bleached endosome may also contribute to this slow regain of fluorescence. However, the kinetics of fluorescence recovery observed in these conditions are dramatically slower than the one observed with the B2R/β-arrestin2-YFP interaction.
9. Typically, over the 3 min time period of the FRAP experiment, a slow and linear decrease of fluorescence is observed from nonbleached endosomes as a consequence of scanning (Fig. 2a, b; control endosomes).

Acknowledgments

We are thankful to A-M. Fay and B. Zimmerman for their helpful comments and for critical reading of the manuscript. This work was supported by a Canadian Institutes of Health Research (CIHR) Operating Grant and a CIHR Confocal Maintenance Grant to S.A.L (MOP-74603 and PRG-82673, respectively). B.A. holds a Fellowship award from the McGill University Health Center Research Institute (MUHC-RI), which is a recognized Fonds de la Recherche en Santé du Québec (FRSQ) supported Institute. S.A.L. holds a Canada Research Chair in Molecular Endocrinology.

References

1. Overington, J. P., Al-Lazikani, B., and Hopkins A. L. (2006) How many drug targets are there? *Nat Rev Drug Discov* **5**, 993–6.
2. Lefkowitz, R. J. (1998) G protein-coupled receptors. iii. New roles for receptor kinases and beta-arrestins in receptor signaling and desensitization. *J Biol Chem* **273**, 18677–80.
3. Claing, A., Laporte, S. A., Caron, M. G., and Lefkowitz, R. J. (2002) Endocytosis of G protein-coupled receptors: Roles of G protein-coupled receptor kinases and beta-arrestin proteins. *Prog Neurobiol* **66**, 61–79.
4. Barak, L. S., Ferguson, S. S., Zhang, J., and Caron, M. G. (1997) A beta-arrestin/green fluorescent protein biosensor for detecting G protein-coupled receptor activation. *J Biol Chem* **272**, 27497–500.
5. Zhang, J., Barak, L. S., Anborgh, P. H., Laporte, S. A., Caron, M. G., and Ferguson, S. S. (1999) Cellular trafficking of g protein-coupled receptor/

beta-arrestin endocytic complexes. *J Biol Chem* **274**, 10999–1006.

6. Oakley, R. H., Laporte, S. A., Holt, J. A., Caron, M. G., and Barak, L. S. (2000) Differential affinities of visual arrestin, beta arrestin1, and beta arrestin2 for G protein-coupled receptors delineate two major classes of receptors. *J Biol Chem* **275**, 17201–10.
7. Simaan, M., Bedard-Goulet, S., Fessart, D., Gratton, J. P., and Laporte, S. A. (2005) Dissociation of beta-arrestin from internalized bradykinin B2 receptor is necessary for receptor recycling and resensitization. *Cell Signal* **17**, 1074–83.
8. Gousseva, V., Simaan, M., Laporte, S. A., and Swain, P. S. (2008) Inferring the lifetime of endosomal protein complexes by fluorescence recovery after photobleaching. *Biophys J* **94**, 679–87.
9. Snapp, E. L., Altan, N., and Lippincott-Schwartz, J. (2003) Measuring protein mobility by photobleaching gfp chimeras in living cells. *Curr Protoc Cell Biol* **Chapter 21**: Unit 21.1.

Chapter 24

Disrupting Protein Complexes Using Tat-Tagged Peptide Mimics

Shupeng Li, Sheng Chen, Yu Tian Wang, and Fang Liu

Abstract

Protein–protein interaction is a widely existing phenomenon and is essential for almost all biological processes, extending from the formation of cellular macromolecular structures and enzymatic complexes to the regulation of signal transduction pathways. Proteins interact with each other through the dynamic associations between modular protein domains within different cellular compartments and with distinct temporal dynamics. Disrupting protein interactions has emerged as an effective way to specifically modulate certain signaling pathways. Tat-tagged peptide mimics are a recently developed experimental tool that is used to disrupt specific interactions between protein complexes. TAT, an 11-amino acid protein transduction domain from HIV Tat protein, is tagged to peptides that mimic the functional fragment of protein interaction domains, and facilitates the delivery of peptides into cells to disrupt the associated protein both competitively and selectively. Here we provide a technical description on the utilization of Tat-tagged peptide mimics as a tool to disrupt protein interaction in cultured neurons and in the rat brain.

Key words: Protein–protein interaction, TAT domain, Peptide mimics, Signal transduction, Neuroscience

1. Introduction

A majority of proteins function by associating with other proteins as either a partner molecule or as a component of an assembly of proteins. Protein–protein interactions govern signals involved in cell growth, differentiation, and intercellular communication through dynamic associations between modular protein domains and their cognate binding partners. Within the cell, the life cycle of a protein and the dynamic external/internal environment require that a specific protein changes its composition and association pattern within defined cellular organelles and subdomains.

Louis M. Luttrell and Stephen S.G. Ferguson (eds.), *Signal Transduction Protocols*, Methods in Molecular Biology, vol. 756,
DOI 10.1007/978-1-61779-160-4_24, © Springer Science+Business Media, LLC 2011

Once interacting proteins of interest have been defined and the discrete protein–protein interacting domains within these proteins have been characterized, it is necessary to find appropriate tools to disrupt the interactions to elucidate the biological function and establish the physiological significance of protein–protein interactions. Disrupting protein interactions has also emerged as an effective way to modulate certain signal pathways. Peptides that mimic the functional fragments of interaction domains can disrupt the protein interactions in a competitive manner. However, the bioavailability barriers, such as the plasma membrane of the cell and the blood–brain barrier, limit the intracellular introduction of peptide mimics.

The TAT-based delivery system has been successfully used for intracellular delivery of a broad variety of cargos including peptides and proteins. TAT is an 11-amino acid (Tyr-Gly-Arg-Lys-Lys-Arg-Arg-Gln-Arg-Arg-Arg) protein transduction domain of the HIV-1 transactivator Tat protein (1). TAT is rich in arginine and lysine, thus highly charged and hydrophilic. It is well established that proteins that are fused to the TAT are rapidly and efficiently introduced into cultured cells and into live tissues when systemically administered into intact animals, while retaining their biological activity (2). The mechanism for TAT delivery through biological membranes is not well understood, but most studies agree on the importance of a direct contact between the highly positively charged TAT peptide and the negative residues on the cell surface (2).

The TAT delivery system offers the advantage that it delivers competitive peptide mimics into cells in vitro and in vivo. It can even deliver peptides across the blood–brain barrier into the brain, thus circumventing the bioavailability barrier, and enables potential therapeutic use of peptides and proteins for central nervous system diseases (3, 4). In this chapter, we will use TAT-D2R peptides and TAT-GluR2 peptides that target D2R-DAT interactions and GluR2-GAPDH interactions, respectively, as examples to illustrate how to disrupt protein interactions in cultured neurons and in the rat brain.

2. Materials

2.1. Neuronal Culture Media

1. 0.1% poly D-lysine solution in 0.1 M Borate Buffer, pH 8.4 (Sigma).
2. Neurobasal™ medium (Invitrogen).
3. Heat-inactivated horse serum.
4. Plating medium: 90% Neurobasal™ medium (v/v), 10% heat-inactivated Horse serum (v/v), 0.5% Penicillin/Streptomycin (v/v).

5. Hank's balanced salt solution (HBSS).
6. Sterile 2.5% trypsin solution.
7. B-27 serum-free supplement (Invitrogen).
8. Feeding medium: 98% Neurobasal™ medium, 2% B-27 serum-free supplement, 0.5 mM L-glutamine, 0.5% Penicillin/Streptomycin (v/v).
9. Cytarabine (Ara-C; Calbiochem).
10. Time-pregnant E18 Wistar rats.
11. 10 cm tissue culture dishes.
12. 37°C CO_2 incubator.

2.2. Intracranial Cannulation and Injection

1. 70% EtOH, Betadine scrub, zephiran chloride, 100 mg/ml ketamine, Xylazine supplied as 0.01 mg/ml, Marcaine 1.25%, penicillin G (Penlong/Durapen), wound spray, lacrilube ophthalmic ointment.
2. Scalpel, bulldog clamps, stereotaxic instrument, forceps, spatula, dental cement powder, methylmethacrylate, sterile surgical packs (drapes, gauze, swabs, suture material), precision syringe 30-gauge needle.
3. Guide cannula C313G, injector cannula C313I, cannula dummy C313DC, 0–80 × 1/8 screws, screw driver, drill bit, drill driver (Plastics One).
4. 10 μl Hamilton microsyringe (Hamilton), PE20 polyethylene tubing (20 cm length), infusate incision clamps, syringe pump.

2.3. Jugular Catheterization

1. Catheter consists of silastic tubing (37 mm length, 0.51 mm inside diameter × 0.94 mm outerside diameter; Dow Corning) connected to 65 mm length PE10 tubing. 6 mm heat-shrink tubing is then used to tighten the connection between the silastic tubing and PE10. The latter is heat fused with 170 mm PE20 tubing. The free end of the PE20 tubing is connected to 15 mm heat-shrink tubing which is heated to fit for a 23-gauge needle. The PE 20 tubing is molded onto a polyester fiber mesh (PlasticsOne) and nylon bolt.
2. 1-cc syringe fitted with blunted 22-gauge needle.
3. Small iris scissors, microscissors, two pairs of forceps (one straight, one curved), needle driver/holder, trocar, suture needle and thread, cyanoacrylic glue.
4. Heparin solution: 50 U/ml heparin in 0.9% saline.
5. Brietal (sodium methohexital): 20 mg/ml in 0.9% saline.

2.4. Immunoprecipitation and Western Blotting

1. Lysis Buffer: 0.15 M NaCl, 5 mM EDTA pH 8, 10 mM Tris–HCl pH 7.4, 0.5% deoxycholate, 1% NP-40. Just before using add: 0.5 mM DTT, 1 mM PMSF, and protease inhibitor cocktail (Sigma).

2. Tissue Grinder.
3. Protein A/G agarose beads (Santa Cruz).
4. Refrigerated centrifuge.
5. Rocking/rotating platform.
6. 2× SDS sample buffer.
7. SDS-PAGE and nitrocellulose transfer equipment.
8. Nitrocellulose membrane.
9. Blocking buffer: 3% BSA in phosphate-buffered saline (PBS) or 5% non-fat dry milk in 0.1% Tween20 in PBS.
10. Primary antibody buffer: 1% bovine serum albumin (BSA), 0.1% Tween20, in PBS.
11. Wash buffer: 0.1% Tween20 in PBS.
12. Primary antibodies against protein(s) of interest.
13. Horseradish peroxidase-conjugated secondary antibody against the primary antibody species.
14. ECL reagents and X-ray film.

3. Methods

TAT-tagged peptides disrupt the physical interaction between proteins both in vitro and in vivo. After delivery into either the cell or cellular organelles by TAT, the TAT peptide competitively binds to the sites where the two proteins interact with each other and blocks the binding of the original partner. Solubility of the peptide depends on the amino acid composition. For most hydrophilic peptides, the peptide can be dissolved in saline. Basic peptides may first be dissolved in 20% acetic acid and then diluted to the desired concentration with saline. For acidic peptides, 10% ammonium bicarbonate may be used to dissolve the peptide. If the peptide is very hydrophobic, dissolve the peptide in a very small amount of dimethylsulfoxide (DMSO) first. For Cys-containing peptides, use dimethylformamide (DMF) instead of DMSO. Some peptides tend to aggregate. If this occurs add 6 M guanidine HCl to disperse and then proceed with the necessary dilutions.

For in vivo tests, TAT-peptides can be administrated via either intraperitoneal injection or intravenous injection. Local injection is also a choice for those that have area selective actions. For brain injection, pretest cannulation is necessary if either multiple injections are performed or if behavioral tests are planned.

3.1. Primary Neuronal Culture

1. One day before culture, coat culture plates/coverslips with 0.1% poly D-lysine solution. Swirl the plate to ensure that the

coating mix covers the entire bottom of the plate. Leave the dishes in the 37°C/5% CO_2 incubator overnight.

2. On the second day, thoroughly wash the plates twice with sterile water; remove the final wash and add 10 ml plating medium to each plate and leave the plates in incubator to balance the temperature and pH of the medium.
3. Anesthetize rats via inhalant anesthetic (e.g., isoflurane) and then sacrifice the rat via cervical dislocation at the embryo age day 18. Lift the peritoneum of the dam and cut through it to open the abdominal cavity. Care should be taken not to injure the embryos or internal organs. Remove the embryos and place them into a sterile 10 cm tissue culture dish filled with cold HBSS on ice. Separate the individual embryos and remove each embryo from its amniotic sac, decapitate, and then place the heads into a separate 10 cm tissue culture dish containing HBSS. Under a dissecting microscope, remove the skin and cut along the scalp in the midline and open the calvarium with fine scissors. Deflect the calvarium with a blunt spatula and remove the brain.
4. Hippocampus and cortex dissection:
   (a) To isolate the hippocampus, place the brain such that the dorsal surface with the brain stem and cerebellum faced up. Gently cut through the longitudinal fissure and separate the two hemispheres through the basal ganglia. Place the spatula in the lateral ventricle underneath the hippocampus. Make cuts superiorly and inferiorly to free both ends of the hippocampus. Roll out the hippocampus with the spatula and cut the hippocampus from the cortex junction.

   (b) To remove the cortex, place the brain ventral side up. Place the spatula in the medial aspect of the ventral cortex and midbrain and cut the cortices off. Place cortices or hippocampi in 10-ml sterile tube containing ice-cold HBSS.
5. Tissue digestion and cell plating:
   (c) Dilute 160 μl 2.5% trypsin into 2 ml HBSS to get a final concentration of 0.2% trypsin. Digest tissue chunks in trypsin for 15 min at 37°C. Inactivate the trypsin by washing tissue twice with plating medium containing serum.

   (d) Triturate the tissue 10–15 times through a fire-polished Pasteur pipette. Wait 3 min to allow undispersed tissue to settle down, then collect and centrifuge the supernatant for 7 min at $200 \times g$.

   (e) Resuspend the cell pellet with plating medium, mix an aliquot with Trypan Blue and count the cell density in a hemacytometer. Plate 1×10^6 live cells per 10 cm culture plate.

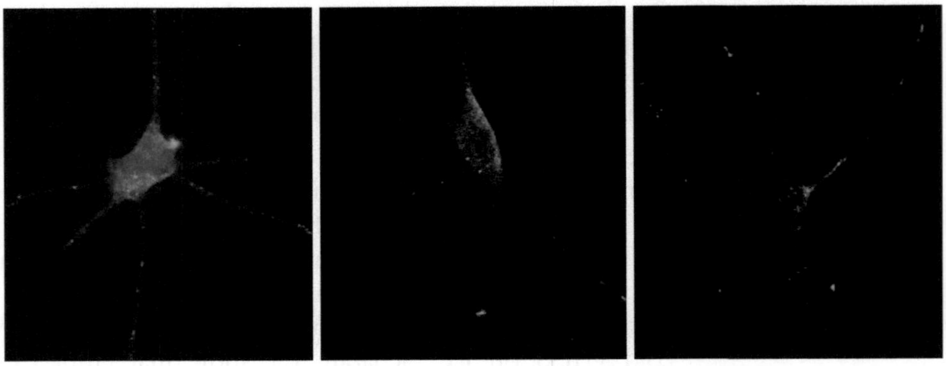

Fig. 1. Transduction of TAT peptide into cultured cortical neurons. Visualization of intereuronal accumulation of FITC-conjugated TAT peptide (100, 10, and 1 μM) 30 min after application in cortical cultures by confocal fluorescence microscopy.

6. Change half of the plating medium to feeding medium 2 days after plating and twice a week thereafter. 5 or 10 μM Ara-C is added on day 6 in culture and left in medium for 24 h to inhibit the growth of nonneuronal cells.
7. Neurons will be ready for experiments 10–12 days after plating. TAT-peptides can be used to treat the neurons at final concentrations from 1 to 100 μM. TAT peptide accumulation is detectable in neurons within 10 min of application, peaks during the next 20 min, and remained detectable for 5 h after washing the peptide from the bath (3). Figure 1 shows the intraneuronal accumulation of a FITC-conjugated TAT peptide 30 min after application to cultured cortical neurons.

3.2. Brain Cannulation, Jugular Vein Catheterization, and Intracerebral Microinjection

Animals are handled and acclimatized to facility for at least 1 week prior to cannulation to reduce the effects of stress generated during transportation.

3.2.1. Brain Cannulation

1. The surgical area and stereotaxic frame are wiped with Zephiran chloride prior to use. All instruments, cannulae, dummy cannulae, and mounting screws are sterilized by immersion in Zephiran chloride for 20 min. They are then rinsed thoroughly in sterile saline prior to use. Instrument beakers and glassware are autoclaved prior to use.
2. The rat is weighed and anesthetized using ketamine/xylazine (10/75 mg/kg) (see Note 1) and the animal's scalp is shaved. The scalp area is scrubbed with Betadine surgical scrub. Soapy residue is removed using 70% ethanol and the area is then painted with Betadine solution and allowed to dry. Marcaine (0.1 ml) is infiltrated subcutaneously along the incision site. Lacrilube is applied to the animal's eyes to prevent drying of the corneas.

3. The animal is carefully secured into the stereotaxic frame using the ear bars and incisor bar. Care is taken not to rupture the tympanic membrane with the earbars. The animal's body temperature is maintained by a thermoregulator.
4. A 2 cm midline incision is made through the scalp so that the bregma and lambda skull landmarks are visible. Bulldog clamps are applied to retract fascia and periosteum. Scrape surface bleeding.
5. Cannulation coordinates are located. For intracerebroventricular (ICV) cannulae, the following coordinates are used: AP: −1.0 mm from bregma; LM: −1.4 mm; DV: −3.6 mm from dura. The incisor bar is set at −3.3 mm to keep the skull flat. A tap drill is used to drill 3–4 holes for the stainless steel mounting screws that will be used to secure the headcap to the animal's skull. It is important that these screws are not positioned on a suture line and are not too close to each other. Screws are then inserted approximately halfway into the skull (5).
6. Cannula holes are drilled using a dental drill equipped with a carbide drill bit. The dura is punctured using a 30-gauge needle.
7. Cannulae are slowly lowered into the brain until the final dorsal/ventral coordinates are reached. The skull is dried and cannulae and screws are cemented in place with dental cement (see Note 2).
8. Once the cement has hardened, the stereotaxic carrier is removed and dummy cannulae are inserted into the guides.
9. Bulldog clamps are removed, sutures are employed if necessary, and the animal is removed from the frame (Fig. 2).

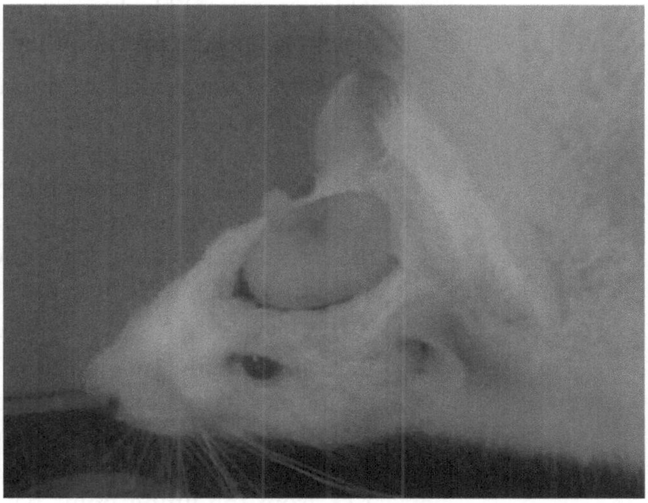

Fig. 2. Image of the operative site after brain cannulation performed on a rat.

10. The animal is transferred to a recovery area, placed on a thermostatically controlled hot water blanket, and given 0.1 ml of Penlong/Durapen (300,000 units of penicillin G) (see Note 3).

3.2.2. Jugular Vein Catheterization

An indwelling catheter is surgically implanted into the right external jugular vein (6). The catheter passes subcutaneously from the animal's back to the jugular vein where the tubing is inserted. The catheter exits between the scapulae and is attached to a modified 22-gauge cannula for peptide administration. A polyethylene assembly is used to mount the catheter on the animal's back. The catheter is flushed daily with 0.1 ml of a sterile heparin-saline solution (50 U/ml) to maintain patency. (see Note 4).

1. It is recommended that the animal weigh at least 300 g. Check the catheter for patency, pressure test for leaks, and ensure that the insertion tip has suitable length to reach the right atrium. The catheter and all other surgical instruments must be sterilized in a 1.5% solution of zephiran chloride.
2. After the animal is anesthetized, shave the right side between the ventral region of the neck and the scapulae. Swab these areas with alcohol and Betadine solution.
3. Make an oblique incision through the skin midway between the right scapulae and the middle line of the neck. Clear the superficial fascia and layers of muscle by blunt dissection. Care should be taken not to damage the underlying nerve and blood vessels.
4. Locate the jugular vein and isolate it from the rest of surrounding fascia. Pass a suture beneath the vein.
5. Place the animal on its abdomen and make an incision between the scapulae. Clear a subcutaneous space by blunt dissection. This space should be large enough to accommodate the mesh assembly and the excess catheter tubing.
6. Insert the forceps into the dorsal incision and direct it toward the ventral incision, running under the right forearm. The forceps should point upward to avoid tissue damage during this process. Punch through the connective tissue at the ventral incision and firmly grab a trocar. Pull the trocar back to the dorsal incision and leave the trocar tunnel through both incisions, which will allow the catheter to be fed through the trocar (see Note 5).
7. Return the animal to its side. Place the catheter so that all tubing lies flat and has no twisting stress. Flush the catheter with sterile saline and make sure there are no bubbles in the line. Lift the distal part of jugular vein and make an incision 1/3 from the lifting point between the grasping point and exposed proximal end using microscissors (see Note 6).

8. Hold the distal part with a suture locked with forceps, pinch the incision point of the vein with curved fine tweezers, grasp the silastic tip of the catheter and insert into the jugular incision. The tubing should feed in freely all the way to the heat shrink to facilitate securing the catheter in place. Verify the patency of the catheter by pulling back the syringe to draw blood into the PE10 tubing. Then, push the blood back into the vein with a small amount of saline (0.1 ml). Adjust the remaining PE20 tubing to lie flatly.
9. Tie the catheter to the jugular vein with two sutures, at both ends of the heat shrink connection across the incision. The distal end of the vein may also be tied to ensure that the suture does not slip off onto the silastic, as it will occlude the tubing. Anchor the PE10 tubing to deep muscle with a single suture. The heat shrink is then secured to underlying tissue using a single drop of cyanoacrylate adhesive on its underside.
10. Close the superficial muscle layer with 1–2 sutures. This serves as additional protection should the animal scratch at the incision. Close the skin with interrupted sutures.
11. Turn the animal onto its abdomen and cap the catheter with a filled silastic plug. Feed the excess PE 20 tubing into the subcutaneous pocket in a looping fashion, usually encircling the incision. Insert the mesh assembly and make sure it lies flat, centered between the scapulae, and above all tubing. Suture the skin around the nylon bolt, making sure neither to wrinkle the skin nor catch the mesh in the sutures (see Note 7).
12. Clean all incisions with saline and dress with wound spray. Inject 0.1 ml Penlong intramuscularly, 3 ml saline subcutaneously (for fluid replacement) and an appropriate dose of buprenorphine for postoperation analgesia. Place the animal on a thermoregulated heating blanket to recover.
13. Return the animal to its home cage when it regains consciousness. It is necessary to house these animals singly to prevent catheter damage.
14. Maintain catheter patency by flushing daily with 0.1 ml heparinized saline (50 U/ml).
15. Catheter patency may be verified by the injection of 0.15 ml Brietal solution. If the catheter is venous, the animal should rapidly lose consciousness. This lapse is extremely brief so care should be taken to avoid startle as the animal regains consciousness. Flush the catheter to leave heparinized saline in the tubing (see Note 8).

3.2.3. Brain Microinjection of TAT Peptides

1. Set up the injection syringe: Using a 10 μL Hamilton microsyringe, pull saline into PE20 tubing connected to a 30-gauge

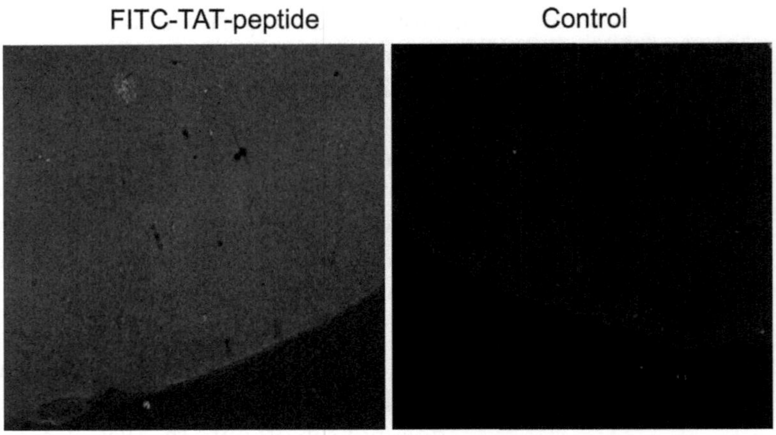

Fig. 3. Delivery of TAT peptide into the brain of an intact animal. Detection of fluorescence in the rat cerebral cortex 1 h after injection of FITC-conjugated TAT peptide. Brain sections from animals injected with FITC-TAT peptide, but not the control, exhibited strong fluorescence in the cortex.

needle; pull a small bubble before dipping the connector into the peptide solution. Load peptide into the PE20 tubing (see Note 9).

2. Gently restrain the animal and remove the cannula cap. Connect and snap lock the injector (which is linked to the PE tube filled with peptide) into the indwelling cannula. The animal is then put in a small cage where it can move freely without the danger of disconnection of injector and PE tube. Turn on the syringe pump and infuse the peptide with the speed 1 μl/min. Monitor the bubble movement (it should begin instantly) and mark both air/saline borders of the bubble, taking care to avoid bubble shrink. After injection, leave the injector in place for another 1 min to allow the peptide infusion, and then gently remove the injector.
3. To confirm the delivery of TAT peptide into the brain of an intact animal, Wistar rats can be injected with a FITC-conjugated TAT peptide. Coronal brain sections taken 1 h after injection can be examined by confocal microscopy for fluorescent peptide uptake. As shown in Fig. 3, sections from animals injected with FITC-TAT peptide, but not control, should exhibit strong fluorescence in the cortex.

3.3. Immunoprecipitation

To confirm the disruptive effects of TAT-tagged peptides, we perform immunoprecipitation to examine the composition of the targeted protein complex as described here.

1. Homogenize harvested neurons or rat brain tissues in ice-cold lysis buffer using a tissue grinder.
2. Centrifuge the lysate at 15,000 × g for 15 min at 4°C to pellet debris.

3. Transfer the supernatant into a fresh 1.5 mL capped microcentrifuge tube. Discard the pellet.
4. Perform a protein assay to determine the protein concentration of the extract. If desired, the extract can be stored at −70°C until use.
5. Add 25 μl of protein A/G agarose beads to an extract that contains 500–1,000 μg of total protein.
6. Rotate the beads and the extract mixture at 4°C for 30 min.
7. Centrifuge at 1,000 × *g* for 5 min at 4°C to pellet the beads. Transfer the supernatant to a fresh eppendorf tube. Discard the pelleted beads, which are used to clear proteins that bind nonspecifically.
8. Add primary antibody to each sample tube.
9. Rotate the beads and antibody at 4°C for 3 h to bind the primary antibody to the protein of interest.
10. Add 25 μl of protein A/G agarose beads.
11. Rotate at 4°C overnight to bind the primary antibody/protein complex to the beads.
12. Centrifuge at 1,000 × *g* for 5 min to pellet the beads and discard the supernatant.
13. Wash the beads three times in lysis buffer at 4°C, resuspending the beads each time in same volume of lysis buffer as the original sample volume.
14. After the third wash, spin down the beads and remove supernatant. Add 25 μl of 2× SDS sample buffer, heat at 100°C for 5 min.

3.4. Western Blotting

1. Prepare an 8 or 10% SDS-polyacrylamide gel.
2. Load the gel with co-immunoprecipitation samples. Run at a constant voltage of 100 V until the bromophenol blue dye is at the bottom of the gel.
3. Transfer the separated proteins in the gel to a nitrocellulose filter using constant current of 400 mA for 1–2 h at 4°C. Larger size proteins may take longer to transfer.
4. After transfer, block the membrane with 3% BSA or 5% milk blocking buffer at room temperature for 1 h.
5. Primary Antibody: After blocking, gently wash the membrane with Tween20/PBS wash buffer and put the membrane in BSA/Tween20/PBS antibody buffer containing the primary antibody at an appropriate dilution. Incubate the membrane overnight at 4°C on a shaker (see Note 10).
6. After the primary antibody incubation, wash the blot three times for 10 min with Tween20/PBS wash buffer.

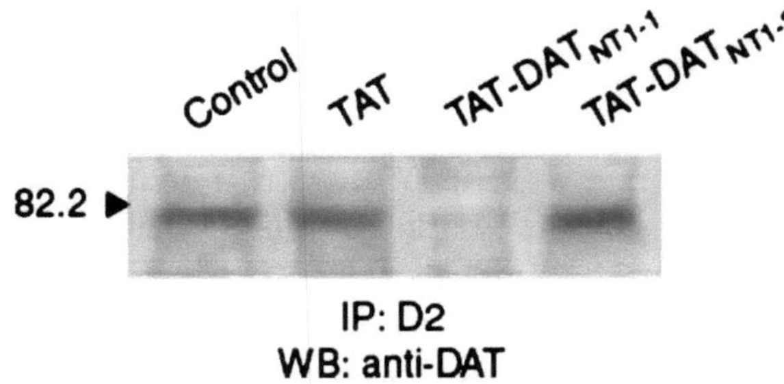

Fig. 4. Disruption of protein interactions by TAT-peptides. Neurons treated with TAT-DAT_{NT1-1} but not TAT-DAT_{NT1-2} or TAT alone (10 μM, 30 min) exhibited a significant decrease in the co-immunoprecipitation of DAT with the D2 receptor. Figure originally published in: Lee F. et al. (2007) EMBO J **26**, 2127–2136.

7. Secondary Antibody: Incubate the membrane with the secondary antibody diluted in Tween20/PBS wash buffer for 2 h at room temperature on a shaker.
8. After the secondary incubation, wash the membrane six times with Tween20/PBS wash buffer.
9. Detect signals on X-ray film after developing the blot using an Amersham ECL kit (Fig. 4).

4. Notes

1. Adequacy of anesthesia is confirmed using the pedal withdrawal reflex and animal is brought to the surgical area and placed on the draped stereotaxic frame.
2. It is important that the surface of the cement headcap is smooth and free of rough edges that may cause irritation.
3. The animal's postoperation recovery is monitored daily (general condition and incision site) and appropriate veterinary care/treatment is provided if necessary.
4. All needle tubing inserted into the heat shrink end of the catheter must be carefully blunted and filed to avoid rough edges.
5. The catheter must be fed in a dorsal-central direction; i.e., the insertion tip is introduced into the trocar from the dorsal end. Once a suitable length of catheter is visible at the ventral site, the trocar may be removed simply by pulling it our through the ventral incision.
6. All drug solutions intended for intravenous use should be filtered through a 22 μm filter for sterilization.
7. This procedure, while cumbersome, ensures that there is no stress exerted on the catheter, thereby protecting its integrity.

8. If the catheter becomes difficult to flush, it is possible to carefully pull the heat shrink out of the nylon bolt and plug directly into PE 20 tubing with a 26-gauge needle.
9. There should be no lag time of bubble movement when pulling solution into the tubing. Mark the initial position of both ends of the bubble to check movement and detect pressure-induced volume changes.
10. The optimal dilution for a specific antibody will vary and must be determined empirically. The primary antibody should be different from the one used in co-immunoprecipitation.

Acknowledgments

We thank Kathleen M. Coen and Zhaoxia Li for their excellent technical assistance. We appreciate Dr. Paul J. Fletcher for critical reading and comments on the manuscript.

References

1. Schwarze, S. R., Ho, A., Vocero-Akbani, A., and Dowdy, S. F. (1999) In vivo protein transduction: delivery of a biologically active protein into the mouse. *Science* **285**, 1569–72.
2. Rapoport, M., and Lorberboum-Galski, H. (2009) TAT-based drug delivery system--new directions in protein delivery for new hopes? *Expert Opin Drug Deliv* **6**, 453–63.
3. Aarts, M., Liu, Y., Liu, L., Besshoh, S., Arundine, M., Gurd, J. W., Wang, Y. T., Salter, M. W., and Tymianski, M. (2002) Treatment of ischemic brain damage by perturbing NMDA receptor-PSD-95 protein interactions. *Science* **298**, 846–50.
4. Brebner,, K., Wong, T. P., Liu, L., Liu, Y., Campsall, P., Gray, S., Phelps, L., Phillips, A. G., and Wang, Y. T. (2005) Nucleus accumbens long-term depression and the expression of behavioral sensitization. *Science* **310**, 1340–3.
5. Erb, S., Funk, D., and Lê, A. D. (2003) Prior repeated exposure to cocaine potentiates locomotor responsivity to central injections of corticotropin-releasing factor (CRF) in rats. *Psychopharmacology (Berl)* **170**, 383–9.
6. Corrigall, W. A., and Coen, K. M. (1989) Nicotine maintains robust self-administration in rats on a limited-access schedule. *Psychopharmacology (Berl)* **99**, 473–8.

Chapter 25

Protein-Fragment Complementation Assays for Large-Scale Analysis, Functional Dissection and Dynamic Studies of Protein–Protein Interactions in Living Cells

Stephen W. Michnick, Po Hien Ear, Christian Landry, Mohan K. Malleshaiah, and Vincent Messier

Abstract

Protein-fragment Complementation Assays (PCAs) are a family of assays for detecting protein–protein interactions (PPIs) that have been developed to provide simple and direct ways to study PPIs in any living cell, multicellular organism, or in vitro. PCAs can be used to detect PPI between proteins of any molecular weight and expressed at their endogenous levels. Proteins are expressed in their appropriate cellular compartments and can undergo any posttranslational modification or degradation that, barring effects of the PCA fragment fusion, they would normally undergo. Assays can be performed in any cell type or model organism that can be transformed or transfected with gene expression DNA constructs. Here we focus on recent applications of PCA in the budding yeast, *Saccharomyces cerevisiae*, that cover the gamut of applications one could envision for studying any aspect of PPIs. We present detailed protocols for large-scale analysis of PPIs with the survival-selection dihydrofolate reductase (DHFR), reporter PCA, and a new PCA based on a yeast cytosine deaminase reporter that allows for both survival and death selection. This PCA should prove a powerful way to dissect PPIs. We then present methods to study spatial localization and dynamics of PPIs based on fluorescent protein reporter PCAs.

Key words: Protein-fragment complementation assay, Protein–protein interactions, Dihydrofolate reductase, Cytosine deaminase, Green fluorescent protein, Luciferase reporter

1. Introduction

In the Protein-fragment Complementation Assay (PCA) strategy, protein–protein interactions (PPIs) are measured by fusing each of the proteins of interest to complementary N- or C-terminal peptides of a reporter protein that has been *rationally* dissected

Louis M. Luttrell and Stephen S.G. Ferguson (eds.), *Signal Transduction Protocols*, Methods in Molecular Biology, vol. 756, DOI 10.1007/978-1-61779-160-4_25, © Springer Science+Business Media, LLC 2011

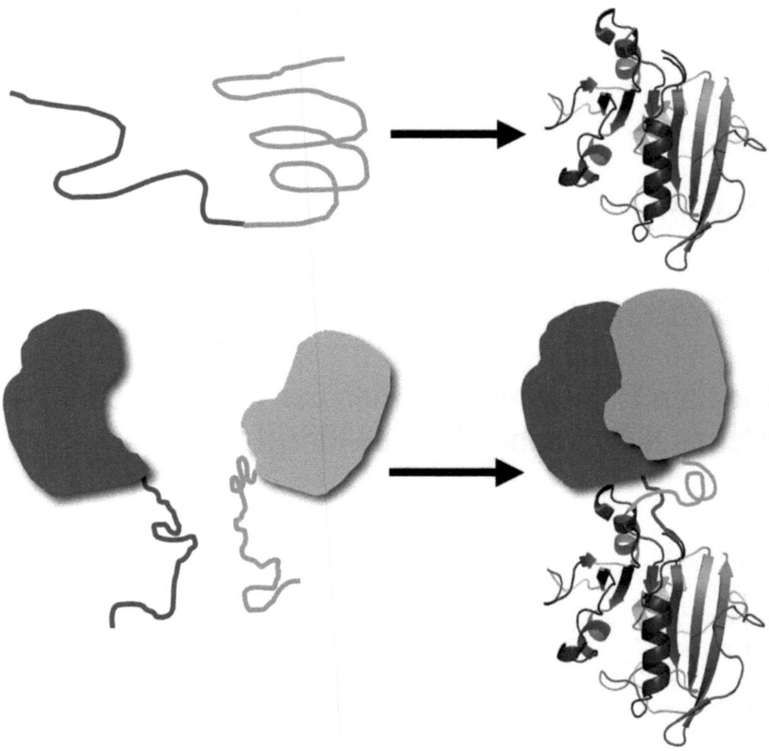

Fig. 1. Conceptual basis of Protein-fragment Complementation Assays. The spontaneous unimolecular folding of a protein from its nascent polypeptide (*upper panel*) can be made into a protein–protein interaction-dependent bimolecular process by fusing two interacting proteins to one or the other complementary N- or C-terminal peptides into which a protein has been dissected (*lower panel*). PPI-mediated folding of a reporter protein from its complementary fragments results in reconstitution of reporter protein activity.

using protein engineering strategies (1–3). The reporter protein fragments are brought into proximity by interaction of the two interacting proteins, allowing them to fold together into the three-dimensional structure of the reporter protein, thus reconstituting the activity of the reporter (Fig. 1). PCAs have been created with many different reporter proteins and thus provide for different types of readouts, depending on the desired application. This generality means that PCA is not a single reporter assay, but rather a toolkit. PCAs have also been developed to study spatial and temporal changes in PPIs under different conditions and also survival-selection assays that provide a simple readout for large-scale systematic analyses of protein interaction networks or directed evolution experiments (reviewed in (4)). Finally, there are two unique features of PCAs we must note. First, by nature of the fact that interactions between two proteins must occur in such a way that the reporter protein can fold, PCAs can provide structural and topological details of how a PPI is formed or if such complexes undergo conformation changes under specific conditions (5, 6). Second, contrary to intuition, most PCAs are fully reversible, allowing for direct studies of the dynamics of both formation and disruption of PPIs.

1.1. General Considerations in Using PCA

Measuring PPI in living cells by any method entails that one reconsider any suppositions that we may have about the nature of a PPI, most importantly if it has only been studied with in vitro methods by indirect methods such as affinity or immunopurification. PCAs detect direct binary or indirect proximal interactions between proteins and thus, if it is assumed that the there is such an interaction based on experiments that only suggest association of proteins in a complex, it is possible that no interaction will be detected. Our advice is "life is short, experiment." However, we can make some general statements about what to consider when setting up any PCA experiment in order to maximize the probability of a successful outcome.

First we consider the sensitivity of PCAs. Like any analytical technique, the sensitivity of the assay depends on the sensitivity of the detection method and background signal that may arise from cells. Regardless of the properties of the reporters, the range of signal detectable will depend in all cases on the quantity of complexes formed, which in turn is determined by the abundance of the proteins studied and their affinity for each other. We have only explored these parameters in great detail for the dihydrofolate reductase (DHFR) PCA (see Subheading 3.1). We have demonstrated that for this simple survival-selection assay, the number of complexes needed to support survival under the selection conditions was as low as approximately 25 per cell for a complex for which the dissociation constant was in the range of 1 nM (5). We recently showed that we could generalize this result across a proteome, demonstrating that the distribution of detected interactions covered the range of protein abundances down to the range of less than 100 molecules per cell (6). We have also shown that an upper limit of the dissociation constant for detection of PPI is likely in the range of 10–100 μM for the DHFR (7) and OyCD (see Subheading 3.2) PCAs (8). These observations suggest that PPI can be detected by PCA within ranges of protein abundances and complex affinities that are commonly observed. However, PPI may or may not be detected depending on the PCA reporter used. For instance, a PPI studied with a fluorescent protein PCA reporters might not be detected if the abundance of complexes is lower than necessary to reconstitute enough fluorescent proteins (see Subheading 3.3). In this case, signal will not be high enough to overcome background fluorescence of cells in the range of wavelengths over which the fluorophore emits. On the other hand, there are no background issues for luciferase-based PCAs and thus detection is limited only by the sensitivity of the detector used (see Subheading 3.4). Finally, an issue of particular importance to studies in yeast where the complementary PCA fragments are fused to gene open reading frames by homologous recombination is whether the genes are hetero- or homozygous for the fusions in diploid cells. In this case, the untagged proteins (A and B) will compete for binding with those that are tagged (A′ and B′), resulting in a reduced number of reconstituted PCA reporter proteins and thus, reporter signal. Only the A′B′ complex (out of the four possible AB, AB′, A′B, and A′B′) results in a reconstituted PCA reporter protein, leading to a fourfold reduction in signal. The number of reconstituted complexes necessary for signal detection in assays performed in diploid cells (6) is therefore much lower than what is expected from the abundance of the interacting partners alone.

A second set of considerations in using PCA is how the fusion of complementary PCA reporter fragments could affect the proteins of interest and the ability to detect PPI. First, as with any fusion construct, it is critical to test the fusions in established functional assays in order to assure that the tags themselves do not impair the function of the protein or lead to gain of function. One should also not assume that a functional fusion protein with a particular tag ensures that other PCA tags will lead to functional fusions. Different tags may have different effects. Second, we can ask if the orientation of fusion (N or C terminus) or identity of the fragment may affect the outcome of a PCA experiment. This can only be determined empirically. We have tested all possible combinations and permutations of tagging individual test proteins that are known to interact (eight total per protein pair) and found that in some cases it made no difference how the proteins were tagged while for others, only an individual arrangement worked (unpublished results).

As we described above, PCAs are sensitive to whether the complementary N- and C-terminal fragments can find each other in space and this depends on the distances between the termini of the interacting proteins to which the fragments are fused. To assure that PCA can occur, we typically insert a 10–15 amino acid flexible polypeptide linker consisting of the sequences $(Gly.Gly.Gly.Gly.Ser)_n$ between the proteins of interest and the PCA reporter protein fragments. We chose the $(Gly.Gly.Gly.Gly.Ser)_n$ linker because it is the most flexible possible and we have empirically observed that linkers of these lengths are sufficiently long to allow for fragments to find each other and fold, regardless of the sizes of the interacting proteins to which the fragments are fused (9).

1.2. DHFR PCA Survival-Selection for Large-Scale Analysis of PPIs

The DHFR PCA was previously developed for *Escherichia coli*, plant protoplasts and mammalian cell lines (2, 5, 10, 11) and has recently been adapted for large-scale screening of PPIs in yeast (6). The principle of the DHFR PCA survival-selection

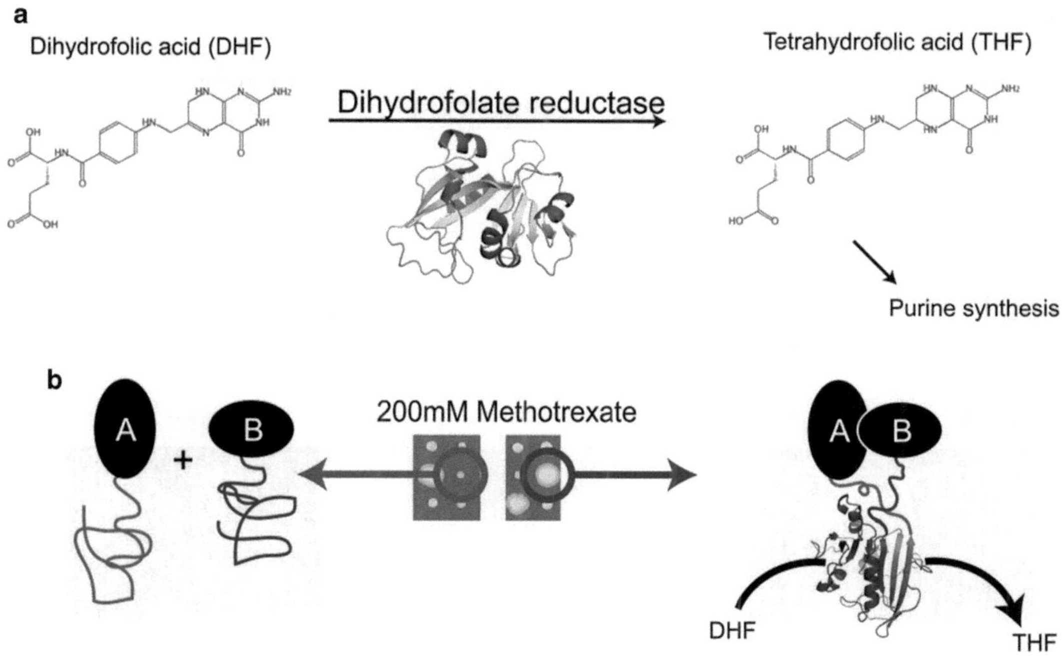

Fig. 2. Basis of the DHFR PCA strategy. (**a**) DHFR catalyzes the reduction of dihydrofolate to tetrahydrofolate, which is required for nucleotide, and in some cases, amino acid synthesis. This reaction can be inhibited by an antifolate, methotrexate. (**b**) In the DHFR PCA strategy, the two proteins of interest are fused to complementary fragments of a mutant DHFR protein that is insensitive to methotrexate. The PCA fragments are inactive unless the proteins of interest interact. If so, the DHFR fragments are brought together in space and fold into the native structure, thus reconstituting the activity of the mutant DHFR and allowing cells to proliferate in the presence of methotrexate.

assay is that cells lacking endogenous DHFR activity, achieved here by inhibiting the *S. cerevisiae* scDHFR with methotrexate, are enabled to proliferate by simultaneously expressing PCA fragments of a methotrexate-resistant DHFR mutant that are fused to interacting proteins or peptides. If the proteins interact and thus allow refolding of the DHFR reporter, cells that are grown in the presence of methotrexate can proliferate (Fig. 2) (5). To adapt the DHFR PCA for high-throughput screening in *S. cerevisiae,* we created a double mutant (L22F and F31S) that is 10,000 times less sensitive to methotrexate than wild-type scDHFR, while retaining full catalytic activity (12). The assay can be used with strains harboring yeast expression vectors of the target protein open reading frame (ORF) fused to the PCA fragment coding sequence. It is also sensitive enough to be used with genomic recombinant strains, expressing proteins fused to the PCA fragment under the control of their endogenous promoters. We created two universal oligonucleotide cassettes encoding each complementary DHFR PCA fragment and two unique antibiotic resistance enzymes to allow for selection of haploid strains that have been successfully transformed and recombined with one or the other homologous recombination

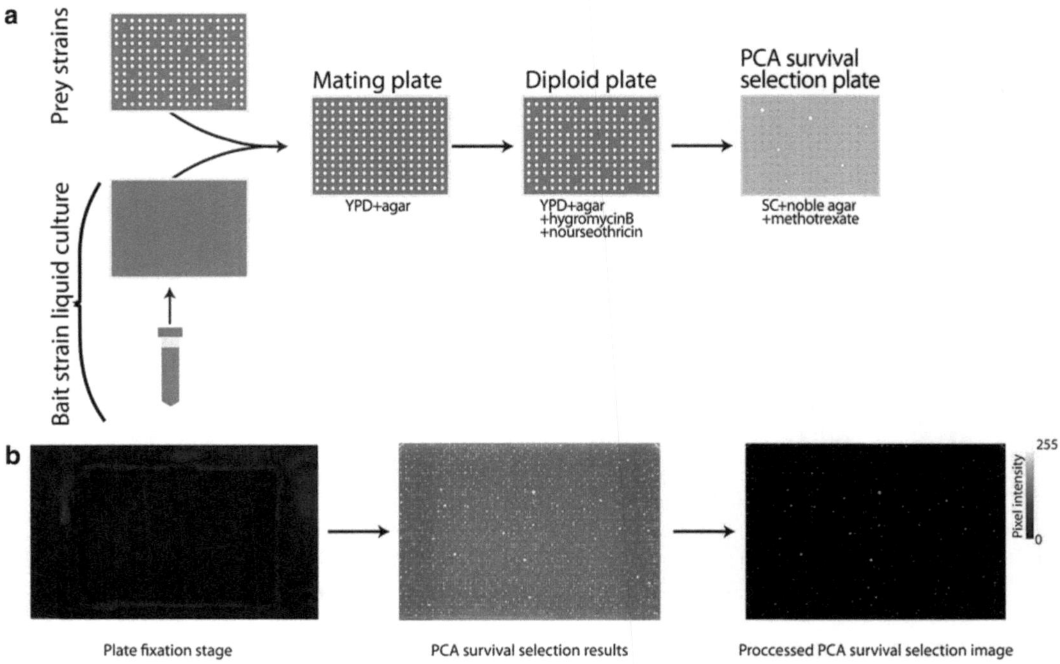

Fig. 3. Use of the DHFR screen for define the yeast protein interactome. (**a**) The DHFR PCA screen is performed as shown in schematic. The bait reporter strain is incubated in liquid culture. The prey reporter strains are printed on solid medium and incubated to be used on multiple assay plates. The mating plate is produced by sequentially printing the bait strain and the prey strains on solid agar containing rich medium, allowing strains to mate and colonies to grow. Resulting haploids and diploid mixture strains are transferred onto solid agar plates containing diploid selective medium. The resulting diploid strains can be transferred onto plates containing PCA survival-selection medium (containing methotrexate). **(b)** The resulting PCA survival-selection plate, here a 6,144 density plate grown for 2 weeks, can be imaged using a black velvet-covered plate fixation platform and a digital camera. The integrated pixel density is computed using pixel intensity, represented here as a color- or gray-coded scale, integrated on the area of each colony.

cassettes (6). The resulting universal templates were used to create homologous recombination cassettes for most budding yeast genes by PCR using 5′ and 3′ oligonucleotides consisting of 40-nucleotide sequences homologous to the 3′ end of each ORF (prior to the Stop codon) and a region approximately 20 nucleotides from the stop codon. Below are protocols to perform DHFR PCA at a large-scale with recombinant strains or with yeast transformed with expression plasmids (Fig. 3).

1.3. A Life and Death Selection PCA Based on the Prodrug-Converting Cytosine Deaminase for Dissection of Protein–Protein Interactions

Another valuable PCA strategy is based on an optimized mutant form of the reporter enzyme yeast cyosine deaminase (OyCD). The choice of yCD as a reporter was based on its role in a pyrimidine salvage pathway and the availability of a prodrug 5-fluorocytosine (5-FC), which is converted to 5-fluorouracil (5-FU) by yCD. Bacteria and yeast can convert cytosine to uracil and use it for the synthesis of UTP and TTP, which are required for cell survival (13). In *S. cerevisiae*, yCD is encoded by the *FCY1* gene and is the enzyme that catalyzes this reaction. In addition to deaminating cytosine, yCD can also deaminate 5-FC to 5-FU. 5-FU will be further processed by enzymes of the pyrimidine salvage pathway to 5-FUTP, a toxic compound that causes cell death. These particular properties of yCD make it an ideal reporter for a life and death selection PCA (Fig. 4a) (8).

The OyCD PCA allows a death and survival assay to be performed without changing the reporter system. In a two-step selection process, we can engineer mutant forms of a protein in order to dissect its different functions; disrupting interactions with one partner, while retaining interaction with others (Fig. 4b). For example, protein A interacts with both protein B and protein C. First, we can screen for mutant forms of protein A that disrupt interaction with protein B. Second, we select for protein A mutants that still interact with protein C. Using OyCD PCA, neither of these selection steps requires replica plating. In addition, no expensive reagents or equipment is required. Specific mutants can be obtained in about 4 weeks.

Both the survival and death selection assays are performed in *fcy1* deletion strains. For the survival-selection assay, uracil must be removed from the selection medium. Only cells that have OyCD PCA activity will be able to synthesize uracil and survive. For the death selection assay, cells are grown in a selection medium in the presence of 5-FC. In this death assay, cells that have OyCD PCA activity will be sensitive to 5-FC.

1.4. Visualizing the Location of Protein–Protein Interactions with GFP Family Fluorescent Protein PCAs

Originally described by Lynne Regan's group for GFP (14–16), we and others have described different color and behavioral variants (17–22). Notably, and unlike other PCAs, those based on these fluorescent proteins are irreversible, which can be both useful (trapping and visualizing rare and transient complexes) but also require care in interpretation of turnover or localization of interacting proteins (15, 18).

It is important that the kinetics of relocalization of protein interactions observed with fluorescence PCAs be confirmed by immunofluorescence or by monitoring the localization of the same proteins fused to full-length fluorescent proteins. Fluorescent protein PCAs are also limited to the temporal range of dynamics that can be studied. Because different variants of these proteins take minutes to hours to fold and mature, they are obviously not appropriate for studying most dynamic processes in a quantitative way, though many important slower processes can be studied. PCAs based on luciferase enzyme reporters are, like the DHFR PCA, fully reversible and can be used to capture kinetics on the second time scale (23, 24).

As we previously demonstrated, protein–protein interactions that occur within a specific biochemical pathway can be modulated in predicted ways by conditions or molecules that activate or inhibit the pathway. We, and others, have shown that at least changes in the formation of complexes can be detected with the

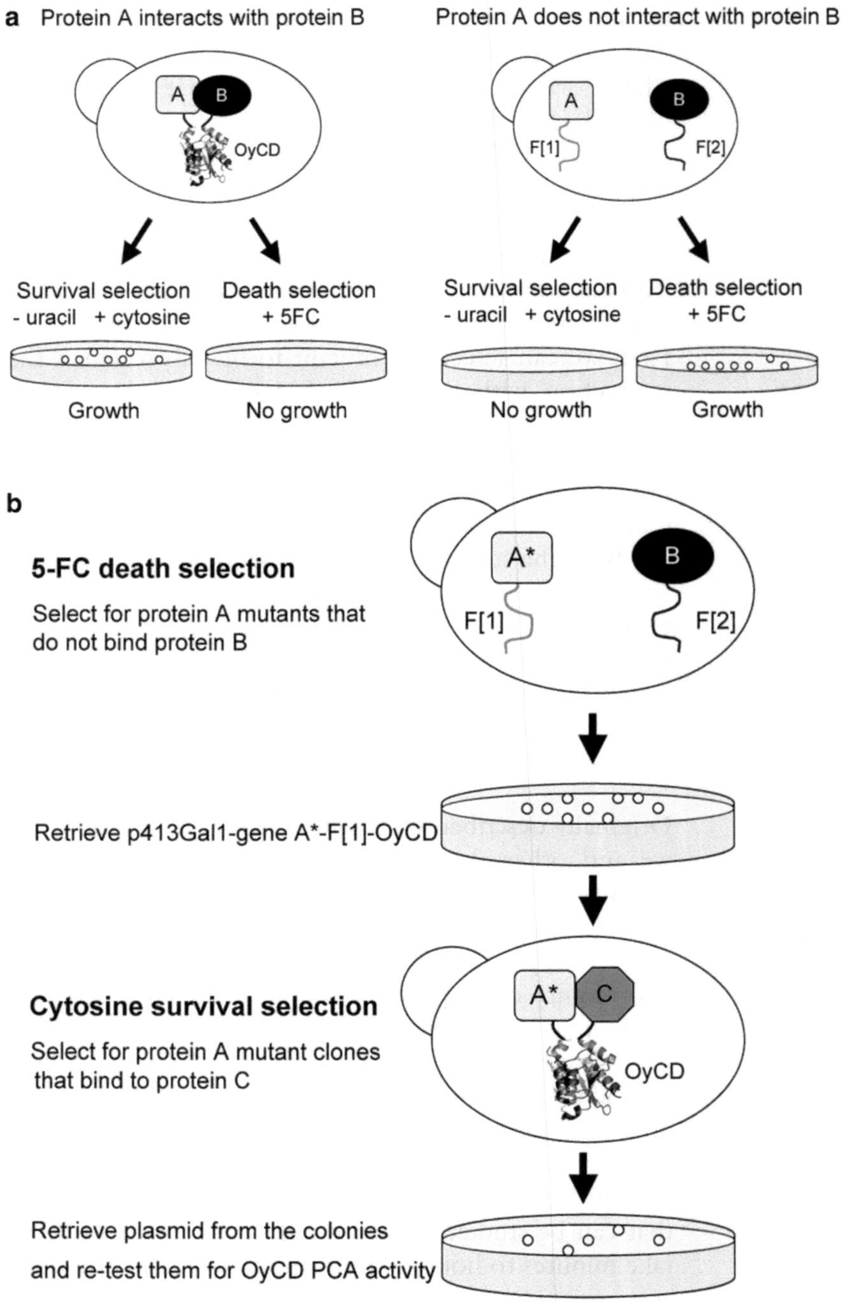

Fig. 4. Basis of the OyCD dual selection PCA. (**a**) The OyCD PCA can serve as a reporter for formation of a protein–protein interaction provided that the reconstituted reporter enzyme supports growth under one condition (survival assay) or no growth under another condition (death assay). In the case where the two test proteins do not interact, the reverse scenarios are observed. (**b**) Screen for mutants of protein A that do not bind to protein B but retain binding to protein C using sequential death followed by survival-selection OyCD PCA. The first death selection screen consists of screening the library of protein A mutants fused to OyCD-F[1] (A*-F[1]) with protein B fused to OyCD-F[2] (B-F[2]) and identifying clones that show loss of OyCD PCA activity (growth in the presence of 5-FC). The second survival-selection screen consists of screening A*-F[1] clones harvested from the first death selection screen against protein C fused to OyCD-F[2] (C-F[2]) to identify clones that show OyCD PCA activity using the life assay (growth in presence of cytosine).

GFP and YFP PCAs (21, 22). Further, the subcellular location of stable complexes and changes in their locations following perturbation can also be detected in intact living cells with the YFP PCA (19, 21, 22). It is this ability to detect the location and intracellular movements of protein complexes that make fluorescent protein-based PCAs unique. Because GFP/YFP-based PCAs do not require additional substrates or cofactors for emission of fluorescence, they are particularly simple to implement. We have shown that protein–protein interactions can be monitored by fluorescence microscopy, flow cytometry, and spectroscopy using GFP- and YFP-based PCAs (19, 21, 22). We have applied these assays to the detection and quantification of protein interactions, localization of complexes in living cells, and cDNA library screening in mammalian cells (19–22, 25, 26). In addition, we have used the YFP-based PCA to detect protein interactions in specific subcellular compartments of *S. Cerevisiae*, such as cytoplasm, nucleus, plasma membrane, and the bud neck (Fig. 5) (30). In the following protocol, we describe methods for studying PPI with the "Venus" mutant of YFP (31).

1.5. Studying Dynamics of Protein–Protein Interactions with Luciferase Reporter PCAs

It has been a major challenge to measure and quantify the dynamics of protein complexes in their native state within living cells. PCAs using *Renilla* luciferase (*R*luc) and *Gaussia* luciferase (*G*luc) have been designed specifically to investigate the dynamics of assembly and disassembly of protein complexes. We have applied these assays to the detection and quantification of protein interactions in mammalian cells as well as yeast. These assays are sensitive enough to detect interactions among proteins expressed at endogenous levels in vivo and to study dynamic changes in both the formation and disruption of protein–protein interactions over seconds without altering the kinetics of binding (23, 24). Both of these luciferases catalyze the oxidation of substrate coelenterate luciferins (coelenterazines) in a reaction that emits blue light (at a peak of 480 nm) and requires no cofactors (32). The substrates readily diffuse through cell membranes and into all cellular compartments, enabling quantitative analysis in live cells. *R*luc and *G*luc are monomeric proteins of 312 (36 kDa) and 185 amino acids (19.9 kDa). Gluc PCA has some advantage in that the reporter protein is smaller and has ten times higher activity to native coelanterizine than Rluc. However, at present, Rluc has the advantage that stable substrates (e.g., benzyl-coelenterizine) can be used with this reporter allowing for easier handling and integration of signal over longer times. In contrast to fluorescent protein-based PCAs, both Rluc and Gluc are fully reversible; a prerequisite to study signaling events by the dynamics of protein complex assembly and disassembly (23, 24). Both Rluc and Gluc PCAs provide for extremely high signal-to-background ratio due to lack of any cellular luminescence and can easily be measured

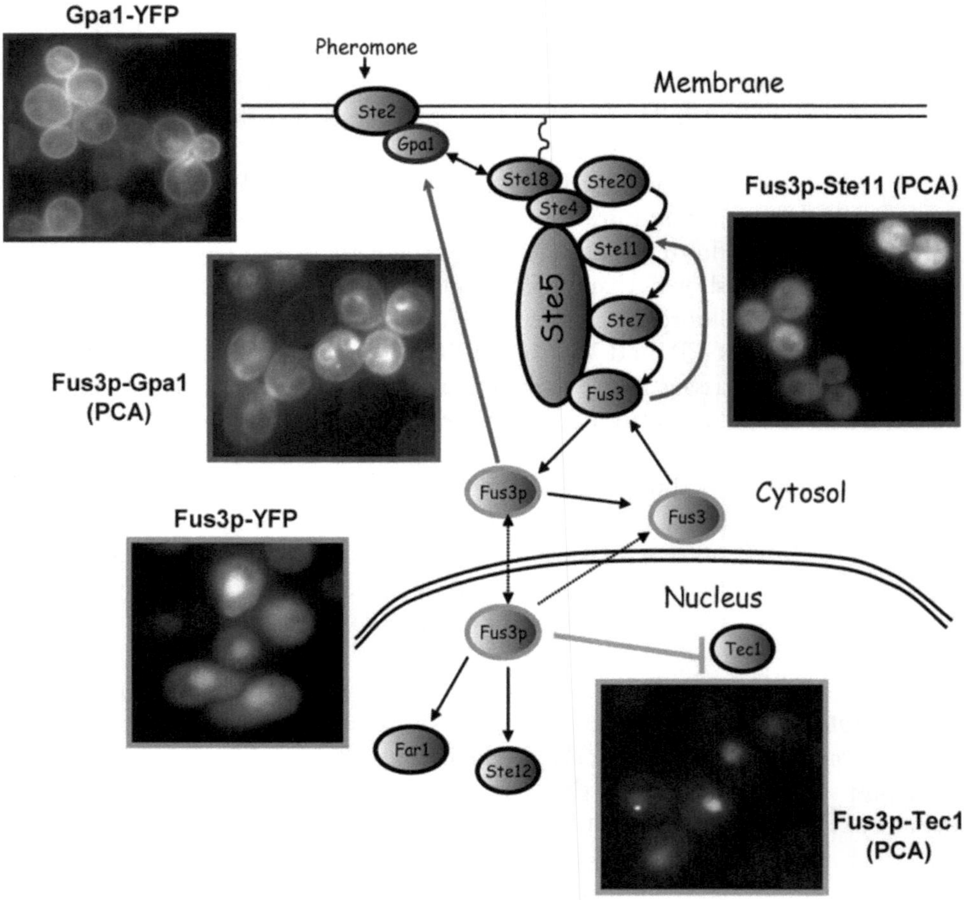

Fig. 5. Venus YFP PCA allows for detection of the location of protein complexes within living cells. This illustration uses the yeast pheromone response mitogen activated protein kinase pathway for visualization of protein complexes in different regions within cells. Images show the location of interactions of Fus3p with Gpa1 (27) to the membrane, with Ste11 (28) to the cytoplasm and with Tec1 (29) to the nucleus. As controls for different localizations, Gpa1 fused to full-length Venus YFP protein is shown to be at the membrane while Fus3-Venus YFP is found in both cytoplasm and the nucleus. Cells containing Fus3-venus YFP were treated with 1 μM alpha-factor pheromone for 2–3 h to induce its translocation to the plasma membrane and nucleus.

spectroscopically on whole cell populations or by imaging single cells. Finally, the luciferase PCAs allow for accurate measurements of time (for time constants greater than 10 s) and dose dependence of pharmacologically induced alterations of protein complexes.

2. Materials

2.1. Reagents and Solutions

2.1.1. DHFR PCA Survival-Selection for Large-Scale Analysis of PPIs

1. Glycerol stocks of *MAT***a** recombinant strains in which ORFs are fused to the complementary DHFR PCA F[1,2] fragment (Open Biosystems).
2. Glycerol stocks of *MATα* recombinant strains in which ORFs are fused to the complementary DHFR F[3] PCA fragment (Open Biosystems).

3. 3% Agar-solidified YPD medium in Nunc omniplates.
4. 3% Agar-solidified YPD medium with 100 μg/mL nourseothricin for *MAT*a recombinant strains (WERNER BioAgents) or 250 μg/mL hygromycin B for *MATα* recombinant strains (Wisent Corporation) in Nunc omniplates.
5. 3% Agar-solidified YPD medium with both 100 μg/mL nourseothricin (WERNER BioAgents) and 250 μg/mL hygromycin B (Wisent Corporation) in Nunc omniplates.
6. 4% Noble Agar (purified agar; Bioshop) solidified synthetic complete (SC) medium with 200 μg/mL methotrexate (prepared from 10 mg/mL methotrexate in DMSO stock solution) in Nunc omniplates.
7. Anti-DHFR polyclonal antibody that specifically recognizes an epitope in the N-terminal F[1,2] fragment (Sigma D1067; working dilution 1:6,000) (Sigma-Aldrich).
8. Anti-DHFR polyclonal antibody that specifically recognizes an epitope in the C-terminal F[3] fragment (Sigma D0942; working dilution1:5,000) (Sigma-Aldrich).

2.1.2. Cytosine Deaminase Life and Death Selection PCA for Dissection of PPIs

1. BY4741, BY4742, or BY4743 strains with a deletion in the *FCY1* gene (*fcy1*Δ) that are resistant to G418 (33).
2. Synthetic complete medium with the appropriate amino acid drop out according to the chosen expression plasmids.
3. Genes of interest fused to the OyCD fragments in yeast expression vectors.
4. Sorbitol Buffer: 1 M sorbitol, 1 mM EDTA, 10 mM Tris, 100 mM Lithium Acetate, pH 8.0.
5. PLATE Solution: 40% PEG 3350, 100 mM Lithium Acetate, 10 mM Tris, 0.4 mM EDTA, pH 7.5.
6. Dimethylsulfoxide (DMSO).
7. Sterile distilled water.
8. G418 (Wisent).
9. Cytosine.
10. 5-Fluorocytosine.
11. Agar (Bioshop).
12. Noble agar (Bioshop).
13. DH5α or MC1061 *E. coli* electro-competent cells.
14. Luris Broth (LB) medium.
15. DNeasy Tissue Kit (Qiagen).
16. Antibodies against yCD fragments: Anti-yCD polyclonal (Biogenesis).
17. 10 mg/ml stock solution of cytosine: Dissolve 100 mg of cytosine in 10 ml of distilled water. Vortex the solution and

incubate at 37°C to make it dissolve. Filter the solution and store at room temperature. It is better to make this solution fresh and use it within a week.

18. 10 mg/ml stock solution of 5-fluorocytosine (5-FC): Dissolve 100 mg of 5-FC in 10 ml of distilled water. Vortex the solution and incubate at 37°C to make it dissolve. Filter the solution and use it right away or aliquot in sterile tubes and store at –20°C.
19. Control plates: Make SC plates for selection of clones harboring the expression plasmids. We used the p413Gal1 and p415Gal1 expression vectors, therefore our control plates contain SC medium without histidine and leucine, with 2% agar, 2% raffinose, and 2% galactose.
20. Cytosine survival-selection plates: Make SC plates without uracil and selection for the expression plasmids. We used the p413Gal1 and p415Gal1 expression vectors, therefore our selection plates contain SC medium without uracil, histidine and leucine, with 3% Noble agar, 2% raffinose, 2% galactose and cytosine (we use 100 μg/ml of 5-FC for our proteins of interest).
21. 5-FC death selection plates: Make SC plates with 5-FC and selection for the expression plasmids. We used the p413Gal1 and p415Gal1 expression vectors, therefore our selection plates contain SC medium without histidine and leucine, with 2% Noble agar, 2% raffinose, 2% galactose and 5-FC (we use 100 μg/ml of 5-FC for our protein of interests).

2.1.3. Fluorescence Protein PCA to Visualize PPI Location

1. Competent *MAT*a or diploid yeast (34).
2. SD medium: 6.7 g/L yeast nitrogen base, without amino acids.
3. SD agar: SD medium with 2% agar.
4. 10× amino acid mix -histidine, -leucine, -lysine: adenine sulphate (0.4 g/mL), uracil (0.2 g/mL), L-tryptophan (0.4 g/mL), L-arginine HCl (0.2 g/mL), L-tyrosine (0.3 g/mL), L-phenylalanine (0.5 g/mL), L-glutamic acid (1.0 g/mL), L-asparagine (1.0 g/mL), L-valine (1.5 g/mL), L-threonine (2.0 g/mL), L-serine (3.75 g/mL), methionine (0.2 g/mL) (do not include when growing diploid yeast).
5. SC agar: SD agar, 2% glucose, 1× amino acids.
6. Low Fluorescence Medium (LFM): 1× low fluorescence yeast nitrogen base, 2% glucose, 1× amino acids (35).
7. 20% glucose solution.
8. PLATE solution: 40% Polyethylene glycol 3,350, 100 mM LiOAc, 10 mM Tris, 0.4 mM EDTA, pH 7.5.
9. DMSO.
10. Poly-L-lysine MW 30,000–70,000 (Sigma) or Concanavalin A (Sigma).

2.1.4. Luciferase Reporter PCAs to Study Dynamics of PPIs

1. Competent *MAT*a or diploid yeast (34).
2. SD medium: 6.7 g/L yeast nitrogen base, without amino acids.
3. SD agar: SD medium with 2% agar.
4. 10× amino acid mix -histidine, -leucine, -lysine: adenine sulphate (0.4 g/mL), uracil (0.2 g/mL), L-tryptophan (0.4 g/mL), L-arginine HCl (0.2 g/mL), L-tyrosine (0.3 g/mL), L-phenylalanine (0.5 g/mL), L-glutamic acid (1.0 g/mL), L-asparagine (1.0 g/mL), L-valine (1.5 g/mL), L-threonine (2.0 g/mL), L-serine (3.75 g/mL), methionine (0.2 g/mL) (do not include when growing diploid yeast).
5. SC agar: SD agar, 2% glucose, 1× amino acids.
6. Low Fluorescence Medium (LFM): 1× low fluorescence yeast nitrogen base, 2% glucose, 1× amino acids (35).
7. 20% glucose solution
8. PLATE solution: 40% polyethylene glycol 3,350, 100 mM LiOAc, 10 mM Tris, 0.4 mM EDTA, pH 7.5.
9. DMSO.
10. Poly-L-lysine MW 30,000–70,000 (Sigma) or concanavalin A (Sigma).
11. Coelenterazine and Benzyl-Coelenterazine (Nanolight).

2.2. Equipment

2.2.1. DHFR PCA Survival-Selection for Large-Scale Analysis of PPIs

1. Pintool: robotically manipulated 96 pintool (0.787 mm flat round-shaped pins, #FP3N; V&P Scientific Inc.); 384 pintool (0.457 mm flat round-shaped pins, custom #FP1N, V&P Scientific Inc.); and 1,536 pintool (0.457 mm flat round-shaped pins, custom #FP1N; V&P Scientific Inc.), or manually manipulated 96 pintool (1.58 mm, 1 μL slot pins, VP 408Sa; V&P Scientific Inc.).
2. Plate Imaging: Minimum 4.0 Mega pixel digital camera (e.g., Powershot A520; Canon), stationary arm (70 cm mini repro, Industria Fototecnica Firenze), and plate-shooting platform.

2.2.2. Cytosine Deaminase Life and Death Selection PCA for Dissection of PPIs

1. Genepulser II electroporator system (Bio-Rad) or Electroporator 2510 (Eppendorf).
2. Electroporation cuvette with 1 mm wide slot (Sigma).
3. Glass spreader.
4. 100 mm Petri dishes.
5. Shaking incubators, preset to 30 and 37°C.
6. Incubator, preset to 30 and 37°C.

2.2.3. Fluorescence Protein PCA to Visualize the Location of PPIs

1. Fluorescence microscope, e.g., Nikon Eclipse TE2000U inverted microscope (Nikon) with a CoolSnap HQ Monochrome CCD camera (Photometrics).

2. 96-well black, glass bottom plate (Molecular Machines).
3. 6-well culture plate or Petri dish.
4. Appropriate sterile tubes to grow yeast.
5. Spectrophotometer Spectra MAX GEMINI XS (Molecular Devices).

2.2.4. Luciferase Reporter PCAs to Study Dynamics of PPIs

1. Luminescence microplate reader, e.g., LMax II^{384} Luminometer (Molecular Devices).
2. Luminescence microscope, e.g., Nikon Eclipse TE2000U inverted microscope with a CoolSnap HQ Monochrome CCD camera (Photometrics).
3. 96-well white plates (Molecular Machines).
4. 6-well culture plate or Petri dish.
5. Appropriate sterile tubes to grow yeast.
6. Spectrophotometer.

3. Methods

3.1. DHFR PCA Survival-Selection for Large-Scale Analysis of PPIs

The general strategy for performing a screen is to generate an array of "prey" strains as indexed colonies grown in a regular grid on agar and then mate them with individual "bait" strains of the opposite mating type to select for diploids and then transfer these to a methotrexate-containing plate for survival-selection (Fig. 3). The choice of whether to use the *MAT*a or *MATα* strains as bait or prey is arbitrary. Here we describe a procedure in which the *MAT*a strains are bait and *MATα* are the prey strains. Baits can also be expressed as fusions to DHFR PCA fragments from expression plasmids available from our lab and transformed into appropriate strains.

3.1.1. Colony Plating and Culture

1. Incubate individual bait strains picked from glycerol stocks in a 45 mL liquid culture of strain selective media (YPD with 100 μg/mL nourseothricin for *MAT*a recombinant strains or 250 μg/mL hygromycin B for *MATα* recombinant strains) and allow culture to reach saturation at 30°C.
2. Print prey strains picked from glycerol stocks onto a 35 mL agar-solidified omniplate of strain selective media (3% agar + YPD with 100 μg/mL nourseothricin for *MAT*a recombinant strains or 250 μg/mL hygromycin B for *MATα* recombinant strains) using four 96 manual or robotic pintool prints for a total of 384 prints per plate and incubate 16 h at 30°C (see Notes 1 and 2).
3. Centrifuge a saturated culture of bait strain at 500 × *g* for 5 min and resuspend in 15 mL of YPD. The bait culture must

be saturated to print enough cells for efficient mating on solid phase. The pintool must be cleaned between each cell transfer. We soak the pins twice in a solution of 10% bleach containing glass beads followed by a 10% bleach wash and two sterile water bath washes.

4. Transfer bait strain suspensions into an empty omniplate.
5. Print the bait strain suspension from the empty omniplate to an omniplate containing 35 mL of solid agar containing rich medium (YPD + 3% agar) at the same density as the prey strains using a pintool appropriate for the desired colony array density.
6. Transfer prey strains onto the bait strain in an omniplate containing 35 mL of solid agar containing rich medium (YPD + 3% agar) using a pintool appropriate for the desired colony array density. Allow mating to occur by incubating the plate for 16 h at 30°C (see Note 3).
7. Transfer the mixed haploid and diploid colonies from step 6 to an omniplate containing 35 mL of diploid selective medium (3% agar + YPD with both 100 μg/mL nourseothricin and 250 μg/mL hygromycin B) using a pintool appropriate for the desired colony array density. Incubate for 16 h at 30°C (see Note 4).
8. Transfer diploid selected strains onto an omniplate containing 35 mL of solid noble agar with synthetic minimal medium plus methotrexate (4% noble agar + SC + 2% glucose + 200 μg/mL methotrexate) using a pintool appropriate for the desired colony array density. Incubate at 30°C and acquire pictures of the colony array every 96 h for approximately 2 weeks (see Note 5).

3.1.2. Anticipated Results and Controls

Additional troubleshooting advice can be found in Table 1. To evaluate a DHFR PCA screen, both positive controls (known PPIs) and negative controls (fragments alone or noninteracting protein partners) should be tested on every plate. These noninteracting protein partner colonies exhibit background growth that should stop after a few days of incubation on methotrexate-containing plates. Colonies containing interacting baits and preys will continue to grow. The PCA fragment fusions expressed alone should not result in cell proliferation because the individual PCA fragments have no activity. Thus if individual colonies do grow for unknown reasons, they should not be considered for further analysis. The most critical controls are those for spontaneous PCA, i.e., cases where a protein PCA fragment fusion interacts with the complementary fragment alone. We found in our own screen that about 5% of bait or prey protein-expressing strains would grow in the presence of methotrexate when mated to a strain harboring an expression vector encoding the complementary fragment alone (6). These complementary DHFR PCA

Table 1
Trouble shooting large-scale DHFR PCA screen

Step	Problem	Possible reason	Solution
Subheading 3.1.1, Steps 1–2	Strains are not growing or i ncomplete prey array growth	Erroneous haploid selection	Verify protocol for appropriate culture conditions
		Low glycerol viability	Strains can be streaked on solid agar-selective medium Petri dishes prior to inoculation to increase viability
		Technical problem	Verify that all pins of the pin tool touch glycerol stocks and the recipient omniplate
Subheading 3.1.1, Step 7	Low number or no colonies on diploid selective plates	Erroneous haploid strains type	Verify mating type of haploid strains
		Technical problem	Pin tool alignment might have changed. No modifications to the pin tool positioning should be done between transfers
Subheading 3.1.1, Step 8	No colony growth on DHFR PCA survival-selective medium	Erroneous selective conditions	Use heteromeric complex $SspB_{YGMF}$:$SspB_{LSLA}$ as a positive control to validate DHFR PCA activity
		Erroneous DHFR PCA	Verify by a strain diagnostic PCR the complementarity of PCA fragments
			Verify DHFR PCA fragment recombinant insertion by genomic sequencing
		DHFR PCA fragment expression	Verify DHFR PCA fragment expression by western blot
	All colonies grow at the same rate on DHFR PCA survival-selective medium	Erroneous selective conditions	Use DHFR PCA fragment controls alone as negative control
		Methotrexate solubility	Verify methotrexate solubility under conditions used. Stock solution should not exceed 10 mg/mL in DMSO and final concentration in solid agar plates should not exceed 200 mg/mL

fragment expression vectors are available upon request form our lab. Other controls can be included to test how the PCA screen performs. For instance, we have used the engineered heteromeric $SspB_{YGMF}$:$SspB_{LSLA}$ interaction as a positive control to validate DHFR PCA activity as suggested in the troubleshooting section. Another elegant control to examine the range of dissociation constants for which the DHFR PCA is sensitive is to use a complex for which single-point mutations are known by other methods disrupt the interaction to different degrees. To this end, we have used in our own work, mutants of the Ras-binding domain of Raf (7). A potential source of false positives in a PCA screen can be through trapping of nonspecific complexes due to irreversible folding of the DHFR fragments. However, we have used the adenosine 3′,5′-monophosphate-dependent dissociation of the yeast protein kinase A complex as a control (24) to show that the DHFR PCA is fully reversible, and thus the trapping of complexes is unlikely (6). Another control one could use is a condition-dependent PPI. In our own work we have used the FK506-binding protein that binds rapamycin and then binds the Target of Rapamycin (TOR) (2). All of these reagents are available upon request.

3.1.3. Analysis of Large-Scale DHFR PCA Screens

The goal of this process is to turn the size of the colonies on the selection plate into binary data that will represent PPIs. First, the digital images have to be transformed into tables containing colony intensities. Second, these colony intensities have to be turned into protein–protein interaction confidence scores.

Image Acquisition and Processing

Plates are imaged using a black velvet-covered plate fixation platform and a basic digital camera. The image must be processed to remove plate sides, allowing image analysis to be performed only on the region containing colonies. Images should be corrected for nonuniform illumination as described in (http://www.mathworks.com/products/image/demos.html?file=/products/demos/shipping/images/ipexrice.html) and small objects, corresponding to bubbles, gel background and other anomalies should be removed using the imopen function.

Image Analysis

Several bioinformatics tools are available to perform colony size measurements from digital images of high-density plates (36–38). Alternatively, tools developed for analysis of spotted DNA microarrays can be modified to estimate the sizes of the colonies spaced on regular grids (39). Globally, the analysis consists of measuring the number of pixels per colony position. In cases where high-density plates are used (above 1,536 position grid), more involved analyses methods have to be utilized to separate adjacent colonies that may touch each other (6). However, because protein–protein interactions are rare, most colonies will have a very slow growth rate and this problem is mostly negligible at lower densities. Thus, when lower densities are used for the screens, simple macros can be implemented in publicly accessible image analysis software such as ImageJ (http://www.rsb.info.nih.gov/ij/). In this case, digital images of plates are first converted to 8-bit grayscale format and colonies are measured by positioning the measurement tool on a colony center and estimating the integrated pixel intensity in an area that corresponds to the maximal colony size allowed. The process is iterated over all the grid positions and then all the plates, and the grid positions and intensity values are exported to text files for further processing in your preferred spreadsheet or statistical analysis software (e.g., the ImageJ scripts that we use are available at our Web site: http://michnick.bcm.umontreal.ca). It is important to note that colonies should always have the same positions on the images. If this is not the case, some of the tools cited above include a step that positions the analysis grid onto the colony positions prior to colony size measurements.

Statistical Analysis of Raw Colony Data: From Continuous to Binary Data

A PCA screen based on survival assay will only be useful if there is a confidence score attached to each of the putative interactions. Raw colony intensity data are continuously distributed, i.e., they cover a wide range of values and cannot be directly turned into "yes" or "no" binary scores. Further, not all the colonies that can grow due to protein-fragment complementation will do so at exactly the same rate. As described above, every PCA experiment should include a set of positive controls consisting of pairs of baits and preys that interact with each other, and negative controls, consisting of pairs or baits and preys that do not interact with each other. These will be used for quality control in order to detect mis-positioning of the grid and batch effects (variation in media, incubation, drug concentration) that affect global growth rate of the different plates. Finally, the positive controls can provide a first, visual analysis of the data, whereby the growth rate of the positive controls indicate roughly the intensity threshold above which we expect strains with interacting bait-prey pairs to grow. Beyond these "qualitative" controls, a statistical analysis should be used to separate the interacting pairs from the noninteracting pairs.

The statistical analysis globally includes two steps. First, it has to be determined whether there is a significant difference in growth rates among the plates before applying a global analysis to the data. If there is significant variation, the data should be normalized such that all the plates have the same average colony size. Alternatively, data could be transformed into relative scores, such as Z-scores, whereby each data point is transformed to become the number of standard deviations that data point is from the average of the plate. We found that combining the Z-score and the raw intensity worked best for our large-scale screen (6). Then continuous values must be turned into binary values by setting a threshold of intensity above which proteins are inferred to interact, and establishing a confidence score for this particular threshold. One way to assign confidence values to PCA interactions is to benchmark the intensity values against a set of data containing interactions that should be detected in the screen (a set of real positives) and others that should not (a set of real negatives). The real positives set can be derived from a set of known and well-supported interactions. The real negative set has, however, to be approximated because it is impossible to show that two proteins never interact. Sets of proteins that are most likely not interacting can be used for this purpose, for instance proteins that are not localized in the same cell compartments and that have negatively correlated expression profiles (40). One can then predict, for a given intensity threshold, what should be the proportion of true-positive interactions and false-positive interactions. In order to decide on the threshold, the ratio of true-positive interactions divided by the total number of inferred positives (true positives + predicted false positives) – known as the Positive Predictive Value (PPV) – is calculated as a function of threshold of intensities. For instance, at a PPV of 95%, one expects 5% of positives to be false. Lower and higher thresholds can be used depending on how stringent one wants the analysis to be. It is important to note that the estimated PPV is only accurate if the relative occurrence of positives and negatives in the reference sets is similar to that of the real positives and negatives (41). In the case of a genome-wide, comprehensive screen, this fraction corresponds to a very low prior probability of finding interactions among all pairwise possibilities. On the other hand, a small-scale screen of a specific biological process will contain a greater proportion of real positives than a random screen. The reference set therefore needs to be tailored for the actual screen being performed, i.e., the space of the interactome that is covered. For a formal treatment of these issues, refer to Jansen et al. (41). Beyond these statistical considerations, analysis such as Gene Ontology enrichment and visualization of interaction clusters should be used to further assess the confidence in the data set being produced. For instance, the matrix of binary interactions can be clustered to identify groups or complexes of interacting proteins. Finally, sets of true positives and negatives are not a panacea and the functional and evolutionary characterization of protein–protein interactions is the only way to provide a definitive answer as to whether an interaction is functionally relevant or not for the cell (42).

3.2. Cytosine Deaminase Life and Death Selection PCA for Dissection of PPIs

The proteins of interest are fused to the N-terminal of OyCD fragment 1 or fragment 2 (protein A-OyCD-F(1), protein B-OyCD-F[2] and protein C-OyCD-F[2]). For some proteins, protein–protein interactions can only be detected when they are fused at the C terminus of OyCD fragment 1 (e.g., OyCD-F[1]-protein A). OyCD-F[1] corresponds to amino acid residues 1–77 of yCD with an A23L point mutation. OyCD-F[2] corresponds to amino acid residues 57–158 of yCD with the following point mutations: V108I, I140L, T95S, and K117E (8). The proteins of interest and OyCD fragments are separated by a 15 amino acid flexible polypeptide linker $(Gly.Gly.Gly.Gly.Ser)_3$. The nucleotide sequences encoding these fusion proteins are cloned into yeast expression vectors. We used the p413Gal1 and p415Gal1 expression plasmids (43). Before proceeding with a screen, verify that interactions of your proteins of interest are detected by OyCD PCA. Titrate the amount of cytosine and 5-FC required for detecting your interactions. We normally try a range between 50 and 1,000 μg/ml of cytosine or 5-FC. For many proteins, 100 μg/ml of either substrate is sufficient for detecting OyCD PCA activity. Use well-known interacting and noninteracting proteins as controls. For the Two-Step OyCD PCA screen, a library of your gene of interest can be generated by methods such as error-prone PCR. For example, if the goal is to engineer mutant forms of protein A that can specifically disrupt interaction with protein B while preserving interaction with protein C, an ideal library of protein A would carry 1–3 mutations per clone.

3.2.1. Death Selection Screen

1. Thaw competent yeast cells on ice.
2. Mix 10 μl of cells with 1 μg of the library encoding mutant forms of protein A (protein A*) in BY4741 *fcy1Δ* strain that already carries a plasmid-expressing protein B (Fig. 4b). 60 μl of PLATE solution and 8 μl DMSO (see Note 6).
3. Heat shock yeast at 42°C for 20 min.
4. Plate half of the transformation on the control plates to select for the presence of both expression plasmids (p413Gal1-gene A*-OyCD-F[1]+p415Gal1-gene B-OyCD-F[2]). These plates serve as controls for reporting the efficiency of the transformation. Plate the other half of the transformation on 5-FC death selection plates (see Notes 7 and 8).
5. Incubate plates at 30°C for 2–3 days. Compare the number of colonies obtained on the 5-FC death selection plates to the control plates (see Note 9).
6. Colonies that grow on 5-FC selection plates are pooled and harvested for DNA extraction with a Qiagen DNeasy Tissue Kit or a genomic DNA purification protocol using phenol-choroform) in order to recover the plasmids that express protein A* (see Note 10).
7. Digest the extracted DNA with enzyme(s) that cut in the plasmids coding for expression of protein B-OyCD-F[2] but not the plasmids coding for expression of protein A*-OyCD-F[1] library.

We use AflII, BspmI, HpaI, MunI, NarI or XcmI since they cut in p415Gal1 and not in p413Gal1 plasmid or the gene of interest. This step is not required if the two expression plasmids do not have the same antibiotic resistance gene.

8. Use 2 μl of extracted DNA for electroporation into electrocompetent *E. coli* cells. We use the MC1061 *E. coli* strain since it has higher transformation efficiency than the DH5α strain. Plate the *E. coli* on LB plates with appropriate antibiotic selection. We used LB with ampicillin for the p41XGal1 plasmids.
9. Pool *E. coli* colonies and extract the plasmid DNA using your mini-prep kit of choice (see Note 11).

3.2.2. Survival-Selection Screen

1. Transform the library encoding for mutant forms of protein A (protein A*) retrieved after the death selection screen in BY4741 *fcy1*Δ strain that already carry a plasmid expressing protein C (see Note 12) (Fig. 4b).
2. Plate half of the transformation on the control plates to select for the presence of both expression plasmids (p413Gal1-gene A*-OyCD-F[1]+p415Gal1-gene C-OyCD-F[2]). These plates serve as control for reporting the efficiency of the transformation. Plate the other half of the transformation on Cytosine survival-selection plates (see Notes 13 and 14).
3. Incubate plates at 30°C for 3–7 days (see Note 15).
4. If the screen resulted in less than 50 colonies, inoculate each yeast colony separately in 5 ml of selection medium and harvest cells for DNA extraction with a Qiagen DNeasy Tissue Kit or a genomic DNA purification protocol using phenol-chloroform. If over 50 colonies were obtained, pool all of the colonies and extract DNA from the pooled cells (see Note 16).
5. Digest the extracted DNA with enzyme(s) that cut in the plasmid expressing protein C-OyCD-F[2] but not the protein A*-OyCD-F[1] library. We use AflII, BspmI, HpaI, MunI, NarI or XcmI since they cut in p415Gal1 and not in p413Gal1 plasmid or the gene of interest. This step is not required if the two expression plasmids do not have the same antibiotic resistance gene.
6. Use 2 μl of extracted DNA for electroporation into electrocompetent MC1061 *E. coli* cells. Plate the *E. coli* on LB plates with the appropriate antibiotic selection.
7. For samples obtained from a single yeast colony in step 5, inoculate one or two *E. coli* colonies for plasmid DNA extraction. For samples obtained from pooled yeast colonies in step 5, inoculate over 90 *E. coli* colonies for plasmid DNA extraction (see Note 17).

8. Digest the isolated plasmids with appropriate restriction enzymes or perform diagnostic PCR to confirm the presence of gene A*-OyCD-F[1].
9. Re-transform individually the purified plasmids expressing protein A mutants in BY4741 *fcy1Δ* strain carrying a plasmid expressing protein B and C, respectively, and test for OyCD PCA activity.
10. Send the purified plasmids expressing protein A mutants for sequencing in order to identify the mutation(s). Trouble-shooting advice can be found in Table 2.

3.3. Fluorescence Protein PCA to Visualize the Location of PPIs

The proteins to test for interaction are fused to the N- and C-terminal fragments of an enhanced YFP (e.g., Venus YFP (27)), either 5′ or 3′ of the YFP fragments (protein A-vYFP-F[1], vYFP-F[1]-protein A, protein B-vYFP-F[1], vYFP-F[1]-protein B). vYFP-F[1] (N-terminal) corresponds to amino acids 1–158, and vYFP-F[2] (C-terminal) corresponds to amino acids 159–239 of Venus YFP. The fusions are subcloned into yeast expression vectors p413ADH for the vYFP-F[1] fusion and p415ADH for the vYFP-F[2] fusion (36). We typically insert a 10 amino acid flexible polypeptide linker consisting of $(Gly.Gly.Gly.Gly.Ser)_2$ between the protein of interest and the vYFP fragments.

3.3.1. Co-transformation of Competent Yeast

1. Thaw competent yeast cells on ice.
2. Mix 10 μl of cells with 1 μl (~250 ng) of each yeast expression plasmid (e.g., p413ADH and p415ADH (36)) encoding the Venus YFP PCA fusion partners (protein A fused to vYFP-F[1] and protein B fused to vYFP-F[2]), 60 μl of PLATE solution and 8 μl DMSO.
3. Heat shock yeast at 42°C for 20 min (see Note 18).
4. Centrifuge at 1,100 × *g* for 3 min. Remove supernatant and resuspend cells in 500 μl SD medium without amino acids or glucose.
5. Plate 20 μl of cell suspension per well on SC agar (SD agar + 2% glucose + 1× amino acids (-his, -leu, -lys for *MAT*a; -his, -leu, -lys, -met for diploids)) in a six-well plate.
6. Incubate at 30°C for 48–72 h.

3.3.2. Preparation of Cells for Fluorescence Microscopy

1. Inoculate a fresh colony for each sample into 3 ml of SC medium (-his, -leu, -lys for *MAT*a; -his ,-leu, -lys, -met for diploids) and grow overnight at 30°C with shaking.
2. The following day, measure the OD_{600} of the overnight culture and inoculate a fresh culture of LFM (-his, -leu, -lys for Mat A; -his, -leu, -lys, -met for diploids) with enough cells to obtain an OD_{600} of approximately 0.1–0.3 at the time of analysis (see Note 19).

Table 2
Troubleshooting an OyCD PCA screen

Step	Problem	Possible reason	Solution
Subheading 3.2.1, Step 5	Less than 10% of colonies died on the 5-FC selection plates	Too many yeast plated on the selection plate	Plate less than 1,000 cells per 100 mm Petri dish Increase 5-FC concentration
Subheading 3.2.1, Step 5	No *E. coli* colonies or very few colonies	Electro-competent *E. coli* cells not very competent	Use freshly prepare electro-competent MC1061 *E. coli* cells
Subheading 3.2.2, Step 4	Several hundreds of colonies grew on the cytosine selection plates	Too many yeast plated on the selection plate	Plate less than 1,000 cells per 100 mm Petri dish Decrease cytosine concentration
	Small colonies form around the initial large colony after 4 days of incubation	Uracil can diffuse out of cells that have OyCD PCA activity and allow for cells that do not have OyCD PCA activity to grow	Pick only the large colony at the center
Subheading 3.2.2, Step 6	No *E. coli* colonies or very few colonies	Electro-competent *E. coli* cells not very competent	Use freshly prepare electro-competent MC1061 *E. coli* cells

Table 3
Troubleshooting vYFP PCA experiments

Step	Problem	Possible reason	Solution
Subheading 3.3.1, Step 6	No colonies after transformation	DNA or cells used is insufficient	Increase quantity of cells and DNA. Increase the volume of cells plated on the Petri dish or six-well plate
	Too many colonies after transformation	Plated too many of cells	Dilute cells before plating on the Petri dish or six-well plate
Subheading 3.3.2, Step 3	Fusion protein is not functioning correctly	Fragment fusion interferes with protein expression/function/stability	Fuse the PCA fragment to the other end of the protein

3. Coat the wells of a glass bottom 96-well plate with a solution of 1 mg/ml poly-L-Lysine, or 50 μg/ml concanavilin A for 10 min, rinse with distilled water and allow to dry. Transfer 70 μl of cell suspension to each well. Wait 10 min to allow the cells to settle in the wells. Acquire images with a fluorescence microscope equipped with a CCD camera, using a YFP filter cube and ~750 ms of exposure time (see Note 20).

3.3.3. Anticipated Results and Controls

Additional troubleshooting advice can be found in Table 3. The fluorescence intensity of the reassembled Venus YFP PCA varies with the expression levels and the interaction dissociation constants for the protein pairs attached to the PCA fragments. In the case of our simplest positive control (GCN4 leucine zipper pair fused to the PCA fragments: Zip-vYFP-F[1] + Zip-vYFP-F[2]), the reconstituted PCAs represent approximately 10–20% of the activity of the full-length Venus YFP. The PCA fusions expressed alone should not result in detectable fluorescence (compared to nontransformed cells) because the individual PCA fragments have no activity. For each study, positive (known interaction) and particularly negative (noninteracting proteins) controls should always be performed in parallel. A PCA response should not be observed if noninteracting proteins are used as PCA partners.

3.4. Luciferase Reporter PCAs to Study Dynamics of PPIs

The proteins to test for interaction are fused to the coding sequences for N- and C-terminal fragments of Rluc or Gluc, either 5′ or 3′ of the luciferase fragments (e.g., protein A-Rluc-F[1], Rluc-F[1]-protein A, protein B-Rluc-F[2], Rluc-F[2]-protein B). Rluc-F[1] (N-terminal) corresponds to amino acids 1–110, and Rluc-F[2] (C-terminal) corresponds to amino acids 111–312 of Rluc (24). Similarly Gluc-F[1] corresponds to amino acids 1–63 and Gluc-F[1] to amino acids 64–185 of Gluc (23).

The fusions are subcloned into yeast expression vectors, e.g., p413ADH for the Rluc-F[1] or Gluc-F[1] fusion and p415ADH for the Rluc-F[2] or Gluc- F[2] fusion (43). In yeast, fragments can also be fused to the genes of interest at their chromosomal loci using a homologous recombination method (13). For this purpose, the PCA fragments are cloned in to nonexpression vectors that provide a selection marker (e.g., antibiotic resistance). For example, pAG25-Rluc-F[1] and pAG32-Rluc-F[2] plasmids are used for the Rluc PCA fragment fusions.

3.4.1. Co-transformation of Competent Yeast

1. Thaw competent yeast cells on ice.
2. Mix 10 μl of cells with 1 μl (~250 ng) of each yeast expression plasmid (e.g., p413ADH and p415ADH) (43) encoding the Rluc or Gluc PCA fusion partners (protein A fused to Rluc-F-[1] or Gluc-F-[1] and protein B fused to Rluc-F F[2]) or Gluc- F[2], 60 μl of PLATE solution, and 8 μl DMSO.
3. Heat shock yeast at 42°C for 20 min (see Note 18).
4. Centrifuge at 1,100 × *g* for 3 min. Remove the supernatant and resuspend cells in 500 μl S.D. medium without amino acids or glucose.
5. Plate 20 μl of cell suspension per well on SC agar (SD agar + 2% glucose + 1× amino acids (-his, -leu, -lys for *MAT*a; -his, -leu, -lys, -met for diploids)) in a six-well plate.
6. Incubate at 30°C for 48–72 h.

3.4.2. Fusion of PCA Fragments at the Native Chromosomal Loci

1. PCR amplify the Rluc or Gluc PCA fragment cassettes containing the PCA fragment followed by a terminator and an antibiotic selection marker (constructs are available from our laboratory upon request).
2. Transform the PCR product into suitable competent cells:
   (a) Mix 10 μl of thawed competent cells with 10 μl (~1–2 μg) of each PCR amplified cassette DNA encoding the Rluc or Gluc PCA fragments along with a resistance marker.
   (b) Add 85 μl of PLATE solution.
   (c) Incubate for 30 min at room temperature. Add 9.5 μl DMSO followed by heat shock at 42°C for 20 min.
   (d) Centrifuge at 1,100 × *g* for 3 min and remove supernatant.
   (e) Resuspend cells in 500 μl YPD medium and incubate at 30°C with shaking for 4 h.
   (f) Centrifuge the cells, remove supernatant and resuspend cells in 200 μl of YPD.
   (g) Plate 60 μl per well in six-well plate or the entire 200 μl on a Petri dish that contains the selection antibiotic.

(h) Incubate the plates at 30°C for 48–72 h, after which the colonies can be verified by colony PCR methods.

3.4.3. Preparation of Cells for Bioluminescence Assay

1. Inoculate a fresh colony for each sample into 3 ml of SC medium (-his, -leu, -lys for *MATa*; -his ,-leu, -lys, -met for diploids) and grow overnight at 30°C with shaking. For cells with fragments fused at chromosomes, grow them in SC medium with suitable antibiotic.
2. The following day, measure the OD_{600} of the culture and inoculate a fresh culture of LFM (–his, -leu, -lys for Mat A; –his, -leu, -lys, -met for diploids), or LFM complete with suitable antibiotics, with enough cells to obtain an OD_{600} of approximately 0.1 to 0.3 at the time of analysis (see Note 21).
3. Transfer 160–180 μl of cell suspension (cells equivalent to 0.1–0.3 OD_{600}) to each well of a 96-well white plate. Manually add or inject 20–40 μl of suitable substrate using the Luminometer injector and initiate the bioluminescence analysis. Optimize the signal integration times depending on the bioluminescence signal strength. For real-time kinetics experiments, add or inject the substrate, immediately initiate the bioluminescence readings with the optimized signal integration time continuously for the desired period. Then, background correct the bioluminescence signals to obtain the net signal. Afterward, normalize the data to total protein concentration in cell lysates if desired (see Note 22).

3.4.4. Anticipated Results and Controls

Additional troubleshooting advice can be found in Table 4. The luminescence intensity of the reassembled Rluc and Gluc PCAs vary with the strength of interaction between the protein pairs attached to the PCA fragments. In the case of our simplest positive control (GCN4 leucine zipper pair fused to the PCA fragments: e.g., Zip-Rluc-F[1] + Zip-Rluc-F[2]), the reconstituted PCAs represent approximately 10–30% of the activity of the full-length Rluc or Gluc enzymes. The PCA fusions expressed alone should not result in detectable luminescence (compared to nontransfected cells) because the individual PCA fragments have no activity. For each study, positive (known interaction) and particularly negative (noninteracting proteins) controls should always be performed in parallel. A PCA response should not be observed if noninteracting proteins are used as PCA partners.

4. Notes

1. For the prey strain, step 2 can be repeated from the 384 prints to be transferred to a maximum of four other 1,536 pintool prints per omniplate to achieve a density of up to 6,144 colonies.

Table 4
Troubleshooting Rluc or Gluc luciferase PCAs

Step	Problem	Possible reason	Solution
Subheading 3.4.1, Step 6	No colonies after transformation	Not enough DNA or cells	Increase quantity of cells and DNA. Increase the number of cells plated on the Petri dish or six-well plate
	Too many colonies after transformation	Too many cells plated	Dilute cells before plating on the Petri dish or six-well plate
Subheading 3.4.3, Step 3	Fusion protein is not functioning correctly	Fragment fusion interferes with protein expression/function/stability	Fuse the PCA fragment to the other end of the protein
	Poor Luminescence signal	Signal integration time is too short	Optimize the signal Integration times
		Not enough substrate	Increase the substrate concentrations
		Not enough cells used	Increase the number of cells used per assay
	No or low signal modulation after stimulus or Inhibitor treatment	Number of cells and signal integration times are not optimal	Optimize the number of cells and signal integration times
		Stimulus or Inhibitor concentration are too low or duration of treatment is not long enough	Try different stimulus or inhibitor treatment times and or concentrations
		Off the optimal signal detection time	Peak signal occurs immediately after addition of colelantrezines. Try optimizing the beginning of signal integration after substrate addition
		Signal-to-background ratio is low	If the signal is very low, find an optimal way to extract the meaningful signal from background. Test appropriate positive and negative controls for the interaction you are studying

2. Endogenous recombinant strain screen setup (steps 1–2): 8 h to 3 days (depending on the screen density achieved).
3. Transferring haploids for mating (steps 3–6): 6 min or more per bait strains (depending on screen density and robotic routine efficiency) + 16 h incubation.
4. Diploid cell selection (step 7): 5 min or more per bait (depending on screen density and robotic routine efficiency) + 16 h incubation.
5. DHFR PCA survival-selection (step 8): 5 min or more per bait (depending on screen density and robotic routine efficiency) + 2 weeks incubation (maximum).
6. Make sure that the efficiency of the transformation gives enough colonies to cover six times the size of the library in order to have good coverage of potential mutants. For example, if the size of the library of protein A* is 5,000 clones, make sure to obtain more than 30,000 clones.
7. Test the efficiency of your competent yeast cells to determine how many cells to plate per 100 mm Petri dish. Do not plate more than 5,000 cells per 100 mm Petri dish.
8. Make a glycerol stock of the pooled yeast colonies obtained on the control plates as a backup source or for future screens if required.
9. We should expect 10–50% fewer colonies on the 5-FC death selection plates compared to the control plates. This variability depends on the pair of interaction chosen and the number of mutations per clone in the library.
10. Yeast cell pellets can be stored at −20°C for months.
11. *E. coli* cell pellets can be stored at −20°C for months.
12. Make sure that the efficiency of the transformation gives enough colonies to cover six times the size of the library in order to have a good coverage of potential mutant clones.
13. Test the efficiency of your competent yeast cells to have an idea how much cells to plate per 100 mm Petri dish. Do not plate more than 2,000 cells per 100 mm Petri dish.
14. Make glycerol stock of the pooled yeast colonies obtained on the control plates as a backup source or for future screens if required.
15. We can expect to obtain from a few to hundreds of colonies at this step. This variability depends mostly on the pair of interaction that was chosen and the complexity of the library.
16. Make glycerol stocks of the single or pooled yeast colonies as a backup source.
17. Make glycerol stocks of the single or pooled bacterial colonies as a backup source.

18. Shorter or longer incubation times at higher or lower temperatures can result in decreased efficiency of transformation.
19. It is particularly important for the cells to be in the log phase of growth in order to avoid including dead and unhealthy cells. These cells are highly autofluorescent and thus would confound quantitative analysis. Cells in the lag phase can be used if they are appropriate to study a particular interaction(s) as long as the condition of the cells is verified by brightfield microscopy.
20. It is best to use a 60× or 100× objective to discriminate subcellular structures. Bright field or phase contrast images can be acquired for each field of view to compare the morphology of the yeast with fluorescent PCA signal. Specific functional assays to further characterize a protein–protein interaction might be performed here.
21. It is particularly important for the cells to be in the log phase of growth in order to avoid including dead and unhealthy cells.
22. Specific functional assays to further characterize a protein–protein interaction might be performed here. For example, incubation of cells with agents, such as specific enzyme or transport inhibitors, can be performed for various amount of time, prior to the Luminometric analysis.

References

1. Michnick, S. W., Remy, I., Campbell-Valois, F. X., Vallée-Bélisle, A., and Pelletier, J. N. (2000). Detection of protein–protein interactions by protein fragment complementation strategies. *Methods Enzymol* **328**, 208–30.
2. Pelletier, J. N., Campbell-Valois, F. X., and Michnick, S. W. (1998) Oligomerization domain-directed reassembly of active dihydrofolate reductase from rationally designed fragments. *Proc Natl Acad Sci U S A* **95**, 12141–6.
3. Pelletier, J. N., and Michnick, S. W. (1997) A Protein Complementation Assay for Detection of Protein–Protein Interactions in vivo. *Protein Engineering* **10**, 89.
4. Michnick, S. W., Ear, P. H., Manderson, E. N., Remy, I., and Stefan, E. (2007) Universal strategies in research and drug discovery based on protein-fragment complementation assays. *Nat Rev Drug Discov* **6**, 569–82.
5. Remy, I., and Michnick, S. W. (1999) Clonal Selection and In Vivo Quantitation of Protein Interactions with Protein Fragment Complementation Assays. *Proc Natl Acad Sci U S A* **96**, 5394–5399.
6. Tarassov, K., Messier, V., Landry, C. R., Radinovic, S., Serna Molina, M. M., Shames, I., Malitskaya, Y., Vogel, J., Bussey, H., and Michnick, S. W. (2008) An in vivo map of the yeast protein interactome. *Science* **320**, 1465–70.
7. Campbell-Valois, F. X., Tarassov, K., and Michnick, S. W. (2005) Massive sequence perturbation of a small protein. *Proc Natl Acad Sci U S A* **102**, 14988–93.
8. Ear, P. H., and Michnick, S. W. (2009) A general life-death selection strategy for dissecting protein functions. *Nat Methods* **6**, 813–6.
9. Remy, I., and Michnick, S. W. (2001) Visualization of biochemical networks in living cells. *Proc Natl Acad Sci U S A* **98**, 7678–83.
10. Pelletier, J. N., Arndt, K. M., Plückthun, A., and Michnick, S. W. (1999) An in vivo library-versus-library selection of optimized protein–protein interactions. *Nat Biotechnol* **17**, 683–90.
11. Subramaniam, R., Desveaux, D., Spickler, C., Michnick, S. W., and Brisson, N. (2001) Direct visualization of protein interactions in plant cells. *Nat Biotechnol* **19**, 769–72.

12. Ercikan-Abali, E. A., Waltham, M. C., Dicker, A. P., Schweitzer, B. I., Gritsman, H., Banerjee, D., and Bertino, J. R. (1996) Variants of human dihydrofolate reductase with substitutions at leucine-22: effect on catalytic and inhibitor binding properties. *Mol Pharmacol* **49**, 430–7.
13. Ghaemmaghami, S., Huh, W. K., Bower, K., Howson, R. W., Belle, A., Dephoure, N., O'Shea, E. K., and Weissman, J. S. (2003) Global analysis of protein expression in yeast. *Nature* **425**, 737–41.
14. Ghosh, I., Hamilton, A. D., and Regan, L. (2000) Antiparallel leucine zipper-directed protein reassembly: application to the green fluorescent protein. *J Am Chem Soc* **122**, 5658–9.
15. Magliery, T. J., Wilson, C. G., Pan, W., Mishler, D., Ghosh, I., Hamilton, A. D., Regan, L. (2005) Detecting protein–protein interactions with a green fluorescent protein fragment reassembly trap: scope and mechanism. *J Am Chem Soc* **127**, 146–57.
16. Wilson, C. G., Magliery, T. J., and Regan, L. (2004) Detecting protein–protein interactions with GFP-fragment reassembly. *Nat Methods* **1**, 255–62.
17. Cabantous, S., Terwilliger, T. C., and Waldo, G. S. (2005) Protein tagging and detection with engineered self-assembling fragments of green fluorescent protein. *Nat Biotechnol* **23**, 102–7.
18. Hu, C. D., Chinenov, Y., and Kerppola, T. K. (2002) Visualization of interactions among bZIP and Rel family proteins in living cells using bimolecular fluorescence complementation. *Mol Cell* **9**, 789–98.
19. MacDonald, M. L., Lamerdin, J., Owens, S., Keon, B. H., Bilter, G. K., Shang, Z., Huang, Z., Yu, H., Dias, J., Minami, T., Michnick, S. W., and Westwick, J. K. (2006) Identifying off-target effects and hidden phenotypes of drugs in human cells. *Nat Chem Biol* **2**, 329–37.
20. Nyfeler, B., Michnick, S. W., and Hauri, H. P. (2005) Capturing protein interactions in the secretory pathway of living cells. *Proc Natl Acad Sci U S A* **102**, 6350–5.
21. Remy, I. and Michnick, S. W. (2004) Regulation of apoptosis by the Ft1 protein, a new modulator of protein kinase B/Akt. *Mol Cell Biol* **24**, 1493–504.
22. Remy, I., Montmarquette, A., and Michnick, S. W. (2004) PKB/Akt modulates TGF-beta signalling through a direct interaction with Smad3. *Nat Cell Biol* **6**, 358–65.
23. Remy, I., and Michnick, S. W. (2006) A highly sensitive protein–protein interaction assay based on Gaussia luciferase. *Nat Methods* **3**, 977–9.
24. Stefan, E., Aquin, S., Berger, N., Landry, C. R., Nyfeler, B., Bouvier, M., and Michnick, S. W. (2007) Quantification of dynamic protein complexes using Renilla luciferase fragment complementation applied to protein kinase A activities in vivo. *Proc Natl Acad Sci U S A* **104**, 16916–21.
25. Benton, R., Sachse, S., Michnick, S. W., and Vosshall, L. B. (2006) Atypical membrane topology and heteromeric function of Drosophila odorant receptors in vivo. *PLoS Biol* **4**, e20.
26. Ding, Z., Liang, J., Lu, Y., Yu, Q., Songyang, Z., Lin, S. Y., and Mills, G. B. (2006) A retrovirus-based protein complementation assay screen reveals functional AKT1-binding partners. *Proc Natl Acad Sci U S A* **103**, 15014–9.
27. Metodiev, M. V., Matheos, D., Rose, M. D., and Stone, D. E. (2002) Regulation of MAPK function by direct interaction with the mating-specific Galpha in yeast. *Science* **296**, 1483–6.
28. Choi, K. Y., Satterberg, B., Lyons, D. M., and Elion, E. A. (1994) Ste5 tethers multiple protein kinases in the MAP kinase cascade required for mating in S. cerevisiae. *Cell* **78**, 499–512.
29. Chou, S., Huang, L., and Liu, H. (2004) Fus3-regulated Tec1 degradation through SCFCdc4 determines MAPK signaling specificity during mating in yeast. *Cell* **119**, 981–90.
30. Manderson, E. N., Malleshaiah, M., and Michnick, S. W. (2008) A Novel Genetic Screen Implicates Elm1 in the Inactivation of the Yeast Transcription Factor SBF. *PLoS ONE* **3**, e1500.
31. Nagai, T., Ibata, K., Park, E. S., Kubota, M., Mikoshiba, K., and Miyawaki, A. (2002) A variant of yellow fluorescent protein with fast and efficient maturation for cell-biological applications. *Nat Biotechnol* **20**, 87–90.
32. Tannous, B. A., Kim, D. E., Fernandez, J. L., Weissleder, R., and Breakefield, X. O. (2005) Codon-optimized Gaussia luciferase cDNA for mammalian gene expression in culture and in vivo. *Mol Ther* **11**, 435–43.
33. Giaever, G., Chu, A. M., Ni, L., Connelly, C., Riles, L., Véronneau, S., Dow, S., Lucau-Danila, A., Anderson, K., André, B., Arkin, A. P., Astromoff, A., El-Bakkoury, M., Bangham, R., Benito, R., Brachat, S., Campanaro, S., Curtiss, M., Davis, K., Deutschbauer, A., Entian, K. D., Flaherty, P., Foury, F., Garfinkel, D. J., Gerstein, M., Gotte, D., Güldener, U., Hegemann, J. H., Hempel, S., Herman, Z., Jaramillo, D. F., Kelly, D. E., Kelly, S. L., Kötter, P., LaBonte, D., Lamb, D. C., Lan, N., Liang, H., Liao, H., Liu, L., Luo, C., Lussier, M.,

Mao, R., Menard, P., Ooi, S. L., Revuelta, J. L., Roberts, C. J., Rose, M., Ross-Macdonald, P., Scherens, B., Schimmack, G., Shafer, B., Shoemaker, D. D., Sookhai-Mahadeo, S., Storms, R. K., Strathern, J. N., Valle, G., Voet, M., Volckaert, G., Wang, C. Y., Ward, T. R., Wilhelmy, J., Winzeler, E. A., Yang, Y., Yen, G., Youngman, E., Yu, K., Bussey, H., Boeke, J. D., Snyder, M., Philippsen, P., Davis, R. W., and Johnston, M. (2002) Functional profiling of the Saccharomyces cerevisiae genome. *Nature* **418**, 387–91.

34. Knop, M., Siegers, K., Pereira, G., Zachariae, W., Winsor, B., Nasmyth, K., and Schiebel, E. (1999) Epitope tagging of yeast genes using a PCR-based strategy: more tags and improved practical routines. *Yeast* **15**, 963–72.
35. Sheff, M. A. and Thorn, K. S. (2004) Optimized cassettes for fluorescent protein tagging in Saccharomyces cerevisiae. *Yeast* **21**, 661–70.
36. Collins, S. R., Schuldiner, M., Krogan, N. J., and Weissman, J. S. (2006) A strategy for extracting and analyzing large-scale quantitative epistatic interaction data. *Genome Biol* 7, R63.
37. Linggi, B. and Carpenter, G. (2006) ErbB receptors: new insights on mechanisms and biology. *Trends Cell Biol* **16**, 649–56.
38. Memarian, N., Jessulat, M., Alirezaie, J., Mir-Rashed, N., Xu, J., Zareie, M., Smith, M., and Golshani, A. (2007) Colony size measurement of the yeast gene deletion strains for functional genomics. *BMC Bioinformatics* **8**, 117.
39. Dudley, A. M., Janse, D. M., Tanay, A., Shamir, R., and Church, G. M. (2005) A global view of pleiotropy and phenotypically derived gene function in yeast. *Mol Syst Biol* **1**, 2005.0001.
40. Collins, S. R., Kemmeren, P., Zhao, X. C., Greenblatt, J. F., Spencer, F., Holstege, F. C., Weissman, J. S., and Krogan, N. J. (2007) Toward a comprehensive atlas of the physical interactome of Saccharomyces cerevisiae. *Mol Cell Proteomics* **6**, 439–50.
41. Jansen, R., and Gerstein, M. (2004) Analyzing protein function on a genomic scale: the importance of gold-standard positives and negatives for network prediction. *Curr Opin Microbiol* 7, 535–45.
42. Levy, E. D., Landry, C. R., and Michnick, S. W. (2009) How perfect can protein interactomes be? *Sci Signal* **2**, pe11.
43. Mumberg, D., Müller, R., and Funk, M. (1995) Yeast vectors for the controlled expression of heterologous proteins in different genetic backgrounds. *Gene* **156**, 119–22.

Index

Louis M. Luttrell and Stephen S.G. Ferguson (eds.), *Signal Transduction Protocols*, Methods in Molecular Biology, vol. 756, DOI 10.1007/978-1-61779-160-4, © Springer Science+Business Media, LLC 2011

MIX
Papier aus verantwortungsvollen Quellen
Paper from responsible sources
FSC® C105338

If you have any concerns about our products,
you can contact us on
ProductSafety@springernature.com

In case Publisher is established outside the EU,
the EU authorized representative is:
Springer Nature Customer Service Center GmbH
Europaplatz 3, 69115 Heidelberg, Germany

Printed by Libri Plureos GmbH
in Hamburg, Germany